JavaScript
编程思想 从ES5到ES9

柯霖廷 编著

清华大学出版社
北京

内 容 简 介

本书从基础到高级，主要阐释 JavaScript 编程各层面的语法、观念、实战示例与习题。其内容深入浅出，主要内容包括：表达式与运算符，数据类型，条件和循环语句，函数与方法，处理数值、字符串、数组、自定义对象与日期和时间，DOM 的事件处理和元素实例，Reflect、Proxy 和 Intl 对象，window.navigator、window.document 和 BOM 的多个对象实例，类，错误处理，数据的验证与传输，响应式机制与源代码加密。

本书含有充足且具有创意的实战示例，利于读者学习、理解和实际运用。另外，各章末的习题也是用来演练学习成果的良方，可起到事半功倍的奇效。本书既适合业界的程序开发者阅读，也可作为大中专院校与培训机构的教学参考书。

本书封面贴有清华大学出版社防伪标签，无标签者不得销售。
版权所有，侵权必究。侵权举报电话：010-62782989　13701121933

图书在版编目（CIP）数据

JavaScript 编程思想：从 ES5 到 ES9 / 柯霖廷编著. —北京：清华大学出版社，2019
ISBN 978-7-302-52661-2

Ⅰ. ①J… Ⅱ. ①柯… Ⅲ. ①JAVA 语言—程序设计 Ⅳ. ①TP312.8

中国版本图书馆 CIP 数据核字（2019）第 055895 号

责任编辑：栾大成
封面设计：王　翔
责任校对：闫秀华
责任印制：杨　艳

出版发行：清华大学出版社
　　　　网　　址：http://www.tup.com.cn，http://www.wqbook.com
　　　　地　　址：北京清华大学学研大厦A座　　邮　　编：100084
　　　　社 总 机：010-62770175　　　　　　　　邮　　购：010-62786544
　　　　投稿与读者服务：010-62776969，c-service@tup.tsinghua.edu.cn
　　　　质 量 反 馈：010-62772015，zhiliang@tup.tsinghua.edu.cn

印 装 者：清华大学印刷厂
经　　销：全国新华书店
开　　本：190mm×260mm　　　　印　张：40.25　　　　字　数：644 千字
版　　次：2019 年 5 月第 1 版　　　　　　　　　　　　　印　次：2019 年 5 月第 1 次印刷
定　　价：118.00 元

产品编号：077875-01

前　言

现今仍然被使用于地球上的编程语言繁多，可分别实现特定任务、功能与运作逻辑。有些编程语言仅被内建于特定集成开发环境（IDE, integrated development environment），并不够普及。现代流行的编程语言，例如 JavaScript、Python、Java、C、C++、C#、PHP、SQL、R 等，被内建于由不同软件业者提供的集成开发环境里，进而蔚为风潮！

为了培养稳定流畅的编程思维逻辑和源代码的衔接能力，程序开发者在初学阶段逐渐学习多种编程语言，肯定是必经之路！一开始挑选现代较为流行的编程语言，可获得相当多的学习资源，进而大幅降低所耗费的心力与时间。

JavaScript 编程语言从 20 多年前问世以来，逐渐风行于全球各国！如今被内建于不同软件平台上的许多集成开发环境中，进而触及更为宽广、更多层次的开发领域。因此，很多较为资深的程序开发员通过 JavaScript 编程语言来实现多种应用程序。

希望长年累月的高级程序开发员、行之有年的 JavaScript 编程工作者、对 JavaScript 编程有兴趣者，甚至涉世未深的初学者，都能通过本书极为丰富的实战示例，提升自己的编程技术能力！

本书分为 21 章，从基础到高级，主要阐释 JavaScript 编程各层面的语法、观念、实战示例与习题。其内容深入浅出，包括如下所有的实战主题：

- 表达式（expression）与运算符（operator）
- 数据类型（data type）
- 条件（condition）和循环（loop）语句
- 函数（function）与方法（method）
- 处理数值（number）、字符串（string）、数组（array）、自定义对象（custom object / user-defined object）与日期和时间（date and time）
- DOM（document object model）的事件处理器（event handler）和元素实例（element instance）
- Reflect、Proxy 和 Intl 对象（object）
- window.navigator、window.document 和 BOM（browser object model）对象实例（object instance）
- 类（class）
- 错误处理（error handling）
- 数据的验证与传输（data validation and transmission）
- 响应式机制（responsive mechanism）与源代码加密（source-code encryption）

另外，本书的源代码示例文档也可扫描下方的二维码下载。若下载时遇到问题，请将标题为【下载问题——JavaScript 编程思想：从 ES5 到 ES9】的电子邮件发送至 741376828@qq.com，后续交由专员为您排困解难。

由于编者水平有限，书中难免存在疏漏之处，恳请广大读者批评指正。

柯霖廷

2019 年 3 月

目 录

第 1 章 ECMAScript 简介 1
   1.1 概述 1
      1.1.1 ECMAScript 各版本 1
      1.1.2 关于 JavaScript 2
      1.1.3 其他脚本语言 2
   1.2 语法的实现 2
      1.2.1 源代码 2
      1.2.2 语句 3
      1.2.3 表达式 5
      1.2.4 子程序 10
      1.2.5 注释 11
      1.2.6 关键字 11
   1.3 开 发 环 境 12
      1.3.1 浏览器 13
      1.3.2 Node.js 14
      1.3.3 其他 JavaScript Shell 14
   1.4 练习题 15

第 2 章 表达式与运算符 17
   2.1 操作数 17
      2.1.1 常量（ES6）... 17
      2.1.2 变量（ES6）... 18
      2.1.3 子表达式 23
      2.1.4 函数的返回值 25
   2.2 运算符 26
      2.2.1 算术运算符（ES7）... 27
      2.2.2 赋值运算符 30
      2.2.3 比较运算符 33
      2.2.4 逻辑运算符 34
      2.2.5 条件运算符 36
      2.2.6 类型运算符 37

        2.2.7　按位运算符 41
        2.2.8　括号运算符 46
        2.2.9　扩展运算符（ES6） 49
        2.2.10　逗号运算符 51
        2.2.11　删除运算符 51
        2.2.12　运算符的优先级（ES6） 53
    2.3　练习题 55

第 3 章　数据类型 57
    3.1　数值类型 57
        3.1.1　2^n 进制的字面量（ES6） 57
        3.1.2　数值的比较（ES6） 58
        3.1.3　数值的正负符号（ES6） 59
        3.1.4　数值的截断（ES6） 60
        3.1.5　数值的特殊格式（ECMA-402） 63
        3.1.6　整数值的安全范围（ES6） 66
    3.2　布尔类型 68
    3.3　数组类型 70
    3.4　对象类型 73
    3.5　字符串类型 75
        3.5.1　一般字符串 75
        3.5.2　格式化字符串（ES6） 77
        3.5.3　日期与时间格式的字符串（ES6） 80
    3.6　集合与地图类型 82
        3.6.1　集合类型（ES6） 82
        3.6.2　地图类型（ES6） 84
    3.7　数据类型的转换（ES6） 85
    3.8　练习题 88

第 4 章　条件和循环语句 90
    4.1　条件语句 90
        4.1.1　if 语句 90
        4.1.2　switch 语句 94
    4.2　循环语句 96
        4.2.1　for 相关语句（ES6） 96
        4.2.2　while 相关语句 104
        4.2.3　break 与 continue 语句 106
    4.3　练习题 108

第 5 章 函数与方法 110

5.1 函数的定义 110
5.1.1 不同形式的函数（ES6） 110
5.1.2 函数名称（ES6） 124
5.1.3 参数（ES6） 125
5.1.4 主体 132
5.1.5 返回数据与 void 关键字（ES6） 135
5.1.6 定义的位置（ES6） 137
5.1.7 函数的调用形式（ES6） 139

5.2 生成器 142
5.2.1 迭代器协议与生成器（ES6） 142
5.2.2 生成器函数的定义和用法（ES6） 143

5.3 搭配 Promise 对象 144
5.3.1 Promise 对象的用法（ES6、ES9） 144
5.3.2 聚集多个 Promise 对象（ES6） 148
5.3.3 异步函数与等待表达式（ES8） 154

5.4 练习题 156

第 6 章 处理数值 159

6.1 Number 对象的内置属性 159
6.1.1 最大的正数和安全整数（ES6） 159
6.1.2 最小的正数和安全整数（ES6） 160
6.1.3 正负无穷值 162
6.1.4 非数值的判断（ES6） 162
6.1.5 浮点数运算的误差值（ES6） 166

6.2 Number 对象的内置函数 167
6.2.1 转换为特定进制的数码字符串 167
6.2.2 处理小数格式 170
6.2.3 转换为数值 172
6.2.4 判断是否为整数或有限数（ES6） 175

6.3 Math 对象 177
6.3.1 Math 对象的常量属性 177
6.3.2 Math 对象的函数（ES6） 178

6.4 练习题 189

第 7 章 处理字符串 190

7.1 String 对象 190
7.1.1 子字符串的索引值 190

	7.1.2	特定模式的子字符串的搜索（ES6）	192
	7.1.3	子字符串的获取	194
	7.1.4	子字符串的替换	196
	7.1.5	字符串的大小写转换	198
	7.1.6	不同字符串的连接	199
	7.1.7	字符串的重复连接	200
	7.1.8	字符串的分割	200
	7.1.9	特定字符和 Unicode 数码的双向转换（ES6）	202
	7.1.10	重复填充子字符串于扩充后的字符串中（ES8）	204
7.2	将冗长的字符串分割为多行（ES6）		205
7.3	字符串的扩展运算（ES6）		207
7.4	字符串的插值格式化（ES6）		207
7.5	原始字符串（ES6）		210
7.6	正则表达式与黏性匹配（ES6）		211
7.7	万国码字面量（ES6）		214
7.8	练习题		216

第 8 章 处理数组 217

8.1	Array 对象		217
	8.1.1	创建特定数组的副本（ES6）	217
	8.1.2	创建来自可迭代对象的新数组（ES6）	220
	8.1.3	数组元素数据所构成的字符串	223
	8.1.4	数组元素的放入和取出	224
	8.1.5	新增或删除数组的多个元素	226
	8.1.6	合并多个数组	227
	8.1.7	切割数组	228
	8.1.8	寻找符合特定条件的数组元素（ES6）	229
	8.1.9	自我复制并覆盖数组的部分元素（ES6）	230
	8.1.10	判断数组各元素是否符合特定条件	232
	8.1.11	数组部分元素的填充（ES6）	233
	8.1.12	筛选出符合特定条件的数组元素（ES6）	235
	8.1.13	循环访问数组各元素	236
	8.1.14	判断是否为数组的实例	237
	8.1.15	访问并渐次处理数组各元素的数据	239
	8.1.16	反转数组各元素的顺序	244
	8.1.17	数组各元素的重新排序	245
	8.1.18	判断特定元素的存在性（ES7）	246
8.2	数组的扩展运算（ES6）		248
8.3	数组元素的匹配（ES6）		249

| 8.4 | 数据类型化的按位数组（ES6） | 250 |
| 8.5 | 练习题 | 252 |

第 9 章　处理自定义对象 ... 253

- 9.1 自定义对象的属性名称、属性数据与副本 ... 253
  - 9.1.1 对象属性的名称或数据所构成的数组（ES8） ... 253
  - 9.1.2 精细创建特定对象实例的副本 ... 255
- 9.2 自定义对象实例之间的相同性与合并 ... 258
  - 9.2.1 判断两个对象实例的数据是否完全相同（ES6） ... 258
  - 9.2.2 合并多个对象实例的所有成员（ES6） ... 260
- 9.3 对象实例的简短自定义语法（ES6） ... 261
- 9.4 自定义对象实例的动态成员名称（ES6） ... 262
- 9.5 对象实例的匹配（ES9） ... 262
- 9.6 练习题 ... 264

第 10 章　处理日期和时间 ... 265

- 10.1 处理日期 ... 265
  - 10.1.1 获取日期相关数据 ... 265
  - 10.1.2 设置日期相关数据 ... 267
  - 10.1.3 带有日期的格式化字符串 ... 268
- 10.2 处理时间 ... 270
  - 10.2.1 获取时间相关数据 ... 270
  - 10.2.2 设置时间相关数据 ... 272
  - 10.2.3 带有时间的格式化字符串 ... 274
  - 10.2.4 处理纪元时间至今的毫秒数（ES6） ... 275
- 10.3 练习题 ... 277

第 11 章　DOM 的事件处理（一） ... 278

- 11.1 鼠标事件 ... 278
  - 11.1.1 单击和双击事件 ... 278
  - 11.1.2 上下文菜单事件 ... 281
  - 11.1.3 鼠标按下与松开事件 ... 282
  - 11.1.4 鼠标指针相关进出事件 ... 284
  - 11.1.5 鼠标移动事件 ... 287
- 11.2 键盘事件 ... 289
  - 11.2.1 按压与按下按键事件 ... 289
  - 11.2.2 松开按键事件 ... 289
- 11.3 网页内容的装卸事件 ... 290
  - 11.3.1 出现错误事件 ... 290

11.3.2	加载和页面显示事件	292
11.3.3	卸载和页面隐藏事件	294
11.3.4	先于卸载事件	295
11.3.5	网址散列变化事件	296
11.3.6	滚动事件	298

11.4 表单事件 299
11.4.1	内容变化事件	299
11.4.2	获取和失去焦点相关事件	300
11.4.3	输入事件	302
11.4.4	无效事件	303
11.4.5	重置事件	304
11.4.6	搜索事件	306
11.4.7	选定文本事件	307
11.4.8	提交事件	308

11.5 练习题 309

第 12 章 DOM 的事件处理（二） 311

12.1 拖动事件 311
12.1.1	正在拖动事件	311
12.1.2	拖动结束事件	313
12.1.3	拖动进入事件	315
12.1.4	拖动离开事件	316
12.1.5	拖动悬停事件	318
12.1.6	拖动开始事件	319
12.1.7	放下事件	321

12.2 剪贴板事件 323
12.2.1	复制事件	324
12.2.2	剪切事件	325
12.2.3	粘贴事件	325

12.3 视频和音频事件 325
12.3.1	加载相关事件	326
12.3.2	清空事件	328
12.3.3	播放结束事件	329
12.3.4	异常相关事件	330
12.3.5	播放与暂停相关事件	331
12.3.6	播放速率变化事件	332
12.3.7	播放位置变化相关事件	333
12.3.8	音量变化事件	335
12.3.9	缓冲等待事件	335

12.4 动画及过渡事件 336
12.4.1 动画相关事件 336
12.4.2 过渡结束事件 338
12.5 其他事件 339
12.5.1 details 元素实例的切换事件 340
12.5.2 鼠标滚轮事件 341
12.5.3 触摸相关事件 342
12.5.4 接收服务器数据相关事件 344
12.6 练习题 346

第 13 章 Reflect 对象 347
13.1 Reflect 对象介绍（ES6）......347
13.2 间接应用特定函数（ES6）......349
13.3 创建特定对象的实例（ES6）......351
13.4 精细定义新属性（ES6）......353
13.5 删除特定属性（ES6）......354
13.6 获取特定属性的数据（ES6）......355
13.7 返回特定属性的描述器（ES8）......356
13.8 返回特定对象的原型（ES6）......359
13.9 判断特定属性的存在性（ES6）......360
13.10 判断与设置特定对象的扩展性（ES6）......361
13.11 简易定义新属性（ES6）......363
13.12 重新设置特定对象的原型（ES6）......365
13.13 返回与列举特定对象的自定义属性（ES6）......367
13.14 练习题 370

第 14 章 Proxy 对象、Intl 对象和 navigator 对象实例 371
14.1 Proxy 对象 371
14.1.1 创建访问代理机制的构造函数（ES6）......372
14.1.2 确定被代理的特定对象（ES6）......373
14.1.3 自定义代理函数（ES6）......373
14.1.4 调试访问代理机制（ES6）......375
14.2 Intl 对象 376
14.2.1 精确比较语言敏感的字符串（ECMA-402）......376
14.2.2 语言敏感的日期与时间格式（ECMA-402）......378
14.2.3 语言敏感的数值格式（ECMA-402）......379
14.2.4 返回规范化语言环境名称（ECMA-402）......380
14.3 window.navigator 对象实例 381
14.3.1 获取浏览器相关信息 381

14.3.2　获取当前地理定位相关数据..........383
　14.4　练习题..........385

第 15 章　window.document 对象实例..........386

　15.1　返回焦点所在的元素..........386
　15.2　附加事件处理器至特定元素..........388
　15.3　收养特定节点..........390
　15.4　返回所有锚点元素构成的集合..........392
　15.5　返回当前网址的相关属性..........393
　15.6　在当前网页中动态生成 HTML 源代码..........394
　15.7　内含特定服务器相关数据的 cookie..........396
　15.8　返回当前网页编码字符集的名称..........398
　15.9　创建代表新属性的节点..........398
　15.10　创建代表新注释的节点..........399
　15.11　创建代表新片段或新元素的节点..........400
　15.12　创建代表新文本的节点..........402
　15.13　返回当前网页的根元素..........404
　15.14　访问当前网址的域名..........405
　15.15　返回所有 embed 元素构成的集合..........406
　15.16　返回所有 form 元素实例构成的集合..........407
　15.17　返回特定身份识别码的元素实例..........408
　15.18　返回被设置带有特定 CSS 类名的所有元素实例的集合..........411
　15.19　返回特定标签名称的所有元素实例的集合..........414
　15.20　判断当前网页是否存在焦点..........416
　15.21　返回当前网页的 head 元素实例..........417
　15.22　返回当前网页所有 image 元素实例的集合..........418
　15.23　创建当前网页或者子网页里的特定节点实例的副本..........420
　15.24　获取当前网页的最近被修改的日期和时间..........423
　15.25　返回当前网页中的所有超链接元素实例的集合..........425
　15.26　返回特定 CSS 选择器名称对应的元素实例或集合..........426
　15.27　返回和处置当前网页的加载状态..........429
　15.28　返回跳转前的网址..........430
　15.29　解除已被附加的事件处理器..........432
　15.30　返回当前网页中的所有 script 元素实例构成的集合..........433
　15.31　访问当前网页的标题文本..........434
　15.32　练习题..........435

第 16 章　DOM 的元素实例..........437

　16.1　设置焦点跳转至特定元素实例上的快捷键..........437

16.2	创建特定元素实例的动画效果（Web Animations）	439
16.3	添加新元素实例和访问特定元素实例的所有属性	441
16.4	使得特定元素实例失去和获取焦点	445
16.5	访问子节点或子元素的实例	447
16.6	访问被应用在特定元素实例的所有 CSS 类名	450
16.7	模拟鼠标单击特定元素实例的动作	454
16.8	获取特定元素的尺寸、坐标与可定位的上层元素	455
16.9	比较两个元素之间的位置关系	458
16.10	判断是否存在特定子元素或可被编辑	461
16.11	访问特定元素实例的文本被书写的方向	463
16.12	返回头尾的子节点实例或子元素实例	464
16.13	访问或删除特定元素实例的特定属性	467
16.14	访问特定元素实例的常见属性的数据	469
16.15	判断是否存在任何子节点实例	472
16.16	在特定子节点实例之前新增另一子节点实例	473
16.17	判断两个节点实例的内容是否完全相同	475
16.18	返回下一个或上一个兄弟节点实例	478
16.19	返回特定节点实例的相关数据	479
16.20	合并多个相邻的文本子节点实例	481
16.21	返回父节点实例	484
16.22	删除或替换子节点实例	486
16.23	获取滚动条的相关数据	488
16.24	练习题	491

第 17 章　BOM 的多个对象实例 ... 492

17.1	window 对象实例	492
17.1.1	显示多种对话框与搜索特定文本	492
17.1.2	滚动至坐标或滚动特定距离	496
17.2	screen 对象实例	498
17.3	history 对象实例	499
17.4	location 对象实例	501
17.5	练习题	503

第 18 章　类 ... 504

18.1	类的定义和继承	504
18.1.1	类的定义（ES6）	505
18.1.2	类的继承（ES6）	507
18.2	类的静态成员	508
18.2.1	静态成员的概念和定义（ES6）	509

		18.2.2　静态成员的运用（ES6）......511
	18.3　类的设置器和取得器......511
		18.3.1　设置器和取得器的概念和定义（ES6）......512
		18.3.2　设置器和取得器的运用（ES6）......513
	18.4　练习题......514

第 19 章　错误处理......516

19.1　异常错误的种类......516
		19.1.1　语法错误......517
		19.1.2　数据类型错误......518
		19.1.3　评估错误......520
		19.1.4　范围错误......522
		19.1.5　引用错误......524
		19.1.6　网址在编码或解码上的错误......526
		19.1.7　逻辑错误......527
	19.2　处置特定异常错误......529
		19.2.1　试验与捕获特定异常错误......529
		19.2.2　抛出自定义的异常错误......530
	19.3　调试机制......531
		19.3.1　严格模式......531
		19.3.2　源代码的断点设置和逐句执行......535
	19.4　练习题......536

第 20 章　数据的验证与传输......538

20.1　HTML 表单的内置验证......538
		20.1.1　必填验证......538
		20.1.2　字符个数和数值范围的验证......540
	20.2　自定义的验证......542
		20.2.1　文本字段的模式验证......544
		20.2.2　JavaScript 源代码实现的验证......545
	20.3　异步数据传输......548
		20.3.1　AJAX 的工作原理......549
		20.3.2　AJAX 的编程方式......552
	20.4　练习题......561

第 21 章　响应式机制......563

21.1　通过 CSS 语法的版本......563
		21.1.1　页面的元信息......565
		21.1.2　媒体查询......565

21.2	通过 JavaScript 语法的版本	567
	21.2.1 简易判断窗口尺寸的版本	572
	21.2.2 直接变更 CSS 规则的版本	575
21.3	源代码的加密	579
	21.3.1 改写 HTML 与 CSS 成为 JavaScript 源代码	580
	21.3.2 JavaScript 源代码的全数加密	585
21.4	练习题	586

附录　练习题答案 588

第1章

ECMAScript 简介

本章内容主要介绍 ECMAScript 的历史、版本以及语法特性和多种集成开发环境,让读者对 ECMAScript 有一个初步的认识,以便为后续学习打好基础。

1.1 概　　述

成立于1961年之极具影响力的国际组织ECMA(European Computer Manufacturers Association,欧洲制造商协会),现今专门制定**信息**和**通信系统**的标准与技术报告,以促进和规范信息通信技术与消费电子产品。

ECMA 至今已经主动贡献了超过 400 个标准和 100 个技术报告,其中三分之二以上的部分,已在**国际上得到了广泛的使用**。其中,由 Ecma 制定的 ECMA-262(ISO/IEC 16262)的 ECMAScript,是具有商标的脚本编程语言(script programming language)的规范。

1.1.1 ECMAScript 各版本

从 1997 年 6 月的第 1 版,到 2018 年 6 月的第 9 版(ECMAScript 9),已问世的 ECMAScript 各版本,主要被用来标准化 JavaScript 脚本编程语言。目前业界的应用主要以 ECMAScript 5/6 为主、ECMAScript 7/8/9 为辅。从 ECMAScript 6 开始的各版本存在如表 1-1 所示的昵称。

表 1-1　ECMAScript 的各版本

正式名称	版本昵称	缩写昵称
ECMAScript 2015	ECMAScript 6	ES2015、ES6
ECMAScript 2016	ECMAScript 7	ES2016、ES7
ECMAScript 2017	ECMAScript 8	ES2017、ES8
ECMAScript 2018	ECMAScript 9	ES2018、ES9

1.1.2　关于 JavaScript

可简称为 JS 的 JavaScript，是一个动态高级解释脚本编程语言（dynamic high-level interpreting scripting programming language）。对于万维网而言，HTML、CSS 与 JavaScript 并列为网页制作的核心编程语言。

JavaScript 主要用来实现网页程序的用户交互机制（interactive mechanism），并通过 ECMAScript 规范，被创建和内置于网页浏览器（web browser）的 JavaScript 引擎中。

JavaScript 支持事件驱动（event-driven）、函数调用（function invocation / call）、面向对象编程（OOP，object-oriented programming）等特性，并具备处理文本、数组、日期、正则表达式（regular expression）和文档对象模型（DOM，document object model）的应用程序编程接口（API，application programming interface）。

JavaScript 引擎早期仅实现于网页浏览器当中，现今则被实现于其他类型的软件里，例如网页服务器、数据库服务器、文字处理器，以及用来编写移动或桌面应用程序的集成开发环境（IDE，integrated development environment）等软件里。

1.1.3　其他脚本语言

通过 ECMAScript 规范实现的编程语言，除了 JavaScript 之外，主要还有 Adobe 的 ActionScript 和 Microsoft 的 JScript，其语法非常相似于 JavaScript。

JScript 被用于 Internet Explorer 浏览器，ActionScript 主要被用于 Adobe Animate CC 集成开发环境中。

1.2　语法的实现

计算机编程语言的语法是一种规则，用来定义此编程语言表面形式的符号组合，以正确组成此语言的源代码（source code），进而堆砌成为片段或文件。

1.2.1　源代码

源代码（source code）通常以纯文本方式，存在于文档里，并作为输入数据，由汇编器（assembler）、编译器（compiler）或解释器（interpreter），转换成计算机可理解的二进制机器代码（machine code）。

通过经常阅读他人编写的源代码，计算机程序员可增进其编程技术的成熟度。因此，开发者互相分享源代码，可算是对彼此的一种贡献。

移植特定软件到其他计算机平台上，通常是异常困难的。然而，若有此软件的源代码，移植任务就会变得简单许多。

1.2.2 语句

在计算机编程当中，语句（statement）是命令式编程语言的语法单元，以表示欲指示计算机进行的一连串动作。

JavaScript 编程语言的语句大致分为简单语句（simple statement）和复合语句（compound statement）两种形式。

1. 简单语句

简单语句的末尾应该加上分号，例如：

（1）
```
let v01, v02 ;
```
变量 v01 与 v02 的声明。

（2）
```
var list01 = ['apple', 'banana', 'cherry'] ;
```
数组变量 list01 的声明与数据设置。

（3）
```
profile = {name: 'Alex', gender: 'male', age: '40'} ;
```
对象变量 profile 的数据设置。

（4）
```
document.writeln ('<h3>world peace</h3>') ;
```
写入带有文本 "world peace" 的 h3 元素实例到当前网页里。

（5）
```
confirm ('Are you sure to delete it?') ;
```
在网页上，显示带有 "Are you sure to delete it?" 信息并带有确认按钮和取消按钮的模式对话框。

（6）
```
username.style.color = 'RoyalBlue' ;
```
设置 id 为 username 元素实例的颜色样式为宝蓝色。

（7）
```
break ;
```
中断循环语句（loop statement），或确认切换语句（switch statement）的分支（branch）。

（8）
```
continue ;
```
在循环语句中，终止当前的循环，并立即进行下一次的循环。

（10）
```
debugger ;
```

设置调试**断点**（breakpoint），以暂停后续源代码的执行，并启动调试机制。

（11）

```
"use strict" ;
```

限制以严格模式执行源代码。

（12）

```
return result ;
```

终止函数的执行，并返回变量 result 的数据。

2. 复合语句

复合语句大都带有一对大括号，例如：

（1）

```
let result = 0 ;
alert ("You're welcome!") ;
{
  let num01 = 75, num02 = 64 ;
  result = num01 + num02 ;
}
```

在前述大括号中的源代码，连同大括号在内，可被称为复合语句。其中，变量 num01 与 num02 只有在大括号里才允许被访问。

（2）

```
with (send_button.style)
{
  color = 'Gold' ;
  backgroundColor = 'DodgerBlue' ;
  fontSize = '1.5em' ;
}
```

with 复合语句包含一对大括号，并且可以简化对象实例属性的访问语法。在此，【send_button.style.color = 'Gold' ;】的语句，在复合语句【with (send_button.style)】的大括号里，可被简化为【color = 'Gold' ;】。

（3）

```
if (score >= 60)
{
  passer_count++ ;
  saying = 'You have passed!' ;
}
```

if 条件复合语句可包含一对大括号，当小括号里的条件【(score >= 60)】为真时，大括号里的各语句就会被执行。

（4）

```
for (let i = 0 ; i < 10 ; i++)
{
  count++ ;
  sum += count * i ** 2 ;
```

}
```

for 循环复合语句可包含一对大括号，并借助小括号【(let i = 0 ; i < 10 ; i++)】中的 3 个子语句，使得大括号中的各语句，可被迭代而执行 10 次。

（5）
```
switch (choice)
{
  case 1:
    grade = 'A+' ;
    break ;
  case 2:
    grade = 'A' ;
    break ;
  case 3:
    grade = 'B' ;
    break ;
  case 4:
    grade = 'C' ;
    break ;
  default:
    grade = '@_@' ;
}
```

switch 复合语句包含一对大括号，并通过小括号【(choice)】中变量 choice 的数值，来决定并执行大括号中的特定 case 或 default 分支的子语句。

### 1.2.3 表达式

编程语言中的表达式（expression）内含运算符（operator）以及代表操作数（operand）的常量（constant）、变量（variable）、函数返回值（function return value）与子表达式（subsidiary expression）。

一般情况下，表达式的结果数据通常是原始数据类型（primitive data type）之一，例如数字（number）、字符串（string）或布尔型（boolean）。

JavaScript 编程语言的表达式大致可分为以下几种。

**1. 算术表达式**（arithmetic expression）

例如：

（1）v01 = v02 + 5 ;

将变量 v02 的数值，加上 5 之后的结果值，赋给变量 v01。

（2）v03 = v02 ** 3 ;

将变量 v02 的数值，进行 3 次方运算的结果值，赋给变量 v03。

（3）v04 = 6 * (v03 + 10) ;

将 6 乘以【变量 v03 加上 10】之后的积，赋给变量 v04。

（4）v05 = v04 % 2 + v04 / 2 ;

将变量 v04 除以 2 的余数，加上变量 v04 除以 2 的结果值，赋给变量 v05。

（5）c01 = c01 + 1 ;
```

将变量 c01 的数值，加上 1 的结果值，赋给变量 c01 本身。

（6）c01 += 1；

将变量 c01 的数值**增加 1** / **递增**（increment）。

（7）c01++；

先返回 c01 的数值，再递增变量 c01。

（8）++c01；

先递增变量 c01 的数值，再返回 c01。

（9）c02 = c02 - 1；

将变量 c02 的数值，**减去 1** / **递减**（decrement）的结果值，赋给变量 c02 本身。

（10）c02 -= 1；

将变量 c02 的值**减去 1** / **递减**的结果值，赋给变量 c02 本身。

（11）c02--；

先返回 c02 的数值，再递减变量 c02。

（12）--c02；

先递减变量 c02 的数值，再返回 c02。

2. 字符串表达式（string expression）

例如：

（1）

```
let subject = 'Alex' ;
let object = 'Jasper' ;
let greeting = "'have a \"nice\" day'" ;
let message = '' ;

message = subject + ' said ' + greeting + ' to ' + object ;
message = `${subject} said ${greeting} to ${object}` ;
```

在一对单引号或双引号内的文本是代表一种常量（constant）的字符串字面量（string literal）。在字符串字面量里的单引号或双引号应该冠上反斜杠（back slash）【\】，成为【\'】或【\"】。多个字符串字面量可通过加法运算符【+】，结合成为新的字符串。

通过一对反引号（back quote）【`】，可创建模板字面量（template literal），并内含语法【${特定变量的名称}】，以动态解析特定变量的数据，成为新字符串的一部分。

在前述源代码里，最后两行具有相同的效果。

（2）

```
let sentence = 'Alice really \
lovingly loves \
lovely beloved of \
Jason very much.' ;
```

借助反斜杠【\】来分割较长的字符串字面量。

在各行中，字符串片段的最后一个字符必须就是反斜杠【\】，不可以再有包括空白字符在内的其他字符。

前述源代码被执行之后,变量 sentence 的数据会是字符串 "Alice really lovingly loves lovely beloved of Jason very much."。

3. 关系与逻辑表达式(relational and logical expression)

例如:

(1)
```
x > y
```
检验若变量 x 的数值大于 y 的数值,则返回布尔值 true。

(2)
```
x <= y && x <=z
```
检验若变量 x 的数值同时小于或等于变量 y 和 z 的数值,则返回布尔值 true。

(3)
```
a > 0 || b > 0 || c > 0
```
检验若变量 a 的数值大于 0、变量 b 的数值大于 0,或是变量 c 的数值大于 0,则返回布尔值 true。

(4)
```
! (v01 > 60 || v02 >= 90) && v03 >= 80
```
检验若【并非变量 v01 的数值大于 60,或是变量 v02 的数值大于或等于 90】,并且【变量 v03 的数值大于或等于 80】,则返回布尔值 true。

(5)
```
name == 'admin' && /^\w{6,}$/.test(password)
```
检验若变量 name 的数据等于字符串字面量 "admin",并且变量 password 的数据是由【0~9】、【大小写 a~z】与【下画线_】所构成的**最少** 6 个字符的字符串,则返回布尔值 true。

4. 主要表达式(primary expression)

例如:

(1)
```
let circle_area = function (r)
{
  let result ;
  result = Math.PI * r ** 2
  console.log (`The circle area of radius ${r} = ${result}`) ;
  return result ;
} ;
```
在所谓的函数表达式(function expression)中,通过关键字 function,定义匿名函数(anonymous function),并以变量的名称 circle_area,作为匿名函数的别名(alias)。

(2)
```
let Cubic = class
```

```
  {
  constructor (l, w, h)
  {
    this.length = l ;
    this.width = w ;
    this.height = h ;
  }

  volume (l = this.length, w = this.width, h = this.height)
  {
    return l * w * h ;
  }

  surface_area (l = this.length, w = this.width, h = this.height)
  {
    return 2 * (l * w + w * h + h * l) ;
  }
} ;
```

在所谓的类表达式（class expression）中，通过关键字 class，定义匿名类（anonymous class），并以变量的名称 Cubic 作为匿名类的别名。

（3）

```
let number_list = [15, 30, 75, 90, 180] ;
console.log(number_list[2]) ;
number_list[5] = 770 ;
```

这 3 个语句均可被视为主要表达式。在等号右侧，没有变量名称的一对中括号，是用来定义数组实例 [15, 30, 75, 90, 180]，并赋给等号左侧的数组变量 number_list。在等号左侧，变量名称衔接的一对中括号，则用来访问数组变量 number_list 中特定索引值（2 与 5）对应的元素值（75 与 770）。

（4）

```
var user_input = 'z1 = x1 ^ 2 + y1 * 3 + 6' ;
var pattern = /[a-zA-Z]\d/g ;
var matches = user_input.match (pattern) ;
console.log (matches) ;
```

这 4 个语句均可被视为主要表达式。在此，通过一对斜杠符号【/】，创建用来匹配特定模式【由字母开头，后接一个数字的文本】的**正则表达式字面量**（regular-expression literal）【/[a-zA-Z]\d/g】，并赋给等号左侧的变量 pattern。

5. 箭头函数表达式（arrow function expression）

箭头函数表达式存在如下几种形式。

（1）

```
(参数列) => { ... } ;
```

通过带有参数列的一对小括号，衔接代表箭头符号的【=>】，与内含主体源代码的一对大括号，以定义没有名称、但可带有多个参数的箭头函数（arrow function）。

（2）

```
单一参数名称 => { ... } ;
```

通过单一参数，衔接代表箭头符号的【=>】，与内含主体源代码的一对大括号，以定义没有名称、仅有一个参数的箭头函数。

（3）

```
单一参数名称 =>单一语句；
```

通过单一参数，衔接代表箭头符号的【=>】，和单一语句的主体源代码，以定义没有名称、仅有一个参数和语句的箭头函数。

（4）

```
() => { ... } ;
```

通过不带任何参数的一对小括号，衔接代表箭头符号的【=>】，与内含主体源代码的一对大括号，以定义没有名称和参数的箭头函数。

（5）

```
() => 单一语句 ;
```

通过不带任何参数的一对小括号，衔接代表箭头符号的【=>】，和单一语句的主体源代码，以定义没有名称与参数、仅有单一语句的箭头函数。

6. 左侧表达式（left-hand expression）

例如：

（1）

```
profile.name = 'Jasper' ;
```

通过点号【.】，将字符串字面量 'Jasper'，赋给变量 profile 的实例属性 name。

（2）

```
profile['age'] = 28 ;
```

借助一对中括号，将整数值 28 赋给变量 profile 的实例属性 age。

（3）

```
super(10, 15) ;
```

通过关键字 super 与一对小括号，调用父类（parent class）的构造函数（constructor），并传入参数值 10 和 15。

（4）

```
super.cylinder_volume(10, 15) ;
```

借助关键字 super、点号与一对小括号，调用在父类中作为成员函数（member function）的函数 cylinder_volume()，并传入参数值 10 和 15。

（5）

```
x = x + 3 ;
```

在等号的左侧，设置变量 x，并将变量 x 的数值加上 3 的结果值，赋给变量 x 本身。

（6）

```
[a, b] = [7, 13] ;
```

在等号的左侧，编写内含变量 a 与 b 的一对中括号，并将右侧另一对中括号里的整数值 7 与 13，分别赋给变量 a 与 b。

（7）

```
[c, d] = [a + 3, b + 7] ;
```

在等号的左侧，编写内含变量 c 和 d 的一对中括号，并将右侧另一对中括号里的 a + 3 与 b + 7 的结果值，分别赋给变量 a 与 b。

1.2.4 子程序

在计算机编程中，子程序（subroutine）也可被称为函数（function）或方法（method），即是可重复被调用（call / invocation），以重新执行特定任务的一连串程序指令单元，进而节省开发与维护的成本，并提高质量与可靠性。

JavaScript 编程语言中的子程序，主要被称为函数，可通过以下语法来加以定义。

（1）

```
function sphere(r)
{
  values = {} ;
  values.volume = 4 / 3 * Math.PI * Math.pow(r, 3) ;
  values.surface_area = 4 * Math.PI * r * r ;

  return values ;
}
```

上述源代码是定义函数的标准语法。通过调用【sphere(10)】，可计算并返回【内含半径为 10 的球体积和球体表面积】相关数据的对象实例。

（2）

```
display = function ()
{
  console.log('Variable-type displaying.') ;
} ;
```

上述源代码是所谓的函数表达式，使得变量 display 具备函数的特征。通过调用【display()】，可在网页浏览器的调试工具【Console】面板中，显示【Variable-type displaying】的信息。

（3）

```
sphere_volume = r => { return 4 / 3 * Math.PI * Math.pow(r, 3) } ;
```

上述源代码中的简洁语法，等同于定义了函数 sphere_volume()。通过调用【sphere_volume(30)】，可计算并返回半径为 30 的球体积的结果值。

1.2.5 注释

在源代码里,注释(comment)是用来辅助程序员,加以理解各源代码片段的意义和用途,却不会被计算机加以执行的文本。

JavaScript 编程语言的注释方式有如下两种。

(1) 在注释文本的行首,加上紧连的两个斜杠符号【//】。例如:

```
// calculate the summary values for bonus table.
```

(2) 在注释文本的开头,加上紧连的斜杠符号与星号【/*】,并在注释文本的末尾,加上紧连的星号与斜杠符号【*/】。例如:

```
/*
    ca(r) calculates circle area of radius r.
    ssa(r) calculates sphere surface area of radius r.
    cv(r, h) calculates cylinder volume of radius r and height h.
*/
```

1.2.6 关键字

在计算机编程语言中,被称为保留字(reserved word)的关键字(keyword),不能或不应该作为常量、变量、属性、函数 / 方法的标识符(identifier)/ 名称(name)的词汇。JavaScript 编程语言的关键字如表 1-2 所示。

表 1-2 JavaScript 的关键字

关键字	关键字	关键字	关键字	关键字
abstract	arguments	await	boolean	break
byte	case	catch	char	class
const	continue	debugger	default	delete
do	double	else	enum	eval
export	extends	false	final	finally
float	for	function	goto	if
implements	import	in	Infinity	instanceof
int	interface	let	long	NaN
native	new	null	package	private
protected	public	return	short	static
super	switch	synchronized	this	throw
throws	transient	true	try	typeof
undefined	var	void	volatile	while
with	yield			

1.3 开发环境

目前，知名的 JavaScript 集成开发环境（IDE，integrated development environment）有如下几种。

1. 存在免费版本的

- Visual Studio Community 或 Visual Studio Code
  https://www.visualstudio.com/zh-hans/downloads
- Eclipse
  http://www.eclipse.org/downloads/eclipse-packages
- IntelliJ IDEA
  https://www.jetbrains.com/idea
- NetBeans IDE
  https://netbeans.org
- Sublime Text
  https://www.sublimetext.com
- Atom IDE
  https://ide.atom.io
- Brackets
  http://brackets.io
- Cloud9（在线系统）
  https://aws.amazon.com/cn/cloud9
- Codeanywhere（在线系统）
  https://codeanywhere.com

2. 仅有付费版本的

- WebStorm
  https://www.jetbrains.com/webstorm
- Komodo IDE
  https://www.activestate.com/komodo-ide

JavaScript 编程语言主要用于开发网页相关应用程序，所以，网页浏览器成为执行与调试 JavaScript 源代码的主要软件。

因此，上述各个集成开发环境所产生的内含 JavaScript 源代码的网页应用，最终仍然在网页浏览器或其他内置 JavaScript 引擎的软件上，被进行必要的调试与执行。

1.3.1 浏览器

网页浏览器（web browser）是用来检索与呈现网络上信息资源（文本信息、图像、音频、视频、动画）的软件。现今较流行的网页浏览器为 Google Chrome、Mozilla Firefox、Opera、Apple Safari、与 Microsoft Edge。

上述的网页浏览器皆内置 JavaScript 调试工具。以 Google Chrome 浏览器为例，启动 Chrome 并按下 Ctrl + Shift + I / J 快捷键之后，即可看到如图 1-1 所示的开发者工具调试窗格。

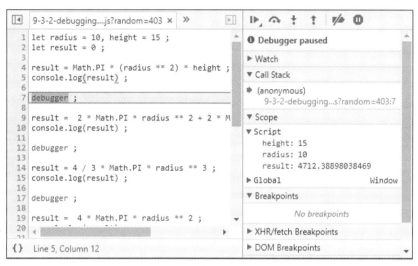

图 1-1

为了能顺利显示其 JavaScript 源代码于调试窗口中，可执行如下操作步骤。

步骤01 在 Chrome 浏览器，打开配书源代码示例文档中的网页文档【js_tester.html】。

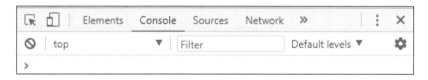

步骤02 按 Ctrl + Shift + I 快捷键，启动【开发者工具】，并可看到默认显示的 Console（控制台）窗口。

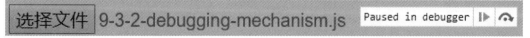

步骤03 请先按 Ctrl+ R 快捷键，以刷新页面，再单击网页中的【选择文件】按钮，并选择 JavaScript 源代码文档【js / 19-3-2-debugging- mechanism.js】之后，可在按钮右侧，明确看到被选定的文档名称和【Paused in debugger】信息。

步骤04 观察调试窗口的变化，从【Console】面板，自动切换至【Source】面板中。

被读取的【19-3-2-debugging-mechanism.js】内含的源代码片段【debugger ;】用来设置调试断点（debugging breakpoint），使得浏览器启动调试机制，进而自动切换至【Source】面板中，并在第 1 个断点上，暂停 JavaScript 源代码的执行，如图 1-2 所示 。

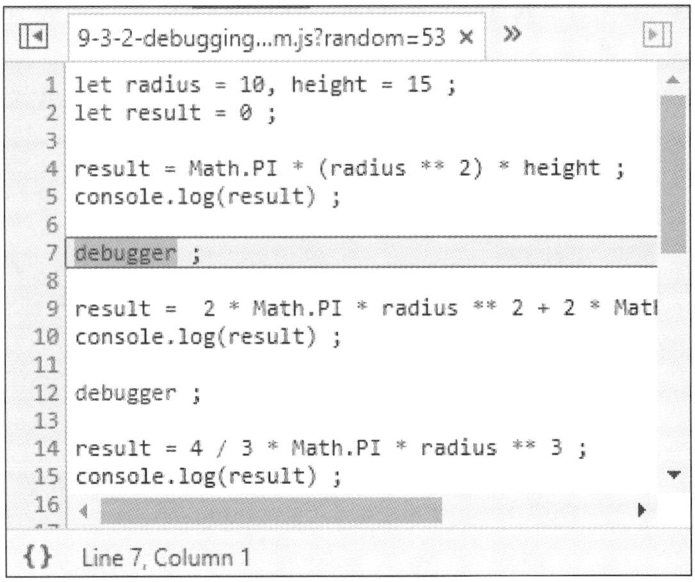

```
9-3-2-debugging...m.js?random=53
1  let radius = 10, height = 15 ;
2  let result = 0 ;
3
4  result = Math.PI * (radius ** 2) * height ;
5  console.log(result) ;
6
7  debugger ;
8
9  result =  2 * Math.PI * radius ** 2 + 2 * Math
10 console.log(result) ;
11
12 debugger ;
13
14 result = 4 / 3 * Math.PI * radius ** 3 ;
15 console.log(result) ;
16
{ } Line 7, Column 1
```

图 1-2

关于 JavaScript 源代码的相关调试，笔者将深入说明于本书 19.3 节。

关于操作扩展名亦为【.js】的【其他】示例文档，请完成至前述第 3 步即可。至于操作**扩展名为**【.html】的示例文档，则直接通过浏览器，**加载**并**浏览**其网页内容。

接着，观察特定扩展名为【.html】的示例文档，出现在浏览器中的网页内容，以及在【**Console**】**面板**里的对应信息，可进一步理解各个 HTML、CSS 和 JavaScript 源代码片段的工作原理。

1.3.2　Node.js

如同【ASP.NET】并非代表扩展名（file extension）为【.NET】的文档名称一样，Node.js 实际上亦不是代表 JavaScript 源代码的文档名称，而是开源且跨平台的**运行期环境**（run-time environment），并用于执行服务器端的 JavaScript 源代码，以生成动态的网页内容。

因此，对于主要借助 JavaScript 编程语言，整合前端（front-end）与后端（back-end）Web 应用程序开发的任务而言，Node.js 如今已然成为基本要素之一。此外，Node.js 具备事件驱动（event-driven）的运行架构，可实现数据异步的传递与接收，进而优化 Web 应用程序的处理能力和扩充灵活性。

1.3.3　其他 JavaScript Shell

通过特定 JavaScript Shell 的协助，可在不刷新特定网页内容的情况下，辅助开发与调试 JavaScript 源代码。较有名的 JavaScript Shell 有如下几种：

- Node.js

  https://nodejs.org
- JSDB

  http://www.jsdb.org
- JavaLikeScript

  http://javalikescript.free.fr
- GLUEscript

  http://gluescript.sourceforge.net
- jspl

  http://jspl.msg.mx
- ShellJS

  http://documentup.com/shelljs/shelljs

1.4 练 习 题

1. 截至 2018 年，ECMAScript 总共有几个版本？
2. 可被并列为网页制作的核心编程语言，除了 JavaScript 之外，还包括什么语言？
3. 网页浏览器相对应的英文短语是什么？
4. 除了 JavaScript 之外，通过 ECMAScript 规范实现的编程语言主要还有哪两个？
5. 依序翻译**在编程语言中**的专有名词：machine code、source code、programming language、syntax、subroutine、variable、constant 和 invocation。
6. 依序翻译**在编程语言中**的专有名词：assembler、compiler 和 interpreter。
7. 在如下的源代码片段里，有哪些**简单语句**？

```
with (send_button.style)
{
  color = 'Gold' ;
  backgroundColor = 'DodgerBlue' ;
  fontSize = '1.5em' ;
}
```

8. 在如下的源代码片段里，有哪些**简单语句**？

```
for (let i = 0 ; i < 10 ; i++)
{
  count++ ;
  sum += count * i ** 2 ;
}
```

9. 在如下的源代码片段里，哪些应该是**变量的名称**？

```
message = subject + ' said ' + greeting + ' to ' + object ;
```

10. 在源代码中的表达式里，哪些可以作为**操作数**？

11. 在如下源代码片段里，存在哪些不同的运算符？

```
! (v01 > 60 || v02 >= 90) && v03 >= 80
```

12. 在如下源代码片段里，存在哪些不同的常量？

```
let number_list = [15, 30, 75, 90, 180] ;
console.log (number_list[2]) ;
number_list[5] = 770 ;
```

13. 在如下源代码片段里，存在哪些不同的常量？

```
let user_input = 'z1 = x1 ^ 2 + y1 * 3 + 6' ;
let pattern = /[a-zA-Z]\d/g ;
let matches = user_input.match(pattern) ;
console.log(matches) ;
```

14. 在 JavaScript 语言里，如何调用如下被定义的函数，进而将半径为 50 的**球体积**和**球体表面积**，赋给变量 result？

```
function sphere(r)
{
 values = {} ;
 values.volume = 4 / 3 * Math.PI * Math.pow(r, 3) ;
 values.surface_area = 4 * Math.PI * r * r ;
 return values ;
}
```

15. 在 JavaScript 语言里，下列哪些可以作为**变量**的**名称**？

```
catchup、extends、finally、super、class、import、export 和 throw
```

16. 特定网页被浏览于 Google Chrome 浏览器窗口内时，有什么方式可以快速显示出浏览器**开发者工具**的 **Console 面板**？

第 2 章

表达式与运算符

本章内容主要介绍构成表达式的要素，包括常量、变量、子表达式、函数的返回值等形式的操作数，以及多种不同的运算符。

2.1 操 作 数

编程语言中的表达式主要由操作数（operand）和运算符（operator）构成。其中，操作数可以是常量（constant）、变量（variable）、子表达式（subsidiary expression）或是特定函数（function）的返回值（return value）。

2.1.1 常量（ES6）

在各种编程语言中，常量（constant）是指固定而明确的数据。例如：

（1）整数（integer）常量：25、770、-110。
（2）浮点数（floating-point number）常量：25.625、-23.71、Math.PI、Math.LN2、Math.SQRT2。
（3）字符串（string）的字面量（literal）：'Alex'、"Happily Ever After"、 '=.=+'。
（4）正则表达式（regular expression）的字面量：/\w+\s*.(\w\d+){1,3}/g、/^\w{2}\d{5}\s*$/。
（5）布尔（boolean）常量：true、false。
（6）原始（primitive）常量：NaN、null、undefined。
（7）编程人员在源代码中，自行声明（declare / declaration）的常量，可简称为自声明常量。例如在源代码片段【const num01 = 157, num02 = 268;】中，num01 和 num02 即是自声明常量。

关于自声明常量，可参考如下示例。

【2-1-1-constants.js】

```
const E = 299792458 ;

console.log('Speed of light is greater than 2.9E8?', E > 2.9e8) ;
```

【相关说明】

```
const E = 299792458 ;
```

声明数值为 299792458 的常量 E。

```
console.log('Speed of light is greater than 2.9E8?', E > 2.9e8) ;
```

在浏览器调试工具的【Console】面板里显示出【Speed of light is greater than 2.9E8? true】的信息。其中，2.9e8 代表 2.9×10^8，也就是 290000000。所以，【E > 2.9e8】等价于【E > 290000000】，会返回布尔值 true，这是因为自声明常量 E 的数值为 299792458，大于 290000000。

在上述简短的源代码中，常量 E 刻意被声明成整数常量 299792458，因此常量 E 在源代码的**运行期间**，不可被变更为其他数据。

2.1.2 变量（ES6）

在各种编程语言中，变量（variable）的标识符（identifier）/名称（name）用来识别特定存储位置中的可变数据。所谓的可变，是指在运行期间，其数据是可被变更的。

在 JavaScript 语言中，变量的作用范围（scope）大致可如下区分：

- 全局（global）范围
- 局部（local）范围
- 块（block）范围

下面通过示例，介绍如何运用全局范围和局部范围的变量。

【2-1-2-e1-global-and-local-scope-variables.js】

```
let start = 123 ;
const STEP = 3 ;
var result = 0 ;

function some_steps(step_count = 1)
{
  var output ; // or let output ;

  output = start + STEP * step_count ;

  console.log(`in function: start = ${start}, STEP = ${STEP}, result = ${result}`) ;
  console.log(`in function: output = ${output}\n\n`) ;

  return output ;
}

result = some_steps(5) ;
```

```
console.log(`outer: start = ${start}, step = ${STEP}, result = ${result}`) ;
console.log(`outer: output = ${output}`) ;
```

【相关说明】

```
let start = 123 ;
const STEP = 3 ;
var result = 0 ;
```

声明全局范围的变量 start 与 result，以及全局范围的自声明常量 STEP。既然是全局范围的变量和常量，即可让后续的任意表达式，加以访问。STEP 是一个常量，所以在后续的源代码里，无法被赋予新数据。

```
function some_steps(step_count = 1)
{
  var output ; // or let output ;

  output = start + STEP * step_count ;

  console.log(`in function: start = ${start}, step= ${STEP}, result = ${result}`) ;
  console.log(`in function: output = ${output}\n\n`) ;

  return output ;
}
```

定义函数 some_steps()，并在内部声明局部范围的变量 output。既然是局部范围的，变量 output 只能在函数 some_steps() 内部被访问。变量 output 接着被赋予【带有全局范围变量 start 和常量 STEP】的表达式的结果值。这个函数内部的多条语句，访问了全局范围的变量 start、常量 STEP，以及局部范围的变量 output。

```
result = some_steps(5) ;
console.log(`outer: start = ${start}, step= ${STEP}, result = ${result}`) ;
```

访问全局范围的变量 start、result 与常量 STEP。

```
console.log(`outer: output = ${output}`) ;
```

因为访问了函数 some_steps() 内的局部范围的变量 output，所以会显示出错误信息【Uncaught ReferenceError: output is not defined】。

关于块范围的理解，可参考如下示例。

【2-1-2-e2-block-scope-variables.js】

```
// different heights measured in meters.
{
  // height of the building.
  let height = 100 ;

  {
    // height of one floor.
    let height = 4 ;

    {
      // height of a room.
      let height = 3 ;
```

```
      {
        // height of a desk.
        let height = 1 ;

        console.log('height of a desk =', height) ;
      }

      console.log('height of a room =', height) ;
    }

    console.log('height of one floor =', height) ;
  }

  console.log('height of the building =', height) ;
}
console.log('') ;

for (var m = 1; m < 10; m++)
{
  for(var n = 1; n < 10; n++)
  {
    console.log(`kernel count = (${m},${n})`) ;
  }

  console.log(`inner count = (${m},${n})\n\n`) ;
}

console.log(`outer count = (${m},${n})`) ;
console.log('') ;

for (let i = 1; i < 10; i++)
{
  for(let j = 1; j < 10; j++)
  {
    console.log(`kernel count = (${i},${j})`) ;
  }

  console.log(`inner count = (${i},${j})\n\n`) ;
}
// console.log(`outer count = (${i},${j})`) ;
```

【相关说明】

```
// different heights measured in meters.
{
  // height of the building.

  let height = 100 ;
```

此为第 1 层大括号内的块范围变量 height，所以第 1 层大括号之外的语句是访问不到的。声明块范围的变量，必须通过关键字 let 才行；由关键字 var 声明的变量，只会成为全局范围或局部范围的变量。在此，可将此层块范围的变量 height 视为**建筑物**的高度。

```
    {
        // height of one floor.
        let height = 4 ;
```

此为第 2 层大括号内的块范围变量 height，因此第 2 层大括号之外的语句是访问不到的。在此，可将此层块范围的变量 height，视为建筑物特定**楼层**的高度。

```
        {
            // height of a room.
            let height = 3 ;
```

此为第 3 层大括号内的块范围变量 height，因此第 3 层大括号之外的语句是访问不到的。在此，可将此层块范围的变量 height，视为特定楼层中特定**房间**的高度。

```
            {
                // height of a desk.
                let height = 1 ;
```

此为第 4 层大括号内的块范围变量 height，因此第 4 层大括号之外的语句是访问不到的。在此，可将此层块范围的变量 height，视为特定房间中特定**桌子**的高度。

```
                console.log('height of a desk =', height) ;
            }
            console.log('height of a room =', height) ;
        }
        console.log('height of one floor =', height) ;
    }
    console.log('height of the building =', height) ;
}
```

通过此源代码片段，可供读者调试出**每层**块范围变量 height 的对应值。

```
console.log('') ;

for (var m = 1; m < 10; m++)
{
  for(var n = 1; n < 10; n++)
  {
    console.log(`kernel count = (${m},${n})`) ;
  }
  console.log(`inner count = (${m},${n})\n\n`) ;
}
console.log(`outer count = (${m},${n})`) ;
```

测试此源代码片段，可看出通过关键字 var 声明的变量 m 与 n，在带有大括号的 for 循环语句结束之后，依然可以被访问到。这也就意味着，关键字 var 声明的，无法成为块范围的变量。

```
console.log('') ;

for (let i = 1; i < 10; i++)
{
  for(let j = 1; j < 10; j++)
  {
```

```
    console.log(`kernel count = (${i},${j})`) ;
  }
  console.log(`inner count = (${i},${j})\n\n`) ;
}
// console.log(`outer count = (${i},${j})`) ;
```

测试此源代码片段，可看出通过关键字 let 声明的变量 m 与 n，只能被访问于 for 循环语句大括号内的块范围里。这也就意味着，关键字 let 声明的，**可以**成为**块范围**的变量。复原最后一行成为注释的语句，并进行测试之后，可在网页浏览器的调试工具【Console】面板中，看到因为访问不到块范围变量 i 与 j 的错误信息。

关于特殊变量的理解，可参考如下示例。

【2-1-2-e3-special-variables.js】

```
document.body.style.backgroundColor = 'RoyalBlue' ;

console.log(document.body.style.backgroundColor) ;

document.body.style.fontSize = '3em' ;

console.log(document.body.style.fontSize) ;

file_selector.style.color = 'Gold' ;

console.log(file_selector.style.color) ;

file_selector.style.zoom = 2 ;
```

【相关说明】

```
file_selector.style.zoom = 2 ;
```

此语句中的子属性 zoom，因为可被设置新数据，所以算是**变量**的一种；然而，此语句中的子属性 style，则无法被设置新数据，也因此算是**常量**的一种。

通过 JavaScript 语法，可访问文档对象模型（DOM, document object model）中的特定元素实例（element instance）。元素实例可内含**常量**属性，只可被读取其数据，而不允许被写入新数据。下面举出 4 个元素实例中的常量属性：

- file_selector.style
- document.body.style
- document.body
- window.document（可被简写为 document）

以下可变更数据的是字符串（string）类型的属性，亦是网页浏览器内置的，所以可认为是网页程序在运行时的**特殊变量**：

- document.body.style.backgroundColor
- document.body.style.fontSize
- file_selector.style.color
- file_selector.style.zoom

2.1.3 子表达式

一般而言,特定表达式可再分割成为子表达式(subsidiary expression)。关于子表达式的运用,请参考如下示例。

【2-1-3-subsidiary-expressions.js】

```js
var r = 10 ;
var volume = 4 / 3 * Math.PI * Math.pow(r, 3) ;

circumference = r => 2 * Math.PI * r ;

circle_area = r => Math.PI * Math.pow(r, 2) ;

sphere_volume = r => 4 / 3 * Math.PI * Math.pow(r, 3) ;

cylinder_volume = (r, h) => circle_area(r) * h ;

cylinder_surface_area = (r, h) => 2 * circle_area(r) + circumference(r) * h ;

rounded_circle_area = circle_area(10).toFixed(3) ;
rounded_cylinder_volume = cylinder_volume(10, 20).toFixed(3) ;
rounded_cylinder_surface_area = cylinder_surface_area(10, 20).toFixed(3) ;

console.log(rounded_circle_area) ;
console.log(rounded_cylinder_volume) ;
console.log(rounded_cylinder_surface_area) ;
```

【相关说明】

```js
var r = 10 ;
```

此语句内含左侧表达式。

```js
var volume = 4 / 3 * Math.PI * Math.pow(r, 3) ;
```

此主要表达式内含如下子表达式:

- 左侧表达式:【var volume =】和【Math.】
- 算术表达式:【4 / 3 * Math.PI * Math.pow(r, 3)】
- 主要表达式:【Math.PI】和【Math.pow(r, 3)】

```js
circumference = r => 2 * Math.PI * r ;
```

此主要表达式内含如下子表达式:

- 左侧表达式:【circumference =】和【Math.】
- 箭头函数表达式:【r => 2 * Math.PI * r】
- 算术表达式:【2 * Math.PI * r】
- 主要表达式:【Math.PI】

```js
circle_area = r => Math.PI * Math.pow(r, 2) ;
```

此表达式内含如下子表达式:

- 左侧表达式：【circle_area =】和【Math.】
- 箭头函数表达式：【r => Math.PI * Math.pow(r, 2)】
- 算术表达式：【Math.PI * Math.pow(r, 2)】
- 主要表达式：【Math.PI】和【Math.pow(r, 2)】

```
sphere_volume = r => 4 / 3 * Math.PI * Math.pow(r, 3) ;
```

此表达式内含如下子表达式：

- 左侧表达式：【sphere_volume =】和【Math.】
- 箭头函数表达式：【r => 4 / 3 * Math.PI * Math.pow(r, 3)】
- 算术表达式：【4 / 3 * Math.PI * Math.pow(r, 3)】
- 主要表达式：【Math.PI】和【Math.pow(r, 3)】

```
cylinder_volume = (r, h) => circle_area(r) * h ;
```

此表达式内含如下子表达式：

- 左侧表达式：【cylinder_volume =】
- 箭头函数表达式：【(r, h) => circle_area(r) * h】
- 算术表达式：【circle_area(r) * h】
- 主要表达式：【circle_area(r)】

```
cylinder_surface_area = (r, h) => 2 * circle_area(r) + circumference(r) * h ;
```

此表达式内含如下子表达式：

- 左侧表达式：【cylinder_surface_area =】
- 箭头函数表达式：【(r, h) => 2 * circle_area(r) + circumference(r) * h】
- 算术表达式：【2 * circle_area(r) + circumference(r) * h】
- 主要表达式：【circle_area(r)】和【circumference(r)】

```
rounded_circle_area = circle_area(10).toFixed(3) ;
```

此表达式内含如下子表达式：

- 左侧表达式：【rounded_circle_area =】和【circle_area(10).】
- 主要表达式：【circle_area(10).toFixed(3)】、【circle_area(10)】和【toFixed(3)】

```
rounded_cylinder_volume = cylinder_volume(10, 20).toFixed(3) ;
```

此表达式内含如下子表达式：

- 左侧表达式：【rounded_cylinder_volume =】和【cylinder_volume(10, 20).】
- 主要表达式：【cylinder_volume(10, 20).toFixed(3)】、【cylinder_volume(10, 20)】和【toFixed(3)】

```
rounded_cylinder_surface_area = cylinder_surface_area(10, 20).toFixed(3) ;
```

此表达式内含如下子表达式：

- 左侧表达式:【rounded_cylinder_surface_area =】和【cylinder_surface_area(10, 20).】
- 主要表达式:【cylinder_surface_area(10, 20).toFixed(3)】、【cylinder_surface_area(10, 20)】和【toFixed(3)】

```
console.log(rounded_circle_area) ;
console.log(rounded_cylinder_volume) ;
console.log(rounded_cylinder_surface_area) ;
```

如上 3 个语句排除分号【;】的其余部分,皆为主要表达式。

2.1.4 函数的返回值

在多种编程语言中,关键字 return 开头的语句,除了会终止当前函数内源代码的执行,并返回调用函数的子表达式之外,亦会使得前述子表达式,被取代成为返回值(return value),成为原始表达式的操作数。关于返回值的理解,可参考如下示例。

【2-1-4-function-return-values.js】

```
// cv stands for "cubic volume".
function cv01(l, w, h)
{
  let cubic_volume = l * w * h ;

  console.log(`cubic volume of (${l}, ${w}, ${h}) = ${cubic_volume}`) ;

  // return ;
}

function cv02(l, w, h)
{
  let cubic_volume = l * w * h ;

  return cubic_volume ;
}

let result01 = cv01(3, 5, 10) ;

let result02 = cv02(3, 5, 10) ;

let result03 = cv02(1, 3, 5) + cv02(2, 4, 6) ;

console.log(`result01 = ${result01}`) ;
console.log(`result02 = ${result02}`) ;
console.log(`result03 = ${result03}`) ;
```

【相关说明】

```
// cv stands for "cubic volume".
function cv01(l, w, h)
{
  let cubic_volume = l * w * h ;
  console.log(`cubic volume of (${l}, ${w}, ${h}) = ${cubic_volume}`) ;
```

```
  // return ;
}
```

- 在上述函数内的末尾处，并未放置 return 语句，或是 return 语句仅仅衔接分号【;】。所以，此段源代码定义了未带返回值的函数 cv01()。
- 函数 cv01()可在网页浏览器的调试工具【Console】面板里，显示出长、宽、高可能不同的立方体积。

```
function cv02(l, w, h)
{
  let cubic_volume = l * w * h ;
  return cubic_volume ;
}
```

- 此源代码片段定义了**具有返回值**的函数 cv02()，因为在上述函数内的末尾处，放置了语句【return cubic_volume ;】。
- 函数 cv02()亦可计算出长、宽、高可能不同的立方体积。

```
let result01 = cv01(3, 5, 10) ;
```

- 此语句调用了函数 cv01(3, 5, 10)，显示出长、宽、高各为 3、5、10 的立方体积。
- 因为函数 cv01()并无返回值，所以变量 result01 会被赋予原始常量 undefined。

```
let result02 = cv02(3, 5, 10) ;
```

此语句调用了函数 cv02(3, 5, 10)，并将长、宽、高各为 3、5、10 的立方体积结果值，返回到此语句，成为新的操作数 150，使得此语句等价于【let result02 = 150 ;】，进而让变量 result02 的数值变成 150。

```
let result03 = cv02(1, 3, 5) + cv02(2, 4, 6) ;
```

此语句调用了函数 cv02(1, 3, 5)和 cv02(2, 4, 6)，并将长、宽、高各为 1、3、5 与 2、4、6 的个别立方体积结果值，返回到此语句，成为新的操作数 15 与 48，使得此语句等价于【let result02 = 15 + 48 ;】，进而让变量 result02 的数值变成 63。

```
console.log(`result01 = ${result01}`) ;
```

此语句显示出变量 result01 的数据为 undefined。

```
console.log(`result02 = ${result02}`) ;
console.log(`result03 = ${result03}`) ;
```

这两个语句分别显示出变量 result02 的数值 150，以及变量 result03 的数值 63。

2.2 运算符

各种计算机编程语言均支持一系列的运算符，并和参与其中的操作数，构成特定形态的表达式，以满足计算机各种复杂的演算、分析与归纳。

2.2.1 算术运算符（ES7）

JavaScript 语言的算术运算符如表 2-1 所示。

表 2-1　JavaScript 的运算符

运算符分类	运算符
加法运算符（addition operator）或一元正号运算符（unary plus operator）	+
减法运算符（subtraction operator）或一元负号运算符（unary negation operator）	-
乘法运算符（multiplication operator）	*
除法运算符（division operator）	/
求余运算符（remainder operator）	%
求幂运算符（exponentiation operator）	**
递增运算符（increment operator）	++
递减运算符（decrement operator）	--

下面通过示例介绍各种算术运算符的运用。

【2-2-1-arithmetic-operators.js】

```
let num01 = 125, num02 = 10, num03 = 5 ;
let result = 0 ;

result = num01 % num02 ;
console.log(result) ;

result = num01 / num02 ;
console.log(result) ;

result = num01 * num02 ;
console.log(result) ;

result = num01 + num02 - num03 ;
console.log(result) ;

result = ++num01 ;
console.log(result, num01) ;

result = num01++ ;
console.log(result, num01) ;

++num01 ;
num01++ ;
console.log(num01) ;

num01 += 1 ;
console.log(num01) ;

num01 = num01 + 1 ;
console.log(num01) ;

result = --num02 ;
```

```
console.log(result, num02) ;

result = num02-- ;
console.log(result, num02) ;

--num02 ;
num02-- ;
console.log(num02) ;

num02 -= 1 ;
num02 = num02 - 1 ;
console.log(num02) ;

result = num03 ** 3 ** 2 ;
console.log(result) ;

result = num03 ** (3 ** 2) ;
console.log(result) ;

result = (num03 ** 3) ** 2 ;
console.log(result) ;
```

【相关说明】

```
let num01 = 125, num02 = 10, num03 = 5 ;
let result = 0 ;
```

这两个语句声明了具有初始数据的变量 num01、num02、num03 与 result。

```
result = num01 % num02 ;
```

此语句使得变量 result，被赋予了【变量 **num01 的数值**除以**变量 num02 的数值**】的余数。

```
result = num01 / num02 ;
```

此语句使得变量 result，被赋予了【变量 **num01 的数值**除以**变量 num02 的数值**】的结果值。

```
result = num01 * num02 ;
```

此语句使得变量 result，被赋予了【变量 **num01 的数值**乘以**变量 num02 的数值**】的结果值。

```
result = num01 + num02 - num03 ;
```

此语句使得变量 result，被赋予了【变量 num01 的数值加上变量 num02 的数值，再减去变量 num03 的数值】的结果值。

```
result = ++num01 ;
```

此语句被执行之前，变量 num01 的数值为 125。

因为运算符 ++ 出现在变量 num01 的左侧，也就意味着 num01 的数值要先递增为 126，再执行【result = num01】。

所以，result 的数值最终为 126。

```
result = num01++ ;
```

此语句被执行之前，变量 num01 的数值已经变成 126。

因为运算符 ++ 出现在变量 num01 的右侧,也就意味着要先执行【result = num01】,之后 num01 的数值才递增为 127。

所以,result 的数值最终仍然为 126。

```
++num01 ;
num01++ ;
```

这两个语句被执行之前,变量 num01 的数值已经变成 127。因为表达式 ++num01 与 num01++ 均单独出现在语句中,所以可简单视为变量 num01 的数值,被进行了 2 次递增,变成 129。

```
num01 += 1 ;
```

此语句被执行之前,变量 num01 的数值已经变成 129。此语句等同于递增变量 num01 的数值,使得 num01 的数值成为 130。

```
num01 = num01 + 1 ;
```

此语句被执行之前,变量 num01 的数值已经变成 130。此语句亦等同于递增变量 num01 的数值,成为 131。

```
result = --num02 ;
```

此语句被执行之前,变量 num02 的数值为 10。因为运算符 -- 出现在变量 num02 的**左侧**,也就意味着 num02 的数值要先递减为 9,再执行【result = num02】。所以,result 的数值最终为 9。

```
result = num02-- ;
```

此语句被执行之前,变量 num02 的数值已经变成 9。因为运算符 -- 出现在变量 num02 的**右侧**,也就意味着要先执行【result = num02】,之后 num02 的数值才递减为 8。所以,result 的数值最终仍然为 9。

```
--num02 ;
num02-- ;
```

这两个语句被执行之前,变量 num02 的数值已经变成 8。因为运算符 --num02 与 num02-- 均单独出现在语句中,所以可简单视为变量 num01 的数值,被进行了两次递减,变成 6。

```
num02 -= 1 ;
```

此语句被执行之前,变量 num02 的数值已经变成 6。此语句等同于递减变量 num02 的数值,成为 5。

```
num02 = num02 - 1 ;
```

此语句被执行之前,变量 num02 的数值已经变成 5。此语句亦等同于递减变量 num02 的数值,成为 4。

```
result = num03 ** 3 ** 2 ;
```

在此,变量 num03 的数值是 5。因为求幂运算符 ** 具有右结合的特征,所以【3 ** 2】要先被评估成为 3^2,也就是 9,然后再评估【num03 ** 9】,结果值为 5^9,也就是 1953125。

```
result = num03 ** (3 ** 2) ;
```

在此，借助小括号运算符()的辅助，可明确得知【(3 ** 2)】会优先被评估成为 3^2，也就是 9，然后评估【num03 ** 9】的结果值为 5^9，也就是 1953125。

```
result = (num03 ** 3) ** 2 ;
```

在此，使用小括号运算符()的辅助，可明确得知【(num03 ** 3)】会优先被评估成为 5^3，也就是 125，然后再评估【125** 2】的结果值为 125^2，也就是 15625。

2.2.2 赋值运算符

JavaScript 编程语言的赋值运算符（assignment operator）存在基本形式的等号【=】，以及如表 2-2 所示的复合形式。

表 2-2 JavaScript 的赋值运算符

赋值运算符	用法	含义
=		a = b
+=	a += b	a = a + b
-=	a -= b	a = a - b
*=	a *= b	a = a * b
/=	a /= b	a = a / b
%=	a %= b	a = a % b
<<=	a <<= b	a = a << b
>>=	a >>= b	a = a >> b
>>>=	a >>>= b	a = a >>> b
&=	a &= b	a = a & b
^=	a ^= b	a = a ^ b
|=	a |= b	a = a | b

关于赋值运算符的运用，可参考如下示例。

【2-2-2-assignment-operators.js】

```
let a = 18, b = 5, c = 2 ;
let result = 0 ;

result = a + b + c ;
console.log(result) ;

// a = a + b ;
a += b ;
console.log(a) ;

// a = a - b ;
a -= b ;
console.log(a) ;

// a = a * b ;
a *= b ;
```

```
console.log(a) ;

// a = a / b ;
a /= b ;
console.log(a) ;

// a = a % b ;
a %= b ;
console.log(a) ;

// b = b ** c ;
b **= c ;
console.log(b) ;

// b = b << c ;
b <<= c ;
console.log(b) ;

// b = b >> c ;
b >>= c ;
console.log(b) ;

// b = b >>> c ;
b >>>= c ;
console.log(b) ;

// b = b & c ;
b &= c ;
console.log(b) ;

// b = b ^ c ;
b ^= c ;
console.log(b) ;

// b = b | c ;
b |= c ;
console.log(b) ;
```

【相关说明】

```
let a = 18, b = 5, c = 2 ;
let result = 0 ;
```

这两个语句声明了具有初始数据的变量 a、b、c 与 result。

```
result = a + b + c ;
```

此语句里的赋值运算符【=】，使得变量 result 的数值，成为变量 a、b、c 各数值的总和。

```
// a = a + b ;
a += b ;
```

此语句里的赋值运算符【+=】，使得变量 a 的数值，成为【变量 a 本身的数值加上变量 b 的数值】的结果值。

```
// a = a - b ;
```

```
a -= b;
```

此语句里的赋值运算符【-=】，使得变量 a 的数值，成为【变量 a 本身的数值减去变量 b 的数值】的结果值。

```
// a = a * b;
a *= b;
```

此语句里的赋值运算符【*=】，使得变量 a 的数值，成为【变量 a 本身的数值乘以变量 b 的数值】的结果值。

```
// a = a / b;
a /= b;
```

此语句里的赋值运算符【/=】，使得变量 a 的数值，成为【变量 a 本身的数值除以变量 b 的数值】的结果值。

```
// a = a % b;
a %= b;
```

此语句里的赋值运算符【%=】，使得变量 a 的数值，成为【变量 a 本身的数值除以变量 b 的数值】的余数。

```
// b = b ** c;
b **= c;
```

此语句里的赋值运算符【**=】，使得变量 b 的数值，成为【**变量 b 本身数值**的 c 幂次】（b^c）的结果值。

```
// b = b << c;
b <<= c;
```

此语句里的赋值运算符【<<=】，使得变量 b 的二进制数值，向**左偏移变量 c** 所代表的**比特位（bit）个数**。

```
// b = b >> c;
b >>= c;
```

此语句里的赋值运算符【>>=】，使得变量 b 的二进制数值，在保留正负号的前提下，向右偏移变量 c 所代表的比特位（bit）个数。所谓的保留正负号就是：

- 若该二进制数值为负值，则其最左侧的符号位（sign bit）是 1，并在向右偏移的同时，保持其符号位为 1。
- 若该二进制数值为正值，则其最左侧的符号位是 0，并在向右偏移的同时，保持其符号位为 0。

```
// b = b >>> c;
b >>>= c;
```

此语句里的赋值运算符【>>>=】，使得变量 b 的数值，成为【变量 b 的二进制数值，向右偏移变量 c 所代表的比特位（bit）个数，并在其左侧补上相同个数的二进制 0】之后的结果值。

在其左侧补上相同个数的二进制 0，也就意味着，一开始无论其最左侧的符号位是 0（正值）或是 1（负值），后续皆在向右偏移的同时，保持符号位成为 0。

```
// b = b & c;
```

```
b &= c ;
```

此语句里的赋值运算符【&=】，使得变量 b 的数值，成为【变量 b 本身的二进制数值与变量 c 的二进制数值，进行按位与（bitwise and）运算】之后的结果值。

```
// b = b ^ c ;
b ^= c ;
```

此语句里的赋值运算符【^=】，使得变量 b 的数值，成为【变量 b 的二进制数值与变量 c 的二进制数值，进行按位异或（bitwise exclusive or) 运算】之后的结果值。

```
// b = b | c ;
b |= c ;
```

此语句里的赋值运算符【^=】，使得变量 b 的数值，成为【变量 b 的二进制数值与变量 c 的二进制数值。进行按位或（bitwise or）运算】之后的结果值。

2.2.3 比较运算符

在各编程语言中，比较运算符（comparison operator）是用来决定其**两侧**操作数相等或不相等的关系。JavaScript 编程语言的比较运算符如表 2-3 所示。

表 2-3 JavaScript 语言的比较运算符

比较运算符	用法	含义	判断后的返回值
==	a == b	判断变量 a 与 b 的数值，是否相等	• 为真，返回 true • 为假，返回 false
===	a === b	判断变量 a 与 b 的数值，是否相等，以及其数据类型是否也相同	
!=	a != b	判断变量 a 与 b 的数值，是否并不相等	
!==	a !== b	判断变量 a 与 b 的数值，是否并不相等，或者其数据类型是否也不同	
>	a > b	判断变量 a 的数值，是否大于变量 b 的数值	
>=	a >= b	判断变量 a 的数值，是否大于或者等于变量 b 的数值	
<	a < b	判断变量 a 的数值，是否小于变量 b 的数值	
<=	a <= b	判断变量 a 的数值，是否小于或者等于变量 b 的数值	

关于比较运算符的综合运用，可参考如下示例。

【2-2-3-comparison-operators.js】

```
let v01 = 100 , v02 = 250, v03 = 500 ;
let s01 = '100', s02 = '250', s03 = '500' ;

console.log(v01 == s01) ;
console.log(v01 === s01) ;

console.log(v02 > s01) ;
console.log(v02 < s03) ;

console.log(v03 >= s03) ;
console.log(v03 <= s03) ;
```

【相关说明】

```
let v01 = 100 , v02 = 250, v03 = 500 ;
```

此语句声明了初始数值为整数常量的变量 v01、v02 和 v03。

```
let s01 = '100', s02 = '250', s03 = '500' ;
```

此语句声明了初始数据为字符串字面量的变量 s01、s02 与 s03。

```
console.log(v01 == s01) ;
```

判断变量 v01 与 s01 的数据是否相同,并显示判断为真的返回值 true,于浏览器的调试工具【Console】面板中。

```
console.log(v01 === s01) ;
```

判断变量 v01 与 s01 的数据是否相同,以及其数据类型是否也相同,并显示判断为假的返回值 false,于浏览器的调试工具【Console】面板中。

```
console.log(v02 > s01) ;
```

判断变量 v02 的数值是否大于变量 s01 的数据,并显示判断为真的返回值 true,于浏览器的调试工具【Console】面板中。

```
console.log(v02 < s03) ;
```

判断变量 v02 的数值是否小于变量 s03 的数据,并显示判断为真的返回值 true,于浏览器的调试工具【Console】面板中。

```
console.log(v03 >= s03) ;
```

判断变量 v03 的数值是否大于**或者**等于变量 s03 的数据,并显示判断为真的返回值 true 于浏览器调试工具的【Console】面板中。

```
console.log(v03 <= s03) ;
```

判断变量 v03 的数值是否小于**或者**等于变量 s03 的数据,并显示判断为真的返回值 true,于浏览器的调试工具【Console】面板中。

2.2.4 逻辑运算符

在各种编程语言中,逻辑运算符(logical operator)主要用来串联带有比较含义的表达式。JavaScript 编程语言的逻辑运算符如表 2-4 所示。

表 2-4 JavaScript 语言的逻辑运算符

逻辑运算符	用法	含义	判断后的返回值
&&	e1 && e2	判断表达式 e1 与 e2,是否皆为真	• 为真,返回 true • 为假,返回 false
||	e1 || e2	判断表达式 e1 或 e2,是否**其中之一**为真	
!	! e1	判断表达式 e1,是否**并非**为真	
in	e1 in e2	判断 e1 所代表的属性名称,是否存在于 e2 所代表的对象实例中	

关于逻辑运算符的综合运用，可参考如下示例。

【2-2-4-logical-operators.js】

```
let v01 = 10, v02 = 50, v03 = 60 ;
let values = [10, 20, 30, 40, 50] ;
let person = {name: 'Gary', gender: 'male', age: '25'} ;

// result = true && true ;
result = v01 < v02 && v02 < v03 ;
console.log(result) ;

// result = false || true ;
result = v01 > v02 || v03 > v02 ;
console.log(result) ;

// result = ! false ;
result = ! (v01 > v02) ;
console.log(result) ;

result = 2 in values ;
console.log(result) ;

result = 'length' in values ;
console.log(result) ;

result = 'round' in Math ;
console.log(result) ;

result = 'gender' in person ;
console.log(result) ;
```

【相关说明】

```
let v01 = 10, v02 = 50, v03 = 60 ;
```

此语句声明了初始数值为整数常量的变量 v01、v02 与 v03。

```
let values = [10, 20, 30, 40, 50] ;
```

此语句声明了初始数据为数组实例的变量 values。

```
let person = {name: 'Gary', gender: 'male', age: '25'} ;
```

此语句声明了初始数据为对象实例的变量 person。

```
// result = true && true ;
result = v01 < v02 && v02 < v03 ;
```

【v01 < v02】为真，所以返回 true；【v02 < v03】为真，所以返回 true。因此，【v01 < v02 && v02 < v03】亦为真而返回 true，使得变量 result 的数据成为布尔值 true。

```
// result = false || true ;
result = v01 > v02 || v03 > v02 ;
```

【v01 > v02】为假，所以返回 false；【v03 > v02】为真，所以返回 true。因此，【v01 > v02 || v03 > v02】亦为真而返回 true，使得变量 result 的数据成为布尔值 true。

```
// result = ! false ;
result = ! (v01 > v02) ;
```

【v01 > v02】为假，所以返回 false。因此，【! (v01 > v02)】为真而返回 true，使得变量 result 的数据成为布尔值 true。

```
result = 2 in values ;
```

变量 values 的数据是数组实例 [10, 20, 30, 40, 50]，所以 values[0]可访问到数组实例的元素值 10；values[4]可访问到数组实例的元素值 50。

在数组实例名称 values 右侧的中括号[]里面，例如 values[3]，存在作为索引值（index value）的整数值 3，以访问其索引值 3 所代表的特定元素 40。

在此，因为索引值 2（第 3 个）对应到 values[2]所代表的第 3 个元素值 30，所以【2 in values】为真而返回 true，使得变量 result 的数据成为布尔值 true。

```
result = 'length' in values ;
```

变量 values 的数据是数组实例，因此变量 values 具有属性 length，可用来获取 values.length 所代表元素个数的整数值 5。所以，【'length' in values】为真而返回 true，使得变量 result 的数据成为布尔值 true。

```
result = 'round' in Math ;
```

内置的对象 Math 具有函数 round()，所以【'round' in Math】为真而返回 true，进而使得变量 result 的数据成为布尔值 true。

```
result = 'gender' in person ;
```

变量 person 的数据是对象实例{name: 'Gary', gender: 'male', age: '25'}，所以属性 gender 可用来访问到 person.gender 所代表属性的数据'male'。所以，【'gender' in person】为真而返回 true，使得变量 result 的数据成为布尔值 true。

2.2.5 条件运算符

从 C 语言开始，许多后继的编程语言，均支持条件运算符（conditional operator）/三元运算符（ternary operator），并用来简化特定形式的 if 语句。关于条件运算符的运用，可参考如下示例。

【2-2-5-conditional-operator.js】

```
let score, passed ;

score = 58 ;

passed = score >= 60 ? 'yes' : 'no' ;
console.log(passed) ;

score = 77 ;

passed = score >= 60 ? 'yes' : 'no' ;
console.log(passed) ;
```

```
score = 60 ;

if (score >= 60) passed = 'yes' ;
else passed = 'no' ;

console.log(passed) ;
```

【相关说明】

```
let score, passed ;
```

此语句声明了变量 score 与 passed。

```
score = 58 ;
```

此语句使得变量 score，被赋予整数值 58。

```
passed = score >= 60 ? 'yes' : 'no' ;
```

对于表达式【score >= 60 ? 'yes' : 'no'】而言，若【score >= 60】为真，则返回字符串'yes'；否则返回字符串'no'。在此，其为假，所以返回字符串'no'，进而使得变量 passed 的数据，成为字符串'no'。

```
score = 77 ;
```

此语句使得变量 score，被赋予整数值 77。

```
passed = score >= 60 ? 'yes' : 'no' ;
```

在此，表达式【score >= 60 ? 'yes' : 'no'】返回字符串'yes'，进而使得变量 passed 的数据，成为字符串'yes'。

```
score = 60 ;
```

此语句使得变量 score，被赋予整数值 60。

```
if (score >= 60) passed = 'yes' ;
else passed = 'no' ;
```

对于此条件语句而言，若【score >= 60】为真，则变量 passed 会被赋予字符串'yes'；否则会被赋予字符串'no'。在此，其为真，所以返回字符串'yes'，进而使得变量 passed，被赋予字符串'yes'。

2.2.6 类型运算符

类型运算符（typeof operator）用来返回特定操作数（operand）的数据类型（data type）。关于类型运算符的综合运用，可参考如下示例。

【2-2-6-typeof-operator.js】

```
let num01 = 33, num02 = 1.414 ;

console.log(typeof num01) ;
console.log(typeof 33) ;
console.log(typeof num02) ;
console.log(typeof 1.414) ;
```

```
console.log('') ;

console.log(typeof Math.PI) ;
console.log(typeof NaN) ;
console.log(typeof Infinity) ;
console.log('') ;

console.log(typeof '') ;
console.log(typeof "") ;
console.log(typeof "Hello, Earth!") ;
console.log('') ;

console.log(typeof true) ;
console.log(typeof false) ;
console.log(typeof (num01 > num02)) ;
console.log('') ;

console.log(typeof undefined) ;
console.log(typeof num03) ;
console.log('') ;

console.log(typeof function() {}) ;
console.log(typeof Array.isArray) ;
console.log('') ;

console.log(typeof Object()) ;
console.log(typeof new Object()) ;
console.log(typeof {}) ;
console.log('') ;

console.log(typeof Array()) ;
console.log(typeof new Array()) ;
console.log(typeof []) ;
console.log('') ;

console.log(typeof null) ;
console.log('') ;

console.log(typeof String('test')) ;
console.log(typeof new String('test')) ;
console.log('') ;

console.log(typeof Number(123)) ;
console.log(typeof new Number(123)) ;
console.log('') ;

console.log(typeof Date()) ;
console.log(typeof new Date()) ;
```

【相关说明】

```
let num01 = 33, num02 = 1.414 ;
```

此语句声明了具有初始数值的变量 num01 与 num02。

```
console.log(typeof num01) ;
```

表达式【typeof num01】会得出变量 num01 的数据类型名称，在此为字符串'number'。

```
console.log(typeof 33) ;
```

表达式【typeof 33】会得出常量 33 的数据类型名称，在此为字符串'number'。

```
console.log(typeof num02) ;
```

表达式【typeof num02】会得出变量 num02 的数据类型名称，在此为字符串'number'。

```
console.log(typeof 1.414) ;
```

表达式【typeof 1.414】会得出常量 1.414 的数据类型名称，在此为字符串'number'。

```
console.log(typeof Math.PI) ;
```

表达式【typeof Math.PI】会得出内置对象 Math 的常量属性 PI 的数据类型名称，在此为字符串'number'。

```
console.log(typeof NaN) ;
```

表达式【typeof NaN】会得出原始常量 NaN 的数据类型名称，在此为字符串'number'。

```
console.log(typeof Infinity) ;
```

表达式【typeof Infinity】会得出内置常量 Infinity 的数据类型名称，在此为字符串'number'。

```
console.log(typeof '') ;
```

表达式【typeof ''】会得出**空**字符串''的数据类型名称，在此为字符串'string'。

```
console.log(typeof "") ;
```

表达式【typeof ""】会得出**空**字符串""的数据类型名称，在此为字符串'string'。

```
console.log(typeof "Hello, Earth!") ;
```

表达式【typeof "Hello, Earth!"】会得出字符串"Hello, Earth!"的数据类型名称，在此为字符串'string'。

```
console.log(typeof true) ;
```

表达式【typeof true】会得出布尔值 true 的数据类型名称，在此为字符串'boolean'。

```
console.log(typeof false) ;
```

表达式【typeof false】会得出布尔值 false 的数据类型名称，在此为字符串'boolean'。

```
console.log(typeof (num01 > num02)) ;
```

表达式【typeof (num01 > num02)】会得出子表达式【num01 > num02】返回值的数据类型名称，在此为字符串'boolean'。

```
console.log(typeof undefined) ;
```

值得关注的是，原始常量 undefined 的数据类型亦是 undefined。表达式【typeof undefined】会得出原始常量 undefined 的数据类型名称，在此为字符串'undefined'。

```
console.log(typeof num03) ;
```

在此，num03 未被声明为变量或函数名称，所以 num03 处于未定义（undefined）的状态。表达式【typeof num03】会得出未被声明的 num03 的数据类型名称，在此为字符串'undefined'。

```
console.log(typeof function() {});
```

表达式【typeof function() {}】会得出匿名函数的数据类型名称，在此为字符串'function'。

```
console.log(typeof Array.isArray);
```

表达式【typeof Array.isArray】会得出内置对象 Array 的函数 isArray()的数据类型名称，在此为字符串'function'。

```
console.log(typeof Object());
```

因为 Object 对象的构造函数 Object()，会返回 Object 对象的实例；也因此表达式【typeof Object()】会得出 Object 对象实例的数据类型名称，在此为字符串'object'。

```
console.log(typeof new Object());
```

表达式【typeof new Object()】也会得出 Object 对象实例的数据类型名称，在此亦为字符串'object'。此外，JavaScript 语言尚未界定【Object()】与【new Object()】的明显区别。

```
console.log(typeof {});
```

表达式【typeof {}】会得出空对象实例{}的数据类型名称，在此亦为字符串'object'。而在 JavaScript 编程语言中，如下 3 个语句是等价的：

- obj = {};
- obj = new Object();
- obj = Object();

```
console.log(typeof Array());
```

因为 Array 对象的构造函数 Array()，会返回 Array 对象的实例；也因此表达式【typeof Array()】会得出 Array 对象实例的数据类型名称，在此竟然也为字符串'object'。

```
console.log(typeof new Array());
```

表达式【typeof new Array()】**亦**会得出 Array 对象实例的数据类型名称，在此亦为字符串'object'。此外，JavaScript 编程语言尚未界定【Array()】与【new Array()】的明显区别。

```
console.log(typeof []);
```

表达式【typeof []】会得出**空**数组实例[]的数据类型名称，在此为字符串'object'。而在 JavaScript 语言中，如下 3 个语句是等价的：

- arr = [];
- arr = new Array();
- arr = Array();

在 JavaScript 语言里，将数组实例（array instance）的数据类型视为 object。

```
console.log(typeof null);
```

【typeof null】表达式会得出原始常量 null 的数据类型名称，在此为字符串'object'。

```
console.log(typeof String('test')) ;
```

内置对象 String 的构造函数 String('test')，只会返回字符串'test'。因此，表达式【typeof String('test')】等同于获得字符串'test'的数据类型名称，在此为字符串'string'。

```
console.log(typeof new String('test')) ;
```

内置对象 String 的构造函数 String('test')会返回字符串'test'。再经过 new 关键字的处理，会返回内含字符串'test'的 String 对象实例。因此，表达式【typeof new String('test')】等同于获得 String 对象实例的数据类型名称，在此为字符串'object'。

```
console.log(typeof Number(123)) ;
```

内置对象 Number 的构造函数 Number(123)，会返回整数常量 123。因此，【typeof Number(123)】表达式等同于获得整数常量 123 的数据类型名称，在此为字符串'number'。

```
console.log(typeof new Number(123)) ;
```

内置对象 Number 的构造函数 Number(123)，会返回整数常量 123。再经过 new 关键字的处理，会返回内含整数常量 123 的 Number 对象实例。因此，表达式【typeof new Number(123)】等同于获得 Number 对象**实例**的数据类型名称，在此为字符串'object'。

```
console.log(typeof Date()) ;
```

内置 Date 对象的构造函数 Date()，会返回当前日期与时间的字符串。因此，表达式【typeof Date()】等同于获得当前日期与时间的字符串的数据类型名称，在此为字符串'string'。

```
console.log(typeof new Date()) ;
```

内置 Date 对象的构造函数 Date()，会返回当前日期与时间的字符串。再经过 new 关键字的处理，会返回内含当前日期与时间的 Date 对象实例。因此，表达式【typeof new Date()】等同于获得 Date 对象实例的数据类型名称，在此为字符串'object'。

2.2.7 按位运算符

JavaScript 语言的按位运算符（bitwise operator），在被运算之前，其两侧操作数（operand）的数据，会被转换成为 32 个比特位的数据。关于按位运算符的综合运用，可参考如下示例。

【2-2-7-bitwise-operators.js】

```
let num01 = 56, num02 = 77 ;
let num03 = 124, num04 = -3 ;

console.log(num01.toString(2)) ;
console.log(num02.toString(2)) ;
console.log(num03.toString(2)) ;
console.log('') ;

console.log(num01 == 0b111000) ;
console.log(num01 == 0b00111000) ;
```

```
console.log('') ;

console.log(num02 == 0b1001101) ;
console.log(num02 == 0b01001101) ;
console.log('') ;

console.log(num03 == 0b1111100) ;
console.log(num03 == 0b01111100) ;
console.log('') ;

/*
 00111000
&01001101
----------
 00001000
*/
result = num01 & num02 ;

console.log(result) ;
console.log(result.toString(2)) ;
console.log('') ;

/*
 00111000
|01001101
----------
 01111101
*/
result = num01 | num02 ;

console.log(result) ;
console.log(result.toString(2)) ;
console.log('') ;

/*
 00111000
^01001101
----------
 01110101
*/
result = num01 ^ num02 ;

console.log(result) ;
console.log(result.toString(2)) ;
console.log('') ;

/*
 ~ 01111100
------------
   10000011 (negative value)
=
 -01111100
+ 00000001 (because of 2's complement)
------------
  -01111101
```

```
*/
result = ~ num03 ;

console.log(result) ;
console.log(result.toString(2)) ;
console.log('') ;

/*
<< 001111100
------------
   011111000
*/
result = num03 << 1 ;

console.log(result) ;
console.log(result.toString(2)) ;
console.log('') ;

/*
>> 001111100
------------
   000111110
*/
result = num03 >> 1 ;

console.log(result) ;
console.log(result.toString(2)) ;
console.log('') ;

/*
but -3 actually is
    11111111111111111111111111111101 in 32-bit 2's complement.
So,
>>> 11111111111111111111111111111101
-------------------------------------
    01111111111111111111111111111110
*/
result = num04 >>> 1 ;

console.log(num04.toString(2)) ;
console.log(result) ;
console.log((2 ** 31 - 1) - 1) ;
console.log(result.toString(2)) ;
```

【相关说明】

```
let num01 = 56, num02 = 77 ;
let num03 = 124, num04 = -3 ;
```

这两个语句声明了初始数据为整数值的变量 num01、num02、num03 与 num04。

```
console.log(num01.toString(2)) ;
console.log(num02.toString(2)) ;
console.log(num03.toString(2)) ;
```

这 3 个语句中的【.toString(2)】，可分别将变量 num01、num02 与 num03 的整数值，转换成为字符串类型的**二进制**数码。

```
console.log(num01 == 0b111000) ;
```

【num01 == 0b111000】可用来判断变量 num01 的数值，是否等于二进制整数【111000】。在此，其返回 true。加上【0b】在二进制数码的左侧，可使得此代码被视为二进制数值。

在此亦可得知，浏览器中的 JavaScript 引擎，仍然将**最左比特位**（left-most bit）为 1 的数值，视为正值（positive value）。因此，若要变更成为负值（negative value），则修改为【- 0b111000】。

```
console.log(num01 == 0b00111000) ;
```

【num01 == 0b00111000】可用来判断变量 num01 的数值，是否等于二进制整数【00111000】。在此，其返回 true。

在此可得知，浏览器中的 JavaScript 引擎将【0b00111000】与【0b111000】，视为相同的二进制数值。

```
console.log(num02 == 0b1001101) ;
console.log(num02 == 0b01001101) ;
```

【num02 == 0b1001101】可用来判断变量 num02 的数值，是否等于二进制整数【1001101】。在此，其返回 true。

【num02 == 0b01001101】可用来判断变量 num02 的数值，是否等于二进制整数【01001101】。在此，其返回 true。

在此可得知，JavaScript 引擎将【0b1001101】与【0b01001101】，视为相同的二进制数值。

```
console.log(num03 == 0b1111100) ;
console.log(num03 == 0b01111100) ;
```

【num03 == 0b1111100】可用来判断变量 num03 的数值，是否等于二进制整数【0b1111100】。在此，其返回 true。

【num03 == 0b01111100】可用来判断变量 num03 的数值，是否等于二进制整数【0b01111100】。在此，其返回 true。

由此可得知，JavaScript 引擎将【0b1111100】与【0b01111100】，视为相同的二进制数值。

```
/*
 00111000
&01001101
----------
 00001000
*/
result = num01 & num02 ;
```

【num01 & num02】可得到【变量 num01 与 num02 的二进制数据，进行**按位和**（bitwise and）运算】之后的结果值。

```
/*
 00111000
|01001101
----------
 01111101
*/
```

```
result = num01 | num02 ;
```

【num01 | num02】可得到【变量 num01 与 num02 的二进制数据，进行**按位或**（bitwise or）运算】之后的结果值。

```
/*
 00111000
^01001101
----------
 01110101
*/
result = num01 ^ num02 ;
```

【num01 ^ num02】可得到【变量 num01 与 num02 的二进制数据，进行**按位异或**（bitwise exclusive or）运算】之后的结果值。

```
/*
 ~ 01111100
------------
   10000011  (negative value)
=
 -01111100
+ 00000001  (because of 2's complement)
------------
 -01111101
*/
result = ~ num03 ;
```

【~ num03】会得到【变量 num03 的二进制数据，进行 **2 的补码**（two's complement）】之后的结果值。

```
/*
<< 001111100
------------
   011111000
*/
result = num03 << 1 ;
```

【num03 << 1】会得出【变量 num03 的 2 进位数据，向**左**偏移 1 个比特位】之后的结果值。

```
/*
>> 001111100
------------
   000111110
*/
result = num03 >> 1 ;
```

【num03 >> 1】会得出【变量 num03 的 2 进位数据，向**右**偏移 1 个比特位】之后的结果值。

```
/*
but -3 actually is
    11111111111111111111111111111101 in 32-bit 2's complement.
So,
>>> 11111111111111111111111111111101
------------------------------------
    01111111111111111111111111111110
```

```
*/
result = num04 >>> 1 ;
```

【num04 >>> 1】会得出【变量 num04 的 2 进位数据,向**右**偏移 1 个比特位的同时,在其最左侧的符号位(sign bit),填入 0】之后的结果值。

此语句被执行之前,变量 num04 的数值为负整数-3。此语句被执行之后,变量 num04 的数值却变成非常大的正整数 2147483646,可见其符号位被填入了 0。

```
console.log((2 ** 31 - 1) - 1) ;
```

【(2 ** 31 - 1) - 1】会评估出(2^{31} - 1) - 1,也就是正整数 2147483646。

2.2.8　括号运算符

括号运算符包含:

- 小括号 / 圆括号(parentheses, round brackets)运算符():除了用于变更特定表达式的运算优先级之外,亦被用于 if、switch、for、while 等语句和函数的调用。
- 中括号 / 方括号(brackets, square brackets)运算符[]:主要用于数组(array)或字符串(string)相关的定义和访问。
- 大括号 / 花括号(braces, curly brackets)运算符{}:用于语句的分组(grouping)、函数的主体构造,以及对象(object)的定义。

关于括号运算符的综合运用,可参考如下示例。

【2-2-8-brackets-operators.js】

```
let a = 10, b = 5, c = 3 ;
let result = 0 ;

result = a - b * c ;
console.log(result) ;

result = (a - b) * c ;
console.log(result) ;

result = a ** c ** 2 ;
console.log(result) ;

result = (a ** c) ** 2 ;
console.log(result) ;
console.log('') ;

///
let now = new Date() ;

console.log(now) ;
console.log(now.toLocaleString()) ;
console.log('') ;
```

```
    let obj = new Object() ;

    obj.name = 'Jasper' ;
    obj.gender = 'male' ;
    obj.age = 28 ;

    console.log(obj) ;
    console.log('') ;
    ///
    function display(choice, message)
    {
      if (choice == 1) alert(message) ;
      else if (choice == 2) confirm(message) ;
    }

    display(1, 'Hello, Earth!') ;
    display(2, 'Hello, are you human?') ;
    ///
    let fruits = ['apple', 'banana', 'cherry', 'durian'] ;

    console.log(fruits[1]) ;
```

【相关说明】

```
    let a = 10, b = 5, c = 3 ;
    let result = 0 ;
```

这两个语句声明了具有初始数值的变量 a、b、c 和 result。

```
    result = a - b * c ;
```

表达式【a - b * c】会先被计算出【b * c】的结果值，再计算出整个表达式的结果值。

```
    result = (a - b) * c ;
```

因为小括号运算符的缘故，表达式【(a - b) * c】会先被计算出【a - b】的结果值，再计算出整个表达式的结果值。

```
    result = a ** c ** 2 ;
```

因为求幂运算符**具有**右结合**的特征，所以表达式【a ** c ** 2】会先被计算出【c ** 2】的结果值，再计算整个表达式的结果值。

```
    result = (a ** c) ** 2 ;
```

因为小括号运算符的缘故，表达式【(a ** c) ** 2】会先被计算出【a ** c】的结果值，再被计算出整个表达式的结果值。

```
    let now = new Date() ;
```

此语句使得变量 now，具有初始数据为【内含当前日期与时间】的 Date 对象实例。

```
    console.log(now) ;
```

借助此语句，可在网页浏览器的调试工具【Console】面板中，显示出标准格式的日期与时间。例如：【Fri Dec 29 2017 01:16:16 GMT+XX00 (XX 标准时间)】。

```
console.log(now.toLocaleString()) ;
```

【now.toLocalString()】可将日期与时间，以精简的本地格式，显示出来。例如：【2017/12/29 上午 1:16:16】。

```
let obj = new Object() ;
```

因为小括号运算符的缘故，使得 Object 对象的构造函数 Object()被调用。此语句使得变量 obj 的初始数据，成为 Object 对象的**空**实例（empty instance）。

```
obj.name = 'Jasper' ;
obj.gender = 'male' ;
obj.age = 28 ;
```

这 3 个语句分别设置了变量 obj 的新属性 name、gender、age 和其个别数据'Jasper'、'male'、28。

```
console.log(obj) ;
```

此语句可显示出变量 obj 的数据，也就是对象实例{name: "Jasper", gender: "male", age: 28}。

```
function display(choice, message)
```

因为小括号运算符的缘故，网页浏览器或其他软件中的 JavaScript 引擎，可认出关键字 function 开头的此源代码片段，即是用来定义【带有参数 choice 和 message，而且其名称为 display】的函数。

```
{
  if (choice == 1) alert(message) ;
  else if (choice == 2) confirm(message) ;
}
```

借助大括号运算符{}，JavaScript 引擎可认出大括号{ ... }里的源代码，即是函数 display()的主体结构。

在函数 display()的主体结构里，仍然可以看到小括号运算符，出现在 if 语句中。另外，亦可看到小括号运算符，出现在内置函数 alert()与 confirm()的调用语法中。

```
display(1, 'Hello, Earth!') ;
display(2, 'Hello, are you human?') ;
```

这两个语句，通过传入不同参数值，而重复调用了函数 display()。

```
let fruits = ['apple', 'banana', 'cherry', 'durian'] ;
```

借助中括号运算符[]，在此语句里的等号右侧，即是内含 4 个字符串元素的数组实例。

```
console.log(fruits[1]) ;
```

此语句通过数组变量的名称 fruits，以及中括号运算符里的元素索引值 1，访问到 fruits[1]所代表的字符串元素'banana'。

2.2.9 扩展运算符（ES6）

简单来说，由 3 个英文句点【...】构成的扩展运算符（spread operator），可被用来【卸除】特定数组的中括号或者特定对象的大括号。关于扩展运算符的综合运用，可参考如下示例。

【2-2-9-spread-operator.js】

```js
var greetings = ['Hi', 'Howdy', 'Hey, man', 'G\'day mate'] ;

var extended_greetings = ['Long time no see', 'Nice to see you', 'Hiya' , ... greetings] ;

console.log(extended_greetings) ;
console.log('') ;

var number_texts = ['one', 'two', 'three']
var number_digits = '123' ;

var numbers = [... number_texts, ... number_digits]
console.log(numbers) ;
console.log('') ;

///
let birthday = new Date(1999, 11, 25, 20, 30) ;
let now = new Date() ;

console.log(birthday.toLocaleString()) ;
console.log(now.toLocaleString()) ;
console.log('') ;

///
let arr01 = [1, 2, 3] ;
let arr02 = [10, 20, 30] ;
let arr03 = [... arr01, ... arr02, 100, 200, 300] ;

let obj01 = {name: 'orange', amount: 10} ;
let obj02 = {name: 'durian', amount: 5, origin: 'Thai'} ;
let obj03 = {... obj01, ... obj02} ;

console.log(arr03) ;
console.log(obj03) ;
console.log('') ;
```

【相关说明】

```js
var greetings = ['Hi', 'Howdy', 'Hey, man', 'G\'day mate'] ;
```
此语句声明了初始数据为数组实例的变量 greetings。

```js
var extended_greetings = ['Long time no see', 'Nice to see you', 'Hiya' , ... greetings] ;
```
此语句声明了初始数据为另一数组实例的变量 extended_greetings。扩展运算符【...】使得变量 extended_greetings 的数组实例，带有变量 greetings 的数组实例中的所有元素。

```
var number_texts = ['one', 'two', 'three']
```

此语句亦声明了初始数据为另一数组实例的变量 number_texts。

```
var number_digits = '123' ;
```

此语句声明了初始数据为字符串'123'的变量 number_digits。

```
var numbers = [... number_texts, ... number_digits]
```

扩展运算符【...】使得变量 numbers 的初始数据，成为合并【变量 number_texts 与 number_digits 的个别数组实例】的新数组实例，进而使得变量 numbers 的数据，成为数组实例["one", "two", "three", "1", "2", "3"]。

```
let birthday = new Date(1999, 11, 25, 20, 30) ;
```

此语句声明了【初始数据为内含日期与时间 1999/12/25 20:30:00 的 Date 对象实例】的变量 birthday。

需留意的是，构造函数 Date()小括号中的第 2 个参数值为 11，其实是代表 12 月份的含义。换句话说，此参数值若为 0，则代表 1 月份。

```
let now = new Date() ;
```

此语句声明了【初始数据为**当前日期与时间**的 Date 对象实例】的变量 now。

```
console.log(birthday.toLocaleString()) ;
```

此语句会显示出【1999/12/25 下午 8:30:00】的信息。

```
console.log(now.toLocaleString()) ;
```

此语句显示出当前日期与时间的信息。

```
let arr01 = [1, 2, 3] ;
let arr02 = [10, 20, 30] ;
```

这两个语句声明了【初始数据为不同数组实例】的变量 arr01 与 arr02。

```
let arr03 = [... arr01, ... arr02, 100, 200, 300] ;
```

此语句声明了变量 arr03，并借助扩展运算符【...】，使得其【初始数据为合并变量 arr01 与 arr02 的数组实例，再衔接子数组实例[100, 200, 300]】的新数组实例[1, 2, 3, 10, 20, 30, 100, 200, 300]。

```
let obj01 = {name: 'orange', amount: 10} ;
let obj02 = {name: 'durian', amount: 5, origin: 'Thai'} ;
```

这两个语句声明了【初始数据为对象实例】的变量 obj01 与 obj02。

```
let obj03 = {... obj01, ... obj02} ;
```

此语句声明了变量 obj03，并借助扩展运算符【...】，使得其【初始数据为合并变量 obj01 与 obj02 对象实例】的新对象实例{name: "durian", amount: 5, origin: "Thai"}。

值得注意的是，变量 obj01 与 obj02 均存在属性 name 与 amount。经过扩展运算之后，变量 obj03 的属性 name 与 amount 却只有一个，而且其属性的数据，均个别与变量 obj02 的属性 name 和 amount

的数据，是相同的。

2.2.10 逗号运算符

逗号运算符（comma operator）主要用来并联多个赋值表达式或操作数。关于逗号运算符的运用，可参考如下示例。

【2-2-10-comma-operator.js】

```
let a = 11, b = 21, c = 31 ;
let d, e, f ;

d = e = a + b, f = b + c ;

console.log(d, e, f) ;
```

【相关说明】

```
let a = 11, b = 21, c = 31 ;
```

此语句声明了初始数据均为整数的变量 a、b 与 c。

```
let d, e, f ;
```

此语句声明了未设置初始数据的变量 d、e 与 f。

```
d = e = a + b, f = b + c ;
```

此语句主要由两个表达式【d = e = a + b】与【f = b + c】构成。

- 第 1 个表达式先被计算出【a + b】的结果值，再赋给变量 d 和 e。
- 第 2 个表达式先被计算出【b + c】的结果值，再赋给变量 f。

修改此语句中的逗号运算符【,】，成为代表语句结束的分号【;】，并不影响其结果值。只是，原本的单一语句，就变成在同一行的两个语句了。

```
console.log(d, e, f) ;
```

通过传入变量 d、e 和 f 的数值，作为函数 console.log() 的参数值，可让这些变量的数值，显示在同一行的信息里，例如【32 32 52】。

2.2.11 删除运算符

JavaScript 语言中的删除运算符（delete operator），仅用来删除特定对象实例的特定属性。关于删除运算符的运用，可参考如下示例。

【2-2-11-delete-operator.js】

```
let person = {name: 'Ivory', gender: 'female', age: '30'} ;
let colors = ['RoyalBlue', 'GreenYellow', 'Gold', 'Cyan'] ;

var num01 = 123 ;
```

```
let num02 = 456 ;
num03 = 789 ;

console.log(num01, num02, num03) ;
console.log('') ;

delete person.name ;
delete colors[1] ;

delete num01 ;
delete num02 ;
delete num03 ;

console.log(person.name) ;
console.log(person) ;
console.log('') ;

console.log(colors[1]) ;
console.log(colors) ;
console.log('') ;

console.log(num01) ;
console.log(num02) ;
console.log(num03) ;
```

【相关说明】

```
let person = {name: 'Ivory', gender: 'female', age: '30'} ;
```

此语句声明了初始数据为对象实例的变量 person。

```
let colors = ['RoyalBlue', 'GreenYellow', 'Gold', 'Cyan'] ;
```

此语句声明了初始数据为数组实例的变量 colors。

```
var num01 = 123 ;
let num02 = 456 ;
num03 = 789 ;
```

无论有无通过关键字 var 或 let 进行声明,这 3 个语句分别声明了初始数据为不同整数的变量 num01、num02 与 num03。

```
delete person.name ;
```

此语句使用关键字 delete,并配合点运算符【.】,可删除变量 person 的对象实例的属性 name。

```
delete colors[1] ;
```

此语句使用关键字 delete,并配合中括号运算符[],可在变量 colors 数组实例中,仅删除索引值为 1 的元素数据'GreenYellow',但是此元素占用的缓存空间,仍然保留在其数组实例中。

```
delete num01 ;
delete num02 ;
delete num03 ;
```

通过这 3 个语句,试图删除变量 num01、num02 与 num03;然而实际上,仅有未通过 var 或 let 关键字,加以声明的变量 num03,可以被成功删除。

```
console.log(person.name) ;
```

【person.name】会返回 undefined，即代表变量 person 的属性 name，已被删除了。

```
console.log(person) ;
```

此语句只会产生{gender: "female", age: "30"}的信息。由此可见，属性 name 和其数据'Ivory'已被删除了。

```
console.log(colors[1]) ;
```

colors[1]在此返回 undefined，即代表变量 colors 中的索引值为 1 的数据，已经被清除了。

```
console.log(colors) ;
```

此语句会产生["RoyalBlue", empty, "Gold", "Cyan"]的信息，可看出 colors[1]对应的元素数据已经被清除了，才会被标记为 empty。

```
console.log(num01) ;
console.log(num02) ;
```

查看这两个语句所产生的信息，可得知变量 num01 与 num02 并未被删除！那是因为这两个变量，是借助 var 或 let 关键字，来加以声明的。

```
console.log(num03) ;
```

执行此语句时，会产生【num03 is not defined 参考错误（reference error）】的信息。由此可知，变量 num03 确实被删除了。

2.2.12 运算符的优先级（ES6）

在特定表达式中，运算符的优先级（operators precedence）可用来决定各子表达式的执行顺序。在相邻的子表达式中，其运算符优先级较高的子表达式，会先被执行。关于各运算符的优先级，如表 2-5 所示。

表 2-5 运算符的优先级

优先级	用途	关联性	运算符
20	变更优先级	无	(…)
19	访问属性或函数	由左至右	… . …
	访问子元素或定义数组实例的元素个数	由左至右	[…]
	创建特定对象【非】默认的实例	无	new … (…)
	调用特定函数	由左至右	… (…)
18	创建特定对象默认的实例	由右至左	new …
17	评估后再递增	无	… ++
	评估后再递减		… --

（续表）

优先级	用途	关联性	运算符
16	逻辑非运算	由右至左	! …
	按位非运算		~ …
	正值		+ …
	负值		- …
	先递增再评估		++ …
	先递减再评估		-- …
	返回数据类型		typeof …
	立即调用匿名函数		void …
	删除特定数据		delete …
	等待评估后的数据		await …
15	求幂运算	由右至左	… ** …
14	乘法运算	由左至右	… * …
	除法运算		… / …
	求余运算		… % …
13	加法运算	由左至右	… + …
	减法运算		… - …
12	按位左移运算	由左至右	… << …
	按位右移运算		… >> …
	按位无符号右移运算		… >>> …
11	比较小于的关系	由左至右	… < …
	比较小于或等于的关系		… <= …
	比较大于的关系		… > …
	比较大于或等于的关系		… >= …
	判断是否存在特定属性		… in …
	判断是否为特定对象的实例		… instanceof …
10	判断数据是否相等	由左至右	… == …
	判断数据是否不相等		… != …
	判断数据【和】类型是否都相同		… === …
	判断数据【或】类型是否不相同		… !== …
9	按位与运算	由左至右	… & …
8	按位异或运算	由左至右	… ^ …
7	按位或运算	由左至右	… | …
6	逻辑与运算	由左至右	… && …
5	逻辑或运算	由左至右	… || …
4	条件运算	由右至左	… ? … : …

(续表)

优先级	用途	关联性	运算符
3	赋值运算	由右至左	… = … … += … … -= … … *= … … /= … … %= … … <<= … … >>= … … >>>= … … &= … … ^= … … |= …
2	返回生成器中的特定数据 返回子生成器中的特定数据	由右至左	yield … yield * …
1	扩展运算	无	… , …
0	逗号运算	由左至右	… , …

2.3 练 习 题

1. 在 JavaScript 语言里，下列哪些项目是常量？

770、Math.PI、'Nice Day'、"good"、/\w\s\d/g、/.\w.\d/、TRUE、FALSE、Undefined、Null。

2. 在 JavaScript 语言里，应该通过什么语法声明名称为 love_you_forever 而代表着整数值 201314 的常量？

3. 在 JavaScript 语言里，应该通过什么语法声明名称为 love_me_longer 而代表着浮点数值 2591.8 的常量？

4. 在如下 JavaScript 源代码片段里，哪些是**全局**变量？哪些是**局部**变量？

```
let value01 = 10 ;
var value02 = 30 ;

function func01(data, identity)
{
  let result ;

  result = 21 + data + 2 * identity ;
  return result ;
}

var str01 = 'Finished', str02 = 'Error' ;

function func02(amount, price)
```

```
{
  let output ;

  output = (value01 + value02 + price) * amount ;
  return output ;
}
```

5. 在如下 JavaScript 源代码片段里，请**至少**列举出其中的 **5 个**表达式。

```
sphere_volume = r => 4 / 3 * Math.PI * Math.pow(r, 3) ;
```

6. 编写带有上底宽度、下底宽度和高度 3 个参数，可计算并返回**梯形面积**的**函数**定义。

7. 编写带有**长轴**长度、**短轴**长度 2 个参数，可计算并返回**椭圆面积**的**函数**定义。

8. 如下源代码被执行之后，变量 result 的结果数据是什么？

```
let result, n01 = 10, n02 = 20 ;
result = n01++ + ++n02 % 6 ;
```

9. 如下源代码被执行之后，变量 result 的结果数据是什么？

```
let result, n03 = 30, n04 = 40 ;
result = n03-- - --n04 % 6 ;
```

10. 如下源代码被执行之后，变量 result 的结果数据是什么？

```
let result, n05 = 50, n06 = 60 ;
result = (n05 / 5) ** 3 + (n06 / 10) ** 2 ;
```

11. 如下源代码被执行之后，变量 result 的结果数据是什么？

```
let result, n07 = 70, n08 = 80 ;
result = (n07 / 10) ** 2 + (n08 / 20) ** 0.5 ;
```

12. 通过编写条件运算符【? ... :】相关的源代码来实现下述功能。

已知两个变量 score 和 rating。当 score 的数值低于 60 时，变量 rating 的数据为字符串'failed'；当 score 的数值大于或等于 60 而小于 80 时，变量 rating 的数据为字符串'passed'；当 score 的数值大于或等于 80 而小于或等于 100 时，变量 rating 的数据为字符串'nice'；当 score 的数值超过 100 时，变量 rating 的数据为字符串'error'。

13. 至少列举类型运算符 typeof 表达式可返回的 5 种数据类型。

第 3 章

数 据 类 型

不同的编程语言存在不同的数据类型（data type），各种数据类型可用来描述容纳特不同数据的数据结构（data structure）。特定表达式被评估时，可返回特定类型的数据。

3.1 数 值 类 型

JavaScript 语言的数值，不仅包括整数（integer），也包含带有小数部分的浮点数（floating-point number）。

3.1.1 2^n 进制的字面量（ES6）

2^n 进制主要是指二进制、八进制与十六进制。在 JavaScript 语言中，可通过 2^n 进制的字面量（literal），来表示 2^n 进制的**数字常量**（number constant）/ **数值**（number value）。

【3-1-1-nth-power-of-2-based-literals.js】

```
var number01 = 111 ;
var number02 = parseInt('111', 16) ;
var number03 = 0x111 ;
var number04 = 0111 ; // cannot use in strict mode.
var number05 = parseInt('111', 8) ;
var number06 = 0o111 ; // new literal in ES6
var number07 = parseInt('111', 2) ;
var number08 = 0b111 ; // new literal in ES6

console.log(number01) ;
console.log(number02) ;
console.log(number03) ;
```

```
console.log(number04) ;
console.log(number05) ;
console.log(number06) ;
console.log(number07) ;
console.log(number08) ;
```

【相关说明】

```
var number01 = 111 ;
```

111 代表十进制的数值 111。

```
var number02 = parseInt('111', 16) ;
```

parseInt('111', 16) 返回十六进制数值 111 所对应的十进制数值 273。

```
var number03 = 0x111 ;
```

0x111 代表十六进制的数值 111。因此，变量 number03 的初始数值，会成为十六进制数值 111 所对应的十进制数值 273。

```
var number04 = 0111 ; // cannot use in strict mode.
```

0111 代表八进制的数值 111。因此，变量 number04 的初始数值，会成为八进制数值 111 所对应的十进制数值 73。

```
var number05 = parseInt('111', 8) ;
```

parseInt('111', 8) 会返回八进制数值 111 所对应的十进制数值 73。

```
var number06 = 0o111 ; // new literal in ES6
```

0o111 代表八进制的数值 111。因此，变量 number06 的初始数值，会成为八进制数值 111 所对应的十进制数值 73。

```
var number07 = parseInt('111', 2) ;
```

parseInt('111', 2) 会返回二进制数值 111 所对应的十进制数值 7。

```
var number08 = 0b111 ; // new literal in ES6
```

0b111 代表二进制的数值 111。因此，变量 number08 的初始数值，会成为二进制数值 111 所对应的十进制数值 7。

3.1.2　数值的比较（ES6）

要比较两个数值类型的操作数（operand）的数值（number value）是相等、不相等、大于或小于的关系，必须通过带有特定比较运算符的表达式来实现。关于数值的比较，可参考如下示例。

【3-1-2-number-comparisons.js】

```
console.log(3 + 0.1416 === 3.1416) ;
console.log(0.5 + 0.25 === 0.75) ;
console.log('') ;

console.log(0.1 + 0.02 === 0.12) ;
```

```
a = 0.1 ;
b = 0.02 ;
c = 0.12 ;
console.log(a + b === c) ;
console.log(Math.abs(a + b - c) < Number.EPSILON) ;
```

【相关说明】

```
console.log(3 + 0.1416 === 3.1416) ;
console.log(0.5 + 0.25 === 0.75) ;
```

在这两个语句中的比较表达式里，比较运算符【===】两侧操作数的数值，被判断为均相等，以及其数据类型均为浮点数（floating-point number）。

```
console.log('') ;

console.log(0.1 + 0.02 === 0.12) ;
```

在这个语句中，虽然比较运算符【===】两侧的数值，看起来是相等的，但是判断的结果却为假而返回布尔值 false。

计算机的浮点数表示法，无法完整表示大部分的浮点数，所以计算机进行浮点数的运算时，常常会出现小数部分的截断误差（truncation error）。看似简单浮点数的比较运算，对计算机来说，反而是困难的任务。

```
a = 0.1 ;
b = 0.02 ;
c = 0.12 ;
```

这 3 个语句声明了初始数值为不同浮点数的变量 a、b 与 c。

```
console.log(a + b === c) ;
```

虽然比较运算符【===】两侧的数值，看起来是相等的，但是判断的结果却为假而返回布尔值 false。

```
console.log(Math.abs(a + b - c) < Number.EPSILON) ;
```

【Number.EPSILON】是内置对象 Number 中的常量属性 EPSILON，用来表示浮点数运算之后，可被容忍的截断误差值。因此，欲正确比较浮点数，表达式【a + b === c】应该被改写成为【Math.abs(a + b - c) < Number.EPSILON】。

3.1.3 数值的正负符号（ES6）

在 JavaScript 语言中，数值可以是负数、正数或零值，而且零值还可以分为正零和负零。关于数值的正负符号的判断与诠释，可参考如下示例。

【3-1-3-number-sign-deteminations.js】

```
console.log(Math.sign(0)) ;
console.log(Math.sign(-0)) ;

console.log(Math.sign(123)) ;
```

```
console.log(Math.sign(-123)) ;

console.log(Math.sign(null)) ;

console.log(Math.sign(NaN)) ;
console.log(Math.sign('Hello')) ;
console.log(Math.sign(undefined)) ;
```

【相关说明】

函数 Math.sign(特定变量名称)的返回值，可用来判断特定变量的如下特征：

- 其返回值为 0，代表特定变量的数值亦为 0。
- 其返回值为-0，代表特定变量的数值亦为-0。
- 其返回值为 1，代表特定变量的数值为正数，例如 123、23.8、1250.75。
- 其返回值为-1，代表特定变量的数值为负数，例如-123、-29.15、-2500.88。
- 其返回值为 NaN（not a number），代表特定变量的数据，无法被转换成为数值，例如 NaN、undefined、'Hello'、'125a'。

```
console.log(Math.sign(0)) ;
```

Math.sign(0)的返回值为 0。

```
console.log(Math.sign(-0)) ;
```

Math.sign(-0)的返回值为-0。

```
console.log(Math.sign(123)) ;
```

Math.sign(123)的返回值为 1。

```
console.log(Math.sign(-123)) ;
```

Math.sign(-123)的返回值为-1。

```
console.log(Math.sign(null)) ;
```

Math.sign(null)的返回值为 0。

```
console.log(Math.sign(NaN)) ;
```

Math.sign(NaN)的返回值为 NaN。其中，NaN 的含义表示**并非一个数值**（not a number）。

```
console.log(Math.sign('Hello')) ;
```

Math.sign('Hello')的返回值为 NaN。

```
console.log(Math.sign(undefined)) ;
```

Math.sign(undefined)的返回值为 NaN。

3.1.4　数值的截断（ES6）

数值的截断（truncation），是指对于特定浮点数，先进行**无条件**或**四舍五入**的**数值修约**，再去除其小数，转换成为整数的过程。关于数值的截断的综合运用，可参考如下示例。

【3-1-4-number-truncations.js】

```
v01 = 12.5 ;
v02 = 12.3 ;
v03 = 0.56 ;
v04 = -0.83 ;

console.log(parseInt(v01)) ;
console.log(Math.trunc(v01)) ;
console.log(Math.floor(v01)) ;
console.log(Math.round(v01)) ;
console.log(Math.ceil(v01)) ;
console.log('') ;

console.log(parseInt(v02)) ;
console.log(Math.trunc(v02)) ;
console.log(Math.floor(v02)) ;
console.log(Math.round(v02)) ;
console.log(Math.ceil(v02)) ;
console.log('') ;

console.log(parseInt(v03)) ;
console.log(Math.trunc(v03)) ;
console.log(Math.floor(v03)) ;
console.log(Math.round(v03)) ;
console.log(Math.ceil(v03)) ;
console.log('') ;

console.log(parseInt(v04)) ;
console.log(Math.trunc(v04)) ;
console.log(Math.floor(v04)) ;
console.log(Math.round(v04)) ;
console.log(Math.ceil(v04)) ;
console.log('') ;

console.log(0 === -0) ;
console.log(+0 === -0) ;
```

【相关说明】

```
v01 = 12.5 ;
v02 = 12.3 ;
v03 = 0.56 ;
v04 = -0.83 ;
```

这 4 个语句声明了初始数据为不同浮点数的变量 v01、v02、v03 与 v04。

```
console.log(parseInt(v01)) ;
```

调用内置函数 parseInt(v01)，会返回变量 v01 的数值 12.5 的整数部分 12。

```
console.log(Math.trunc(v01)) ;
```

调用内置函数 Math.trunc(v01)，会返回变量 v01 的数值 12.5 的整数部分 12。

```
console.log(Math.floor(v01));
```

调用内置函数 Math.floor(v01)，会返回**小于**但**接近**变量 v01 的数值 12.5 的整数值 12。

```
console.log(Math.round(v01));
```

调用内置函数 Math.round(v01)，会返回对变量 v01 的数值 12.5 进行**四舍五入**之后的整数值 13。

```
console.log(Math.ceil(v01));
```

调用内置函数 Math.ceil(v01)，会返回**大于**但**接近**变量 v01 数值 12.5 的整数值 13。

```
console.log(parseInt(v02));
```

调用内置函数 parseInt(v02)，会返回变量 v02 的数值 12.3 的整数部分 12。

```
console.log(Math.trunc(v02));
```

调用内置函数 Math.trunc(v02)，会返回变量 v02 的数值 12.3 的整数部分 12。

```
console.log(Math.floor(v02));
```

调用内置函数 Math.floor(v02)，会返回**小于**但**接近**变量 v02 的数值 12.3 的整数值 12。

```
console.log(Math.round(v02));
```

调用内置函数 Math.round(v02)，会返回对变量 v02 的数值 12.3 进行**四舍五入**之后的整数值 12。

```
console.log(Math.ceil(v02));
```

调用内置函数 Math.ceil(v02)，会返回**大于**但**接近**变量 v02 的数值 12.3 的整数值 13。

```
console.log(parseInt(v03));
```

调用内置函数 parseInt(v03)，会返回变量 v03 的数值 0.56 的整数部分 0。

```
console.log(Math.trunc(v03));
```

调用内置函数 Math.trunc(v03)，会返回变量 v03 的数值 0.56 的整数部分 0。

```
console.log(Math.floor(v03));
```

调用内置函数 Math.floor(v03)，会返回**小于**但**接近**变量 v03 的数值 0.56 的整数值 0。

```
console.log(Math.round(v03));
```

调用内置函数 Math.round(v03)，会返回对变量 v03 的数值 0.56 进行**四舍五入**之后的整数值 1。

```
console.log(Math.ceil(v03));
```

调用内置函数 Math.ceil(v03)，会返回**大于**但**接近**变量 v03 的数值 0.56 的整数值 1。

```
console.log(parseInt(v04));
```

调用内置函数 parseInt(v04)，会返回变量 v04 的数值-0.83 的整数部分 0。

```
console.log(Math.trunc(v04));
```

调用内置函数 Math.trunc(v04)，会返回变量 v04 的数值-0.83 的**负整数**部分-0。

```
console.log(Math.floor(v04));
```

调用内置函数 Math.floor(v04)，会返回**小于**但**接近**变量 v04 的数值-0.83 的整数值-1。

```
console.log(Math.round(v04)) ;
```

调用内置函数 Math.round(v04)，会返回对变量 v04 的数值-0.83 进行**四舍五入**之后的整数值-1。

```
console.log(Math.ceil(v04)) ;
```

调用内置函数 Math.ceil(v04)，会返回大于但接近变量 v04 的数值-0.83 的负整数值-0。

```
console.log(0 === -0) ;
```

【0 === -0】返回布尔值 true，所以 0 与 -0 的数值相等、数据类型相同。

```
console.log(+0 === -0) ;
```

【+0 === -0】返回布尔值 true，所以+0 与 -0 的数值相等、数据类型相同。

3.1.5 数值的特殊格式（ECMA-402）

在此，数值的特殊格式，是指千分位（thousands）或者货币（currency）形式的数值表示方式。关于数值的特殊格式的综合运用，可参考如下示例。

【3-1-5-number-formattings.js】

```
var nf_en = Intl.NumberFormat("en") ;
var nf_de = Intl.NumberFormat("de") ;

var number01 = 2533591.8 ;

console.log(nf_en.format(number01)) ;
console.log(nf_de.format(number01)) ;
console.log('') ;

console.log(number01.toString()) ;
console.log(number01.toLocaleString()) ;
console.log('') ;

console.log(number01.toLocaleString('en')) ;
console.log(number01.toLocaleString('de')) ;
console.log('') ;

// currency number format
var cnf_cn = Intl.NumberFormat('cn', {style: 'currency', currency: 'cny'}) ;
var cnf_jp = Intl.NumberFormat('jp', {style: 'currency', currency: 'jpy'}) ;
var cnf_en = Intl.NumberFormat('en', {style: 'currency', currency: 'usd'}) ;
var cnf_uk = Intl.NumberFormat('gb', {style: 'currency', currency: 'gbp'}) ; // Great Britain
 Pound
var cnf_de = Intl.NumberFormat('de', {style: 'currency', currency: 'eur'}) ;

var price01 = 25324700.56 ;

console.log(cnf_cn.format(price01)) ;
console.log(cnf_jp.format(price01)) ;
console.log(cnf_en.format(price01)) ;
```

```
console.log(cnf_uk.format(price01)) ;
console.log(cnf_de.format(price01)) ;
```

【相关说明】

```
var nf_en = Intl.NumberFormat("en") ;
```

Intl.NumberFormat 对象的构造函数 Intl.NumberFormat("en")，会返回【以字符串形式，表示英文（en, English）数值格式（number format）】的 Intl.NumberFormat 对象实例。因此，后续源代码，可借助其数据为 Intl.NumberFormat 对象实例的变量 nf_en，将特定数值，表示成为**英文**数值格式化之后的字符串。

```
var nf_de = Intl.NumberFormat("de") ;
```

Intl.NumberFormat("de")会返回【以字符串形式，表示德文 / 德国（de, DENIC eG for Germany）数值格式】的 Intl.NumberFormat 对象实例。因此，后续源代码，可借助其数据为 Intl.NumberFormat 对象实例的变量 nf_de，将特定数值，表示成为**德文**数值格式化之后的字符串。

```
var number01 = 2533591.8 ;
```

此语句声明了初始数据为 2533591.8 的变量 number01。

```
console.log(nf_en.format(number01)) ;
```

因为变量 nf_en 的数据为 Intl.NumberFormat 对象实例，所以存在可调用的函数 format()。nf_en.format(number01)会返回【将变量 number01 的数值 2533591.8，表示成为**英文**数值格式化】之后的字符串"2,533,591.8"。

```
console.log(nf_de.format(number01)) ;
```

nf_en.format(number01)会返回【将变量 number01 的数值 2533591.8，表示成为**德文**数值格式化】之后的字符串"2.533.591,8"。

```
console.log('') ;
console.log(number01.toString()) ;
```

number01.toString()仅会返回【将变量 number01 的数值 2533591.8，转换成为字符串】之后的"2533591.8"。

```
console.log(number01.toLocaleString()) ;
```

number01.toString()仅会返回【将变量 number01 的数值 2533591.8，表示成为**本地语言**的数值格式化】之后的"2,533,591.8"。在此，本地语言为中文（cn, China）。

```
console.log('') ;
console.log(number01.toLocaleString('en')) ;
```

console.log(number01.toLocaleString('en'))和如下源代码片段之一，存在着异曲同工之妙：

- var nf_en = Intl.NumberFormat("en") ;
  console.log(nf_en.format(number01));
- console.log(Intl.NumberFormat("en").format(number01)) ;

```
console.log(number01.toLocaleString('de')) ;
```

number01.toLocaleString('de')和如下源代码片段之一，存在着异曲同工之妙：

- var nf_de = Intl.NumberFormat("de") ;
  console.log(nf_de.format(number01)) ;
- console.log(Intl.NumberFormat("de").format(number01)) ;

```
console.log('') ;

// currency number format
var cnf_cn = Intl.NumberFormat('cn', {style: 'currency', currency: 'cny'}) ;
```

Intl.NumberFormat 对象的构造函数 Intl.NumberFormat('cn', {style: 'currency', currency: 'cny'})，会返回【以字符串形式，表示中文／中国（cn, China）与人民币（cny, China Yuan）的数值格式】的 Intl.NumberForma 对象实例。后续源代码，可借助其数据为 Intl.NumberFormat 对象实例的变量 cnf_cn，将特定数值，表示成为**中文**与人民币的数值格式化之后的字符串。

```
var cnf_jp = Intl.NumberFormat('jp', {style: 'currency', currency: 'jpy'}) ;
```

Intl.NumberFormat('jp', {style: 'currency', currency: 'jpy'})，会返回【以字符串形式，表示日文（jp, Japan）与日币（jpy, Japan Yuan）的数值格式】的 Intl.NumberFormat 对象实例。后续源代码，可借助其数据为 Intl.NumberFormat 对象实例的变量 cnf_jp，将特定数值，表示成为**日文**与日币的数值格式化之后的字符串。

```
var cnf_en = Intl.NumberFormat('en', {style: 'currency', currency: 'usd'}) ;
```

Intl.NumberFormat('en', {style: 'currency', currency: 'usd'})，会返回【以字符串形式，表示英文（en, English）与美元（usd, United States Dollar）的数值格式的 Intl.NumberFormat 对象实例。后续源代码可借助其数据为 Intl.NumberFormat 对象实例的变量 cnf_en，将特定数值，表示成为**英文**与美元的数值格式化之后的字符串。

```
var cnf_uk = Intl.NumberFormat('gb', {style: 'currency', currency: 'gbp'}) ; // Great Britain
Pound
```

Intl.NumberFormat('gb', {style: 'currency', currency: 'gbp'})，会返回【以字符串形式，表示英式英文（gb, Great Britain）与英镑（gbp, Great Britain Pound）的数值格式】的 Intl.NumberFormat 对象实例。后续源代码可借助其数据为 Intl.NumberFormat 对象实例的变量 cnf_uk，将特定数值，表示成为**英式英文**与英镑的数值格式化之后的字符串。

```
var cnf_de = Intl.NumberFormat('de', {style: 'currency', currency: 'eur'}) ;
```

Intl.NumberFormat('de', {style: 'currency', currency: 'eur'})会返回【以字符串形式，表示德文／德国（de, DENIC eG for Germany）与欧元（eur, Euro）的数值格式】的对象实例。后续源代码可借助其数据为 Intl.NumberFormat 对象实例的变量 cnf_de，将特定数值，表示成为**德文**与欧元的数值格式化之后的字符串。

```
var price01 = 25324700.56 ;
```

此语句声明了初始数据为 25324700.56 的变量 price01。

```
console.log(cnf_cn.format(price01));
```

因为变量 cnf_cn 是 Intl.NumberFormat 对象实例，所以存在可调用的函数 format()。cnf_cn.format(price01)会返回【将变量 price01 的数值 25324700.56，表示成为**人民币**数值格式化】之后的字符串"CN¥25,324,700.56"。

```
console.log(cnf_jp.format(price01));
```

cnf_jp.format(price01)会返回【将变量 price01 的数值 25324700.56，表示成为**日币**数值格式化】之后的字符串"JP¥25,324,701"。

```
console.log(cnf_en.format(price01));
```

cnf_en.format(price01)会返回【将变量 price01 的数值 25324700.56，表示成为**美元**数值格式化】之后的字符串"$25,324,700.56"。

```
console.log(cnf_uk.format(price01));
```

cnf_uk.format(price01)会返回【将变量 price01 的数值 25324700.56，表示成为**英镑**数值格式化】之后的字符串"£25,324,700.56"。

```
console.log(cnf_de.format(price01));
```

cnf_de.format(price01)会返回【将变量 price01 的数值 25324700.56，表示成为**德文**与**欧元**的数值格式化】之后的字符串"£25,324,700.56"。

关于语言代码（language code），例如 cn、en、gb 等，以及货币代码（currency code），例如 cny、usd、gbp 等，可参考如下网址的内容：

- 语言代码： en.wikipedia.org/wiki/Language_code
- 货币代码： en.wikipedia.org/wiki/ISO_4217

3.1.6 整数值的安全范围（ES6）

简单来说，JavaScript 语言的整数值的安全范围是-(2^{53} - 1) ~ +(2^{53} - 1)。关于整数值的安全范围的理解，可参考如下示例。

【3-1-6-number-safe-ranges.js】

```
console.log(Number.isNaN(NaN));
console.log(Number.isNaN(123));

console.log(Number.isFinite(456));
console.log(Number.isFinite(NaN));

console.log(Number.isFinite(Infinity));
console.log(Number.isFinite(-Infinity));

console.log('');

console.log(Number.isSafeInteger(Math.pow(2, 48) + 100));
console.log(Number.isSafeInteger(Math.pow(2, 53) - 1));
console.log(Number.isSafeInteger(Math.pow(2, 53)));
```

【相关说明】

函数 Number.isNaN(特定变量名称)的返回值，可用来判断特定变量的数值，是否并非一个数值（not a number）：

- 其返回值为 true，代表特定变量的数值，并不是一个数值。
- 其返回值为 false，代表特定变量的数值，是一个数值。

```
console.log(Number.isNaN(NaN));
```

Number.isNaN(NaN)的返回值为 true。

```
console.log(Number.isNaN(123));
```

Number.isNaN(123)的返回值为 false。函数 Number.isFinite(特定变量名称)的返回值，可用来判断特定变量的数值，是否为**有限**（finite）的数值：

- 其返回值为 true，代表特定变量的数值，是计算机可明确表示的数值。
- 其返回值为 false，代表特定变量的数值，**并不是**计算机可明确表示的数值。

```
console.log(Number.isFinite(456));
```

Number.isFinite(456)的返回值为 true。

```
console.log(Number.isFinite(NaN));
```

Number.isFinite(NaN)的返回值为 false。

```
console.log(Number.isFinite(Infinity));
```

Number.isFinite(Infinity)的返回值为 false。

```
console.log(Number.isFinite(-Infinity));
```

Number.isFinite(-Infinity)的返回值为 false。

函数 Number.isSafeInteger(特定变量名称)的返回值，可用来判断特定变量的数值，是否为 IEEE-754 所规范的安全范围 $-(2^{53} - 1) \sim +(2^{53} - 1)$ 里的整数值：

- 返回值为 true，代表特定变量的数值，是安全范围内的整数值。
- 返回值为 false，代表特定变量的数值，并不是安全范围内的整数值。

```
console.log(Number.isSafeInteger(Math.pow(2, 48) + 100));
```

Math.pow(2, 48) + 100 的结果值为 $2^{48} + 100$。

Number.isSafeInteger(Math.pow(2, 48) + 100)的返回值为 true。

```
console.log(Number.isSafeInteger(Math.pow(2, 53) - 1));
```

Math.pow(2, 53) - 1 的结果值为 $2^{53} - 1$。

Number.isSafeInteger(Math.pow(2, 53) - 1)的返回值为 true。

```
console.log(Number.isSafeInteger(Math.pow(2, 53)));
```

Math.pow(2, 53)的返回值为 2^{53}。

Number.isSafeInteger(Math.pow(2, 53))的返回值为 false。

```
console.log(Math.pow(2, 64)) ;
```

Math.pow(2, 64)的返回值，理应为 2^{64}，也就是 18446744073709551**616**；但是，其真实的返回值却为 18446744073709552**000**。所以，其最右侧的 4 位数，明显存在误差。有误差的缘故，就是因为 2^{64} 已经是超越安全范围的整数值了。

3.2 布尔类型

JavaScript 布尔（boolean）类型的数据，只存在布尔值 false（假）和布尔值 true（真）两个组合。其中，false 代表假、否、非、不对、不是等**不成立**的含义；true 则代表真、对、是等**成立**的含义。

【3-2---Boolean-data-type.js】

```
let passed = false, score = 0 ;

score = 80 ;

passed = score > 60 ;

console.log(passed) ;

score = 55 ;

passed = score > 60 ;

console.log(passed) ;
console.log('') ;

///
let conversion ;

conversion = Boolean(-10.8) ;

console.log(conversion) ;

conversion = Boolean(15.6) ;

console.log(conversion) ;
console.log('') ;

///
conversion = Boolean(0) ;

console.log(conversion) ;

conversion = Boolean(null) ;

console.log(conversion) ;
```

```
conversion = Boolean(false) ;

console.log(conversion) ;

conversion = Boolean('') ;

console.log(conversion) ;

conversion = Boolean(undefined) ;

console.log(conversion) ;
console.log('') ;

///
conversion = new Boolean(-10.8) ;

console.log(conversion.valueOf()) ;

conversion = new Boolean(15.6) ;

console.log(conversion.valueOf()) ;

conversion = new Boolean(0) ;

console.log(conversion.valueOf()) ;
```

【相关说明】

```
let passed = false, score = 0 ;
```

此语句声明了初始数据为 false 的变量 passed，以及初始数据为 0 的变量 score。

```
score = 80 ;
```

此语句赋予整数值 80 给变量 score。

```
passed = score > 60 ;
```

在此，【score > 60】是比较运算式，得出来的结果值为布尔值 true，并被赋给变量 passed。

```
console.log(passed) ;
```

```
score = 55 ;
```

此语句赋予整数值 55 给变量 score。

```
passed = score > 60 ;
```

此语句将【score > 60】的结果值 false，赋给变量 passed。

```
console.log(passed) ;
console.log('') ;

///
let conversion ;
```

此语句声明了变量 conversion。

```
conversion = Boolean(-10.8) ;
```

-10.8 是一个非 0 的数值，因此，经过 Boolean 对象的构造函数 Boolean() 处理之后，会返回 true。

```
console.log(conversion) ;

conversion = Boolean(15.6) ;
```

15.6 是一个非 0 的数值，因此，经过 Boolean 对象的构造函数 Boolean() 处理之后，会返回 true。

```
console.log(conversion) ;
console.log('') ;

///
conversion = Boolean(0) ;

conversion = Boolean(null) ;

conversion = Boolean(false) ;

conversion = Boolean('') ;

conversion = Boolean(undefined) ;
```

在前段各语句中，无论是 0、null、false、空字符串"或 undefined，均是具有零值意义的常量，经过构造函数 Boolean() 的处理之后，均会返回 false。

```
console.log(conversion) ;
console.log('') ;

conversion = new Boolean(-10.8) ;

conversion = new Boolean(15.6) ;

conversion = new Boolean(0) ;
```

在前段各语句中，构造函数 Boolean() 左侧被衔接了关键字 new，使得变量 conversion 的数据，成为内含布尔值 true 或 false 的 Boolean 对象实例。

```
console.log(conversion.valueOf()) ;
```

调用 conversion.valueOf() 函数，可返回变量 conversion 内含的布尔值 true 或 false。

3.3 数 组 类 型

数组（array）是用来表示一组【具有**各自数据**】的元素。然而，就 JavaScript 语言来说，其数组实例的数据类型（data type）为**对象**类型。数组实例中各**元素**的数据，可以同时存在**数值**、**布尔值**、**字符串**、**对象**，甚至是子**数组**！关于数组类型的综合运用，可参考如下示例。

【3-3---Array-data-type.js】

```
let numbers = [520, 530, 1314, 2013, 2014] ;
let profile = ['Tommy', 'male', 33, [180, 72]] ;
let newone = new Array(6) ;

console.log(numbers) ;
console.log(numbers[2]) ;

numbers[1] = 2591.8 ;

console.log(numbers) ;
console.log(numbers.length) ;
console.log('') ;

///
console.log(profile) ;
console.log(profile[0]) ;
console.log(profile[2]) ;
console.log(profile[3][0]) ;
console.log(profile[3][1]) ;

profile[3][1] = 70 ;

console.log(profile) ;
console.log(profile.length) ;
console.log(profile[3].length) ;

///
console.log(numbers[numbers.length]) ;
console.log(profile[profile.length]) ;
console.log('') ;
console.log(newone[0]) ;
console.log(newone[newone.length]) ;
```

【相关说明】

```
let numbers = [520, 530, 1314, 2013, 2014] ;
```

此语句声明了初始数据为内含多个**整数元素**的数组实例[520, 530, 1314, 2013, 2014]的变量 numbers。

```
let profile = ['Tommy', 'male', 33, [180, 72]] ;
```

此语句声明了初始数据为数组实例 ['Tommy', 'male', 33, [180, 72]] 的变量 profile。

```
let newone = new Array(6) ;
```

new Array(6)会返回内含 6 个空元素的数组实例。此语句声明了初始数据为内含 6 个空元素的数组实例的变量 newone。

```
console.log(numbers) ;
```

在网页浏览器的调试工具【Console】面板中，显示出[520, 530, 1314, 2013, 2014]的信息。

```
console.log(numbers[2]) ;
```

```
numbers[1] = 2591.8 ;
```

数值 2591.8 被存为在变量 numbers 数组实例中，其索引值为 1（第 2 个）的元素值。

```
console.log(numbers) ;
```

显示出[520, 2591.8, 1314, 2013, 2014]的信息。

```
console.log(numbers.length) ;
```

numbers.length 会返回变量 numbers 数组实例中的元素个数。

```
console.log(profile) ;
```

在网页浏览器的调试工具【Console】面板中，显示出**可展开**的["Tommy", "male", 33, Array(2)]信息。

展开上述信息之后，可看见 Array(2)代表子数组实例[180, 70]。在此，应该为整数值 72 的，却被显示为如下屏幕快照里的整数值 70，是因为调试工具的运行机制，即时响应了随后源代码的执行结果。

```
▼(4) ["Tommy", "male", 33, Array(2)]
    0: "Tommy"
    1: "male"
    2: 33
  ▶ 3: (2) [180, 70]
    length: 4
  ▶ __proto__: Array(0)
```

```
console.log(profile[0]) ;
```

profile[0]会返回在变量 profile 的数组实例中，其索引值为 0（第 1 个）的元素的数据"Tommy"。

```
console.log(profile[2]) ;
```

profile[2]会返回在变量 profile 的数组实例中，其索引值为 2（第 3 个）的元素的数值 33。

```
console.log(profile[3][0]) ;
console.log(profile[3][1]) ;
```

profile[3]会返回在变量 profile 的数组实例中，其索引值为 3（第 4 个）的元素的数据，也就是子数组实例[180,72]。

所以，profile[3][0]会返回在子数组实例[180, 72]中，其索引值为 0（第 1 个）的元素的数值 180。profile[3][1]则会返回在子数组实例[180, 72]中，其索引值为 1（第 2 个）的元素的数值 72。

```
profile[3][1] = 70 ;
```

此语句使得在 profile[3]所代表的子数组实例[180, **72**]中，其索引值为 1（第 2 个）的元素的数值，被修改成为 70。因此，子数组实例变更为[180, **70**]。

```
console.log(profile.length) ;
```

profile.length 会返回在变量 profile 的数组实例中，其**第 1 层**元素的个数 4。

```
console.log(profile[3].length) ;
```

profile[3]代表子数组实例[180, 72]。profile[3].length 会返回子数组实例[180,72]的元素个数 2。

```
console.log(numbers[numbers.length]) ;
```

numbers.length 会返回变量 numbers 的数组实例中的元素个数 5。

在变量 numbers 的数组实例中，若其索引值为 5，则代表第 6 个元素的索引值。

在变量 numbers 的数组实例中，一开始并无第 6 个元素，所以，numbers[numbers.length]如同 numbers[5]的语法，会返回 undefined。

```
console.log(profile[profile.length]) ;
```

profile.length 会返回变量 profile 的数组实例中的元素个数 4。在变量 profile 的数组实例中，若其索引值为 4，则代表第 5 个元素的索引值。在变量 profile 的数组实例中，一开始并无第 5 个元素，所以 profile[profile.length]如同是 profile[4]的语法，会返回 undefined。

```
console.log(newone[0]) ;
console.log(newone[newone.length]) ;
```

因为变量 newone 一开始为内含 6 个空元素的数组实例，所以 newone[0]会返回 undefined，newone.length 会返回变量 newone 的数组实例中的元素个数 6。在变量 newone 的数组实例中，若其索引值为 6，则代表第 7 个元素的索引值。然而，一开始并无第 7 个元素，所以 newone[newone.length]如同是 newone[7]的语法，会返回 undefined。

3.4 对象类型

对象（object）通过属性（property）和方法（method）/函数（function），模拟在真实世界中，具有**身份数据**（identity data）和表现**特征行为**（characteristic behavior）的物体（object）。关于对象类型的理解，可参考如下示例。

【3-4---Object-data-type.js】

```
// let item01 = {} ;
let item01 = new Object() ;

item01.name = 'Tablet PC' ;
item01.price = 1000 ;
item01.origin = 'China' ;
item01['manufacture date'] = '2018/12/15' ;
item01['color'] = 'RoyalBlue' ;
item01[''] = 'secret data...' ; // empty property name

console.log(item01) ;
console.log(item01['']) ;
console.log(item01.color) ;
console.log(item01['manufacture date']) ;
console.log(item01.price) ;

item01.price = 900 ;
```

```
item01['color'] = 'Gold' ;

console.log(item01) ;
```

【相关说明】

```
// let item01 = {} ;
let item01 = new Object() ;
```

通过关键字 new 和 Object 对象的构造函数 Object()的语法，可返回 Object 对象的空实例（empty instance），或称为空的 Object 对象实例。在此，【= new Object()】等同于【= {}】的语法；而且此语句声明了初始数据为空的 Object 对象实例的变量 item01。

```
item01.name = 'Tablet PC' ;
item01.price = 1000 ;
item01.origin = 'China' ;
```

通过变量 item01 衔接点运算符【.】，可动态创建变量 item01 的 Object 对象实例的 3 个新属性 name、price、origin 及其个别的数据'Table PC'、1000、'China'。

```
item01['manufacture date'] = '2018/12/15' ;
item01['color'] = 'RoyalBlue' ;
item01[''] = 'secret data...' ;  // empty property name
```

通过变量 item01 衔接中括号运算符[]，亦可动态创建变量 item01 的 Object 对象实例的 3 个新属性 manufacture date、color、''，以及其个别的数据'2018/12/15'、'RoyalBlue'、'secret data...'。值得注意的是，和衔接点运算符【.】比起来，衔接中括号运算符[]的方式，可创建具有如下特殊名称的属性：

- 带有空格（space）字符的属性名称，例如【manufacture date】。
- 空（empty）的属性名称，例如【''】。

```
console.log(item01) ;
```

此语句使得网页浏览器在调试工具【Console】面板中，显示出如下信息：

```
▼ {name: "Tablet PC", price: 1000, origin: "China", manufacture date:
  "RoyalBlue", …}
    "": "secret data..."
    color: "Gold"
    manufacture date: "2018/12/15"
    name: "Tablet PC"
    origin: "China"
    price: 900
  ▶ __proto__: Object
```

```
console.log(item01['']) ;
```

item01['']返回**空**属性对应的数据"secret data..."。

```
console.log(item01.color) ;
```

item01.color 返回属性 color 对应的数据"RoyalBlue"。

```
console.log(item01['manufacture date']) ;
```

item01['manufacture date']返回属性 manufacture date 对应的数据"2018/12/15"。

```
console.log(item01.price) ;
```

item01.price 返回属性 price 对应的数值 1000。

```
item01.price = 900 ;
```

此语句使得变量 item01 的属性 price，重新被赋予整数值 900。

```
item01['color'] = 'Gold' ;
```

此语句使得变量 item01 的属性 color，重新被赋予字符串'Gold'。

```
console.log(item01) ;
```

再次通过此语句，查看变量 item01 的数据是否被变更了。

3.5　字符串类型

字符串（string）是具有一串字符（character）的文本数据。在各种应用程序的画面当中，显示给用户观看的文本，即是持续由多个字符串，拼凑而成的产物。

3.5.1　一般字符串

在 JavaScript 语言中，字符串的字面量（literal）均以【内含一串字符】的一对单引号或双引号来表示。关于一般字符串的运用，请看下面的示例。

【3-5-1-common-strings.js】

```
let sentence = 'Hi,\n\tlong time no see!\nhow are you today?\n\nBest Regards,\nAlex' ;

console.log(sentence) ;

sentence = "Limit '3' days and \"5\" persons." ;

console.log(sentence) ;

sentence = '\\It costs \'370\' dollars.\\' ;

console.log(sentence) ;

sentence = 'Alex lovingly loves ' +
'lovely beloved of ' +
'Daisy.' ;

console.log(sentence) ;

sentence = 'Alex lovingly loves \
```

```
                lovely beloved of \
                Daisy.' ;

                console.log(sentence) ;
                console.log(sentence[0]) ;
                console.log(sentence[1]) ;
                console.log(sentence[2]) ;
                console.log(sentence[3]) ;
                console.log('') ;

                console.log(sentence.charAt(0)) ;
                console.log(sentence.charAt(1)) ;
                console.log(sentence.charAt(2)) ;
                console.log(sentence.charAt(3)) ;
```

【相关说明】

```
let sentence = 'Hi,\n\tlong time no see!\nhow are you today?\n\nBest Regards,\nAlex' ;
```

此语句声明了初始数据为字符串的变量 sentence。在此字符串中，存在多个换行（new line）字符'\n'，以及间隔至多如同 8 个空格的制表（tab）字符'\t'。

```
console.log(sentence) ;
```

此语句显示出变量 sentence 的如下数据字符串：

Hi,

       long time no see!

how are you today?

Best Regards,

Alex

```
sentence = "Limit '3' days and \"5\" persons." ;
```

此语句赋予新的字符串字面量，给变量 sentence。在此语句中，通过一对双引号【""】，容纳字符串。然而，在此双引号之内，还存在其他单引号与双引号。

容纳字符串的一对双引号【""】或单引号【''】，不可以冲突于其内部的双引号或单引号！其内部发生冲突的引号左侧，必须衔接反斜杠（back slash）字符才行，例如：

- "...\"....\"....'..'...."
- '...".....".....\'..\'....'

```
console.log(sentence) ;
```

此语句显示出变量 sentence 的数据字符串【Limit '3' days and "5" persons.】。

```
sentence = '\\It costs \'370\' dollars.\\' ;
```

此语句将新的字符串字面量，赋给变量 sentence。在此语句中，通过一对单引号【''】容纳字符串。然而，在单引号之内，还存在反斜杠和其他单引号字符。

在字符串内，单引号、反斜杠均是特殊字符，所以必须在其左侧，衔接额外的反斜杠字符，成为【\\】或【\'】，才能正常呈现在信息里。

```
console.log(sentence) ;
```

此语句显示出变量 sentence 的数据字符串【\It costs '370' dollars.\】。

```
sentence = 'Alex lovingly loves ' +
'lovely beloved of ' +
'Daisy.' ;
```

此语句被分成 3 行，在等号右侧，借助加法运算符【+】，进行 3 个字符串的合并运算。

```
console.log(sentence) ;
```

此语句显示出多个字符串被合并之后的新字符串'Alex lovingly loves lovely beloved of Daisy.'。

```
sentence = 'Alex lovingly loves \
lovely beloved of \
Daisy.' ;
```

此语句亦被分成 3 行，在等号右侧，借助反斜杠字符，串接字符串的各个片段。

欲使用此法，在各个反斜杠字符的右侧，不可再加上包括空格（space）在内的任何字符。

```
console.log(sentence) ;
```

此语句亦显示出多个字符串被合并之后的新字符串'Alex lovingly loves lovely beloved of Daisy.'。

```
console.log(sentence[0]) ;
console.log(sentence[1]) ;
console.log(sentence[2]) ;
console.log(sentence[3]) ;
```

变量 sentence 目前的数据为字符串'Alex lovingly loves lovely beloved of Daisy.'。所以，将变量 sentence 的数据字符串，视为内含多个字符的数组实例时，每个元素的数据，即为各个单一字符。这 4 个语句分别显示出在变量 sentence 的数据字符串里，其第 1 个至第 4 个元素的数据字符'A'、'l'、'e'与'x'。

```
console.log(sentence.charAt(0)) ;
console.log(sentence.charAt(1)) ;
console.log(sentence.charAt(2)) ;
console.log(sentence.charAt(3)) ;
```

这 4 个语句较为正统，和前面 4 个语句有异曲同工之妙，亦可显示出在变量 sentence 的数据字符串里，其第 1 个至第 4 个元素的数据字符'A'、'l'、'e'与'x'。

3.5.2 格式化字符串（ES6）

在 JavaScript 语言中，所谓的格式化字符串（formatting string）的正式名称为模板字面量（template literal），可用来嵌入待评估的表达式。当这些表达式被评估完成而返回特定数据时，格式化字符串才能被确认最终的模样。关于格式化字符串的运用，可参考如下示例。

【3-5-2-string-interpolations.js】

```
var users =
[
  {name: 'John', age: '33', gender: 'male'},
  {name: 'Jessica', age: '27', gender: 'female'},
  {name: 'Daisy', age: '33', gender: 'female'},
  {name: 'Sean', age: '24', gender: 'male'}
] ;

var nations = [ 'China', 'Canada', 'America', 'New Zeland'] ;

var days_amount = 5 ;

var flight_message = `${users[1].name} decides to flight to ${nations[0]} after ${days_amount}
  days.` ;

console.log(flight_message) ;

flight_message = `${users[3].name} decides to flight to ${nations[2]} after ${days_amount}
  days.` ;

console.log(flight_message) ;

var items =
[
  {product_id: 15023, price: 330},
  {product_id: 16002, price: 500}
]

var checkout_message = `This product costs ${items[1].price * 0.8}.` ;

console.log(checkout_message) ;

var string01 = 'Hello\nEarth!' ;
var string02 = `Hello\nEarth!` ;
var string03 = String.raw `Hello\nEarth!` ;

console.log(string01) ;
console.log(string02) ;
console.log(string03) ;

number_digits = '1 2 3 4 5 6 7' ;

// string04 = String.raw({raw: 'a b c d e f g'}, '1',' ','2',' ','3',' ','4',' ','5',' ','6','
 ','7') ;
string04 = String.raw({raw: 'a b c d e f g'}, ... number_digits) ;


console.log(string04) ;
```

【相关说明】

```
var users =
[
  {name: 'John', age: '33', gender: 'male'},
```

```
    {name: 'Jessica', age: '27', gender: 'female'},
    {name: 'Daisy', age: '33', gender: 'female'},
    {name: 'Sean', age: '24', gender: 'male'}
] ;
```

此语句声明了初始数据为【内含 4 个对象实例的数组实例】的变量 users。

```
var nations = ['China', 'Canada', 'America', 'New Zeland'] ;
```

此语句声明了初始数据为【内含多个字符串的数组实例】的变量 nations。

```
var days_amount = 5 ;
```

此语句声明了初始数值为整数 5 的变量 days_amount。

```
var flight_message = `${users[1].name} decides to flight to ${nations[0]} after ${days_amount} days.` ;
```

此语句声明了初始数据为格式化字符串的变量 flight_message。格式化字符串必须放入一对反引号（back quote）【``】中。在格式化字符串里，可通过${变量名称}或${可评估的表达式}的语法，将特定变量的数据或者表达式被评估之后的数据，放置于格式化字符串里。

```
console.log(flight_message) ;
```

此语句显示出变量 flight_message 的数据，也就是已经被转换完成的字符串'Jessica decides to flight to China after 5 days.'。

```
flight_message = `${users[3].name} decides to flight to ${nations[2]} after ${days_amount} days.` ;
```

此语句使得变量 flight_message，被赋予另一个格式化字符串。

```
console.log(flight_message) ;
```

此语句显示出变量 flight_message 的数据，也就是已经被转换完成的字符串'Sean decides to flight to America after 5 days.'。

```
var items =
[
    {product_id: 15023, price: 330},
    {product_id: 16002, price: 500}
]
```

此语句声明了初始数据为【内含对象实例的数组实例】的变量 items。

```
var checkout_message = `This product costs ${items[1].price * 0.8}.` ;
```

此语句声明了初始数据为格式化字符串的变量 checkout_message。

```
console.log(checkout_message) ;
```

此语句显示出变量 checkout_message 的数据，也就是已经被转换完成的字符串'This product costs 400.'。

```
var string01 = 'Hello\nEarth!' ;
```

此语句通过单引号【''】，声明了初始数据为一般字符串的变量 string01。

```
var string02 = `Hello\nEarth!` ;
```

此语句通过反引号【``】，声明了初始数据为格式化字符串的变量 string02。

```
var string03 = String.raw `Hello\nEarth!` ;
```

此语句通过内置的 String 对象的函数 raw()，保留其中各字符的原始编码，并声明了初始数据为原始（raw）字符串的变量 string03。也就是说，在字符串里的【\n】，已经被视为一般的字符【\】与【n】，而不再具有换行（new line）的特征。

值得注意的是，raw()明明是一个函数，在此却不能衔接一对小括号，而是衔接一对反引号，以实现出一个原始字符串。

```
console.log(string01) ;
console.log(string02) ;
```

这两个语句显示的信息相同，具体如下：

Hello
Earth!

```
console.log(string03) ;
```

此语句显示的信息是【Hello\nEarth!】，【\n】的两个字符【\】与【n】均被保留了下来。

```
number_digits = '1 2 3 4 5 6 7' ;
```

此语句声明了初始数据为字符串的变量 number_digits。

```
// string04 = String.raw({raw: 'a b c d e f g'}, '1',' ','2',' ','3',' ','4',' ','5',' ','6',' ','7') ;
string04 = String.raw({raw: 'a b c d e f g'}, ... number_digits) ;
```

等号右侧的 String.raw()，会返回**交叉分组**之后的字符串"a1   b2   c3   d4   e5   f6   g7 "。在此，如下语法是等价的：

- String.raw({raw: 'a b c d e f g'}, ... number_digits)
- String.raw({raw: 'a b c d e f g'}, '1',' ','2',' ','3',' ','4',' ','5',' ','6',' ','7')

```
console.log(string04) ;
```

此语句显示出变量 string04 的数据，也就是被**交叉分组**之后的字符串"a1   b2   c3   d4   e5   f6   g7 "。

3.5.3 日期与时间格式的字符串（ES6）

内含日期（date）与时间（time）相关数据的字符串，可被称为日期与时间格式的字符串。关于日期与时间格式的字符串的运用，可参考如下示例。

【3-5-3-date-and-time-strings.js】

```
var dt_cn = new Intl.DateTimeFormat('cn') ;
var dt_en = new Intl.DateTimeFormat('en') ;
var dt_de = new Intl.DateTimeFormat('de') ;
```

```
origin_datetime = new Date('2018-01-23') ;

console.log(origin_datetime.toDateString()) ;
console.log(origin_datetime.toLocaleDateString()) ;

console.log('') ;

dt01 = dt_cn.format(origin_datetime) ;
dt02 = dt_en.format(origin_datetime) ;
dt03 = dt_de.format(origin_datetime) ;

console.log(dt01) ;
console.log(dt02) ;
console.log(dt03) ;
```

【相关说明】

```
var dt_cn = new Intl.DateTimeFormat('cn') ;
```

Intl.DateTimeFormat 对象的构造函数 Intl.DateTimeFormat("cn")，会返回中文 / 中国（cn, China）格式的 Intl.DateTimeFormat 对象实例。后续源代码可借助变量 dt_cn 中的 Intl.DateTimeFormat 对象实例，将其他格式的日期与时间，表示成为中文格式的日期与时间。

```
var dt_en = new Intl.DateTimeFormat('en') ;
```

Intl.DateTimeFormat 对象的构造函数 Intl.DateTimeFormat("en")会返回英文（en, English）格式的 Intl.DateTimeFormat 对象实例。后续源代码可借助变量 dt_en 中的 Intl.DateTimeFormat 对象实例，将其他格式的日期与时间，表示成为英文格式的日期与时间。

```
var dt_de = new Intl.DateTimeFormat('de') ;
```

Intl.DateTimeFormat 对象的构造函数 Intl.DateTimeFormat("de")会返回德文 / 德国（de, DENIC eG for Germany）格式的 Intl.DateTimeFormat 对象实例。后续源代码可借助变量 dt_de 中的 Intl.DateTimeFormat 对象实例，将其他格式的日期与时间，表示成为德文格式的日期与时间。

```
origin_datetime = new Date('2018-01-23') ;
```

此语句声明了变量 origin_datetime，并被初始化为【日期是 2018/01/23】的 Date 对象实例。在此，Date('2018-01-23')或 Date('2018/01/23')均可被解读成功。

```
console.log(origin_datetime.toDateString()) ;
```

origin_datetime.toDateString()会返回字符串'Tue Jan 23 2018'。

```
console.log(origin_datetime.toLocaleDateString()) ;
```

origin_datetime.toLocaleDateString()会返回字符串'2018/1/23'。

```
dt01 = dt_cn.format(origin_datetime) ;
```

因为变量 dt_cn 的数据是 Intl.DateTimeFormat 对象实例，所以存在可调用的函数 format()。dt_cn.format(origin_datetime)会返回【在变量 origin_datetime 的 Date 对象实例中，其内含的日期与时间，被表示成为中文格式】之后的字符串'2018/1/23'。

```
dt02 = dt_en.format(origin_datetime) ;
```

因为变量 dt_en 的数据是 Intl.DateTimeFormat 对象实例，所以存在可调用的函数 format()。dt_en.format(origin_datetime)会返回【在变量 origin_datetime 的 Date 对象实例中，其内含的日期与时间，被表示成为英文格式】之后的字符串'1/23/2018'。

```
dt03 = dt_de.format(origin_datetime) ;
```

因为变量 dt_de 的数据是 Intl.DateTimeFormat 对象的实例，所以存在可调用的函数 format()。dt_de.format(origin_datetime)会返回【在变量 origin_datetime 的 Date 对象实例中，其内含的日期与时间，被表示成为德文格式】之后的字符串'23.1.2018'。

3.6 集合与地图类型

对 JavaScript 语言来说，集合（set）与地图（map）主要可用来简化编程的负担，是从 ECMAScript 2015（ES6）版本开始，才具有的数据类型。

3.6.1 集合类型（ES6）

集合（set）内含其数据均**不重复**的元素，和数学理论中的集合，有着非常相似的含义与原理。换句话说，在特定集合内，只存在其数据不相同的元素。关于集合的运用，可参考如下示例。

【3-6-1-Set-data-type.js】

```
let actions = new Set() ;

actions.add('read') ;
actions.add('write').add('update') ;
actions.add('delete') ;

actions.add('read').add('read').add('delete').add('write').add('update') ;

console.log(actions) ;

console.log(actions.entries()) ;
console.log(actions.keys()) ;
console.log(actions.values()) ;
console.log('') ;

for (let element of actions)
{
  console.log(element) ;
}

console.log('') ;
console.log(actions.size) ;

console.log(actions.has('hide')) ;
```

```
console.log(actions.has('write')) ;
```

【相关说明】

```
let actions = new Set() ;
```

此语句声明了初始数据为【空的 Set 对象实例】的变量 actions。

```
actions.add('read') ;
```

当前变量 actions 的数据为一个 Set 对象实例，因此存在可调用的函数 add()。在变量 actions 的 Set 对象实例中，添加数据为字符串'read'的新元素。

```
actions.add('write').add('update') ;
```

此语句使得在变量 actions 的 Set 对象实例中，连续添加数据为字符串'write'与'update'的两个新元素。

```
actions.add('delete') ;
```

此语句使得在变量 actions 的 Set 对象实例中，添加数据为字符串'delete'的新元素。

```
actions.add('read').add('read').add('delete').add('write').add('update') ;
```

在变量 actions 的 Set 对象实例中，连续添加数据为字符串'read'、'read'、'delete'、'write'与'update'的 5 个新元素。在这 5 个新元素里，存在重复的数据字符串；所以，被添加至变量 actions 的 Set 对象实例中的时候，重复的元素会自动被排除在外。

```
console.log(actions) ;
```

此语句显示出变量 actions 的 Set 对象实例【Set(4) {"read", "write", "update", "delete"}】的信息。在此，亦可看出在这个 Set 对象实例中，的确没有数据相同的元素。

```
console.log(actions.entries()) ;
console.log(actions.keys()) ;
console.log(actions.values()) ;
```

这 3 个语句皆显示出相同的信息【SetIterator {"read", "write", "update", "delete"}】。其中，Set Iterator 具有集合迭代器的含义。

```
for (let element of actions)
{
  console.log(element) ;
}
```

通过循环语句 for ... of 的迭代处理,可在每次循环中,显示出在变量 actions 的 Set 对象实例中，其特定元素的如下数据字符串：

- read
- write
- update
- delete

```
console.log(actions.size) ;
```

actions.size 会返回 Set 对象实例中的元素个数 4。

```
console.log(actions.has('hide')) ;
```

actions.has('hide')会返回 false，意味着在 Set 对象实例中，并无数据字符串为'hide'的元素。

```
console.log(actions.has('write')) ;
```

actions.has('write')会返回 true，代表着在 Set 对象实例中，存在数据字符串为'write'的元素。

3.6.2 地图类型（ES6）

地图（map）内含【**键名（key name）**对应到**值（value）/ 数据（data）**】的组合。地图与对象（object）极为相似，只是节省了对象的累赘和限制。关于地图的运用，可参考如下示例。

【3-6-2-Map-data-type.js】

```javascript
let items = new Map() ;

items.set('slipper', 50) ;
items.set('shoes', 200) ;
items.set('pants', 100).set('shirt', 150) ;

console.log(items) ;

console.log(items.size) ;

console.log(items.entries()) ;
console.log(items.keys()) ;
console.log(items.values()) ;

for (let [product, price] of items)
{
  console.log(`One ${product} costs ${price}.`) ;
}
```

【相关说明】

```
let items = new Map() ;
```

此语句声明了初始数据为【空的 Map 对象实例】的变量 items。

```
items.set('slipper', 50) ;
```

变量 items 的数据，当前为一个 Map 对象实例，因此存在可调用的函数 set()。此语句使得在变量 items 的 Map 对象实例中，被添加键名为 slipper、值为 50 的新元素。

```
items.set('shoes', 200) ;
```

此语句使得在变量 items 的 Map 对象实例中，被添加键名为 shoes、值为 200 的新元素。

```
items.set('pants', 100).set('shirt', 150) ;
```

此语句使得在变量 items 的 Map 对象实例中,连续添加键名分别为 pants 与 shirt、值分别为 100 与 150 的两个新元素。

```
console.log(items) ;
```

此语句显示出在变量 items 的 Map 对象实例【Map(4) {"slipper" => 50, "shoes" => 200, "pants" => 100, "shirt" => 150}】的信息。

```
console.log(items.size) ;
```

items.size 会返回 Map 对象实例中的元素个数 4。

```
console.log(items.entries()) ;
```

此语句显示出 MapIterator {"slipper" => 50, "shoes" => 200, "pants" => 100, "shirt" => 150}的信息。其中，Map Iterator 具有地图迭代器的含义。

```
console.log(items.keys()) ;
```

此语句显示出 MapIterator {"slipper", "shoes", "pants", "shirt"}的信息。

```
console.log(items.values()) ;
```

显示出 MapIterator {50, 200, 100, 150}的信息。

```
for (let [product, price] of items)
{
  console.log(`One ${product} costs ${price}.`) ;
}
```

通过此循环语句 for ... of 的迭代处理，可在每次循环中，显示出在变量 items 的 Set 对象实例中，特定元素的如下数据字符串：

- One slipper costs 50.
- One shoes costs 200.
- One pants costs 100.
- One shirt costs 150.

值得注意的是，在上述循环语句的小括号里，【let [product, price] of items】的语句，使得在每次循环中：

- 变量 product 的数据，成为在变量 items 的 Map 对象实例中，特定元素的键名，例如 slipper。
- 变量 price 的数据，成为在变量 items 的 Map 对象实例中，特定元素的值，例如 50。

3.7 数据类型的转换（ES6）

对于各编程语言来说，特定变量在不同的时间点上，可被赋予不同类型的数据！在 JavaScript 引擎中，通过自动转换机制或者一些内置函数，可使得特定变量的数据，从原来的数据类型，转换成为新的数据类型。

最常见的数据类型的转换，莫过于【其他数据→字符串】和【字符串→数值】。关于数据类型的转换的综合运用，可参考如下示例。

【3-7---data-type-conversions.js】

```
let digital_string = '  123  ';
result = Number(digital_string);

console.log(result);

digital_string = '  0o123';
result = Number(digital_string);

console.log(result);

digital_string = '0x123  ';
result = Number(digital_string);

console.log(result);

result = parseInt(digital_string);

console.log(result);

result = Math.round(digital_string);

console.log(result);
console.log('');

///
digital_string = '35.62';
result = Math.floor(digital_string);

console.log(result);

digital_string = '28.2';
result = Math.ceil(digital_string);

console.log(result);

digital_string = '12.5';
result = Math.round(digital_string);

console.log(result);
///
let value = 53.8125;
result = value.toString(2);

console.log(result);

result = value.toString(8);

console.log(result);

result = value.toString(16);
```

```
console.log(result);
```

【相关说明】

```
let digital_string = ' 123 ';
```

此语句声明了初始数据为字符串' 123 '的变量 digital_string。

```
result = Number(digital_string);
```

通过内置的 Number 对象的构造函数 Number()，将变量 digital_string 的数据字符串' 123 '，转换成十进制整数值 123。在转换过程中，所有空格（space）字符皆会被过滤掉。

```
digital_string = ' 0o123';
```

将字符串' 0o123'，赋给变量 digital_string。

```
result = Number(digital_string);
```

通过内置的 Number 对象的构造函数 Number()，将变量 digital_string 的数据字符串' 0o123'，先视为成八进制整数码 123，再转换成为等价的十进制整数值 83。

```
digital_string = '0x123 ';
```

将字符串'0x123 '赋给变量 digital_string。

```
result = Number(digital_string);
```

通过内置的 Number 对象的构造函数 Number()，将变量 digital_string 的数据字符串'0x123 '，先视为十六进制整数码 123，再转换成为等价的十进制整数值 291。

```
result = parseInt(digital_string);
result = Math.round(digital_string);
```

这两个语句皆可达成【将**字符串**'0x123 '，转换成为等价的十进制**整数值** 291】的任务。

```
digital_string = '35.62';
result = Math.floor(digital_string);
```

这两个语句将字符串'35.62'，转换成为十进制整数值 35。

```
digital_string = '28.2';
result = Math.ceil(digital_string);
```

这两个语句将字符串'28.2'，转换成为十进制整数值 29。

```
digital_string = '12.5';
result = Math.round(digital_string);
```

这两个语句将字符串'12.5'，转换成为十进制整数值 13。

```
let value = 53.8125;
```

此语句声明了初始数值为 53.8125 的变量 value。

```
result = value.toString(2);
```

value.toString(2)会返回转换之后的二进制数码 110101.1101。

```
result = value.toString(8) ;
```

value.toString(8)会返回转换后的八进制数码 65.64。

```
result = value.toString(16) ;
```

value.toString(16)会返回转换后的十六进制数码 35.d。

3.8 练 习 题

1. 在 JavaScript 语言里，应该使用什么语句，才能将浮点数 25.75，分别表示成为二进制、八进制和十六进制数码的字符串？

2. 在 JavaScript 语言里，应该使用什么语句，才能将二进制数码 110111011，**直接**转换成为十六进制数码？

3. 在 JavaScript 语言里，应该使用什么语句，才能直接将二进制数码 101100011100、八进制数码 1275、十六进制数码 51cf 的总和，转换成**二进制**数码？

4. 已知变量 x 与 y，请将 $3x^2 + 2(x-1)^2 y + 2xy^2 + 5y^3$，编写成为 JavaScript 语言中的**算术表达式**。

5. 已知有如下主要表达式：

```
let product = {item01: ['fruit_set', 100], item02: ['sticker_set', 250], item03: ['magnet_set',
    350], item04: ['drink_set', 150], item05: ['pizza_set', 300]} ;
```

请编写【将上述**各个整数值**的**总和**，赋给变量 result】的语句。

6. 说明原始常量 NaN、Infinity 和 undefined 的含义。

7. 执行如下 JavaScript 源代码片段之后，变量 result 的结果值是什么？

```
let num01 = 28.56, num02 = 32.47 ;
let result ;

result = parseInt(num01) + Math.trunc(num01) + Math.floor(num01) + Math.round(num01) +
    Math.ceil(num01) ;

result += parseInt(num02) + Math.trunc(num02) + Math.floor(num02) + Math.round(num02) +
    Math.ceil(num02) ;
```

8. 请编写 JavaScript 源代码，使得变量 price 的数值 72583000，显示成为**货币数值格式**的**字符串**' CN¥72,583,000.00'。

9. 列举两种 JavaScript 语法，来计算并显示【$\sqrt[3]{x}$, where $x = 768$】的数值。

10. 列举两个等价于【price >= 300 && amount < 10】的**比较与逻辑**表达式。

11. 已知如下数组实例：

```
let arr = [[[5, 10], [15, 20]], [[25, 30], [35, 40]], [[45, 50], [55, 60]]] ;
```

访问上述整数值 25、40 和 60 的语法是什么？

12. 已知如下对象实例：

```
let obj = {product:{en: 'browser', cn: '浏览器' }, developer: {en: 'Google', cn: '谷歌'}, price:
    {en: 'free', cn: '免费'}} ;
```

访问字符串'browser'、'谷歌'与'free'的语法是什么？

13. 已知数据为对象实例的变量 profile 和如下主要表达式：

```
let message = profile.name + ' now lives on ' + profile.planet + '.' ;
```

让变量 message 被设置为相同数据字符串的**等价语法**是什么？

14. 欲显示如下**分行**的信息：

```
Apple: 11
Banana: 15
Guava: 23
```

其较为简短的 JavaScript 语法应该是什么？

15. 欲显示当前的日期与时间，其较为简短的 JavaScript 语法应该是什么？

16. 欲声明其数据为如下**集合**实例的变量 components：

```
Set(5) {"window", "pane", "dialogue", "button", "scrollbar"}
```

其较为简短的 JavaScript 语法应该是什么？

17. 欲声明其数据为如下**地图**实例的变量 devices：

```
Map(4) {"mobile phone" => 10, "tablet PC" => 7, "notebook PC" => 3, "desktop PC" => 20}
```

其较为简短的 JavaScript 源代码应该是什么？

18. 已知变量 num 的数值为十进制整数值 201314，请编写【带有**置入变量名称**的模板字面量，并且显示如下信息】的 JavaScript 源代码：

```
变量 num 的：
    十进制数值 = 201314
    二进制数码 = 110001001001100010
    八进制数码 = 611142
    十六进制数码 = 31262
```

第 4 章

条件和循环语句

本章主要介绍 if 和 switch 两种条件语句,以及 for 和 while 两类循环语句,并且进一步介绍 for ... of 语句和迭代器的协作应用,以及 break 与 continue 语句的运用时机。

4.1 条件语句

条件语句(conditional statement)可依据不同的条件,决定并执行相对应的任务。在 JavaScript 语言里,条件语句大致可分为 if 和 switch 语句。

4.1.1 if 语句

JavaScript 语言的 if 语句,和 Java、C、C++、C#、PHP、R 语言等的 if 语句是相同的,并且可分为如下几种语法格式:

(1)

```
if ( ... )
{
   |
}
```

(2)

```
if ( ... )
{
   |
}
else
{
```

```
    |
}
```

（3）

```
if ( … )
{
    |
}
else if ( … )
{
    |
}
else
{
    |
}
```

在同一组 if 语句中，else if 子语句可以超过 1 个，但是 else 子语句最多只能有 1 个。关于 if 语句的综合运用，可参考如下示例。

【4-1-1-if-statements.js】

```
let positives = ['y', 'ye', 'yes', 'yea', 'yeah', 'yeap' ] ;
let user_input ;

user_input = 'YEA' ;
user_input = user_input.toLowerCase() ;

if (positives.indexOf(user_input) != -1)
{
  console.log('As your wish!') ;
}

///
user_input = 'N' ;
user_input = user_input.toLowerCase() ;

if (positives.indexOf(user_input) != -1)
{
  console.log('As your wish!') ;
}
else
{
  console.log('Okay, take it easy.') ;
}

///
let score, message = '' ;
score = 30 ;

if (score > 90)
{
  message = 'Excellent!' ;
}
else if (score > 80)
```

```
{
  message = 'Quite good!' ;
}
else if (score > 70)
{
  message = 'Not bad.' ;
}
else if (score > 60)
{
  message = 'Decent.' ;
}
else if (score > 40)
{
  message = 'Be industrious...' ;
}
else
{
  message = '= =+' ;
}

console.log(message) ;
```

【相关说明】

```
let positives = ['y', 'ye', 'yes', 'yea', 'yeah', 'yeap' ] ;
```

此语句声明了初始数据为【内含字符串的数组实例】的变量 positives。

```
let user_input ;
```

此语句声明了变量 user_input。

```
user_input = 'YEA' ;
```

此语句将字符串'YEA'赋给变量 user_input。

```
user_input = user_input.toLowerCase() ;
```

因为变量 user_input 的数据当前为字符串，所以存在函数 toLowerCase()可被调用，以转换成为全部小写字母（lower-case letter）的字符串'yea'。

```
if (positives.indexOf(user_input) != -1)
{
  console.log('As your wish!') ;
}
```

若表达式【positives.indexOf(user_input) != -1】为真，则代表【positives.indexOf(user_input)】的返回值并不是-1。这也就意味着，变量 user_input 的数据字符串'yea'，的确存在于变量 positives 的数组实例中。

一旦 if 条件语句的小括号里的表达式为真，其大括号里的源代码就会被执行。

```
user_input = 'N' ;
```

此语句重新赋予字符串'N'，给变量 user_input。

```
user_input = user_input.toLowerCase() ;
```

此语句再次调用 user_input.toLowerCase()，以转换成为小写字母（lower-case letter）的字符串 'n'。

```
if (positives.indexOf(user_input) != -1)
{
  console.log('As your wish!') ;
}
else
{
  console.log('Okay, take it easy.') ;
}
```

【positives.indexOf(user_input)】的返回值并不是-1，这也就意味着变量 user_input 的字符串'n'存在于变量 positives 的数组当中。一旦 if 条件语句的小括号里的表达式为真，关键字 if 下方、else 上方的大括号里的源代码，就会被执行；若为假，则关键字 else 下方大括号里的源代码，才会被执行。

```
let score, message = '' ;
```

此语句声明了变量 score，以及初始数据为空字符串的变量 message。

```
score = 30 ;
```

此语句将整数值 30 赋给变量 score。

```
if (score > 90)
{
  message = 'Excellent!' ;
}
```

若变量 score 的数值大于 90，则执行关键字 if 正下方大括号里的源代码。

```
else if (score > 80)
{
  message = 'Quite good!' ;
}
```

若变量 score 的数值小于或等于 90，并且大于 80，则执行此段 else if 大括号里的源代码。

```
else if (score > 70)
{
  message = 'Not bad.' ;
}
```

若变量 score 的数值小于或等于 80，并且大于 70，则执行此段 else if 大括号里的源代码。

```
else if (score > 60)
{
  message = 'Decent.' ;
}
```

若变量 score 的数值小于或等于 70，并且大于 60，则执行此段 else if 大括号里的源代码。

```
else if (score > 40)
{
  message = 'Be industrious...' ;
}
```

若变量 score 的数值小于或等于 60，并且大于 40，则执行此段 else if 大括号里的源代码。

```
else
{
  message = '＝ ＝+' ;
}
```

若变量 score 的数值小于或等于 40，则执行 else 正下方大括号里的源代码。

```
console.log(message) ;
```

通过此语句所显示的信息，即可得知究竟是哪一对大括号里的源代码被执行了。

4.1.2 switch 语句

JavaScript 语言的 switch 语句，和 Java、C、C++、C#、PHP 语言等的 switch 语句，有着相同的语法格式。switch 语句主要依据【相同】数据或【相等】数值的比较条件，决定并执行相对应的源代码片段。在需要比较【大于】或【小于】的场合里，运用 if 语句会更为适合。

switch 语句的语法格式如下：

```
switch ( … )
{
  case … :

    break ;
  case … :

    break ;

  default:

    break ;
}
```

其中，case 子语句可以有多个，但是 default 子语句只能有 1 个。关于 switch 语句的运用，可参考如下示例。

【4-1-2-switch-statements.js】

```
let digital_word, message = '' ;

digital_word = 2091.8 ;

switch (digital_word)
{
  case 530:
    message = '我想你' ;
    break ;

  case 520:
    message = '我爱你' ;
    break ;
```

```
    case 201314:
      message = '爱你一生一世';
      break;

    case 2013:
      message = '爱你一生';
      break;

    case 2014:
      message = '爱你一世';
      break;

    case 2091.8:
      message = '爱你久一点吧!';
      break;

    default:
      message = '+= =';
      // break;
  }
  console.log(message);
```

【相关说明】

```
let digital_word, message = '';
```

此语句声明了变量 digital_word 与 message。

```
digital_word = 2091.8;
```

此语句将 2091.8 赋给变量 digital_word。

```
switch (digital_word)
```

通过 switch 语句，可根据变量 digital_word 的不同数据，执行大括号中特定 case 子语句里的源代码片段。

```
{
  case 530:
    message = '我想你';
    break;
```

若 digital_word 的数值等于 530，则执行【case 530:】之后，至最接近的【break;】为止的源代码片段。

```
  case 520:
    message = '我爱你';
    break;
```

若 digital_word 的数值等于 520，则执行【case 520:】之后，至最接近的【break;】为止的源代码片段。

```
  case 201314:
    message = '爱你一生一世';
```

```
    break ;
```

若 digital_word 的数值等于 201314，则执行【case 201314:】之后，至最接近的【break ;】为止的源代码片段。

```
case 2013:
    message = '爱你一生' ;
    break ;
```

若 digital_word 的数值等于 2013，则执行【case 2013:】之后，至最接近的【break ;】为止的源代码片段。

```
case 2014:
    message = '爱你一世' ;
    break ;
```

若 digital_word 的数值等于 2014，则执行【case 2014:】之后，至最接近的【break ;】为止的源代码片段。

```
case 2091.8:
    message = '爱你久一点吧！' ;
    break ;
```

若 digital_word 的数值等于 2091.8，则执行【case 2091.8:】之后至最接近的【break ;】为止的源代码片段。

```
default:
    message = '+= =' ;
    // break ;
```

若 digital_word 不是以上数值，则执行【default:】之后，至最接近的【break ;】或者 switch 语句末尾为止的源代码片段。其中，default 子语句被放在 switch 语句的末尾，所以【break ;】是可被省略的。

```
}
console.log(message) ;
```

通过此语句所显示的信息，即可得知究竟是哪一个 case 子语句被执行了。

4.2 循 环 语 句

循环语句（loop statement）可用来描述重复执行特定源代码片段的动作。JavaScript 语言支持的循环语句，主要为 for 相关语句和 while 相关语句。

4.2.1 for 相关语句（ES6）

JavaScript 语言的 for 相关语句，可细分如下：

（1）

```
for ( … )
{
  |
}
```

（2）

```
for ( … of … )
{
  |
}
```

（3）

```
for ( … in … )
{
  |
}
```

在 for 语句中，关键字 of 或 in 的存在与否，会影响其迭代方式：

- 没有关键字 of 和 in，仅依据特定变量的数据是否符合特定条件，而被进行迭代任务。
- 存在关键字 of，必须根据特定变量的数组实例中的元素个数，而被进行迭代任务。
- 存在关键字 in，必须根据特定变量的对象实例中的属性个数，而被进行迭代任务。

关于 for 语句的综合运用，可参考如下示例。

【4-2-1-e1-for-loops.js】

```
let fruits = ['apple', 'blueberry', 'cherry', 'durian', 'Fig', 'Grape', 'Haw', 'Kiwi', 'Lichee',
  'Mango', 'Nucleus', 'Orange', 'Pear', 'Raspberry', 'Strawberry', 'Tangerine', 'Watermelon'] ;

let message = '' ;

for (let i = 1; i < fruits.length + 1; i++)
{
  message += `(${i}) ${fruits[i - 1]} ` ;
}

console.log(message) ;
console.log('') ;

message = '' ;

for (let value of fruits)
{
  message += `${value} ` ;
}

console.log(message) ;
console.log('') ;

///
let product = {id: 5685, name: 'Tablet PC', color: 'Gold', Price: '1200', OS: 'Android'} ;
```

```
message = '' ;

for (let str in product)
{
  message += `${str} ` ;
}

console.log(message) ;
```

【相关说明】

```
let fruits = ['apple', 'blueberry', 'cherry', 'durian', 'Fig', 'Grape', 'Haw', 'Kiwi', 'Lichee',
  'Mango', 'Nucleus', 'Orange', 'Pear', 'Raspberry', 'Strawberry', 'Tangerine',
  'Watermelon'] ;
```

此语句声明了初始数据为内含字符串元素的数组实例的变量 fruits。

```
let message = '' ;
```

此语句声明了初始数据为空字符串的变量 message。

```
for (let i = 1; i < fruits.length + 1; i++)
{
  message += `(${i}) ${fruits[i - 1]} ` ;
}
```

因为 fruits.length 的返回值为整数值 17,代表变量 fruits 的数组实例中的元素个数为 17,所以 for 循环语句会迭代 17 次,每次将【(编号) 水果名称】模式的子字符串,衔接到变量 message 的数据字符串里。待此循环语句的迭代结束时,变量 message 的数据,会演变成为如下字符串:

- '(1) apple   (2) blueberry   (3) cherry   (4) durian   (5) Fig   (6) Grape   (7) Haw   (8) Kiwi   (9) Lichee   (10) Mango   (11) Nucleus   (12) Orange   (13) Pear   (14) Raspberry   (15) Strawberry   (16) Tangerine   (17) Watermelon '.

```
console.log(message) ;
console.log('') ;

message = '' ;
```

此语句清空了变量 message 中的字符串。

```
for (let value of fruits)
{
  message += `${value} ` ;
}
```

此语句是用来简化数组实例的各元素的访问语法。通过在关键字 of 的左侧,声明块范围的变量 value,并在其右侧,放入其数据为数组实例的变量 fruits;此语句因此按照迭代的顺序,将变量 fruits 的数组实例中**特定元素的数据**,赋给变量 value。待此循环语句的迭代结束时,变量 message 的数据会演变成为如下字符串:

- apple   blueberry   cherry   durian   Fig   Grape   Haw   Kiwi   Lichee   Mango   Nucleus

Orange   Pear   Raspberry   Strawberry   Tangerine   Watermelon

```
console.log(message) ;
console.log('') ;

let product = {id: 5685, name: 'Tablet PC', color: 'Gold', Price: '1200', OS: 'Android'} ;
```

此语句声明了初始数据为对象实例{id: 5685, name: 'Tablet PC', color: 'Gold', Price: '1200', OS: 'Android'}的变量 product。

```
message = '' ;
```

此语句用来清空变量 message 中的字符串。

```
for (let str in product)
{
  message += `${str} ` ;
}
```

此语法是是用来简化对象实例的各属性的访问语法。通过在关键字 in 的左侧，声明块范围的变量 str，并在其右侧，放入数据为对象实例的变量 product；此语句因此按照迭代的顺序，将变量 product 的对象实例中特定属性的名称，赋给变量 str。待此循环语句的迭代结束时，变量 message 的数据会演变成为如下字符串：

- id   name   color   Price   OS

for ... of 语句必须配合可迭代（iterable）的对象实例，才可迭代其各个属性。关于 for ... of 语句和迭代器的运用，参看如下示例。

【4-2-1-e2-for-of-loops-and-iterators.js】

```
let a01 = [520, 530, 1314, 2013, 2014, 2591.8] ;

let sum = 0 ;

for (let num of a01)
{
  sum += num ;

  console.log(`current number = ${num}`) ;
}

console.log('') ;

let greetings = 'Hello, world, Solar System, Galaxy.' ;

for (let c of greetings)
{
  console.log(c) ;
}

console.log('') ;

function display(... args)
```

```
{
  for (let arg of args)
  {
    console.log(arg) ;
  }

  console.log('') ;

  for (let arg of arguments)
  {
    console.log(arg) ;
  }
}

display('what', 'when', 'where', 'which', 'who') ;

function * gen01()
{
  let top_value = parseInt(30 * Math.random()) ;

  for (let i = 1; i < top_value; i += 3)
  {
    yield i ;
  }
}

let gen02 =
{
  * [Symbol.iterator]()
  {
    let top_value = parseInt(30 * Math.random()) ;

    for (let i = 1; i < top_value; i += 3)
    {
      yield i ;
    }
  }
} ;

for (let num of gen01())
{
  console.log(num) ;
}

console.log('') ;

for (let num of gen02)
{
  console.log(num) ;
}

let number_array = [... gen01()] ;

console.log(number_array) ;
```

```
console.log('') ;

class Planets
{
  * solar_system()
  {
    let planet_list = ['水星 (Mercury)', '金星 (Venus)', '地球 (Earth)' , '火星 (Mars)', '木星
      (Jupiter)', '土星 (Saturn)', '天王星 (Uranus)', '海王星 (Neptune)' , '冥王星 (Pluto)'] ;

    for (let planet of planet_list)
    {
      yield planet ;
    }
  }
}

p01 = new Planets() ;

let a_copy = [... p01.solar_system()] ;

console.log(a_copy) ;
```

【相关说明】

```
let a01 = [520, 530, 1314, 2013, 2014, 2591.8] ;
```

此语句声明了初始数据为数组实例的变量 a01。

```
let sum = 0 ;
```

此语句声明了初始数值为 0 的变量 sum。

```
for (let num of a01)
{
  sum += num ;

  console.log(`current number = ${num}`) ;
}
```

for …of 循环语句需要一个可作为迭代器（iterator）的变量。其数据为数组实例的变量 a01，就具有迭代器的特征。变量 a01 的数组实例中的元素个数为 6，所以此循环语句会迭代 6 次，使得其大括号里的源代码片段，会被执行 6 次。每次迭代而执行的动作是：

- 让变量 num 按照迭代的顺序，被赋予变量 a01 的数组实例中特定元素的数值。
- 让变量 sum 的数值，加上变量 num 的数值。
- 显示变量 num 当前的数值。

```
console.log('') ;

let greetings = 'Hello, world, Solar System, Galaxy.' ;
```

此语句声明了初始数据为字符串'Hello, world, Solar System, Galaxy.'的变量 greetings。

```
for (let c of greetings)
{
  console.log(c) ;
```

}
```

其数据为字符串的变量 greetings，亦具有迭代器的特征。变量 greetings 的字符串里的字符个数为 35，所以此循环语句会迭代 35 次。每次迭代而执行的动作是：

- 让变量 c 按照迭代的顺序，被赋予变量 greetings 的数据字符串里的特定字符。
- 显示变量 c 当前的数据**字符**。

```
console.log('') ;

function display(... args)
```

通过【... args】的语法，定义此函数的参数列，使得被传入的各个参数，被压缩成为参数变量 args 的数组实例中的各个元素。举例来说，在函数 display('what', 'when', 'where', 'which', 'who') 被调用的情况下，参数变量 args 的数据，即是数组实例 ['what', 'when', 'where', 'which', 'who']。

```
{
  for (let arg of args)
  {
    console.log(arg) ;
```

在 for...of 循环语句的每次迭代中，局部变量 arg 的数据，会成为参数变量 args 的数组实例中特定字符串元素。此循环语句会逐一显示如下信息：

- what
  when
  where
  which
  who

```
  console.log('') ;

  for (let arg of arguments)
  {
    console.log(arg) ;
  }
```

此循环语句和上一段的循环语句，具有相同的效果。值得注意的是，JavaScript 引擎支持内置的参数变量 arguments，并且和前述变量 args 有相同的数据。

```
}

display('what', 'when', 'where', 'which', 'who') ;
```

此语句调用了被传入多个数据的函数 display('what', 'when', 'where', 'which', 'who')。

```
function * gen01()
```

此语句用来定义新的生成器函数（generator function），必须在关键字 function 与函数名称之间，放置星号【*】。
```

```
{
    let top_value = parseInt(30 * Math.random()) ;
```

此语句将 parseInt(30 * Math.random()) 返回的 0～29 的随机整数值，赋给变量 top_value。每当此语句重新被执行时，变量 top_value 的数值，都会随机变化。

```
    for (let i = 1; i < top_value; i += 3)
    {
        yield i ;
    }
```

在此 for 语句的每次迭代中，其大括号里的【yield i ;】会被执行，使得变量 i 当前的数值，动态被产生（yield）至最终会被返回的迭代器实例中。另外，此示例源代码每次被重新执行时，都会使得上述的迭代器实例，具有不定个数的元素。在上述的迭代器实例中，各个元素的数值皆小于 30，而且相邻元素的数值的差距均为 3，例如【1 4 7 10 13】或【1 4 7 10 13 16 19 22】等。

```
}
let gen02 =
{
 * [Symbol.iterator]()
    {
        let top_value = parseInt(30 * Math.random()) ;

        for (let i = 1; i < top_value; i += 3)
        {
            yield i ;
        }
    }
} ;
```

此表达式可用来定义具有生成器（generator）特征的变量。因为其语法较为复杂，建议读者可以简单理解成【let 变量名称 = { * [Symbol.iterator]() { ... } } ;】的语法格式。实际上，变量 gen02 与函数 gen01() 的功能是相同的。

```
for (let num of gen01())
{
  console.log(num) ;
}
```

在关键字 of 的右侧，有生成器**函数** gen01()，会返回可作为迭代器的新实例。

```
console.log('') ;

for (let num of gen02)
{
  console.log(num) ;
}
```

在关键字 of 右侧，有生成器**变量** gen02，亦会返回可作为迭代器的新实例。

```
let number_array = [... gen01()] ;
```

若 gen01() 返回的迭代器实例为<1, 4, 7, 10, 13>，则[... gen01()]等价于[... <1, 4, 7, 10, 13>]，并

且可被化简成为[1, 4, 7, 10, 13]。简单来看，扩展运算符【...】具有解开迭代器实例的用途。

```
console.log(number_array) ;
```

此语句显示[1, 4, 7, 10, 13]等类似的信息。

```
console.log('') ;

class Planets
```

关键字 class 在此用来定义名称为 Planets 的类。

```
{
  * solar_system()
  {
    let planet_list = ['水星 (Mercury)', '金星 (Venus)', '地球 (Earth)' , '火星 (Mars)', '木星
      (Jupiter)', '土星 (Saturn)', '天王星 (Uranus)', '海王星 (Neptune)' , '冥王星 (Pluto)'] ;

    for (let planet of planet_list)
    {
      yield planet ;
    }
  }
```

定义生成器函数 solar_system()，会动态生成【内含太阳系九大行星的中英文名称字符串】的迭代器实例。

```
}

p01 = new Planets() ;
```

此语句声明初始数据为 Planets 类实例的变量 p01。

```
let a_copy = [... p01.solar_system()] ;
```

在此，p01.solar_system()会返回迭代器实例<"水星 (Mercury)", "金星 (Venus)", "地球 (Earth)", "火星 (Mars)", "木星 (Jupiter)", "土星 (Saturn)", "天王星 (Uranus)", "海王星 (Neptune)", "冥王星 (Pluto)">。

因为，[... p01.solar_system()]等价于[... <"水星 (Mercury)", "金星 (Venus)", "地球 (Earth)", "火星 (Mars)", "木星 (Jupiter)", "土星 (Saturn)", "天王星 (Uranus)", "海王星 (Neptune)", "冥王星 (Pluto)">]，并可进一步化简成为["水星 (Mercury)", "金星 (Venus)", "地球 (Earth)", "火星 (Mars)", "木星 (Jupiter)", "土星 (Saturn)", "天王星 (Uranus)", "海王星 (Neptune)", "冥王星 (Pluto)"]。

所以，变量 a_copy 的初始数据即为["水星 (Mercury)", "金星 (Venus)", "地球 (Earth)", "火星 (Mars)", "木星 (Jupiter)", "土星 (Saturn)", "天王星 (Uranus)", "海王星 (Neptune)", "冥王星 (Pluto)"]。

4.2.2 while 相关语句

JavaScript 语言的 while 相关语句，有如下两种。

（1）
```
while ( … )
{
   ⋮
}
```

（2）
```
do
{
   ⋮
} while ( … ) ;
```

其中，do … while 语句至少会迭代 1 次，这是因为【判断是否继续迭代】的比较表达式，位于整个语句末尾的小括号里，使得判断的动作，是在每次迭代完成之后。这也就意味着，无论比较表达式是否为真，迭代的动作至少会进行 1 次。关于 while 语句的综合运用，可参考如下示例。

【4-2-2-while-loops.js】

```
let fruits = ['apple', 'blueberry', 'cherry', 'durian', 'Fig', 'Grape', 'Haw', 'Kiwi', 'Lichee',
 'Mango', 'Nucleus', 'Orange', 'Pear', 'Raspberry', 'Strawberry', 'Tangerine',
 'Watermelon'] ;

let product = {id: 5685, name: 'Tablet PC', color: 'Gold', Price: '1200', OS: 'Android'} ;

let message = '', count = 1 ;

while (count < fruits.length + 1)
{
 message += `(${count}) ${fruits[count - 1]} ` ;
 count++ ;
}

console.log(message) ;
console.log('') ;

///
message = '', count = 1 ;

do
{
 message += `(${count}) ${fruits[count - 1]} ` ;
 count++ ;
} while (count < fruits.length + 1) ;

console.log(message) ;
```

【相关说明】

```
let fruits = ['apple', 'blueberry', 'cherry', 'durian', 'Fig', 'Grape', 'Haw', 'Kiwi', 'Lichee',
 'Mango', 'Nucleus', 'Orange', 'Pear', 'Raspberry', 'Strawberry', 'Tangerine',
 'Watermelon'] ;
```

此语句声明了初始数据为内含字符串元素的数组实例的变量 fruits。

```
let product = {id: 5685, name: 'Tablet PC', color: 'Gold', Price: '1200', OS: 'Android'} ;
```

此语句声明了初始数据为对象实例的变量 products。

```
let message = '', count = 1 ;
```

此语句声明了初始数据为**空**字符串的变量 message，以及初始数值为整数 1 的变量 count。

```
while (count < fruits.length + 1)
```

在 while 循环语句的小括号里，每当其比较表达式【count < fruits.length + 1】为真时，其大括号里的源代码片段，就会被执行 1 次。

```
{
  message += `(${count}) ${fruits[count - 1]} ` ;
  count++ ;
}
```

在此，直到【count < fruits.length + 1】为假之前，while 循环语句总共迭代了 17 次，渐次将【(编号) 水果名称】模式的字符串，衔接至变量 message 的数据字符串里，最后演变成为如下字符串：

- (1) apple (2) blueberry (3) cherry (4) durian (5) Fig (6) Grape (7) Haw (8) Kiwi (9) Lichee (10) Mango (11) Nucleus (12) Orange (13) Pear (14) Raspberry (15) Strawberry (16) Tangerine (17) Watermelon

```
console.log(message) ;
console.log('') ;

message = '', count = 1 ;
```

此语句重新设置变量 message 和 count 的新数据。

```
do
{
  message += `(${count}) ${fruits[count - 1]} ` ;
  count++ ;
} while (count < fruits.length + 1) ;
```

在此，上述 do…while 循环语句至少会迭代 1 次，这是因为其比较表达式【count < fruits.length + 1】被安排在整个末尾的小括号里。换言之，其大括号里的源代码片段，至少会被执行 1 次。

4.2.3　break 与 continue 语句

在循环语句中，关键字 break 加上分号而形成的 break 语句【break ;】，可用来终止循环语句的执行；关键字 continue 加上分号而形成的 continue 语句【continue ;】，则可用来直接跳至循环语句的下一次迭代动作。关于 break 与 continue 语句的综合运用，可参考如下示例。

【4-2-3-statements-of-break-and-continue.js】

```
let weekday = ['Monday', 'Tuesday', 'Wednesday', 'Thursday', 'Friday', 'Saturday','Sunday'] ;
```

```
let message = 'The working days of this week: ' ;

let bnum = 3 + parseInt(4 * Math.random()) ;

let cnum01 = parseInt(3 * Math.random()) ;

let cnum02 = 3 + parseInt(4 * Math.random()) ;

for (let i = 0; i < weekday.length; i++)
{
  if (i == bnum) break ;

  message += weekday[i] + ', ' ;
}

message = message.slice(0, -2) ;

console.log(message) ;

///
message = 'Cannot have day-off on: ' ;

for (let i = 0; i < weekday.length; i++)
{
  if (i == cnum01 || i == cnum02) continue ;

  message += weekday[i] + ', ' ;
}

message = message.slice(0, -2) ;

console.log(message) ;
```

【相关说明】

```
let weekday = ['Monday', 'Tuesday', 'Wednesday', 'Thursday', 'Friday', 'Saturday','Sunday'] ;
```

此语句声明了初始数据为【内含字符串元素的数组实例】的变量 weekday。

```
let message = 'The working days of this week: ' ;
```

此语句声明了初始数据为字符串的变量 message。

```
let bnum = 3 + parseInt(4 * Math.random()) ;
```

此语句声明了变量 bnum，并设置其初始数值成为【大于或等于 3 且小于 7】的随机整数值。

```
let cnum01 = parseInt(3 * Math.random()) ;
```

此语句声明了变量 cnum01，并设置其初始数值为【大于或等于 0 且小于 3】的随机整数值。

```
let cnum02 = 3 + parseInt(4 * Math.random()) ;
```

此语句声明了变量 cnum02，并设置其初始数值也为【大于或等于 3 且小于 7】的随机整数值。

```
for (let i = 0; i < weekday.length; i++)
{
```

```
    if (i == bnum) break ;

  message += weekday[i] + ', ' ;
}
```

此 for 循环语句的重点，在于子语句【if (i == bnum) break ;】。若其比较表达式【i == bnum】为真，则执行【break ;】子语句，进而立即终止这个循环语句的执行。

```
message = message.slice(0, -2) ;
```

因为变量 message 的数据为字符串，所以存在内置函数 slice() 可被调用。调用函数 messsage.slice(0, -2)，可使得其数据字符串的末尾的逗号【,】被移除。例如：从字符串'week: Monday, Tuesday, Wednesday,'，变成'week: Monday, Tuesday, Wednesday'。

```
console.log(message) ;

message = 'Cannot have day-off on: ' ;
```

此语句重新设置了变量 message 的数据字符串。

```
for (let i = 0; i < weekday.length; i++)
{
  if (i == cnum01 || i == cnum02) continue ;

  message += weekday[i] + ', ' ;
}
```

此 for 循环语句的重点，在于子语句【if (i == cnum01 || i == cnum02) continue ;】。若其比较表达式【i == cnum01 || i == cnum02】为真，则执行【continue ;】子语句，进而立即进行下一次迭代。

4.3 练 习 题

1. 请将如下条件表达式，分别改写为 if 语句和 switch 语句的版本：

```
let dtype = ['integer', 'float', 'alphabet'][parseInt(3 * Math.random())] ;

let value =
  dtype == 'integer' ? parseInt(100 * Math.random()) :
  dtype == 'float' ? (100 * Math.random()).toFixed(3) :
  dtype == 'alphabet' ? String.fromCharCode(65 + parseInt(26 * Math.random())) : null ;

console.log(value) ;
```

2. 已知如下数组实例：

```
let num_list = [10, 33, 21, 56, 77, 64, 82, 98, 2] ;
```

试着编写通过 for、for...of 和 for...in 语句的版本，将上述数组实例中所有元素的整数值的总和，赋给变量 result，并且显示出来。

3. 已知存在如下对象实例：

```
let chairs = {wood: 15, metal: 23, plastic: 37, others: 60} ;
```

试着编写通过循环语句的版本，将上述对象实例中的数据，显示成为如下分行的信息：

```
wood: 15
metal: 23
plastic: 37
others:60
```

4. 已知存在如下数据为字符串的变量 str：

```
let str = 'Happily ever after.' ;
```

试着编写可颠倒显示成为如下字符串的版本：

```
.retfa reve ylippaH
```

5.完成如下两小题。

（1）定义【通过参数数据的不同，而迭代如下其中一组字符串】的迭代函数 weekday()：

星期日 → 星期一 → 星期二 → 星期三 → 星期四 → 星期五 → 星期六

Sunday → Monday → Tuesday → Wednesday → Thursday → Friday → Saturday

（2） 配合循环语句，调用 weekday()，【分别】显示其中一组字符串。

6. 已知如下数组实例：

```
let base_list = [7, 12, 21, 30, 40, 55] ;
```

试着编写通过 for、while 和 do...while 语句的版本，以计算并显示出上述数组实例中各元素的整数值的平方和。

第 5 章

函数与方法

在特定对象（object）或类（class）的定义语法里，被进一步定义的函数（function），可被称为方法（method）。在 JavaScript 语言中，函数（function）的定义，内含实现特定任务的源代码块。因此，调用函数，就等同于执行特定任务的源代码块。

5.1 函数的定义

在现今的 JavaScript 语言里，函数的定义形式有好几种，其语法格式大致包含关键字 function、函数名称（function name）、在一对小括号里的参数列（parameter / argument list），以及在一对大括号里的主体（body）源代码块。其中，有些定义形式，不具有关键字 function、函数名称或参数列。

5.1.1 不同形式的函数（ES6）

1. 标准函数

标准函数的定义语法如下：

（1）

```
function 函数名称(参数列)
{
    |
}
```

（2）

```
let 作为匿名函数的别名的变量名称 = function (参数列)
{
```

```
    }
  } ;
```

其中，在此可省略关键字 let，或者变更为关键字 var。等号后方的【function (参数列) {…} ;】，因为不具有函数名称，所以可被称为匿名函数（anonymous function）的定义。【作为匿名函数的别名的变量名称】具有函数名称的特征。在此，整个语句可被称为函数表达式（function expression），因此在末尾处，应该加上**分号**【;】。关于标准函数的综合运用，可参考如下示例。

【5-1-1-e1-standard-functions.js】

```
range_calc(5, 15, '+') ;

console.log('') ;

function range_calc(start, end, type)
{
  let result ;

  if ('+-*/'.indexOf(type) == -1)
  {
    console.log('Unsupported calculation...') ;
    return NaN ;
  }
  else
  {
    if ('*/'.indexOf(type) == -1) result = 0 ;

    else result = 1 ;
  }

  for (let i = start; i < end + 1; i++)
    eval(`result ${type}= i ;`) ;

  console.log(`The result of operator ${type} action from ${start} to ${end} = ${result}`) ;

  return result ;
}

let returned = null ;

range_calc(1, 10, '+') ;
range_calc(1, 10, '-') ;

returned = range_calc(1, 10, '*') ;
console.log(returned) ;
console.log('') ;

returned = range_calc(1, 3, '/') ;
console.log(returned) ;
console.log('') ;

///
// if uncomment the following code, there will show 'displaying is not defined' error message.
// displaying(12) ;
```

```
// function-expression variable
let displaying = function (num)
{
  console.log(`The tested number = ${num}`) ;
} ;

displaying(12) ;
```

【相关说明】

```
range_calc(5, 15, '+') ;
```

此语句调用了传入 3 个参数数据的函数 range_calc(5, 15, '+')。

```
console.log('') ;
```

```
function range_calc(start, end, type)
```

此语句通过关键字 function，定义名称为 range_calc 的函数。位于其函数名称右侧小括号内的部分，是以逗号【,】隔开多个参数名称的参数列。参数如同函数内部的局部变量，而此函数有 3 个参数，分别为 start、end 与 type。

```
{
  let result ;
```

在函数大括号内，通过关键字 var 或 let 声明的变量，是这个函数的局部变量。也因此，result 在此为局部变量。

```
  if ('+-*/'.indexOf(type) == -1)
```

此语法是用来判断变量 type 中的字符，是否存在于字符串'+-*/'中。变量 type 当前的数据为字符'+'，使得比较表达式【'+-*/'.indexOf(type) == -1】为假。因此，if 语句大括号里的源代码片段，并不会被执行。

```
  {
    console.log('Unsupported calculation...') ;
    return NaN ;
```

在此，【return NaN ;】被执行时，即是完成此函数的调用任务，并返回 NaN。

```
  }
  else
  {
    if ('*/'.indexOf(type) == -1) result = 0 ;
```

比较表达式【'*/'.indexOf(type) == -1】是用来判断变量 type 中的字符，是否在字符串'*/'里。若前述比较表达式为真，则执行【result = 0 ;】。

```
    else result = 1 ;
```

若上述比较表达式为假，则执行【result = 1 ;】。

```
  }
```

```
        for (let i = start; i < end + 1; i++)
            eval(`result ${type}= i ;`) ;
```

从变量 start 的数值 5，到变量 end 的数值 15，上述 for 语句总共迭代 11 次。在【eval(`result ${type}= i ;`) ;】中的格式化字符串【`result ${type}= i ;`】，会先被转换成为字符串'result += i ;'，然后作为参数数据，传入内置函数 eval()，使得字符串'result += i ;'，最终被转换成为真正可执行的表达式【result += i ;】。

```
        console.log(`The result of operator ${type} action from ${start} to ${end} = ${result}`) ;
```

此语句显示出【The result of operator + action from 5 to 15 = 110】的信息。

```
        return result ;
```

此语句，即是完成了这个函数的调用任务，并返回变量 result 的数值 110。

```
    }

    let returned = null ;
```

此语句声明了初始数据为 null 的变量 returned。

```
    range_calc(1, 10, '+') ;
```

此语句调用了函数 range_calc(1, 10, '+')，进而执行此函数内部的源代码块，最终计算出 1 + 2 + 3 + … + 10 的结果值 55。

```
    range_calc(1, 10, '-') ;
```

此语句调用了函数 range_calc(1, 10, '-')，进而执行此函数内部的源代码，最终计算出-1 - 2 - 3 - … - 10 的结果值-55。

```
    returned = range_calc(1, 10, '*') ;
```

此语句调用了函数 range_calc(1, 10, '*')，进而执行此函数内部的源代码，最终计算出 1 × 2 × 3 × … × 10 的结果值 3628800，并返回给变量 returned。

```
    console.log(returned) ;
```

此语句显示变量 returned 当前的数值 3628800。

```
    returned = range_calc(1, 3, '/') ;
```

此语句调用了函数 range_calc(1, 3, '/')，进而执行此函数内部的源代码，最终计算出 1 ÷ 2 ÷ 3 的结果值 0.16666666666666666，并返回给变量 returned。

```
    console.log(returned) ;
```

此语句显示变量 returned 当前的数值 0.16666666666666666。

```
    // if uncomment the following code, there will show 'displaying is not defined' error message.
    // displaying(12) ;

    // function-expression variable
    let displaying = function (num)
    {
```

```
console.log(`The tested number = ${num}`) ;
} ;
```

在此段语法中,【function (num) { ... }】定义了没有名称的匿名函数。而且,变量名称 displaying 如同是该匿名函数的别名(alias name)。

```
displaying(12) ;
```

此语句调用了参数数值为 12 的函数 displaying(12),使得【The tested number = 12】被显示出来。

2. 匿名函数

匿名(anonymous)函数的语法格式如下:

(1)
```
function (参数列)
{
  |
}
```

这种形式的匿名函数的定义语法,没有可调用的函数名称,因此必须依附于可间接调用这个匿名函数的特殊语法里。

(2)
```
(function (参数列)
{
  |
}
)() ;
```

这种形式的匿名函数的定义语法,也没有可调用的函数名称;但是,却可**立即**调用这个匿名函数。

(3)
```
void function (参数列)
{
  |
}() ;
```

这种形式的匿名函数的定义语法,虽然也没有可调用的函数名称;但是,却也可**立即**调用这个匿名函数。此定义语法,借助关键字 void,替代了在【上一种匿名函数的定义语法】的外侧小括号。关于匿名函数的综合运用,可参考如下示例。

【5-1-1-e2-anonymous-functions.js】

```
(function ()
{
  console.log('Anonymous 01 function is executed!') ;
}
)() ;

void function ()
{
```

```
    console.log('Anomynous function 02 is executed.') ;
} () ;

setTimeout(function () { console.log('Showing after 2 seconds.') ; }, 2000) ;

///
let displaying = function (num = 0)
{
  console.log(`The test number = ${num}`) ;
} ;

displaying(15) ;

setTimeout(displaying, 1000) ;

setTimeout(displaying(21), 1500) ;

setTimeout(() => console.log('The test number = 1800'), 3000) ;
```

【相关说明】

```
(function ()
{
  console.log('Anomynous function 01 is executed.') ;
}
) () ;
```

此语法定义并**立即调用**了匿名函数，进而显示出【Anomynous function 01 is executed.】的信息。这种调用函数的方式格外特别，仅供调用匿名函数 1 次。

```
void function ()
{
  console.log('Anomynous function 02 is executed.') ;
} () ;
```

此语法则借助关键字 void，替代了在【上一个匿名函数的定义语法】的外侧小括号，亦定义并立即调用了匿名函数，进而显示出【Anomynous function 02 is executed.】的信息。此语法仅能调用匿名函数 1 次。

```
setTimeout(function () { console.log('Showing after 2 seconds.') ; }, 2000) ;
```

内置函数 setTimeout()可用来设置每隔多少时间，调用匿名函数 1 次。在这个 setTimeout()的小括号中，第 1 个参数为匿名函数的定义语法【function () { console.log('Showing after 2 seconds.') ; }】，第 2 个参数的数值，代表着每次调用前述匿名函数的间隔时间，在此为 2000 毫秒，也就是 2 秒。换言之，此语句使得第 1 个参数的匿名函数，每隔 2 秒就被调用一次。

```
let displaying = function (num = 0)
{
  console.log(`The test number = ${num}`) ;
} ;
```

这个语法中的【function (num = 0) { ... }】，定义了没有名称的匿名函数。也因此，变量名称 displaying 如同是该匿名函数的别名（alias name）。其中，【(num = 0)】是用来设置参数 num 的默认数值为 0。

```
displaying(15) ;
```

此语句以传入参数数值 15 的方式，调用函数 displaying(15)。

```
setTimeout(displaying, 1000) ;
```

此语句使得每隔 1 秒，就间接调用名称为 displaying 的函数。调用时，没有传入参数数值，所以等同于参数数值为默认数值 0。

```
setTimeout(displaying, 1500) ;
```

此语句使得每隔 1.5 秒，就间接调用名称为 displaying 的函数。

```
setTimeout(() => console.log('The test number = 1800'), 3000) ;
```

此语句使得每隔 3 秒，就间接调用匿名函数【() => console.log('The test number = 1800')】。此匿名函数是借助箭头符号【=>】而实现的箭头函数（arrow function）。

3. 对象函数

在对象（object）的定义语法里，除了可描述代表特定数据的属性（property）之外，亦可描述代表特定行为的方法（method）/函数（function）。JavaScript 语言支持如下对象函数的两种定义形式：

（1）
```
var 变量名称 =
{
  │
  ,
    对象函数的名称: function (参数列)
    {
      │
    },
  │
}
```

（2）
```
var 变量名称 =
{
  │
  ,
    对象函数的名称 (参数列)
    {
      │
    },
  │
}
```

在上述两种对象的定义语法里，均可用来描述对象函数。调用对象函数的语法为【变量名称.对象函数名称(参数列);】。关于对象函数的综合运用，可参考如下两个示例。

【5-1-1-e3-function-similar-properties-part1.js】

```javascript
var r_obj =
{
  circle_area(r)
  {
    return Math.PI * r * r ;
  },
  circumference(r)
  {
    return 2 * Math.PI * r ;
  },
  sphere_volume(r)
  {
    return 4 / 3 * Math.PI * Math.pow(r, 3) ;
  },
  cylinder_volume(r, h)
  {
    return this.circle_area(r) * h ;
  }
}

console.log(r_obj.circle_area(10)) ;

console.log(r_obj['circle_area'](10)) ;

console.log(r_obj.circumference(15)) ;
console.log(r_obj['circumference'](15)) ;

console.log(r_obj.sphere_volume(20)) ;
console.log(r_obj['sphere_volume'](20)) ;

console.log(r_obj.cylinder_volume(10, 20)) ;
console.log(r_obj['cylinder_volume'](10, 20)) ;
```

【相关说明】

```javascript
var r_obj =
{
  circle_area(r)
  {
    return Math.PI * r * r ;
  },
  circumference(r)
  {
    return 2 * Math.PI * r ;
  },
  sphere_volume(r)
  {
    return 4 / 3 * Math.PI * Math.pow(r, 3) ;
  },
  cylinder_volume(r, h)
  {
    return this.circle_area(r) * h ;
  }
```

}

上述冗长的语法，声明了初始数据为【内含多个函数的对象实例】的变量 r_obj。

```
console.log(r_obj.circle_area(10)) ;
```

【r_obj.circle_area(10)】调用了变量 r_obj 内的对象函数 circle_are(10)，以返回半径为 10 的圆面积 314.1592653589793。

```
console.log(r_obj['circle_area'](10)) ;
```

【r_obj['circle_area'](10)】亦调用了变量 r_obj 内的对象函数 circle_are(10)，并返回半径为 10 的圆面积 314.1592653589793。因为变量 r_obj 当前的数据是对象实例，所以可通过中括号运算符[]，加以访问内部的属性或函数。

```
console.log(r_obj.circumference(15)) ;
console.log(r_obj['circumference'](15)) ;
```

这两个语句的效果相同，皆会显示出半径为 15 的圆周长 94.24777960769379。

```
console.log(r_obj.sphere_volume(20)) ;
console.log(r_obj['sphere_volume'](20)) ;
```

这两个语句的效果相同，皆会显示出半径为 20 的球体积 33510.32163829113。

```
console.log(r_obj.cylinder_volume(10, 20)) ;
console.log(r_obj['cylinder_volume'](10, 20)) ;
```

这两个语句的效果也相同，皆会显示出半径为 10、高度为 20 的圆柱体积 6283.185307179587。

【5-1-1-e4-function-similar-properties-part2.js】

```
var cubic_obj =
{
 length: 1,
 width: 1,
 height: 1,
 volume: function ()
  {
   return this.length * this.width * this.height ;
  },
 surface_area()
  {
   with (this)
    return 2 * (length * width + width * height + height * length) ;
  }
}

console.log(cubic_obj.volume()) ;

console.log(cubic_obj.surface_area()) ;

console.log('') ;

cubic_obj.length = 10 ;
cubic_obj.width = 20 ;
cubic_obj.height = 30 ;
```

```
console.log(cubic_obj.volume()) ;

console.log(cubic_obj.surface_area()) ;
```

【相关说明】

```
var cubic_obj =
{
  length: 1,
  width: 1,
  height: 1,
  volume: function ()
  {
    return this.length * this.width * this.height ;
  },
  surface_area()
  {
    with (this)
      return 2 * (length * width + width * height + height * length) ;
  }
}
```

此语法声明了初始数据为【内含多个属性和函数的对象实例】的变量 cubic_obj。

```
console.log(cubic_obj.volume()) ;
```

在变量 cubic_obj 的对象实例中，其 3 个属性 length（长）、width（宽）和 height（高）的数值皆为 1，所以 cubic_obj.volume()会返回长、宽和高皆为 1 的正方体积 1。

```
console.log(cubic_obj.surface_area()) ;
```

cubic_obj.surface_area()则会返回长、宽和高皆为 1 的正方体表面积 6。

```
console.log('') ;

cubic_obj.length = 10 ;
cubic_obj.width = 20 ;
cubic_obj.height = 30 ;
```

这 3 个语句重新设置了在变量 cubic_obj 的对象实例中的属性 length、width 和 height 的数值，分别为 10、20 与 30。

```
console.log(cubic_obj.volume()) ;
```

cubic_obj.volume()在此会返回长、宽和高分别为 10、20 与 30 的长方体积 6000。

```
console.log(cubic_obj.surface_area()) ;
```

cubic_obj.volume()在此会返回长、宽和高分别为 10、20 与 30 的长方体表面积 2200。

4．箭头函数

在特定函数的主体内，若其源代码块较为简短，则可被重新定义成为带有箭头符号【=>】的箭头函数（arrow function），见表 5-1。

表 5-1 箭头函数的语法格式

箭头的左侧语法	箭头	箭头的右侧语法
（参数列） 单一参数名称 ()	=>	• { 被分号隔开的多个语句 } • { return 语句 } • return 以外的单一语句 • 可被评估的单一表达式 • (可被评估的单一表达式) • (单一数组实例) • (单一对象实例)

在箭头函数的定义语法里，并没有函数名称，所以箭头函数也是匿名函数的一种，其定义语法必须依附于【可间接调用匿名函数】的语法中。关于箭头函数的综合运用，可参考如下示例。

【5-1-1-e5-arrow-functions.js】

```
function sphere01(r)
{
 values = {} ;

 values.volume = 4 / 3 * Math.PI * Math.pow(r, 3) ;
 values.surface_area = 4 * Math.PI * r * r ;

 return values ;
}

let sphere02 = (r) => { return { volume: 4 / 3 * Math.PI * Math.pow(r, 3), surface_area:
 4 * Math.PI * r * r } } ;

console.log(sphere02(10)) ;

let sphere03 = r => { return { volume: 4 / 3 * Math.PI * Math.pow(r, 3), surface_area:
 4 * Math.PI * r * r } } ;

console.log(sphere03(10)) ;

let sphere04 = r => ( { volume: 4 / 3 * Math.PI * Math.pow(r, 3), surface_area: 4 * Math.PI
 * r * r } ) ;

console.log(sphere04(10)) ;

let sphere_volume = r => { return 4 / 3 * Math.PI * Math.pow(r, 3) } ;

console.log(sphere_volume(10)) ;

let sphere_surface_area = r => 4 * Math.PI * r * r ;

console.log(sphere_surface_area(10)) ;

// setInterval( function() { console.log( new Date() .getSeconds() ) } , 3000 ) ;
setInterval( () => console.log( new Date() .getSeconds() ) , 3000 ) ;
```

【相关说明】

```
function sphere01(r)
{
 values = {} ;

 values.volume = 4 / 3 * Math.PI * Math.pow(r, 3) ;
 values.surface_area = 4 * Math.PI * r * r ;

 return values ;
}
```

此段语法是标准函数的定义形式之一。也就是【function 函数名称(参数列) { ... }】。在上述函数中，其返回数据为局部变量 values 中的对象实例{volume: 球体积的数值, surface_area: 球体表面积的数值}。

```
let sphere02 = (r) => { return { volume: 4 / 3 * Math.PI * Math.pow(r, 3), surface_area:
 4 * Math.PI * r * r } } ;
```

此段语法则是箭头函数的定义形式之一，也就是【let 作为函数的别名的变量名称 =(参数列) => { ... };】。注意，此函数的返回数据为对象实例{ volume: 球体积的数值, surface_area: 球体表面积的数值}。

```
console.log(sphere02(10)) ;
```

调用 sphere02(10)会返回对象实例{volume: 4188.790204786391, surface_area: 1256.6370614359173}。

```
let sphere03 = r => { return { volume: 4 / 3 * Math.PI * Math.pow(r, 3), surface_area:
 4 * Math.PI * r * r } } ;
```

此段语法是【单一参数】的箭头函数的定义形式之一，也就是【let 作为函数的别名的变量名称 = 唯一的参数名称 => { ⋯ };】。此函数的返回数据为对象实例{volume: 球体积的数值, surface_area: 球体表面积的数值}。

```
console.log(sphere03(10)) ;
```

调用 sphere03(10)会返回对象实例{volume: 4188.790204786391, surface_area: 1256.6370614359173}。

```
let sphere04 = r => ( { volume: 4 / 3 * Math.PI * Math.pow(r, 3), surface_area: 4 * Math.PI
 * r * r } ) ;
```

此段语法为【单一参数、省略关键字 return】的箭头函数的定义形式之一，也就是【let 作为函数的别名的变量名称 = 唯一的参数名称 =>（ 对象实例);】。在箭头运算符【=>】右侧的对象实例外侧，必须加上一对小括号里，才不至于发生错误。此函数的返回数据为对象实例{ volume: 球体积的数值, surface_area: 球体表面积的数值}。

```
console.log(sphere04(10)) ;
```

调用 sphere04(10)会返回对象实例{volume: 4188.790204786391, surface_area: 1256.6370614359173}。

```
let sphere_volume = r => { return 4 / 3 * Math.PI * Math.pow(r, 3) } ;
```

此段语法为【单一参数】的箭头函数的定义形式之一，也就是【作为函数的别名的变量名称 = 唯一的参数名称 => { return 语句 };】。此函数的返回数据为球体积的数值。

```
console.log(sphere_volume(10)) ;
```

调用 sphere_volume(10)会返回球体积 4188.790204786391。

```
let sphere_surface_area = r => 4 * Math.PI * r * r ;
```

此段语法为【单一参数、省略关键字 return、最简化】的箭头函数的定义形式之一,也就是【作为函数的别名的变量名称 = 唯一的参数名称 => 可被评估的单一表达式 ;】。注意,在箭头运算符【=>】右侧的小括号已经被省略了。此函数的返回数据为球体表面积的数值。

```
console.log(sphere_surface_area(10)) ;
```

调用 sphere_surface_area(10)会返回球体表面积 1256.6370614359173。

```
// setInterval( function() { console.log( new Date() .getSeconds() ) }, 3000 ) ;
setInterval( () => console.log( new Date() .getSeconds() ) , 3000 ) ;
```

【() => console.log(new Date() .getSeconds()】是箭头函数最简化的定义形式之一,也就是【()=>单一语句】。另外,此语句实现的功能,是每隔 3 秒,显示当前时间的秒数。

5. 块范围的函数

在 JavaScript 语言中,块范围(block scope)是指一对大括号里的范围。在特定块范围里,若遵循特定规则(例如:通过关键字 let,而不是关键字 var,声明特定变量或定义特定函数),则在块范围之外是无法访问到块范围里的变量或函数的。关于块范围的函数的运用,可参考如下示例。

【5-1-1-e6-block-scope-function.js】

```
function cylinder(r, h)
{
  function circle(r)
  {
   var values = {} ;

   values.area = Math.PI * r * r ;
   values.circumference = 2 * Math.PI * r ;

   return values ;
  }

  var values = {} ;

  values.circle = circle(r) ;

  values.volume = values.circle.area * h ;
  values.surface_area = 2 * values.circle.area + values.circle.circumference * h ;

  return values ;
}

let result = cylinder(12, 50) ;

console.log(result.circle) ;
console.log(result.volume) ;
console.log(result.surface_area) ;
```

【相关说明】

```
function cylinder(r, h)
{
  function circle(r)
  {
    var values = {} ;
```

```
values.area = Math.PI * r * r ;
```

此语句将特定圆面积的数值,赋给变量 values 的对象实例的属性 area。

```
    values.circumference = 2 * Math.PI * r ;
```

此语句将特定圆周长的数值,赋给变量 values 的对象实例的属性 circumference。

```
    return values ;
```

此语句返回局部变量 values 的对象实例{area: 圆面积的数值, circumference: 圆周长的数值}。

```
  }
  var values = {} ;
  values.circle = circle(r) ;
```

此语句调用了函数 circle(r),并将其返回值,赋给变量 values 的对象实例的属性 circle。

```
  values.volume = values.circle.area * h ;
```

此语句将特定圆柱体积(圆面积×高)的数值,赋给变量 values 的对象实例的属性 volume。

```
  values.surface_area = 2 * values.circle.area + values.circle.circumference * h ;
```

此语句将特定圆柱体表面积(2×圆面积+圆周长×高)的数值,赋给变量 values 的对象实例的属性 surface_area。

```
  return values ;
```

此语句返回局部变量 values 的对象实例{circle: {area: 圆面积的数值, circumference: 圆周长的数值}, volume: 球体积的数值, surface_area: 球体表面积的数值}。

值得注意的是,在被返回的对象实例中,还存在子对象实例{area: 圆面积的数值, circumference: 圆周长的数值}。

```
}
```

在上述源代码中,可看到两个函数 cylinder()与 circle()。其中,函数 circle()被定义于函数 cylinder()的内部。这就意味着,调用函数 circle(),只能进行于函数 cylinder()的内部!

```
let result = cylinder(12, 50) ;
```

此语句将函数 cylinder(12, 50)所返回的对象实例,赋给变量 result。

```
console.log(result.circle) ;
```

此语句显示出{area: 452.3893421169302, circumference: 75.39822368615503}的信息。

```
console.log(result.volume) ;
```

此语句显示出 22619.46710584651 的信息。

```
console.log(result.surface_area) ;
```

此语句显示出 4674.689868541612 的信息。

5.1.2　函数名称（ES6）

在编程语言中，函数名称（function name）主要用于重复调用特定函数。没有名称的匿名函数，可通过特殊语法，**间接**或**立即**调用特定匿名函数。下面对函数名称的定义和运用，进行举例说明。

【5-1-2-function-names.js】

```
function display01()
{
  console.log('Standard displaying.') ;
}

display02 = function ()
{
  console.log('Variable-type displaying.') ;
} ;

display01() ;

display02() ;

(function ()
 {
   console.log('Anonymous displaying.') ;
 }
)() ;

setTimeout(function () { console.log('Indirect anonymous displaying.') ; }, 1000) ;

setTimeout(() => console.log('Indirect arrow-type anonymous displaying.'), 1500) ;
```

【相关说明】

```
function display01()
{
  console.log('Standard displaying.') ;
}
```

此语法定义了名称为 display01 的函数。

```
display02 = function ()
{
  console.log('Variable-type displaying.') ;
} ;
```

在此语法的等号右侧，定义了没有名称的匿名函数。然而，变量名称 display02，变成此匿名函数的别名。

```
display01() ;
```

此语句调用了名称为 display01 的函数。

```
display02 () ;
```

此语句则调用了名称为 display02 的函数。

```
(function ()
{
  console.log('Anonymous displaying.') ;
}
)() ;
```

这个特殊语法，定义并立即调用了没有名称的匿名函数。

```
setTimeout(function () { console.log('Indirect anonymous displaying.') ; }, 1000) ;
```

此语句在 1 秒之后，间接调用了没有名称的匿名函数【function () { console.log('Indirect anonymous displaying.')】。

```
setTimeout(() => console.log('Indirect arrow-type anonymous displaying.'), 1500) ;
```

此语句则在 1.5 秒之后，间接调用了箭头函数【() => console.log('Indirect arrow-type anonymous displaying.')】。

5.1.3 参数（ES6）

在特定函数的参数列里，可描述多个**参数**（parameter）的名称。参数在函数主体的内部，被视为局部变量（local variable）。例如，在语法【profile(a01 = 'none', a02 = 'none', a03 = 'none', ... others)】中，a01、a02、a03 和 others 是函数 profile() 的**参数**（parameter）。

特定函数被调用时，需要提及函数名称和传入的参数数据（argument）。例如，在语法【profile('name', 'gender', 'age', 'position', 'department')】中，字符串'name'、'gender'、'age'、'position'、'department'，就是被传入函数 profile() 内部的参数数据（argument）。

特别值得注意的是，parameter 和 argument 皆被翻译为参数；但是，读者应该要加以区别：parameter 是指**参数**，argument 是指被传入的参数的数据。

1. 参数列

在特定函数的定义语法中，参数列可内含多个参数的名称与默认数据。关于参数列的综合运用，可参考如下示例。

【5-1-3-e1-parameter-list.js】

```
function profile(a01 = 'none', a02 = 'none', a03 = 'none', ... others)
{
  console.log(arguments) ;
  console.log(arguments.length) ;
  console.log('') ;

  console.log(arguments[0], arguments[1], arguments[2]) ;
  console.log(a01, a02, a03) ;
  console.log('') ;
```

```
    console.log(others) ;
    console.log(arguments[3], arguments[4]) ;
    console.log(others[0], others[1]) ;
}

profile('name', 'gender', 'age', 'position', 'department') ;

console.log('') ;

profile() ;

///
console.log('\n\n') ;

let arg_list = ['model', 'name', 'color', 'weight', 'price'] ;

profile(... arg_list) ;
```

【相关说明】

```
function profile(a01 = 'none', a02 = 'none', a03 = 'none', ... others)
```

在函数 profile() 的参数列里，存在 4 个参数 a01、a02、a03 与 others。其中，前 3 个参数被设置了默认数据，参数 others 则借助扩展运算符【...】而变成数组实例，并且内含从第 4 个开始的被传入的参数数据。

```
{
    console.log(arguments) ;
```

局部变量 arguments 是内置的，所以在任何函数的内部，均是可被访问的。举例来说，若函数 profile('name', 'gender', 'age', 'position', 'department') 被调用，则其内部的局部变量 arguments 的数据，将会是【相似于数组实例】的 Arguments 对象实例 ["name", "gender", "age", "position", "department"]。

```
    console.log(arguments.length) ;
```

若函数 profile('name', 'gender', 'age', 'position', 'department') 被调用，则其 arguments.length 的数值，将会为被传入的参数数据的个数 5。

```
    console.log('') ;

    console.log(arguments[0], arguments[1], arguments[2]) ;
```

若函数 profile('name', 'gender', 'age', 'position', 'department') 被调用，则此语句会显示【name gender age】的信息。

```
    console.log(a01, a02, a03) ;
```

若函数 profile('name', 'gender', 'age', 'position', 'department') 被调用，则此语句会显示【name gender age】的信息。

```
    console.log('') ;

    console.log(others) ;
```

若函数 profile('name', 'gender', 'age', 'position', 'department')被调用，则此语句会显示["position", "department"]的信息。

```
console.log(arguments[3], arguments[4]) ;
console.log(others[0], others[1]) ;
```

若函数 profile('name', 'gender', 'age', 'position', 'department')被调用，则这两个语句会显示相同的信息【position department】。

```
}
```

```
profile('name', 'gender', 'age', 'position', 'department') ;
```

在此，函数 profile()被调用时，被传入了 5 个字符串'name'、'gender'、'age'、'position'与'department'。

```
console.log('') ;
```

```
profile() ;
```

在此，函数 profile()被调用时，并没有被传入任何参数数据。所以，前 3 个参数将会分别采用其默认数据，进而等同于函数 profile('none', 'none', 'none')被调用时的效果。

```
console.log('\n\n') ;
```

```
let arg_list = ['model', 'name', 'color', 'weight', 'price'] ;
```

此语句声明了初始数据为数组实例['model', 'name', 'color', 'weight', 'price']的变量 arg_list。

```
profile(... arg_list) ;
```

借助了扩展运算符【...】，使得此语句等同于调用了函数 profile('model', 'name', 'color', 'weight', 'price')。

2. 参数的默认数据

若特定参数未被设置默认数据（default data），则内置的默认数据会是原始常量 undefined。欲设置特定参数的默认数据，可在参数名称的后方加上【= 默认数据】，成为【参数名称 = 默认数据】。例如，在 function income(identity = 'company', currency = 'USD', days = 30, daily_income = 5000)的语法里：

- 字符串'company'是参数 identity 的默认数据。
- 字符串'USD'是参数 currency 的默认数据。
- 整数值 30 是参数 days 的默认数值。
- 整数值 5000 是参数 daily_income 的默认数值。

调用特定函数，却未传入**特定参数（parameter）的数据**（argument）时，此参数的默认数据，会被视为当前的数据。关于参数的默认数据的理解和运用，可参考如下示例。

【5-1-3-e2-parameter-default-data.js】

```
function income(identity = 'company', currency = 'USD', days = 30, daily_income = 5000)
{
  let result = 0 ;
  let monthly_income = days * daily_income ;
```

```js
    // monthly_income = monthly_income.toLocaleString() ;
    monthly_income = Intl.NumberFormat().format(monthly_income) ;

    result = `The ${identity}'s monthly income = ${monthly_income} ${currency}` ;

    console.log(result) ;
}

income() ;
income('Jasper') ;
income('Paula', 'GBP') ;
income('Eric', 'GBP', 31) ;
income(undefined, 'RMB', undefined, 1000) ;
```

【相关说明】

```js
function income(identity = 'company', currency = 'USD', days = 30, daily_income = 5000)
```

在此语法中，函数 income()共有 4 个参数，每个参数皆设置了默认数据。

```js
{
    let result = 0 ;

    let monthly_income = days * daily_income ;
```

在此段语法中，声明了变量 monthly_income，其初始数值为当前的月薪的数值。

```js
    // monthly_income = monthly_income.toLocaleString() ;
    monthly_income = Intl.NumberFormat().format(monthly_income) ;
```

此语句将变量 monthly_income 的月薪的数值，转换成为具有**千分位隔符**（group seperator）的字符串。

```js
    result = `The ${identity}'s monthly income = ${monthly_income} ${currency}` ;

    console.log(result) ;
}

income() ;
```

此语句则是在没有传入任何**参数数据**的情况下，直接调用函数 income()。因为每个参数皆被设置了默认数据，调用 income()等同于调用 income('company', 'USD', 30, 5000)，进而显示出【The company's monthly income = 150,000 USD】的信息。

```js
income('Jasper') ;
```

此语句仅传入第 1 个参数的数据，即调用了函数 income('Jasper')，其效果等同于调用了 income('Jasper', 'USD', 30, 5000)，进而显示出【The Jasper's monthly income = 150,000 USD】的信息。

```js
income('Paula', 'GBP') ;
```

此语句传入前两个参数的数据，即调用了函数 income('Paula', 'GBP')，其效果等同于调用了 income('Paula', 'GBP', 30, 5000)，进而显示出【The Paula's monthly income = 150,000 GBP】的信息。

```js
income('Eric', 'GBP', 31) ;
```

此语句传入前 3 个参数的数据，即调用了函数 income('Eric', 'GBP', 31)，其效果等同于调用了 income('Eric', 'GBP', 31, 5000)，进而显示出【The Eric's monthly income = 155,000 GBP】的信息。

```
income(undefined, 'RMB', undefined, 1000) ;
```

此语句完整传入 4 个参数的数据，并调用了函数 income(undefined, 'RMB', undefined, 1000)。因为传入的第 1 个和第 3 个数据为 undefined，所以调用 income(undefined, 'RMB', undefined, 1000) 如同调用 income('company', 'RMB', 30, 1000)，进而显示出【The company's monthly income = 30,000 RMB】的信息。

3. 参数的一般配对

参数的一般配对是指，在特定函数参数列的定义中存在对象或数组的描述，此函数被调用时被传入参数的数据为对象或数组。关于参数的一般配对的综合运用，可参考如下示例。

【5-1-3-e3-parameter-common-matches.js】

```
var item01 = {name: 'fruit_set', price: 250} ;

function display01(item)
{
  with (item)
    message = `${name}'s price is ${price} dollars now.` ;

  return message ;
}

function display02({name, price})
{
  message = `${name}'s price is ${price} dollars now.` ;

  return message ;
}

function display03({name: n, price: p})
{
  message = `${n}'s price is ${p} dollars now.` ;

  return message ;
}

console.log(display01(item01)) ;
console.log(display02(item01)) ;
console.log(display03(item01)) ;

///
var item02 = ['bread_set', 120] ;

function display04([name, price])
{
  message = `${name}'s price is ${price} dollars now.` ;

  return message ;
}
```

```
console.log(display04(item02));
```

【相关说明】

```
var item01 = {name: 'fruit_set', price: 250};
```

此语句声明了初始数据为对象实例{name: 'fruit_set', price: 250}的变量 item01。

```
function display01(item)
```

对于此语法而言，若传入的数据是对象实例{name: 'fruit_set', price: 250}，则参数 item 的数据将会是{name: 'fruit_set', price: 250}。

```
{
  with (item)
    message = `${name}'s price is ${price} dollars now.`;

  return message;
}
function display02({name, price})
```

对于此语法而言，若传入的数据是对象实例{name: 'fruit_set', price: 250}，则特殊格式的参数列{name, price}会对应到对象实例{name: 'fruit_set', price: 250}，使得参数 name 的初始数据为 'fruit_set'、参数 price 的初始数据为 250。

```
{
  message = `${name}'s price is ${price} dollars now.`;

  return message;
}
function display03({name: n, price: p})
```

对于此语法而言，若传入的数据是对象实例{name: 'fruit_set', price: 250}，则特殊格式的参数列{name: n, price: p}依然会对应到对象实例{name: 'fruit_set', price: 250}，使得参数 n 的初始数据为 'fruit_set'、参数 p 的初始数据为 250。

```
{
  message = `${n}'s price is ${p} dollars now.`;

  return message;
}
console.log(display01(item01));
console.log(display02(item01));
console.log(display03(item01));
```

这 3 个语句分别调用了不同的函数，但均将变量 item01 的对象实例{name: 'fruit_set', price: 250}，传入而作为各个参数的数据，进而分别显示相同的信息【fruit_set's price is 250 dollars now.】。

```
///
var item02 = ['bread_set', 120];
```

此语句声明了初始数据为数组实例['bread_set', 120]的变量 item02。

```
function display04([name, price])
```

对于此语法而言,若传入的数据是数组实例['bread_set', 120],则参数 name 和 price 的数据分别为字符串'bread_set'和整数值 120。

4. 参数的扩展配对

参数的扩展配对(spreading match),是指在特定函数的参数列的定义语法中,存在扩展运算符和特定参数名称。例如,函数的定义语法 user_cart(id, name, ... items)中的【... items】。

在此,若【user_cart('Alex', 'USA-TX-21532', ['Red apple', 3], ['Durian', 5], ['Grapefruit', 13], ['Watermelon', 2])】被调用,则其参数 id 的数据为字符串'Alex',其参数 name 的数据为字符串'USA-TX-21532',而参数 items 的数据将会是由 3 个一维数组所构成的二维数组实例[['Red apple', 3], ['Durian', 5], ['Grapefruit', 13], ['Watermelon', 2]]。关于参数的扩展配对的综合运用,可参考如下示例。

【5-1-3-e4-parameter-spreading-matches.js】

```
function user_cart(id, name, ... items)
{
  let title = `User ${name} whose id is ${id}` ;

  console.log(title) ;
  console.log(items) ;
  console.log('') ;

  for (let i = 0; i < items.length; i++)
  {
    let item_message = `item ${i}: Product = ${items[i][0]}, Amount = ${items[i][1]}` ;
    console.log(item_message) ;
  }
}
user_cart('Alex', 'USA-TX-21532', ['Red apple', 3], ['Durian', 5], ['Grapefruit', 13],
['Watermelon', 2]) ;
```

【相关说明】

```
function user_cart(id, name, ... items)
{
  let title = `User ${name} whose id is ${id}` ;

  console.log(title) ;
```

此语句显示出【User USA-TX-21532 whose id is Alex】的信息。

```
  console.log(items) ;
```

此语句显示出[['Red apple', 3], ['Durian', 5], ['Grapefruit', 13], ['Watermelon', 2]]的信息。

```
  console.log('') ;

  for (let i = 0; i < items.length; i++)
```

```
        let item_message = `item ${i}: Product = ${items[i][0]}, Amount = ${items[i][1]}` ;
        console.log(item_message) ;
```

在上述 for 循环语句的 4 次迭代中，会显示出如下信息：

- item 0: Product = Red apple, Amount = 3
- item 1: Product = Durian, Amount = 5
- item 2: Product = Grapefruit, Amount = 13
- item 3: Product = Watermelon, Amount = 2

```
    }
}
```

上述源代码定义了带有 3 个参数的函数 user_cart()。其中，最右侧的参数，被冠上扩展运算符【...】。若调用前述函数时，被传入的数据个数超过两个，则第 3 个到最后 1 个数据，均可被整合成为参数 items 的数组实例中的各元素。在此，第 3 个数据['Red apple', 3]、第 4 个数据['Durian', 5]、第 5 个数据['Grapefruit', 13]、第 6 个数据['Watermelon', 2]，均会被整合至参数 items 的数组实例[['Red apple', 3], ['Durian', 5], ['Grapefruit', 13], ['Watermelon', 2]]中。

```
user_cart('Alex', 'USA-TX-21532', ['Red apple', 3], ['Durian', 5], ['Grapefruit', 13],
    ['Watermelon', 2]) ;
```

此语句调用了函数 user_cart()，并传入一连串的数据。

5.1.4　主体

在函数的定义语法中，大括号里的块范围，是函数的主体（body）。在函数的主体里：

- 可声明局部变量。
- 可引用内置但只读（read only）的局部变量 this。
- 可引用内置且含有被传入的各数据的局部变量 arguments。
- 可通过 return 语句，返回特定数据。

关于函数之主体的综合运用，可参考如下示例。

【5-1-4-function-bodies.js】

```
var value01 = -1, value02 = -1 ;
var result = 0 ;

function choice(n, r, h)
{
  switch(n)
  {
    case 1:
      result = Math.PI * (r ** 2) ;
      break ;
    case 2:
```

```
      result = 4 / 3 * Math.PI * (r ** 3) ;
      break ;
    case 3:
      result = Math.PI * (r ** 2) * h ;
      break ;
  }

  console.log(result.toFixed(2)) ;
  console.log('') ;

  console.log(result) ;
  console.log(value01) ;
  console.log(value02) ;
  console.log('') ;

  return result.toFixed(2) ;
}

choice(1, 10) ;
choice(2, 10) ;
choice(3, 10, 15) ;
console.log('') ;

///
function another_choice(n, l, w, h)
{
  this.value = 0 ;

  switch(n)
  {
    case 1:
      this.value = l * w ;
      break ;
    case 2:
      this.value = l * w * h ;
      break ;
    case 3:
      this.value = 2 * (l * w + w * h + h * l) ;
      break ;
  }

  console.log(this.value) ;
}

var pointer = new another_choice(2, 10, 20, 30) ;

console.log('\n\n') ;
console.log(pointer) ;

console.log(pointer.value) ;
```

【相关说明】

```
var value01 = -1, value02 = -1 ;
```

此语句声明了初始数值均为-1 的变量 value01 与 valuc02。

```
var result = 0 ;
```

此语句声明了初始数值为 0 的变量 result。

```
function choice(n, r, h)
{
  switch(n)
  {
    case 1:
      result = Math.PI * (r ** 2) ;
      break ;
    case 2:
      result = 4 / 3 * Math.PI * (r ** 3) ;
      break ;
    case 3:
      result = Math.PI * (r ** 2) * h ;
      break ;
  }
```

此语法根据变量 n 的数值，决定变量 result 的数值，是代表圆面积、球体积，还是圆柱体积。

```
  console.log(result.toFixed(2)) ;
```

此语法显示出变量 result 的精确到第 2 个小数位的数值。

```
  console.log('') ;

  console.log(result) ;
  console.log(value01) ;
  console.log(value02) ;
  console.log('') ;

  return result.toFixed(2) ;
```

此语法返回变量 result 的精确到第 2 个小数位的数值。

```
}

choice(1, 10) ;
```

此语法调用了函数 choice(1, 10)，以计算并显示出半径为 10 的圆面积。

```
choice(2, 10) ;
```

此语法调用了函数 choice(2, 10)，以计算并显示出半径为 10 的球体积。

```
choice(3, 10, 15) ;
```

此语法调用了函数 choice(3, 10, 15)，以计算并显示出半径为 10、高度为 15 的圆柱体积。

```
console.log('') ;

function another_choice(n, l, w, h)
{

  this.value = 0 ;
```

调用函数 another_choice()时：

- 若未冠上关键字 new，例如【var pointer = another_choice(2, 10, 20, 30) ;】，则【this.value】的语法，会使得 value 突然变成全局变量。
- 若冠上关键字 new，例如【var pointer = **new**another_choice(2, 10, 20, 30) ;】，则【this.value】的语法，会使得 value 成为【函数 another_choice()默认对应的新对象实例】中的属性。

```
switch(n)
{
  case 1:
    this.value = l * w ;
    break ;
  case 2:
    this.value = l * w * h ;
    break ;
  case 3:
    this.value = 2 * (l * w + w * h + h * l) ;
    break ;
}
```

此语法根据变量 n 的数值，决定 this.value 的数值，是代表矩形面积、方体积还是方体表面积。

```
  console.log(this.value) ;
}
```

```
var pointer = new another_choice(2, 10, 20, 30) ;
```

在此语法中，调用函数 another_choice()时，冠上了关键字 new，所以变量 pointer 的初始数据，是函数 another_choice()默认对应的新对象实例【another_choice {value: 6000}】。

```
console.log('\n\n') ;
```

```
console.log(pointer) ;
```

此语句显示出【another_choice {value: 6000}】的信息。

```
console.log(pointer.value) ;
```

此语句显示出 pointer.value 的数值 6000。

5.1.5 返回数据与 void 关键字（ES6）

特定函数被调用之后，当执行到函数主体内的 return 语句时，便代表函数此次调用任务结束，并返回特定数据。此外，配合 void 关键字，可使得匿名函数的定义立即被调用。关于函数的返回数据和 void 关键字的运用，可参考如下示例。

【5-1-5-return-data-and-keyword-void.js】

```
let returned = -1 ;

function a_number(start, end)
{
  let number ;
```

```
    if (end <= start) return ;

    number = start + parseInt(Math.random() * (end - start)) ;

    return number ;
}

returned = a_number(50, 100) ;

console.log(returned) ;
console.log('') ;

returned = a_number(100, 10) ;

console.log(returned) ;
console.log('') ;

///
(function ()
{
   console.log('Anomynous function 01 is executed.') ;
}
)() ;

void function ()
{
   console.log('Anomynous function 02 is executed.') ;
}() ;
```

【相关说明】

```
let returned = -1 ;

function a_number(start, end)
{
   let number ;

   if (end <= start) return ;
```

在此，若变量 end 的数值，小于或等于变量 start 的数值，则执行语句【return ;】以完成函数的调用任务，并返回原始常量 undefined。

```
    number = start + parseInt(Math.random() * (end - start)) ;
```

此语句将【变量 start 的数值】到【变量 end 的数值】之间的随机整数值，赋给变量 number。

```
    return number ;
```

在此，通过关键字 return 开头的语句，返回变量 number 的数值。

```
}

returned = a_number(50, 100) ;
```

此语句将函数 a_number(50, 100)返回的 50～100 的随机整数值,赋给变量 returned。

```
console.log(returned) ;

returned = a_number(100, 10) ;
```

此语句调用函数 a_number(100, 10)时,其第一个参数数值 100,并没有小于或等于其第 2 个参数数值 10,所以函数 a_number(100, 10)返回了原始常量 undefined,并赋给变量 returned。

```
console.log(returned) ;
console.log('') ;

(function ()
{
  console.log('Anomynous function 01 is executed.') ;
}
)() ;
```

此特殊语句通过 3 对小括号分别定义没有名称的匿名函数,并随即调用匿名函数。

```
void function ()
{
  console.log('Anomynous function 02 is executed.') ;
}() ;
```

此特殊语法通过关键字 void 和两对小括号,定义并立即调用了没有名称的匿名函数。

5.1.6 定义的位置（ES6）

函数的定义语法,类似于变量的声明语法,可被放置于:

- 全局（global）的位置:也就是源代码第一层级的位置。
- 局部（local）的位置:也就是特定函数的主体内部的位置。
- 块范围里的位置:在特定块范围内部,通过关键字 let 开头的函数表达式(function expression),可定义只能在特定块范围内部,被调用的函数。

关于函数的定义位置的综合运用,可参考如下示例。

【5-1-6-locations-of-function-definitions.js】

```
gfunc() ;

// global function
function gfunc()
{
  lfunc() ;

  // local function
  function lfunc()
  {
    console.log('Local function is executed.') ;
  }

  lfunc() ;
```

```
    console.log('Global function is executed.') ;
}

// Error occurs if bfunc() is invoked here.
// bfunc() ;

{
  // function expression defined with keyword 'let' in a block
  let bfunc = function ()
  {
    console.log('Blocking function is executed.') ;
  }

  bfunc() ;
}
// Error also occurs if bfunc() is invoked here.
// bfunc() ;

gfunc() ;

// Error occurs if lfunc() is invoked here.
// lfunc() ;
```

【相关说明】

```
gfunc() ;
```

此语句调用了被定义在下方的全局函数 gfunc()。

```
// global function
function gfunc()
{
  lfunc() ;
```

此语句调用了被定义在下方的函数 lfunc()。

```
  // local function
  function lfunc()
  {
    console.log('Local function is executed.') ;
  }
```

函数 lfunc()被定义在函数 gfunc()里，成为块范围的函数，因此在函数 gfunc()之外的位置，均无法调用函数 lfunc()。

```
  lfunc() ;
```

此语句调用了被定义在上方的函数 lfunc()。

```
  console.log('Global function is executed.') ;
}
```

函数 gfunc()虽然被定义在被调用的位置之后，但是在 JavaScript 语法中，是可行的。

```
// Error occurs if bfunc() is executed here.
// bfunc() ;
```

这里无法调用函数 bfunc()，所以将此语句，放到注释里。

```
{
  // function in a block
  let bfunc = function ()
  {
    console.log('Blocking function is executed.') ;
  }
```

函数 bfunc() 是通过关键字 let 开头的函数表达式（function expression），被定义在上层大括号里，进而成为块范围的函数。因此，在上层大括号之外的位置，无法调用到函数 bfunc()。

```
  bfunc() ;
```

此语句调用了被定义在上方的函数 bfunc()。

```
}

// Error also occurs if bfunc() is invoked here.
// bfunc() ;
```

这里也无法调用函数 bfunc()。

```
gfunc() ;
```

此语句调用了被定义在上方的全局函数 gfunc()。

```
// Error occurs if lfunc() is executed here.
// lfunc() ;
```

这里亦无法调用函数 lfunc()。

5.1.7 函数的调用形式（ES6）

特定函数的定义语法，原则上并不会一并使得这个函数被调用（invoke / call）；必须有额外的调用语法才行。函数的调用形式，大致可分为以下几种：

- 常态（normally）调用。
- 定期（periodically）调用。
- 递归（recursively）调用。
- 定义后立即（immediately）调用。

关于函数的调用形式的理解，可参考如下示例。

【5-1-7-function-invocations.js】

```
let count = 0 ;

function func01()
{
```

```
  console.log(`[${++count}]: function 01 is executed.`) ;
}

func01() ;

console.log('') ;

///
function func02(n = 1)
{
  console.log(`[${++count}]: function 02 is executed.`) ;

  if (n < 2) return 1 ;
  else return n * func02(n-1) ;
}

let result = func02(10) ;

console.log(result) ;
console.log('') ;

///
function func03()
{
  console.log(`[${++count}]: function 03 is executed.`) ;
}

setTimeout(func03, 1000) ;

let interval = setInterval(func03, 2000) ;

///
(function ()
 {
  console.log('Anonymous 1 function is executed.') ;
 }
)() ;

void function ()
{
  console.log('Anonymous 2 function is executed.') ;
}() ;
```

【相关说明】

```
let count = 0 ;
```

此语句声明了初始数值为 0 的变量 count。

```
function func01()
{
  console.log(`[${++count}]: function 01 is executed.`) ;
}
```

此语法定义了函数 func01()。

```
func01() ;
```

此语句调用了函数 func01() ;

```
console.log('') ;

function func02(n = 1)
{
  console.log(`[${++count}]: function 02 is executed.`) ;

  if (n < 2) return 1 ;
  else return n * func02(n-1) ;
}
```

在函数 func02()的定义语法里,if...else 语句中的【func02(n-1)】,递归调用了函数 func02()本身,这样的函数可称为递归函数(recursive function)。

```
let result = func02(10) ;
```

此语句声明了变量 result,其初始数值为函数 func02(10)的返回值。

```
console.log(result) ;
console.log('') ;

function func03()
{
  console.log(`[${++count}]: function 03 is executed.`) ;
}
```

此语法定义了函数 func03()。

```
setTimeout(func03, 1000) ;
```

此语句使得函数 func03()在延迟 1 秒(1000 毫秒)之后,才间接被调用。

```
let interval = setInterval(func03, 2000) ;
```

此语句则使得函数 func03()每隔 2 秒(2000 毫秒),就被间接调用 1 次。

```
(function ()
{
  console.log('Anonymous 1 function is executed.') ;
}
)() ;
```

此语法是一种【立即可调用的函数表达式 (IIFE, immediately invokable function expression)】,可用来定义并立即调用其中的匿名函数。。

```
void function ()
{
  console.log('Anonymous 2 function is executed.') ;
}() ;
```

此语法定义并立即调用了其中的匿名函数。

5.2 生成器

在 JavaScript 语言里,生成器(generator)是可被迭代的对象实例(object instance),并且可由生成器函数(generator function)加以动态产生。

5.2.1 迭代器协议与生成器(ES6)

在所谓的迭代器协议(iterator protocol)里,描述了【作为迭代器(iterator)的对象实例(instance),必须支持动态产生可被迭代(iterable)的多个元素】的集合。

在 JavaScript 语言中,数组(array)、字符串(string)、集合(set)与地图(map)类型的实例,均被默认为可迭代的。此外,通过自定义的生成器函数(generator function),也可动态产生可被迭代的生成器(generator)。关于迭代器协议与生成器的理解和运用,可参考如下示例。

【5-2-1-iterator-protocol-and-generator.js】

```
let iterator01 =
{
 *[Symbol.iterator]()
  {
   for(let i = 0; i < 10; i++)
   {
    yield parseInt(10 * (i + Math.random())) ;
   }
  }
}

for (let n of iterator01)
{
 console.log(n) ;
}
```

【相关说明】

```
let iterator01 =
{
 *[Symbol.iterator]()
  {
   for(let i = 0; i < 10; i++)
   {
    yield parseInt(10 * (i + Math.random())) ;
```

通过关键字 yield,此 for 循环语句在每次迭代时,均会将表达式【parseInt(10 * (i + Math.random()))】的结果值,产生 (yield) 至【这个生成器最终返回】的数组实例中。

```
   }
  }
```

上述语法依据迭代器协议的规范，定义了作为生成器的变量 iterator01。其中，【Symbol.iterator】可被视为【在特定对象实例（object instance）中，作为内置的迭代器（iterator）】的索引数据（index data），进而使得【*[Symbol.iterator]() { ... }】定义了新的迭代器函数。

为了让特定对象实例具有可被迭代的生成器特征，必须在这个对象实例的定义语法中，再加上前述的迭代器函数的描述。

```
}
```

上述语法将【等号右侧内含迭代器函数】的对象实例，赋给作为生成器的变量 iterator01。

```
for (let n of iterator01)
{
  console.log(n) ;
}
```

在此，通过带有语法【... of 作为生成器的变量名称】的 for 循环语句，可让其大括号里的源代码片段，被迭代数次。

5.2.2　生成器函数的定义和用法（ES6）

通过关键字 function 衔接星号【*】，可用来定义生成器函数（generator function），以动态返回可被迭代（iterable）、可用于扩展（spread）与可用于配对（match）的生成器（generator）。关于生成器函数的定义和用法的综合运用，可参考如下示例。

【5-2-2-generator-function-definitions-and-matches.js】

```
function * iterator02(end)
{
  for(let i = 0; i < end; i++)
  {
    yield parseInt(10 * (i + Math.random())) ;
  }
}

for (let n of iterator02(10))
{
  console.log(n) ;
}

let numbers = [ ... iterator02(20) ] ;

let [a, b, c, ... others] = iterator02(8) ;

console.log(numbers) ;
console.log(a, b, c) ;
console.log(others) ;
```

【相关说明】

```
function * iterator02(end)
{
  for(let i = 0; i < end; i++)
```

```
    {
      yield parseInt(10 * (i + Math.random())) ;
    }
}
```

此语法是生成器函数的标准定义形式。

```
for (let n of iterator02(10))
{
  console.log(n) ;
}
```

通过带有【... of 生成器函数名称()】的 for 循环语句，也可达成迭代的效果。

```
let numbers = [ ... iterator02(20) ] ;
```

此语句声明了初始数据为新数组实例的变量 numbers。在此，扩展运算符【...】衔接生成器函数 iterator02(20)的调用语法，可将生成器函数 iterator02(20)的返回数据，进一步扩展成为新数组实例中的各元素。

```
let [a, b, c, ... others] = iterator02(8) ;
```

若生成器函数 iterator02(8)返回了数组[5, 16, 28, 36, 47, 54, 66, 71]，则此语句就相当于声明了初始数值为 5 的变量 a、初始数值为 16 的变量 b、初始数值为 28 的变量 c，以及初始数据为数组实例[36, 47, 54, 66, 71]的变量 others。

5.3 搭配 Promise 对象

Promise 具有承诺的含义，而 Promise 对象可用来描述特定承诺，是否已经成功（success）完成或者以失败（failure）收场，进而分别进行后续的处置。

5.3.1 Promise 对象的用法（ES6、ES9）

欲捕获 Promise 对象实例发生的异常错误，可通过内置的函数 catch()，间接调用进行错误处理的函数。无论有无发生异常错误，皆可通过内置的函数 finally()，间接调用进行收尾处理的函数。其中，传入函数 catch()或 finally()的单一参数，可以是【匿名函数的定义语法】或者【自定义函数的名称】。其语法大致为：

- 作为 Promise 对象实例的变量名称.**catch**(匿名函数的定义语法或者自定义函数的名称)；
- 作为 Promise 对象实例的变量名称.**finally**(匿名函数的定义语法或者自定义函数的名称)；

欲加以处理 Promise 对象实例的【解决承诺】和【拒绝承诺】的后续动作，可通过内置函数 then()，间接调用不同的两个函数。其语法大致为：

- 作为 Promise 对象实例的变量名称.**then**(**解决**承诺之后加以调用的匿名函数的定义语法或者函数名称, **拒绝**承诺之后加以调用的匿名函数的定义语法或者函数名称)；

运用 Promise 对象的机制，需编写如下各个部分的源代码片段：

（1）定义启动承诺的函数，例如 function promise_executor() { ... }。

（2）调用内置的 Promise 对象的构造函数 Promise()，例如 Promise(promise_executor)，并返回 Promise 对象实例。

（3）将 Promise 对象实例，赋给特定变量，例如【message_promise = new Promise(promise_executor) ;】。

（4）分别定义【解决承诺之后】和【拒绝承诺之后】的处置函数，例如：

- function positive(message) { ... }
- function negative(message) { ... }

（5）调用 Promise 对象实例所对应的内置函数 then()，例如【message_promise.then(positive, negative) ;】。

关于 Promise 对象的细部运用，可参考如下示例。

【5-3-1-Promise-usage.js】

```js
let textfield = document.createElement('input') ;

textfield.id = 't01' ;

textfield.style.fontSize = '2em' ;
textfield.style.paddingLeft = '5px' ;
textfield.style.color = 'RoyalBlue' ;

document.body.innerHTML += '<p></p>' ;

document.body.appendChild(textfield) ;

let message_box = document.createElement('span') ;

message_box.id = 'mbox' ;
message_box.style.fontSize = '1.5em' ;
message_box.style.marginLeft = '15px' ;

document.body.appendChild(message_box) ;

textfield.focus() ;

///
let message_promise ;

function promise_executor(resolve, reject)
{
  if (t01.value != '')
  {
    resolve('at most 12 characters.') ;
  }
  else
  {
```

```
    reject('you have only 5 seconds to input!') ;
  }
}

function positive(message)
{
  mbox.style.color = 'ForestGreen' ;
  mbox.innerHTML = message ;
}

function negative(message)
{
  mbox.style.color = 'Pink' ;
  mbox.innerHTML = message ;
}

function check_error()
{
  message_promise = new Promise(promise_executor) ;

  message_promise.then(positive, negative) ;
}

setTimeout(check_error, 5000) ;
```

【相关说明】

```
let textfield = document.createElement('input') ;
```

此语句声明了变量 textfield，其初始数据为网页上新的 input 元素实例。

```
textfield.id = 't01' ;
```

此语句设置了新 input 元素实例的属性 id 及其数据字符串't01'。在此，这个 input 元素实例的属性 id 的数据字符串为't01'，即是意味着这个 input 元素实例的身份识别码为 t01，可供后续源代码片段加以访问。

```
textfield.style.fontSize = '2em' ;
textfield.style.paddingLeft = '5px' ;
textfield.style.color = 'RoyalBlue' ;
```

这 3 个语句设置了新 input 元素实例的属性 style 中的子属性 fontSize、paddingLeft、color。设置特定元素实例的属性 style，会驱使浏览器变更这个元素实例的外观，例如字体大小（fontSize）、左侧边界内的填充距离（paddingLeft）与字体颜色（color）。

```
document.body.innerHTML += '<p></p>' ;
```

此语句使得无任何子节点的 p 元素实例，被新增至 body 元素实例中。

```
document.body.appendChild(textfield) ;
```

此语句将变量 textfield 的新的 input 元素实例，新增至 body 元素实例中。

```
let message_box = document.createElement('span') ;
```

此语句声明了变量 message_box，其初始数据为网页上新的 span 元素实例。

```
message_box.id = 'mbox' ;

message_box.style.fontSize = '1.5em' ;
message_box.style.marginLeft = '15px' ;
```

这两个语句设置了【在变量 message_box 的新 span 元素实例中，其属性 style】的子属性 fontSize、marginLeft，以改变这个元素实例的字体大小（fontSize）和左侧边界外的间隔距离（marginLeft）。

```
document.body.appendChild(message_box) ;
```

此语句将变量 message_box 的新的 span 元素实例，新增至 body 元素实例中。

```
textfield.focus() ;
```

此语句使得变量 textfield 的 input 元素实例，成为网页上的焦点（focus）。成为网页上的焦点，是指特定 input 元素实例，处于选中状态，以便在**无须**自行单击这个元素实例的情况下，用户可以立即输入文本。

```
let message_promise ;
```

此语句声明了未设置初始数据的全局变量 message_promise。

```
function promise_executor(resolve, reject)
{
  if (t01.value != '')
  {

    resolve('at most 12 characters.') ;
```

此语句调用了内置的局部函数 resolve()，并将字符串'at most 12 characters.'作为其参数数据。

```
  }
  else
  {

    reject('you have only 5 seconds to input!') ;
```

此语句调用了内置的局部函数 reject()，并将字符串'you have only 5 seconds to input!'作为其参数数据。

```
  }
}
```

上述语法定义了具有两个参数的函数 promise_executor()，以用来启动承诺。被传入的两个参数数据，即是【解决承诺之后】和【拒绝承诺之后】的处置函数的名称。

```
function positive(message)
{
  mbox.style.color = 'ForestGreen' ;
  mbox.innerHTML = message ;
}
```

此语法定义了【解决承诺之后】的处置函数 positive()。

```
function negative(message)
{
  mbox.style.color = 'Pink' ;
```

```
    mbox.innerHTML = message ;
}
```

此语法则定义了【拒绝承诺之后】的处置函数 negative()。

```
function check_error()
{
```

```
    message_promise = new Promise(promise_executor) ;
```

此语句在等号右侧，描述了一个 Promise 对象实例，并被传入用来启动承诺的函数名称 promise_executor，再赋给全局变量 message_promise。也因此，这个 Promise 对象实例，间接内含了用来启动承诺的函数 promise_executor()。

```
    message_promise.then(positive, negative) ;
```

在特定承诺被解决或拒绝时，此语句可用来捕捉所发生的异常错误。特定承诺被解决时，调用函数 positive()；特定承诺被拒绝时，则调用函数 negative()。

```
}
```

上述语法定义了函数 check_error()，以创建一个全新的 Promise 对象实例，并间接调用了函数 promise_executor()。

```
setTimeout(check_error, 5000) ;
```

此语句使得函数 check_error()，在 5 秒（5000 毫秒）之后，被调用 1 次。

5.3.2 聚集多个 Promise 对象（ES6）

在一些场合当中，整个承诺可分割为数个细部承诺。而且，解决或拒绝每个细部承诺，最终势必会影响到，是否解决了整个承诺。此时，则可考虑运用 Promise 对象的内置函数 all()，来聚集多个细部承诺对应的个别 Promise 对象实例，以实现前述一连串的机制。其实现的步骤大致如下：

（1）声明一个空（empty）的数组实例，例如【let promise_array = [] ;】，以放入内含处置各细部承诺的各个 Promise 对象实例。

（2）利用 5.3.1 节示例中的方式，在上述空数组实例中，放入代表各细部承诺的 Promise 对象实例。

（3）定义用来处置整个承诺被解决之后和细部承诺被拒绝之后的处置函数，例如：

- function final_success(value_or_list) { ... }
- function final_failure(value_or_list) { ... }

（4）调用【聚集所有细部承诺】的函数 Promise.all()，例如【Promise.all(promise_array)】。

（5）通过整个承诺对应的 Promise 对象实例，进一步调用内置的函数 then()，例如【Promise.all(promise_array).then(final_success, final_failure)】。

关于聚集多个 Promise 对象的综合运用，可参考如下示例。

【5-3-2-Promise-aggregation.js】

```javascript
document.body.innerHTML += '<p></p>' ;

let textfield ;

for (let i = 1 ; i < 5 ; i++)
{
  textfield = document.createElement('input') ;

  textfield.id = 't0' + i ;

  textfield.style.width = '80px' ;

  textfield.style.fontSize = '2em' ;
  textfield.style.paddingLeft = '5px' ;
  textfield.style.marginRight = '5px' ;
  textfield.style.color = 'RoyalBlue' ;

  document.body.appendChild(textfield) ;
}

let message_box = document.createElement('span') ;

message_box.id = 'mbox' ;

message_box.style.fontSize = '1.5em' ;
message_box.style.marginLeft = '15px' ;

document.body.appendChild(message_box) ;

t01.focus() ;

///
let promise_array = [] ;

let filled = 0, blank = 0 ;

let id_no ;

function promise_executor(resolve, reject)
{
  let ref = document.getElementById('t0' + id_no) ;

  console.log(ref) ;

  if (ref.value != '')
  {
    ref.style.borderColor = 'Green' ;

    resolve() ;
  }
  else
  {
    ref.style.borderColor = 'Red' ;
```

```
    reject() ;
  }
}

function check_error()
{
  for (id_no = 1 ; id_no < 5 ; id_no++)
  {
    promise_array[id_no - 1] = new Promise(promise_executor) ;

    promise_array[id_no - 1].then(positive, negative) ;
  }

  Promise.all(promise_array).then(final_success, final_failure) ;
}

function positive(value)
{
  filled++ ;
}

function negative(value)
{
    console.log('value =', value) ;
  blank++ ;
}

function final_success(value_or_list)
{
  mbox.style.color = 'ForestGreen' ;
  mbox.innerHTML = 'All fields are filled.' ;
}

function final_failure(value_or_list)
{
  mbox.style.color = 'Pink' ;
  mbox.innerHTML = `${blank} Blank field is/are left...` ;
}

setTimeout(check_error, 4000) ;
```

【相关说明】

```
document.body.innerHTML += '<p></p>' ;
```

此语句使得在网页上 body 元素实例内，被新增一个空的 p 元素实例，以便让【选择文件】按钮与即将被新增的 input 元素实例之间，存在着间隔距离。

```
let textfield ;
```

此语句声明了没有初始数据的变量 textfield。

```
for (let i = 1 ; i < 5 ; i++)
{
```

```
textfield = document.createElement('input') ;
```

此语句将新的 input 元素实例，赋给变量 textfield。

```
textfield.id = 't0' + i ;
```

此语句设置了各个 input 元素实例的属性 id 的数据，分别成为字符串't01'、't02'、't03'与't04'。这也就意味着，前述各个 input 元素实例的身份识别码，分别成为 t01、t02、t03 与 t04。

```
textfield.style.width = '80px' ;
```

此语句设置了 input 元素实例被呈现在网页上的宽度，成为 80 像素。

```
textfield.style.fontSize = '2em' ;
textfield.style.paddingLeft = '5px' ;
textfield.style.marginRight = '5px' ;
textfield.style.color = 'RoyalBlue' ;
```

这 4 个语句变更了新 input 元素实例显示在网页上的外观参数。

```
document.body.appendChild(textfield) ;
```

通过此语句，为网页上的 body 元素实例，新增变量 textfield 所代表的新 input 元素实例。

```
}

let message_box = document.createElement('span') ;
```

通过此语句，设置变量 message_box 的数据，成为新的 span 元素实例。

```
message_box.id = 'mbox' ;
```

通过此语句，在变量 message_box 所代表的新 span 元素实例中，设置其属性 id 的数据为字符串'mbox'。

```
message_box.style.fontSize = '1.5em' ;
message_box.style.marginLeft = '15px' ;
```

通过这两个语句，设置新 span 元素实例显示在网页上的外观参数。

```
document.body.appendChild(message_box) ;
```

通过此语句，为网页上的 body 元素实例，新增变量 message_box 所代表的新 span 元素实例。

```
t01.focus() ;
```

通过此语句，设置变量 t01 所代表的 input 元素实例，成为网页上的焦点。

```
let promise_array = [] ;
```

此语句声明了初始数据为空数组实例的变量 promise_array，并且准备用来存放 4 个 Promise 对象实例。

```
let filled = 0, blank = 0 ;
```

此语句声明了初始数值均为 0 的变量 filled 与 blank。变量 filled 用来记录已填入文本的 input 元素实例的个数，变量 blank 则是用来记录尚未填入文本的 input 元素实例的个数。

```
    let id_no ;
```

此语句声明了变量 id_no，并且准备用来存放特定 input 元素实例的属性 id 对应的编号值，比如整数值 1、2、3 或 4。

```
function promise_executor(resolve, reject)
{
  let ref = document.getElementById('t0' + id_no) ;
```

此语句声明了局部变量 ref，其初始数据为属性 id 是 t01、t02、t03 或 t04 的其中一个 input 元素实例，即是用来代表特定 input 元素实例。

```
  console.log(ref) ;

  if (ref.value != '')
  {
    ref.style.borderColor = 'Green' ;
```

此语句设置了特定 input 元素实例的边框颜色，成为绿色。

```
    resolve() ;
```

此语句调用了内置的局部函数 resolve()，等同于间接调用了函数 positive()。

```
  }
  else
  {
    ref.style.borderColor = 'Red' ;
```

此语句设置了特定 input 元素实例的边框颜色，成为红色。

```
    reject() ;
```

此语句调用了内置的局部函数 reject()，等同于间接调用了函数 negative()。

```
  }
}
```

上述语法定义了具有两个参数的函数 promise_executor()，主要用来启动承诺。在语法【function promise_executor(resolve, reject)】中，参数 resolve 是对应到【解决承诺之后】的处置函数的名称，而参数 reject 则是对应到【拒绝承诺之后】的处置函数的名称。

```
function check_error()
{
  for (id_no = 1 ; id_no < 5 ; id_no++)
  {
    promise_array[id_no - 1] = new Promise(promise_executor) ;
```

此语句在等号右侧，描述了通过关键字 new，以及传入【用来启动承诺】的函数名称 promise_executor，到 Promise 对象实例的构造函数 Promise()里，以动态产生一个新的 Promise 对象实例。再将这个 Promise 对象实例，设置成为全局变量 promise_array 的数组实例中的元素。

```
      promise_array[id_no - 1].then(positive, negative) ;
```

此语句用来在处置解决承诺或拒绝承诺时，捕捉所发生的异常错误。处置解决承诺时，调用函数 positive()；处置拒绝承诺时，则调用函数 negative()。

```
  }

  Promise.all(promise_array).then(final_success, final_failure) ;
```

此语句是用来在**整个**承诺被解决时，或者**细部**承诺被拒绝时，捕捉所发生的异常错误。整个承诺被解决时，调用函数 final_success()；细部承诺被拒绝时，则调用函数 final_failure()。

```
}
```

上述语法定义了函数 check_error()，以创建**多个**全新的 Promise 对象实例，并间接调用了函数 promise_executor()。

```
function positive(value)
{

  filled++ ;
```

变量 filled 是用来记录被填入文本的 input 元素实例的个数。此语句递增了变量 filled 的数值，代表已被填入文本的 input 元素实例的个数，增加了一个。

```
}
```

上述语法定义了**细部**承诺被解决之后的处置函数 positive()，以计算出被填入文本的 input 元素实例的个数。因为在本示例中，调用函数 resolve() 时，并没有传入任何参数数据；所以函数 positive(value) 的参数 value 的数据，会变成默认的原始常量 undefined。

```
function negative(value)
{

  blank++ ;
```

变量 blank 是用来记录并未被填入文本的 input 元素实例的个数。此语句递增了变量 blank 的数值，代表**逾时**而没有被填入文本的 input 元素实例的个数，增加了一个。

```
}
```

上述语法定义了**细部**承诺被拒绝之后的处置函数 negative()，以计算出**逾时**而没有被填入文本的 input 元素实例的个数。因为在本示例中，调用函数 reject() 时，并没有传入任何参数数据，所以函数 negative(value) 的参数 value 的数据，会变成默认的原始常量 undefined。

```
function final_success(value_or_list)
{
  mbox.style.color = 'ForestGreen' ;
  mbox.innerHTML = 'All fields are filled.' ;
}
```

整个承诺被解决之后，函数 final_success() 就会被调用，并被传入由【函数 resolve(参数的数据) 与 reject(参数的数据) 返回的各个数据】所组成的数组实例。

因为在本示例中，调用函数 resolve() 时，并没有传入任何参数数据，所以函数

final_success(value_or_list)的参数 value_or_list 的数据，会变成内含 4 个默认的原始常量 undefined 的数组实例[undefined, undefined, undefined, undefined]。

```
function final_failure(value_or_list)
{
 mbox.style.color = 'Pink' ;
 mbox.innerHTML = `${blank} Blank field is/are left...` ;
}
```

任何一个细部承诺被拒绝时，函数 final_failure()就会被调用。因为在本示例中，调用函数 reject()时，并没有传入任何参数数据；所以函数 final_failure(value_or_list)的参数 value_or_list 的数据，会变成默认的原始常量 undefined。

```
setTimeout(check_error, 4000) ;
```

此语句使得函数 check_error()，在 4 秒（4000 毫秒）之后，被调用 1 次。

5.3.3 异步函数与等待表达式（ES8）

在 JavaScript 语言中，可通过开头的关键字 async，衔接**标准函数**的定义语法，描述【可返回内含特定数据的 Promise 对象实例】的异步函数（asynchronous function）。

在异步函数的主体语法内，可放入关键字 await 开头的等待表达式（await expression），以便执行其主体内的源代码块时，其执行动作被暂停于等待表达式的所在位置，一直到等待表达式被评估完成，并返回新的 Promise 对象实例为止。关于异步函数与等待表达式的综合运用，可参考如下示例。

【5-3-3-async-function-and-await-expressions.js】

```
function calc(value)
{
 function promise_executor(resolve)
 {
   // call resolve(value) after 2 seconds
   setTimeout(resolve, 2000, value + 100) ;
 }

 return new Promise(promise_executor) ;
}

async function plus(value)
{
 let num01 = calc(20) ;
 let num02 = calc(30) ;

 // return after only 2 seconds.
 return value + await num01 + await num02 ;

 /*
 let num01 = await calc(20) ;
 let num02 = await calc(30) ;
```

```
  // return after (2 + 2) seconds.
  return value + num01 + num02 ;
  */
}

function positive(value)
{
  console.log('The value =', value) ;
}

let result = plus(10) ;

result.then(positive) ;

console.log('Promise object: ', result) ;
```

【相关说明】

```
function calc(value)
{
  function promise_executor(resolve)
  {
    // call resolve(value) after 2 seconds
    setTimeout(resolve, 2000, value + 100) ;
```

此语句是用来模拟，调用内置函数 resolve()的时间差，以便在延迟 2 秒（2000 毫秒）之后，间接调用内置函数 resolve(value + 100)。

```
  }
```

上述语法定义了具有单一参数的函数 promise_executor(resolve)，以用来**启动**承诺。在此，其参数 resolve 可被视为处置解决承诺的函数 resolve()。

```
  return new Promise(promise_executor) ;
```

此语句设置并返回了一个新的 Promise 对象实例。这个新的 Promise 对象实例，是通过关键字 new，以及传入【用来启动承诺】的函数名称 promise_executor，到 Promise 对象实例的构造函数 Promise()里，才被动态产生出来的。

```
}
```

上述语法定义了函数 calc()。

```
async function plus(value)
{
  let num01 = calc(20) ;
  let num02 = calc(30) ;
```

这两个语句均在其等号右侧，调用函数 calc()，并分别传入参数数值20与30，进而将函数 calc(参数数值)返回的新的 Promise 对象实例，分别赋给变量 num01 及 num02。

```
  // return after only 2 seconds.
  return value + await num01 + await num02 ;
```

在此语句中，表达式【value + await num01 + await num02】被顺利计算出来之前，必须等待变量 num01 和 num02 的数值先被返回才行。

这是因为，在变量 num01 与 num02 的左侧，均被冠上关键字 await，使得**同时**被延迟 2 秒之后，num01 才被赋予 20 + 100 的结果值 120，num02 才被赋予 30 + 100 的结果值 130。最后，表达式【value + await num01 + await num02】才被计算出【10 + 120 + 130】的结果值 260。

```
/*
let num01 = await calc(20) ;
let num02 = await calc(30) ;

// return after (2 + 2) seconds.
return value + num01 + num02 ;
*/
```

若将关键字 await，个别放置在调用 calc(20) 与 calc(30) 的语法左侧，则因为**前后分别**被延迟 2 秒，所以总共会延迟 4 秒之后，【return value + num01 + num02】才会返回整数值 260。

```
}
```

上述语法通过关键字 async，定义了异步（asynchronous）函数 plus()。

```
function positive(value)
{
  console.log('The value =', value) ;
}
```

此语法定义了特定承诺被解决之后的处置函数 positive()。

```
let result = plus(10) ;
```

此语句声明了初始数据为【函数 plus(10) 的返回值 260】的变量 result。

```
result.then(positive) ;
```

在特定承诺被解决，此语句可用来捕捉所发生的异常错误。特定承诺被解决时，会调用函数 positive()。

```
console.log('Promise object: ', result) ;
```

此语句用来显示出内含最终结果值 260 的 Promise 对象实例的内容。

5.4 练 习 题

1. 已知有如下标准函数的定义语法：

```
function timing()
{
  console.log(new Date().toTimeString()) ;
}
```

试改写上述函数的定义语法，成为等价的函数表达式。

2. 已知有如下标准函数的定义语法：

```
function today()
{
  console.log(new Date().toDateString()) ;
}
```

试改写上述函数的定义语法，成为等价的 Object 对象实例中的函数（方法）。

3. 已知有如下标准函数的定义语法：

```
function monster_base(name = 'new', HP = 100, MP = 50)
{
  console.log('About this monster:') ;
  console.log(`  name: ${name}`) ;
  console.log(`  HP: ${HP}`) ;
  console.log(`  MP: ${MP}`) ;
}
```

试改写上述函数的定义语法，成为等价的箭头函数表达式。

4. 已知有如下 3 个函数的定义语法：

```
function f01()
{
  console.log('Hello, Earth!') ;
}

function f02()
{
  console.log('Hello, Solar System!') ;

  f01() ;
}

function f03()
{
  console.log('Hello, Galaxy!') ;
}
```

试改写上述各函数的定义语法，使得函数 f01()不能在函数 f03()内部被调用，仅能在函数 f02()内部被调用。

5. 已知有如下标准函数的定义语法：

```
function welcome()
{
  alert('G\'day, and welcome to our official website.') ;
}
```

试改写上述函数的定义语法，成为仅能被调用 1 次的匿名函数。

6. 已知参数 *n* 的数据为小于 100 的正整数，试编写一个递归函数 sum_of_square()，以返回从正整数 1 到变量 *n* 的正整数为止的平方和。

7. 试定义名称为 zodiac01 的生成器，以及名称为 zodiac02 的生成器函数，使得生成器 zodiac01 和生成器函数 zodiac02()，均可迭代如下顺序的字符串：

'Rat'、'Ox'、'Tiger'、'Hare'、'Dragon'、'Snake'、'Horse'、'Sheep'、'Monkey'、'Rooster'、'Dog'、

'Pig'。

8. 通过 Promise 对象实例，编写实现如下功能的版本：

限定用户在 6 秒内，在特定的文本字段里，输入介于 1～100 之间的整数值。并且，无论被输入的整数值，是否在范围内，都要告知用户是否输入正确。

第 6 章

处 理 数 值

JavaScript 语言支持一系列内置的函数（function）/方法（method）与属性（property），以便处理不同类型的数据。

在网页浏览器的 JavaScript 引擎里，内置的函数/方法与属性，均附挂于特定的对象（object）或对象实例（object instance）中。例如：

- 函数 window.parseInt()与属性 window.Infinity 就是被附挂于 window **对象实例**中的函数与属性，并且可分别简写为 parseInt()和 Infinity。
- 函数 Math.pow()与属性 Math.PI 就是被附挂于 Math **对象**中的函数与属性。其完整写法分别为 window.Math.pow()和 window.Math.PI。

在 JavaScript 语言里，整数和带有小数的浮点数的数据类型（data type）均为数值（number value），所以处理数值的函数与属性，主要被附挂于 Number 和 Math 对象里。

6.1　Number 对象的内置属性

Number 对象中的所有属性，均代表特定的数值（number value）或无穷值（infinity）。

6.1.1　最大的正数和安全整数（ES6）

在 Number 对象里，其属性：

- MAX_VALUE 代表可被表示的最大正数值，例如 1.7976931348623157e+308，也就是 $1.7976931348623157 \times 10^{308}$。
- MAX_SAFE_INTEGER 代表可被表示的最大安全整数（maximum safe integer），例如 $2^{53} - 1$,

也就是 9007199254740991。

下面通过示例，进行最大正数和安全整数的调试。

【6-1-1-maximum-values-and-isSafeInteger.js】

```
console.log(Number.MAX_VALUE) ;
console.log(-Number.MAX_VALUE) ;
console.log(Number.MAX_SAFE_INTEGER) ;
console.log(-Number.MAX_SAFE_INTEGER) ;

let answer = Number.isSafeInteger(Number.MAX_SAFE_INTEGER) ;

console.log(answer) ;

answer = Number.isSafeInteger(Number.MAX_SAFE_INTEGER + 1) ;

console.log(answer) ;
```

【相关说明】

```
console.log(Number.MAX_VALUE) ;
```

显示出 1.7976931348623157e+308，也就是 $1.7976931348623157 \times 10^{308}$。

```
console.log(-Number.MAX_VALUE) ;
```

显示出 -1.7976931348623157e+308，也就是 $-1.7976931348623157 \times 10^{308}$。

```
console.log(Number.MAX_SAFE_INTEGER) ;
```

显示出 9007199254740991。

```
console.log(-Number.MAX_SAFE_INTEGER) ;
```

显示出 -9007199254740991。

```
let answer = Number.isSafeInteger(Number.MAX_SAFE_INTEGER) ;
```

函数 Number.isSafeInteger()用来判断，被传入的数据，是否为安全范围内的整数值。在此，Number.isSafeInteger(Number.MAX_SAFE_INTEGER)返回布尔值 true，所以此语句声明了初始数据为布尔值 true 的变量 answer。

```
console.log(answer) ;

answer = Number.isSafeInteger(Number.MAX_SAFE_INTEGER + 1) ;
```

【Number.MAX_SAFE_INTEGER + 1】的结果值为 9007199254740992，已经高于安全范围的最大整数值了。在此，Number.isSafeInteger(Number.MAX_SAFE_INTEGER + 1)返回布尔值 false，所以此语句将布尔值 false，赋给变量 answer。

6.1.2 最小的正数和安全整数（ES6）

在 Number 对象里，其属性：

- MIN_VALUE 代表可被表示的最小正数值，例如 5e-324，也就是 5×10^{-324}。
- MIN_SAFE_INTEGER 代表可被表示的最小安全整数（maximum safe integer），例如-(2^{53} - 1)，也就是-9007199254740991。

下面通过示例，进行最小的正数值和安全整数的调试。

【6-1-2-mininum-values-and-isSafeInteger.js】

```
console.log(Number.MIN_VALUE) ;
console.log(-Number.MIN_VALUE) ;
console.log(Number.MIN_SAFE_INTEGER) ;
console.log(-Number.MIN_SAFE_INTEGER) ;

let answer = Number.isSafeInteger(Number.MIN_SAFE_INTEGER) ;

console.log(answer) ;

answer = Number.isSafeInteger(Number.MIN_SAFE_INTEGER - 1) ;

console.log(answer) ;
```

【相关说明】

```
console.log(Number.MIN_VALUE) ;
```

显示 5e-324，也就是代表 5×10^{-324}。

```
console.log(-Number.MIN_VALUE) ;
```

显示-5e-324，也就是代表-5×10^{-324}。

```
console.log(Number.MIN_SAFE_INTEGER) ;
```

显示-9007199254740991。

```
console.log(-Number.MIN_SAFE_INTEGER) ;
```

显示 9007199254740991。

```
let answer = Number.isSafeInteger(Number.MIN_SAFE_INTEGER) ;
```

函数 Number.isSafeInteger()用来判断，被传入的数据，是否为安全范围内的整数值。在此，Number.isSafeInteger(Number.MIN_SAFE_INTEGER)返回布尔值 true，所以此语句声明了初始数据为布尔值 true 的变量 answer。

```
console.log(answer) ;

answer = Number.isSafeInteger(Number.MIN_SAFE_INTEGER - 1) ;
```

【Number.MIN_SAFE_INTEGER - 1】的结果值为 -9007199254740992，低于安全范围的最小整数值。在此，Number.isSafeInteger(Number.MIN_SAFE_INTEGER)返回布尔值 false，所以此语句将布尔值 false，赋给变量 answer。

6.1.3 正负无穷值

在 Number 对象里，Number.NEGATIVE_INFINITY 代表负无穷值（negative infinity），会返回-Infinity；Number.POSITIVE_INFINITY 代表正无穷值（positive infinity），会返回 Infinity。欲描述无穷值 Infinity，也可直接在源代码里，编写成为 Infinity 或 window.Infinity。关于正负无穷值的调试，可参考如下示例。

【6-1-3-infinities.js】
```
console.log(Number.NEGATIVE_INFINITY) ;
console.log(-Number.NEGATIVE_INFINITY) ;
console.log('') ;

console.log(Number.POSITIVE_INFINITY) ;
console.log(-Number.POSITIVE_INFINITY) ;
```

【相关说明】
```
console.log(Number.NEGATIVE_INFINITY) ;
```
显示-Infinity。

```
console.log(-Number.NEGATIVE_INFINITY) ;
```
显示 Infinity。

```
console.log('') ;
```

```
console.log(Number.POSITIVE_INFINITY) ;
```
显示 Infinity。

```
console.log(-Number.POSITIVE_INFINITY) ;
```
显示-Infinity。

6.1.4 非数值的判断（ES6）

函数 isNaN()和 Number.isNaN()均可用来判断，被传入的参数数据，是否为【非数值 (not a number)】。其不同点在于：

- isNaN()会试图转换被传入的数据，成为数值。若可转换成为数值(number value)，则 isNaN(被传入的数据)返回布尔值 false，代表被传入的数据【并非数值】是假（false）的。这也就意味着，被传入的数据，可被视为数值！
- Number.isNaN()不会进行传入数据的转换动作。换句话说，任何数据类型不是数值（number value）的数据，均会被忽略，使得 Number.isNaN(被传入的数据)返回**让人较难理解**的布尔值 false。其中，大概只有 Number.isNaN(NaN)和 Number.isNaN(0 / 0)会返回布尔值 true。

关于**并非数值**的评估的调试，可参考如下示例。

【6-1-4-NaN-and-isNaN.js】
```
let answer ;

console.log(NaN) ;

answer = isNaN(NaN) ;

console.log(answer) ;

answer = Number.isNaN(NaN) ;

console.log(answer) ;
console.log('\n\n') ;

///
console.log(isNaN(undefined)) ;
console.log(Number.isNaN(undefined)) ;
console.log('') ;

console.log(isNaN(true)) ;
console.log(Number.isNaN(true)) ;
console.log('') ;

console.log(isNaN(false)) ;
console.log(Number.isNaN(false)) ;
console.log('') ;

console.log(isNaN(null)) ;
console.log(Number.isNaN(null)) ;
console.log('\n\n') ;

console.log(isNaN('25.37')) ;
console.log(Number.isNaN('25.37')) ;
console.log('') ;

console.log(isNaN(25.37)) ;
console.log(Number.isNaN(25.37)) ;
console.log('') ;

console.log(isNaN('32one')) ;
console.log(Number.isNaN('32one')) ;
console.log('') ;

console.log('hello: ', isNaN('hello')) ;
console.log('hello: ', Number.isNaN('hello')) ;
console.log('') ;

console.log(isNaN('')) ;
console.log(Number.isNaN('')) ;
console.log('') ;

console.log(isNaN(' ')) ;
console.log(Number.isNaN(' ')) ;
console.log('') ;
```

```
console.log(isNaN(new Date())) ;
console.log(Number.isNaN(new Date())) ;
console.log('') ;
```

【相关说明】

```
let answer ;
```

声明变量 answer。

```
console.log(NaN) ;
```

显示原始常量 NaN，代表非数值（not a number）的含义。

```
answer = isNaN(NaN) ;
```

内置函数 isNaN()可用来判断，传入的数据，是否非数值。若是，则返回 true；反之，则返回 false。

```
console.log(answer) ;
```

```
answer = Number.isNaN(NaN) ;
```

内置的函数 Number.isNaN()和 isNaN()的功能几乎相同，其不同点在于 isNaN()会试图转换传入的数据成为数值，然后再进行判断。

```
console.log(answer) ;
console.log('\n\n') ;
```

```
console.log(isNaN(undefined)) ;
```

显示出 true。

```
console.log(Number.isNaN(undefined)) ;
```

显示出 false。

```
console.log('') ;

console.log(isNaN(true)) ;
console.log(Number.isNaN(true)) ;
```

这两个语句均显示出 false。

```
console.log('') ;

console.log(isNaN(false)) ;
console.log(Number.isNaN(false)) ;
```

这两个语句均显示出 false。

```
console.log('') ;

console.log(isNaN(null)) ;
console.log(Number.isNaN(null)) ;
```

这两个语句均显示出 false。

```
console.log('\n\n');

console.log(isNaN('25.37'));
console.log(Number.isNaN('25.37'));
```

这两个语句均显示出 false。

```
console.log('');

console.log(isNaN(25.37));
console.log(Number.isNaN(25.37));
```

这两个语句均显示出 false。

```
console.log('');

console.log(isNaN('32one'));
```

显示出 true。

```
console.log(Number.isNaN('32one'));
```

显示出 false。

```
console.log('');

console.log('hello: ', isNaN('hello'));
```

显示出【hello: true】的信息。

```
console.log('hello: ', Number.isNaN('hello'));
```

显示出【hello: false】的信息。

```
console.log('');

console.log(isNaN(''));
console.log(Number.isNaN(''));
```

这两个语句均显示出 false。

```
console.log('');

console.log(isNaN(' '));
console.log(Number.isNaN(' '));
```

这两个语句均显示出 false。

```
console.log('');

console.log(isNaN(new Date()));
console.log(Number.isNaN(new Date()));
```

这两个语句均显示出 false。

6.1.5 浮点数运算的误差值（ES6）

在现代的计算机里，进行任何浮点数运算，均会产生微小的误差值。在 JavaScript 语言中，任何浮点数运算的误差值，都不会大于常量属性 Number.EPSILON 所代表的数值 2.220446049250313e-16，也就是 $2.220446049250313 \times 10^{-16}$。关于浮点数运算的误差值的调试，可参考如下示例。

【6-1-5-Number-EPSILON.js】

```
let a = 0.33, b = 0.47, c = 0.8 ;

console.log(a + b == c) ;
console.log(c - a == b) ;
console.log(c - b == a) ;
console.log('') ;

console.log(c - a) ;
console.log(c - b) ;
console.log('') ;

console.log(Number.EPSILON) ;
console.log(c - a - b < Number.EPSILON) ;
console.log(c - b - a < Number.EPSILON) ;
```

【相关说明】

```
let a = 0.33, b = 0.47, c = 0.8 ;
```

声明变量 a、b 与 c，其初始数值分别为浮点数 0.33、0.47 与 0.8。

```
console.log(a + b == c) ;
```

【a + b == c】当前等价于【0.33 + 0.47 == 0.8】，而且返回布尔值 true。

```
console.log(c - a == b)
```

【c - a == b】当前等价于【0.8 - 0.33 == 0.47】，理论上【应该】返回布尔值 true 才对，实际却返回布尔 false！这是因为计算机进行浮点数的运算时，常常会有误差。

```
console.log(c - b == a) ;
```

【c - b == a】当前等价于【0.8 - 0.47 == 0.33】，理论上【应该】返回布尔值 true，实际亦返回布尔值 false。

```
console.log(c - a) ;
```

显示出【c - a】的结果值为 0.47000000000000003，却不是正确的 0.47。

```
console.log(c - b) ;
```

显示出【c - b】的结果值为 0.33000000000000007，却不是正确的 0.33。

```
console.log('') ;

console.log(Number.EPSILON) ;
```

显示出常量属性 Number.EPSILON 的数值为 2.220446049250313e-16，也就是 $2.220446049250313 \times 10^{-16}$，代表着浮点数运算的最大误差。

```
console.log(c - a - b < Number.EPSILON) ;
```

为了正常比较【c - a == b】，可将之改成带有 Number.EPSILON 的等价表达式【c - a - b < Number.EPSILON】。

```
console.log(c - b - a < Number.EPSILON) ;
```

为了正常比较【c - b == a】，可将之改成带有 Number.EPSILON 的等价表达式【c - b - a < Number.EPSILON】。

6.2　Number 对象的内置函数

在 Number 对象中，有些函数 / 方法被附挂在第 1 层，例如【Number.isFinite()】；有些函数则被附挂于第 2 层属性 prototype 里，例如【Number.prototype.toString()】。

【Number.prototype.函数名称()】在实际使用上要被改写成为【变量名称.函数名称()】才行，例如，已知当前数据为特定数值的变量 value，通过改写 Number.prototype.toString(16) 成为 value.toString(16)，代表要将变量 value 中的数值，转换成十六进制数码。

6.2.1　转换为特定进制的数码字符串

在 JavaScript 语言里，显示给用户观看的数码，是十进制的数值（number value）。在转换为其他进制的场合中，编程人员可通过【当前数据为特定数值的变量名称.toString(特定进制的基数值)】的语法，转换特定数值，成为数值所代表的特定进制的数码字符串。关于转换成为特定进制的数码字符串的综合运用，可参考如下示例。

【6-2-1-Number-toString.js】

```
console.log(Number.isFinite(32000)) ;
console.log(Number.isFinite('32000')) ;
console.log('') ;

let value = 32000 ;

console.log(value.toString(2)) ;
console.log(value.toString(4)) ;
console.log(value.toString(8)) ;
console.log(value.toString(16)) ;
console.log('') ;

value = 0b10111100 ;

console.log(value.toString(4)) ;
console.log(value.toString(8)) ;
```

```
console.log(value.toString(10)) ;
console.log(value.toString(16)) ;
console.log('') ;

value = 0o2375 ;

console.log(value.toString(2)) ;
console.log(value.toString(4)) ;
console.log(value.toString(10)) ;
console.log(value.toString(16)) ;
console.log('') ;

value = 0xf25c ;

console.log(value.toString(2)) ;
console.log(value.toString(4)) ;
console.log(value.toString(8)) ;
console.log(value.toString(10)) ;
console.log('') ;
```

【相关说明】

```
console.log(Number.isFinite(32000)) ;
```

函数 Number.isFinite()可判断传入的数据，是否为有限的数值。此语句显示出 true，因为数值 32000 是有限数。

```
console.log(Number.isFinite('32000')) ;
```

此语句显示出 false，因为字符串'32000'并未被主动转换为数值 32000 之后，再进行判断。

```
let value = 32000 ;
```

此语句声明了初始数值为整数值 32000 的变量 value。

```
console.log(value.toString(2)) ;
```

显示出十进制整数值 32000 的二进制数码 111110100000000。

```
console.log(value.toString(4)) ;
```

显示出十进制整数值 32000 的 4 进制数码 13310000。

```
console.log(value.toString(8)) ;
```

显示出十进制整数值 32000 的八进制数码 76400。

```
console.log(value.toString(16)) ;
```

显示出十进制整数值 32000 的十六进制数码 7d00。

```
console.log('') ;

value = 0b10111100 ;
```

此语句将二进制（binary）整数值 10111100，赋给变量 value。其中，0b 开头的一串 0 和 1 的组合，是二进制字面量（binary literal），用来表示二进制的整数值。

```
console.log(value.toString(4)) ;
```

显示出二进制整数值 10111100 的 4 进制数码 2330。

```
console.log(value.toString(8)) ;
```

显示出二进制整数值 10111100 的八进制数码 274。

```
console.log(value.toString(10)) ;
```

显示出二进制整数值 10111100 的十进制数码 188。

```
console.log(value.toString(16)) ;
```

显示出二进制整数值 10111100 的十六进制数码 bc。

```
console.log('') ;

value = 0o2375 ;
```

此语句将八进制（octal）整数值 2375，赋给变量 value。其中，0o 开头的一串 0 至 7 的组合，是八进制字面量（octal literal），用来表示八进制的整数值。

```
console.log(value.toString(2)) ;
```

显示出八进制整数值 2375 的二进制数码 10011111101。

```
console.log(value.toString(4)) ;
```

显示出八进制整数值 2375 的 4 进制数码 103331。

```
console.log(value.toString(10)) ;
```

显示出八进制整数值 2375 的十进制数码 1277。

```
console.log(value.toString(16)) ;
```

显示出八进制整数值 2375 的十六进制数码 4fd。

```
console.log('') ;

value = 0xf25c ;
```

此语句将十六进制（hexadecimal）整数值 f25c，赋给变量 value。其中，0x 开头的一串 0～9 至 a～f 的组合，是十六进制字面量（hexadecimal literal），用来表示十六进制的整数值。

```
console.log(value.toString(2)) ;
```

显示出十六进制整数值 f25c 的二进制代码 1111001001011100。

```
console.log(value.toString(4)) ;
```

显示出十六进制整数值 f25c 的 4 进制代码 33021130。

```
console.log(value.toString(8)) ;
```

显示出十六进制整数值 f25c 的八进制代码 171134。

```
console.log(value.toString(10)) ;
```

显示出十六进制整数值 f25c 的十进制代码 62044。

6.2.2 处理小数格式

浮点数具有小数的部分，可通过【当前数据为特定浮点数值的变量名称】衔接.toExponential（小数位数）、.toFixed（小数位数）和.toPrecision（整数和小数的总位数），以决定浮点数的小数位数。关于处理小数格式的细部运用，可参考如下示例。

【6-2-2-decimal-formattings.js】

```
let result = Math.PI * (100 ** 2) ;

console.log(result) ;
console.log('') ;

console.log(result.toExponential()) ;
console.log(result.toExponential(4)) ;
console.log(result.toExponential(3)) ;
console.log(result.toExponential(2)) ;
console.log('') ;

console.log(result.toFixed()) ;
console.log(result.toFixed(4)) ;
console.log(result.toFixed(3)) ;
console.log(result.toFixed(2)) ;
console.log('') ;

console.log(result.toPrecision()) ;
console.log(result.toPrecision(4)) ;
console.log(result.toPrecision(3)) ;
console.log(result.toPrecision(2)) ;
console.log('') ;

let value = 55.8175 ;

console.log(value.toPrecision()) ;
console.log(value.toPrecision(4)) ;
console.log(value.toPrecision(3)) ;
console.log(value.toPrecision(2)) ;
console.log('') ;
```

【相关说明】

```
let result = Math.PI * (100 ** 2) ;
```

此语句声明了初始数值为【半径是 100 的圆面积】的变量 result。

```
console.log(result) ;
```

显示出半径为 100 的圆面积的数值 31415.926535897932。

```
console.log('') ;

console.log(result.toExponential()) ;
```

result.toExponential()以标准科学记数法，返回圆面积的数值 3.1415926535897932e+4。

```
console.log(result.toExponential(4)) ;
```

result.toExponential(4)以标准科学记数法，返回圆面积的具有 4 位小数的数值 3.1416e+4。

```
console.log(result.toExponential(3)) ;
```

result.toExponential(3)以标准科学记数法，返回圆面积的具有 3 位小数的数值 3.142e+4。

```
console.log(result.toExponential(2)) ;
```

result.toExponential(2)以标准科学记数法，返回圆面积的具有 2 位小数的数值 3.14e+4。

```
console.log('') ;
console.log(result.toFixed()) ;
```

result.toFixed()以标准科学记数法，返回圆面积的无小数的定点表示法 31416。

```
console.log(result.toFixed(4)) ;
```

result.toFixed(4)以定点表示法，返回圆面积的具有 4 位小数的数值 31415.9265。

```
console.log(result.toFixed(3)) ;
```

result.toFixed(3)以定点表示法，返回圆面积的具有 3 位小数的数值 31415.927。

```
console.log(result.toFixed(2)) ;
```

result.toFixed(2)以定点表示法，返回圆面积的具有 2 位小数的数值 31415.93。

```
console.log('') ;
console.log(result.toPrecision()) ;
```

result.toPrecision()通过无特殊设置的小数精度表示法，返回圆面积的数值 31415.926535897932。

```
console.log(result.toPrecision(4)) ;
```

result.toPrecision(4)以小数精度表示法，返回圆面积的总位数 4 位的数值 3.142e+4。

```
console.log(result.toPrecision(3)) ;
```

result.toPrecision(3)以小数精度表示法，返回圆面积的总位数 3 位的数值 3.14e+4。

```
console.log(result.toPrecision(2)) ;
```

result.toPrecision(2)以小数精度表示法，返回圆面积的总位数 2 位的数值 3.1e+4。

```
console.log('') ;
let value = 55.8175 ;
```

此语句声明了初始数值为 55.8175 的变量 value。

```
console.log(value.toPrecision()) ;
```

value.toPrecision()通过无特殊设置的小数精度表示法，返回变量 value 的数值 55.8175。

```
console.log(value.toPrecision(4));
```

value.toPrecision()以小数精度表示法，返回变量 value 的总位数 4 位的数值 55.82。

```
console.log(value.toPrecision(3));
```

value.toPrecision()以小数精度表示法，返回变量 value 的总位数 3 位的数值 55.8。

```
console.log(value.toPrecision(2));
```

value.toPrecision()以小数精度表示法，返回变量 value 的总位数 2 位的数值 56。

函数【toPrecision(总位数)】会通过一般格式或科学记数法，返回特定数值。通过科学记数法，加以返回的时机，是当特定数值的总位数过多时。

6.2.3 转换为数值

在 JavaScript 语言里，可顺利被转换成为数值的数据，大致有以下几种：

- 内含特定原始数值（primitive number value）的 Number 对象实例的变量，例如在语句【let num_obj = new Number(3591.8);】中的变量 num_obj。
- 布尔值 false 或 true。
- 原始常量 undefined、null 或空字符串"。
- 最多内含 1 个小数点纯粹数字 0 ~ 9 所构成的字符串，例如'2591.8'、'1314.35'、'2013'、'775'等。
- 不包含小数点'0x'、'0o'、'0'和'0b'开头并衔接纯粹数字 0 ~ 9 所构成的字符串！例如'0x5c6f'、'0o237'、'0237'、'0b11011101'。这也就意味着不支持【转换带有小数部分的十六进制、八进制及二进制】的字符串代码。

关于转换成为数值的综合运用，可参考如下示例。

【6-2-3-converting-to-numbers.js】

```
let num_obj = new Number(3591.8);

num_obj.second = 5201.53;
num_obj.third = 25.35;

console.log(num_obj);
console.log(typeof num_obj);
console.log('');

console.log(num_obj.valueOf());
console.log(typeof num_obj.valueOf());
console.log('');

console.log(num_obj.second);
console.log(typeof num_obj.second);
console.log('');

console.log(num_obj.third);
console.log(typeof num_obj.third);
```

```
console.log('') ;
///
let answer = Number('333') == parseInt('333') ;

console.log(answer) ;
console.log('') ;

answer = Number('7.7') == parseFloat('7.7') ;

console.log(answer) ;
console.log('') ;

console.log(Number('0b11011')) ;
console.log(parseInt('0b11011')) ;
console.log('') ;

console.log(Number('0o1557')) ;
console.log(parseInt('0o1557')) ;
console.log('') ;

console.log(Number('0xffcc')) ;
console.log(parseInt('0xffcc')) ;
console.log('') ;

console.log(Number('1314love')) ;
console.log(parseInt('1314love')) ;
```

【相关说明】

`let num_obj = new Number(3591.8) ;`

此语句声明了变量 num_obj，其初始数据为内含数值 3591.8 的 Number 对象实例。

`num_obj.second = 5201.53 ;`

此语句在变量 num_obj 的 Number 对象实例中，增加了数值为 5201.53 的属性 second。

`num_obj.third = 25.35 ;`

此语句在变量 num_obj 的 Number 对象实例中，增加了数值为 25.35 的属性 third。

`console.log(num_obj) ;`

显示出【Number {3591.8, second: 5201.53, third: 25.35}】的信息。

`console.log(typeof num_obj) ;`

显示出变量 num_obj 的数据类型为'object'。

`console.log('') ;`

`console.log(num_obj.valueOf()) ;`

显示出变量 num_obj 的 Number 对象实例中的数值 3591.8。

`console.log(typeof num_obj.valueOf()) ;`

显示出在变量 num_obj 的 Number 对象实例中，其数值的数据类型'number'。

```
console.log('') ;
console.log(num_obj.second) ;
```

显示出属性 num_obj.second 的数值 5201.53。

```
console.log(typeof num_obj.second) ;
```

显示出属性 num_obj.second 数值的数据类型'number'。

```
console.log('') ;
console.log(num_obj.third) ;
```

显示出属性 num_obj.third 的数值 25.35。

```
console.log(typeof num_obj.third) ;
```

显示出属性 num_obj.third 的数值的数据类型'number'。

```
console.log('') ;
let answer = Number('333') == parseInt('333') ;
```

此语句声明变量 answer，其初始数值为布尔值 true，来自于表达式【Number('333') == parseInt('333')】的结果值。内置函数 parseInt()与 Number 对象的构造函数 Number()，均可用来将传入的字符串，转换成整数值。

```
console.log(answer) ;
console.log('') ;
answer = Number('7.7') == parseFloat('7.7') ;
```

表达式【Number('7.7') == parseFloat('7.7')】的结果值为布尔值 true，所以此语句将布尔值 true，赋给变量 answer。

```
console.log(answer) ;
console.log('') ;
console.log(Number('0b11011')) ;
```

Number('0b11011')返回二进制数码 11011 对应的十进制数值 27。

```
console.log(parseInt('0b11011')) ;
```

parseInt('0b11011')返回 0，明显无法将二进制数码，转换为对应的数值。

```
console.log('') ;
console.log(Number('0o1557')) ;
```

Number('0o1557')返回八进制数码 1557 对应的十进制数值 879。

```
console.log(parseInt('0o1557')) ;
```

parseInt('0o1557')返回 0，明显无法将八进制数码，转换为对应的数值。

```
console.log('') ;

console.log(Number('0xffcc')) ;
```

Number('0xffcc')返回十六进制数码 ffcc 对应的十进制数值 65484。

```
console.log(parseInt('0xffcc')) ;
```

parseInt('0xffcc')可以返回十六进制数码 ffcc 对应的十进制数值 65484。

```
console.log('') ;

console.log(Number('1314love')) ;
```

Number('1314love')返回 NaN (not a number)，表示无法将字符'1314love'成功转换为数值。

```
console.log(parseInt('1314love')) ;
```

parseInt('1314love')却可以返回数值 1314。

6.2.4　判断是否为整数或有限数（ES6）

通过 Number 对象的函数 isInteger()，可判断传入的数据是否为整数（integer）；而其函数 isFinite()则可判断传入的数据是否为有限数（finite number）。关于判断是否为整数或有限数的调试，可参考如下示例。

【6-2-4-Number-isInteger-and-isFinite.js】

```
console.log(Number.isInteger(Number.MAX_VALUE)) ;
console.log(Number.isInteger(Number.MIN_VALUE)) ;
console.log(Number.isInteger(Math.PI)) ;
console.log(Number.isInteger('777')) ;
console.log(Number.isInteger(777)) ;
console.log('') ;

console.log(Number.isInteger(true)) ;
console.log(Number.isInteger(false)) ;
console.log('') ;

console.log(Number.isInteger(NaN)) ;
console.log(Number.isInteger(Infinity)) ;
console.log('') ;

///
console.log(isFinite(Infinity)) ;
console.log(isFinite(-Infinity)) ;
console.log(isFinite(NaN)) ;
console.log('') ;

console.log(isFinite(Number.MAX_VALUE)) ;
console.log(isFinite(undefined)) ;
console.log(isFinite(true)) ;
console.log(isFinite(false)) ;
console.log(isFinite(null)) ;
```

【相关说明】

```
console.log(Number.isInteger(Number.MAX_VALUE));
```

Number.isInteger()可用来判断传入的数据是否为整数值。此语句显示出 true。

```
console.log(Number.isInteger(Number.MIN_VALUE));
```

显示出 false。

```
console.log(Number.isInteger(Math.PI));
```

此语句显示出 false，因为 Math.PI 返回浮点数值。

```
console.log(Number.isInteger('777'));
```

此语句显示出 false，因为'777'是字符串。

```
console.log(Number.isInteger(777));
```

显示出 true。

```
console.log('');
console.log(Number.isInteger(true));
```

此语句显示出 false，因为传入函数 isInteger()的参数数据，是布尔值 true，并非整数值。

```
console.log(Number.isInteger(false));
```

此语句显示出 false，因为传入函数 isInteger()的参数数据，是布尔值 false，并非整数值。

```
console.log('');
console.log(Number.isInteger(NaN));
console.log(Number.isInteger(Infinity));
```

这两个语句均显示出 false 的信息，因为分别传入函数 isInteger()的数据 NaN（not a number）及 Infinity（无穷大），均为原始常量。

```
console.log('');
console.log(isFinite(Infinity));
console.log(isFinite(-Infinity));
```

内置函数 isFinite()可判断传入的数据，是否为有限的数值。这两个语句均显示出 false 的信息，因为正无穷大（Infinity）与负无穷大（-Infinity）并不是有限数。

```
console.log(isFinite(NaN));
```

此语句显示出 false，因为原始常量 NaN（not a number）并不是数值。

```
console.log('');
console.log(isFinite(Number.MAX_VALUE));
```

显示出 true。

```
console.log(isFinite(undefined));
```

此语句显示出 false，因为原始常量 undefined 并不是数值。

```
console.log(isFinite(true)) ;
console.log(isFinite(false)) ;
console.log(isFinite(null)) ;
```

这 3 个语句均显示出 true 的信息，因为布尔值 true、false 以及原始常量 null，被视为有限数。

6.3 Math 对象

在内置的 Math 对象里，存在数学相关的属性和函数，以便对特定数值，进行数学形式的操作。

6.3.1 Math 对象的常量属性

Math 对象存在数学相关的常量属性，例如自然对数、自然底数（在自然对数函数中的底数 e）、对数、圆周率、平方根等。关于 Math 对象的常量属性的调试，请看下面的例子。

【6-3-1-Math-constant-properties.js】

```
console.log(Math.E) ;
console.log(Math.PI) ;
console.log(Math.LN2) ;
console.log(Math.LN10) ;
console.log(Math.SQRT2) ;
console.log(Math.LOG2E) ;
console.log(Math.LOG10E) ;
console.log(Math.SQRT1_2) ;
```

【相关说明】

```
console.log(Math.E) ;
```

Math 对象的常量属性 E，返回自然底数 e 的数值 2.718281828459045。

```
console.log(Math.PI) ;
```

Math.PI 返回圆周率（pi）的数值 3.141592653589793。

```
console.log(Math.LN2) ;
```

Math.LN2 返回 2 的自然对数值 0.6931471805599453。

```
console.log(Math.LN10) ;
```

Math.LN10 返回 10 的自然对数值 2.302585092994046。

```
console.log(Math.SQRT2) ;
```

Math.SQRT2 返回 2 的平方根值 1.4142135623730951。

```
console.log(Math.LOG2E) ;
```

Math.LOG2E 返回以 2 为底数的 e 的对数值 1.4426950408889634。

```
console.log(Math.LOG10E) ;
```

Math.LOG10E 返回以 10 为底数的 e 的对数值 0.4342944819032518。

```
console.log(Math.SQRT1_2) ;
```

Math.SQRT1_2 返回 $\frac{1}{2}$ 的平方根值 0.7071067811865476。

6.3.2　Math 对象的函数（ES6）

Math 对象存在数学相关的函数，例如求出绝对值的函数、三角函数以及求出对数值、最大值、最小值、幂值等函数。关于 Math 对象的函数的综合调试，可参考如下示例。

【6-3-2-Math-functions.js】

```js
let value = 100 * Math.random() - 100 * Math.random() ;

console.log(value) ;
console.log(Math.abs(value)) ;
console.log('') ;

with (Math)
{
  console.log(abs(value)) ;
  console.log('') ;

  // 反余弦函数 acos()
  console.log(acos(-1)) ;
  console.log(cos(Math.PI)) ;
  // 双曲余弦函数 cosh()
  console.log(cosh(5)) ;
  console.log(acosh(10)) ;
  console.log('') ;

  console.log(asin(1)) ;
  console.log(sin(Math.PI * 0.5)) ;
  console.log(sinh(5)) ;
  console.log(asinh(10)) ;
  console.log('') ;

  console.log(atan(1)) ;
  console.log(atan2(10, 20)) ;
  console.log(tan(Math.PI * 0.25)) ;
  console.log(tanh(1.25)) ;
  console.log('') ;

  console.log(cbrt(-8)) ;
  console.log(cbrt(27)) ;
  console.log(27 ** (1/3)) ;
  console.log(pow(27, 1/3)) ;
  console.log('') ;
```

```
console.log(ceil(32.1)) ;
console.log(ceil(32.7)) ;
console.log(ceil(-32.1)) ;
console.log(ceil(-32.7)) ;
console.log('') ;

console.log(clz32(0b111)) ;
console.log(clz32(0777)) ; // old literal
console.log(clz32(0o777)) ; // new literal
console.log(clz32(0xfff)) ;
console.log('') ;

console.log(exp(1)) ;
console.log(E) ;
console.log(exp(3)) ;
console.log(E ** 3) ;
console.log(pow(E, 3)) ;
console.log('') ;

console.log(expm1(1)) ;
console.log(E - 1) ;
console.log(expm1(3)) ;
console.log(E ** 3 - 1) ;
console.log(pow(E, 3) - 1) ;
console.log('') ;

console.log(floor(15.8)) ;
console.log(floor(15.1)) ;
console.log(floor(-15.8)) ;
console.log(floor(-15.1)) ;
console.log('') ;

// 32-bit single-precision float representation of a number.
console.log(fround(9.875)) ;
console.log(fround(9.876)) ;
console.log(fround(9.125)) ;
console.log(fround(9.126)) ;
console.log('') ;

// hypotenuse of a right triangle 直角三角形的斜边
console.log(hypot(3, 4)) ;
console.log(hypot(6, 8)) ;
console.log(hypot(5, 12)) ;
console.log('') ;

let num_list = Array.from(Array(11).keys()) ;

console.log(hypot(... num_list)) ;
console.log('') ;

// 32-bit integer multiplication
console.log(imul(2.1, 10)) ;
console.log(imul(-5, -8)) ;
console.log(imul(-3, 7)) ;
```

```javascript
console.log('') ;

// natural logarithm (base e) of a number
console.log(log(E)) ;
console.log(log(E ** 7)) ;
console.log(log2(2 ** 15)) ;
console.log(log10(10 ** 8)) ;
console.log('') ;

// natural logarithm (base e) of 1 plus a number.
console.log(log1p(E - 1)) ;
console.log(log1p(E ** 7 - 1)) ;
console.log('') ;

// num_list = Array.from(new Array(10), (val, index) => index * 3) ;
num_list = Array.from(new Array(10), (value, index) => parseInt(100 * Math.random())) ;

console.log(num_list) ;
console.log(max(... num_list)) ;
console.log(min(... num_list)) ;
console.log('') ;

console.log(pow(2, 5)) ;
console.log(2 ** 5) ;
console.log('') ;

console.log(pow(27, 1/3)) ;
console.log(27 ** (1/3)) ;
console.log('') ;

console.log(pow(2, 0.5)) ;
console.log(2 ** 0.5) ;
console.log(SQRT2) ;
console.log('') ;

console.log(pow(0.5, 0.5)) ;
console.log(0.5 ** 0.5) ;
console.log(SQRT1_2) ;
console.log('') ;

console.log(random()) ;
console.log(random()) ;
console.log('') ;

console.log(100 + parseInt(100 * Math.random())) ;
console.log(-50 + parseInt(100 * Math.random())) ;
console.log('') ;

console.log(round(32.48)) ;
console.log(round(32.51)) ;
console.log(round(-32.48)) ;
console.log(round(-32.51)) ;
console.log('') ;

console.log(sign(+77)) ;
```

```
    console.log(sign(77)) ;
    console.log(sign(-77)) ;
    console.log('') ;

    console.log(sign('+66')) ;
    console.log(sign('66')) ;
    console.log(sign('-66')) ;
    console.log('') ;

    console.log(sign(+0)) ;
    console.log(sign(0)) ;
    console.log(sign(-0)) ;
    console.log('') ;

    console.log(sqrt(64)) ;
    console.log(64 ** 0.5) ;
    console.log(pow(64, 0.5)) ;
    console.log('') ;

    console.log(trunc(56.78)) ;
    console.log(trunc(0.779)) ;
    console.log(trunc(-0.36)) ;
    console.log(trunc('-33.66')) ;
}
```

【相关说明】

```
let value = 100 * Math.random() - 100 * Math.random() ;
```

此语句声明了变量 value，其初始数值为两个介于 0 ~ 100 之间的随机数值，加以相减之后的结果值。

```
console.log(value) ;
```

```
console.log(Math.abs(value)) ;
```

Math.abs(value)返回变量 value 数值的绝对值（absolute value）。

```
console.log('') ;
```

```
with (Math)
```

此语句可使得 Math.abs()、Math.cos()、Math.E 等，在下方大括号里，被简化成为 abs()、cos()和 E 等。

```
{
    console.log(abs(value)) ;
```

abs(value)返回变量 value 的数值的绝对值（absolute value）。

```
    console.log('') ;

    // 反三角函数
    console.log(acos(-1)) ;
```

acos(-1)返回-1 的反余弦值（arccosine）。

```
        console.log(cos(Math.PI)) ;

// 双曲正弦函数 sinh、双曲余弦函数 cosh、双曲正切函数 tanh
        console.log(cosh(5)) ;
```

cosh(5)返回 5 的双曲余弦值（hyperbolic cosine）。

```
        console.log(acosh(10)) ;
```

acosh(10) 返回 10 的反双曲余弦值（hyperbolic arccosine）。

```
        console.log('') ;
        console.log(asin(1)) ;
```

asin(1) 返回 1 的反正弦值（arcsine）。

```
        console.log(sin(Math.PI * 0.5)) ;
```

sin(Math.PI * 0.5)返回【Math.PI × 0.5】的正弦值（sine）。

```
        console.log(sinh(5)) ;
```

sinh(5)返回 5 的双曲正弦值（hyperbolic sine）。

```
        console.log(asinh(10)) ;
```

asinh(10)返回 10 的反双曲正弦值（hyperbolic arcsine）。

```
        console.log('') ;
         console.log(atan(1)) ;
```

atan(1)返回 1 的反正切值（arctangent）。

```
        console.log(atan2(10, 20)) ;
```

atan2(10, 20) 返回【10÷20】的反正切值。

```
        console.log(tan(Math.PI * 0.25)) ;
```

tan(Math.PI * 0.25)返回 Math.PI × 0.25 的正切值（tangent）。

```
        console.log(tanh(1.25)) ;
```

tanh(1.25)返回 1.25 的双曲正切值（hyperbolic tangent）。

```
        console.log('') ;
          console.log(cbrt(-8)) ;
```

cbrt(-8)返回-8 的立方根（cube root）值，也就是 $-8^{\frac{1}{3}}$ 的数值。

```
        console.log(cbrt(27)) ;
```

cbrt(27) 返回 $27^{\frac{1}{3}}$ 的数值。

```
        console.log(27 ** (1/3)) ;
```

27 ** (1/3)也返回 $27^{\frac{1}{3}}$ 的数值。

```
console.log(pow(27, 1/3)) ;
```

pow(27, 1/3)返回 27 的 1/3 次方（power）值，也是 $27^{\frac{1}{3}}$ 的数值。

```
console.log('') ;
  console.log(ceil(32.1)) ;
```

ceil(32.1)返回大于且最接近 32.1 的整数值。

```
console.log(ceil(32.7)) ;
```

ceil(32.7)返回大于且最接近 32.7 的整数值。

```
console.log(ceil(-32.1)) ;
```

ceil(-32.1)返回大于且最接近-32.1 的整数值。

```
console.log(ceil(-32.7)) ;
```

ceil(-32.7)返回大于且最接近-32.7 的整数值。

```
console.log('') ;
  console.log(clz32(0b111)) ;
```

clz32(0b111)先将二进制数值 111，转换成为 **32 位**二进制数码 00000000000000000000000000000111 之后，再返回其开头 0 的个数 29。

```
console.log(clz32(0777)) ; // old literal
```

clz32(0777)先将八进制数值 777，转换成为 32 位二进制数码，再返回其开头 0 的个数。

```
console.log(clz32(0o777)) ; // new literal
```

clz32(0o777)也是先将八进制数值 777，转换成为 32 位二进制数码，再返回其开头 0 的个数。

```
console.log(clz32(0xfff)) ;
```

clz32(0x777)先将十六进制数值 fff，转换成为 32 位二进制数码，再返回其开头 0 的个数。

```
console.log('') ;
  console.log(exp(1)) ;
```

exp(1)返回自然底数 e 的 1 次方值。

```
console.log(E) ;
```

E 返回自然底数 e 的数值。

```
console.log(exp(3)) ;
```

exp(3)返回自然底数的 3 次方值。

```
console.log(E ** 3) ;
```

E ** 3 返回自然底数的 3 次方值。

```
console.log(pow(E, 3)) ;
```

pow(E, 3)亦返回自然底数的 3 次方值。

```
console.log('') ;
 console.log(expm1(1)) ;
```

expm1(1)返回【自然底数的 1 次方值，再减去 1】的结果值。

```
console.log(E - 1) ;
```

E - 1 返回【自然底数减去 1】的结果值。

```
console.log(expm1(3)) ;
```

expm1(3)返回【自然底数的 3 次方值，再减去 1】的结果值。

```
console.log(E ** 3 - 1) ;
```

E ** 3 - 1 返回【自然底数的 3 次方值，再减去 1】的结果值。

```
console.log(pow(E, 3) - 1) ;
```

pow(E, 3) - 1 也返回【自然底数的 3 次方值，再减去 1】的结果值。

```
console.log('') ;
 console.log(floor(15.8)) ;
```

floor(15.8)返回小于且最接近 15.8 的整数值。

```
console.log(floor(15.1)) ;
```

floor(15.1)返回小于且最接近 15.1 的整数值。

```
console.log(floor(-15.8)) ;
```

floor(-15.8)返回小于且最接近-15.8 的整数值。

```
console.log(floor(-15.1)) ;
```

floor(-15.1)返回小于且最接近-15.1 的整数值。

```
console.log('') ;
 // 32-bit single-precision float representation of a number.
 console.log(fround(9.875)) ;
```

fround(9.875)返回最接近 9.875 的单精度的浮点数值。

```
console.log(fround(9.876)) ;
```

fround(9.876)返回最接近 9.876 的单精度的浮点数值。

```
console.log(fround(9.125)) ;
```

fround(9.125)返回最接近 9.125 的单精度的浮点数值。

```
console.log(fround(9.126)) ;
```

fround(9.126)返回最接近 9.126 的单精度的浮点数值。

```
console.log('') ;

  // hypotenuse of a right triangle 直角三角形的斜边
  console.log(hypot(3, 4)) ;
```

hypot(3, 4)返回【3 与 4 的平方和】的平方根值。

```
console.log(hypot(6, 8)) ;
```

hypot(6, 8)返回【6 与 8 的平方和】的平方根值。

```
console.log(hypot(5, 12)) ;
```

hypot(5, 12)返回【5 与 12 的平方和】的平方根值。

```
console.log('') ;

  let num_list = Array.from(Array(11).keys()) ;
```

此语句声明了变量 num_list，其初始数据为 Array.from(Array(11).keys())动态产生的数组实例 [0, 1, 2, 3, 4, 5, 6, 7, 8, 9, 10]。

```
  console.log(hypot(... num_list)) ;
```

hypot(... num_list)返回【0、1、…、10 的平方和】的平方根值。

```
console.log('') ;

  // 32-bit integer multiplication
  console.log(imul(2.1, 10)) ;
```

函数 imul()会进行【两个数值的整数部分的 32 位乘法运算】，并返回其结果值。在此，imul(2.1, 10)会进行 2.1 与 10 的整数部分的乘法运算，并返回其结果值。

```
  console.log(imul(-5, -8)) ;
```

imul(-5, -8)返回-5 与-8 的乘法运算的结果值。

```
  console.log(imul(-3, 7)) ;
```

imul(-3, 7)返回-3 与 7 的乘法运算的结果值。

```
console.log('') ;

  // natural logarithm (base e) of a number
  console.log(log(E)) ;
```

log(E)返回 ln(e)的数值。

```
  console.log(log(E ** 7)) ;
```

log(E ** 7)返回 ln(e^7)的数值。

```
  console.log(log2(2 ** 15)) ;
```

log2(2 ** 15)返回 $\log_2(2^{15})$ 的数值。

```
  console.log(log10(10 ** 8)) ;
```

log10(10 ** 8)返回 $\log_{10}(10^8)$ 的数值。

```
console.log('') ;

// natural logarithm (base e) of 1 plus a number.
console.log(log1p(E - 1)) ;
```

log1p(E - 1)返回 ln(1+(e-1))的数值。

```
console.log(log1p(E ** 7 - 1)) ;
```

log1p(E ** 7 - 1)返回 ln(1+(e^7-1))的数值。

```
console.log('') ;

// num_list = Array.from(new Array(10), (val, index) => index * 3) ;
num_list = Array.from(new Array(10), (value, index) => parseInt(100 * Math.random())) ;
```

此语句将内含 10 个 0～100 的随机整数值的数组实例，例如[33, 86, 55, 3, 9, 51, 93, 66, 82, 7, 91]，赋给变量 num_list。

```
console.log(num_list) ;
  console.log(max(... num_list)) ;
```

若 num_list 的数据为数组实例[33, 86, 55, 3, 9, 51, 93, 66, 82, 7, 91]，则 max(... num_list)等价于 max(33, 86, 55, 3, 9, 51, 93, 66, 82, 7, 91)，会返回各数值中的最大值 93。

```
console.log(min(... num_list)) ;
```

若 num_list 的数据为数组实例[33, 86, 55, 3, 9, 51, 93, 66, 82, 7, 91]，则 min(... num_list)等价于 min(33, 86, 55, 3, 9, 51, 93, 66, 82, 7, 91)，会返回各数值中的最小值 7。

```
console.log('') ;

  console.log(pow(2, 5)) ;
```

pow(2, 5)返回 2^5 的数值。

```
console.log(2 ** 5) ;
```

2 ** 5 也返回 2^5 的数值。

```
console.log('') ;

console.log(pow(27, 1/3)) ;
```

pow(27, 1/3)返回 $27^{\frac{1}{3}}$ 的数值。

```
console.log(27 ** (1/3)) ;
```

27 ** (1/3)也返回 $27^{\frac{1}{3}}$ 的数值。

```
console.log('') ;

  console.log(pow(2, 0.5)) ;
```

pow(2, 0.5)返回 $2^{0.5}$ 的数值。

```
console.log(2 ** 0.5) ;
```

2 ** 0.5 也返回 $2^{0.5}$ 的数值。

```
console.log(SQRT2) ;
```

SQRT2 返回 $\sqrt{2}$，等于 $2^{0.5}$ 的数值。

```
console.log('') ;
  console.log(pow(0.5, 0.5)) ;
```

pow(0.5, 0.5)返回 $0.5^{0.5}$，也就是 $\sqrt{\dfrac{1}{2}}$ 的数值。

```
console.log(0.5 ** 0.5) ;
```

0.5 ** 0.5 也返回 $0.5^{0.5}$ 的数值。

```
console.log(SQRT1_2) ;
```

SQRT1_2 返回 $\sqrt{\dfrac{1}{2}}$，也是 $0.5^{0.5}$ 的数值。

```
console.log('') ;
  console.log(random()) ;
  console.log(random()) ;
```

random()会返回大于或等于 0，并且小于 1 的随机浮点数值。

```
console.log('') ;
  console.log(100 + parseInt(100 * Math.random())) ;
```

100 + parseInt(100 * Math.random())的结果值，是介于 100～200 之间的随机整数值。

```
console.log(-50 + parseInt(100 * Math.random())) ;
```

-50 + parseInt(100 * Math.random())的结果值，是介于-50～50 之间的随机整数值。

```
console.log('') ;
  console.log(round(32.48)) ;
```

round(32.48)返回 32.48 的四舍五入至个位的数值 32。

```
console.log(round(32.51)) ;
```

round(32.51)返回 32.51 的四舍五入至个位的数值 33。

```
console.log(round(-32.48)) ;
```

round(-32.48)返回-32.48 的四舍五入至个位的数值-32。

```
console.log(round(-32.51)) ;
```

round(-32.51)返回-32.51 的四舍五入至个位的数值-33。

```
console.log('') ;
  console.log(sign(+77)) ;
```

sign(+77)返回 1，代表+77 是正数。

```
console.log(sign(77)) ;
```

sign(77)返回 1，代表 77 是正数。

```
console.log(sign(-77)) ;
```

sign(-77)返回-1，代表-77 是负数。

```
console.log('') ;
  console.log(sign('+66')) ;
```

sign('+66')返回 1，代表字符串'+66'被转换后的 66 是正数。

```
console.log(sign('66')) ;
```

sign('66')返回 1，代表字符串'66'被转换后的 66 是正数。

```
console.log(sign('-66')) ;
```

sign('-66')返回 1，代表字符串'-66'被转换后的-66 是负数。

```
console.log('') ;
  console.log(sign(+0)) ;
```

sign(+0)返回 0，代表+0 是零值。

```
console.log(sign(0)) ;
```

sign(0)返回 0，代表 0 是零值。

```
console.log(sign(-0)) ;
```

sign(-0)返回-0，代表-0 是负零值。

```
console.log('') ;
  console.log(sqrt(64)) ;
```

sqrt(64)返回 $\sqrt{64}$ 的数值 8。

```
console.log(64 ** 0.5) ;
```

64 ** 0.5 返回 $64^{0.5}$ 的数值，亦等同于 $\sqrt{64}$ 的整数值 8。

```
console.log(pow(64, 0.5)) ;
```

pow(64, 0.5)也返回 $64^{0.5}$ 的数值，亦等同于 $\sqrt{64}$ 的整数值 8。

```
console.log('') ;
  console.log(trunc(56.78)) ;
```

trunc(56.78)返回 56.78 被**截断**（truncate）小数部分的整数值 56。

```
console.log(trunc(0.779)) ;
```

trunc(0.779)返回 0.779 被截断小数部分的整数值 0。

```
console.log(trunc(-0.36)) ;
```

trunc(-0.36)返回-0.36 被截断小数部分的整数值-0。

```
console.log(trunc('-33.66')) ;
```

trunc('-33.66')先转换字符串'-33.66'为数值-33.66，再返回被截断小数部分的整数值-33。

6.4 练 习 题

1. 在 JavaScript 语言里，可处理的最大正整数是什么？安全范围内的最大安全整数又是什么？
2. 在 JavaScript 语言里，Number.MIN_VALUE 代表一个整数还是浮点数？
3. 在 JavaScript 语言里，Number.EPSILON 代表什么数值？
4. 试编写 JavaScript 源代码，以转换并显示浮点数 2013.5625 对应的二进制、八进制和十六进制的数码。
5. 试编写 JavaScript 源代码，以转换并显示十六进制数值 ff285c 对应的二进制、八进制的数码。
6. 试编写 JavaScript 源代码，以转换变量 num03 内含的圆周率数值，成为具有 3 位小数的浮点数。
7. 试编写至少 3 个版本的 JavaScript 源代码，将变量 str01 的字符串'0x57fc'，和变量 str02 的字符串'0b111011'，先分别转换成为数值之后，再进行相加。最后再将其结果值，显示成为十进制数值。
8. 试编写 JavaScript 源代码，以显示出【变量 value01 的不超过 100 的正整数，除以变量 value02 的不超过 10 的正整数的结果值，是否依然为整数】的判断结果，是为真（true）或者为假（false）。
9. 通过 3 种 JavaScript 的语法，表示【$\sqrt[3]{x}, \text{where} \{ x \in N \,|\, 10 \leq x < 70 \}$】。
10. 已知特定直角三角形的斜边的长度为 150，其中一个锐角的角度为 30°。试通过 Math 对象支持的三角函数，计算并显示两个直角边的长度。
11. 试编写 JavaScript 源代码，以显示出介于-100～+100 之间的随机整数。

第 7 章

处理字符串

在应用程序中,字符串(string)经常被处理成为界面上的各种信息。在 JavaScript 语言中,字符串的搜索、获取、替换、连接、分割、填充、扩展、匹配等处理机制,自然是不可或缺的。

7.1　String 对象

在 JavaScript 语言中,处理字符串的各项机制,主要依赖 String 对象中的各函数/方法的协助,以及单引号【'】、双引号【"】、反引号【`】、加号【+】和反斜杠【\】运算符的辅助。

7.1.1　子字符串的索引值

子字符串(sub string)的索引值(index value),是指特定子字符串,在特定字符串里的索引位置。若特定子字符串位于特定字符串中的索引值为 n,则代表在特定字符串里,特定子字符串出现在第 n + 1 个字符(character)开始的位置。在 JavaScript 语言里,索引值是从 0 开始起算的,所以索引值 0 代表第 1 个字符的位置,而索引值 n 代表第 n + 1 个字符的位置。下面通过示例来介绍子字符串索引值的运用。

【7-1-1-sub-string-indexings.js】

```
let sentence = 'I hope you could live happily ever after.'

console.log(sentence.length) ;
console.log('') ;

with (console)
{
  log(sentence.indexOf('happy')) ;
```

```
        log(sentence.indexOf('happily')) ;
        log(sentence.indexOf('live')) ;
        log('') ;

        log(sentence.indexOf('ou')) ;
        log(sentence.indexOf('ou', 10)) ;
        log(sentence.indexOf('ou', 14)) ;
        log('') ;

        log(sentence.lastIndexOf('l')) ;
        log(sentence.lastIndexOf('l', 26)) ;
        log(sentence.lastIndexOf('l', 16)) ;
        log(sentence.lastIndexOf('l', 13)) ;
}
```

【相关说明】

```
let sentence = 'I hope you could live happily ever after.'
```

声明初始数据为字符串的变量 sentence。

```
console.log(sentence.length) ;
```

sentence.length 返回变量 sentence 的数据字符串的字符个数 41。

```
with (console)
{
    log(sentence.indexOf('happy')) ;
```

sentence.indexOf('happy')查找不到子字符串'happy',在变量 sentence 的数据字符串中的索引位置,所以返回-1。

```
log(sentence.indexOf('happily')) ;
```

sentence.indexOf('happily')找寻到子字符串'happily'在变量 sentence 的数据字符串的索引位置 21,也就是第 22 个字符的位置。

```
log(sentence.indexOf('live')) ;
```

sentence.indexOf('live')找寻到子字符串'live',在于变量 sentence 的数据字符串的索引位置 17,也就是第 18 个字符的位置。

```
log('') ;

    log(sentence.indexOf('ou')) ;
```

sentence.indexOf('ou')在变量 sentence 的数据字符串里,找寻到子字符串'ou',存在于其索引位置 8,也就是第 9 个字符的位置。

```
log(sentence.indexOf('ou', 10)) ;
```

sentence.indexOf('ou', 10)在变量 sentence 的数据字符串里,从索引位置 10 开始,向后找寻到字符串'ou',存在于其索引位置 12,也就是第 13 个字符的位置。

```
log(sentence.indexOf('ou', 14)) ;
```

sentence.indexOf('ou', 14)在变量 sentence 的数据字符串里，从索引位置 14 开始，向后查找不到字符串'ou'，存在于剩余的字符串里，所以返回-1。

```
log('') ;

 log(sentence.lastIndexOf('l')) ;
```

sentence.lastIndexOf('l')在变量 sentence 的数据字符串里，找寻到字符'l'，存在于索引位置 27，也就是第 28 个字符的位置。

```
log(sentence.lastIndexOf('l', 26)) ;
```

sentence.indexOf('l', 26)在变量 sentence 的数据字符串里，从索引位置 26 开始，【往前】查找到字符'l'，存在于索引位置 17，也就是第 18 个字符的位置。

```
log(sentence.lastIndexOf('l', 16)) ;
```

sentence.indexOf('l', 16)在变量 sentence 的数据字符串里，从索引位置 16 开始，【往前】查找到字符'l'，存在于索引位置 14，也就是第 15 个字符的位置。

```
log(sentence.lastIndexOf('l', 13)) ;
```

sentence.indexOf('l', 13)在变量 sentence 的数据字符串里，从索引位置 13 开始，【往前】无法查找到字符'l'，存在于剩余的字符串里，所以返回-1。

7.1.2　特定模式的子字符串的搜索（ES6）

若在特定字符串里，搜寻符合特定模式（pattern）的第 1 个子字符串，则可通过【当前数据为字符串的变量名称.search(其数据为特定正则表达式字面量的变量)】来实现。

若欲找出所有符合特定模式的所有子字符串，则可通过语法【当前数据为字符串的变量名称.match(其数据为特定正则表达式字面量的变量)】来达成。

若在特定字符串里，欲判断是否包含特定子字符串，或是以特定子字符串开头或结尾，也有其他相关函数可供调用。关于特定模式的子字符串的搜索，可参考如下示例。

【7-1-2-sub-string-searchings.js】

```
let saying = 'Joanna lovingly loves lovely beloved of Jason.' ;

let re01 = /lov\w+/g ;

let index = saying.search(re01) ;

console.log(index) ;
console.log('') ;

console.log(saying.match(re01).length) ;
console.log(saying.startsWith('Jo')) ;
console.log('') ;

console.log(saying.startsWith('lov')) ;
console.log(saying.startsWith('lov', saying.indexOf('lovingly'))) ;
```

```
console.log('') ;

console.log(saying.endsWith('ed')) ;
console.log(saying.endsWith('on.')) ;
console.log('') ;

console.log(saying.includes('Jason')) ;
```

【相关说明】

```
let saying = 'Joanna lovingly loves lovely beloved of Jason.' ;
```

声明初始数据为字符串的变量 saying。

```
let re01 = /lov\w+/g ;
```

此语句声明了初始数据为正则表达式字面量 (regular-expression literal) 的变量 re01。正则表达式字面量/lov\w+/g 代表所有'lov'开头的子字符串。其中，【\w】代表可以是字符 a～z、A～Z、0～9 或【_】，而【g】代表全局（global），也就是**所有**的含义。

```
let index = saying.search(re01) ;
```

此语句声明了变量 index，其初始数值为 7，代表符合变量 re01 的字面量/lov\w+/的第 1 个子字符串 lovingly，出现在变量 saying 的数据字符串里的索引位置 7。

```
console.log(index) ;
console.log('') ;

console.log(saying.match(re01).length) ;
```

saying.match(re01)会返回数组实例["lovingly", "loves", "lovely", "loved"]，其内含符合变量 re01 的字面量/lov\w+/g 的【所有】子字符串。saying.match(re01).length 会返回此数组实例中的字符串元素的个数 4。

```
console.log(saying.startsWith('Jo')) ;
```

saying.startsWith('Jo')会返回 true，代表变量 saying 的数据字符串，是由子字符串'Jo'开头的。

```
console.log('') ;

console.log(saying.startsWith('lov')) ;
```

saying.startsWith('lov')会返回 false，代表变量 saying 的数据字符串，并不是由子字符串'Jo'开头的。

```
console.log(saying.startsWith('lov', saying.indexOf('lovingly'))) ;
```

saying.indexOf('lovingly')返回索引值 7，代表子字符串'lovingly'，出现在变量 saying 的数据字符串里的索引位置 7。saying.startsWith('lov', saying.indexOf('lovingly'))返回 true，是因为从索引位置 7 开始，到末尾的子字符串'lovingly loves lovely beloved of Jason.'，是由子字符串'lov'开头的。

```
console.log('') ;

console.log(saying.endsWith('ed')) ;
```

saying.endsWith('ed')返回 false，代表变量 saying 的数据字符串，并不是由子字符串'ed'结尾的。

```
console.log(saying.endsWith('on.')) ;
```

saying.endsWith('on.')返回 true,代表变量 saying 的数据字符串,是由子字符串'on.'结尾的。

```
console.log('') ;
```

```
console.log(saying.includes('Jason'))
```

saying.includes('Jason')返回 true,代表变量 saying 的数据字符串,内含子字符串'Jason'。

7.1.3 子字符串的获取

若欲在特定字符串里,获取特定索引位置开始的某一段子字符串,可以使用 String 对象实例支持的 3 个函数之一,即 slice()、substring() 与 substr()。下面通过具体示例来说明这 3 个函数的调用方式。

【7-1-3-sub-string-gettings.js】

```
let saying = 'Do one thing at a time, and do well.' ;

let piece01 = saying.slice(0, 12) ;
let piece02 = saying.substring(0, 12) ;
let piece03 = saying.substr(0, 12) ;

console.log(piece01) ;
console.log(piece02) ;
console.log(piece03) ;
console.log('') ;

///
let start = saying.indexOf(',') ;
let end = saying.indexOf('.') ;

// piece01 = saying.slice(start + 2, end) ;
piece01 = saying.slice(start + 2, -1) ;
piece02 = saying.substring(start + 2, end) ;
piece03 = saying.substr(start + 2, end - start - 2) ;

console.log(piece01) ;
console.log(piece02) ;
console.log(piece03) ;
console.log('') ;

///
piece01 = saying.slice(start + 2) ;
piece02 = saying.substring(start + 2) ;
piece03 = saying.substr(start + 2) ;

console.log(piece01) ;
console.log(piece02) ;
console.log(piece03) ;
console.log('') ;

///
```

```
piece01 = saying.slice(-5, -1) ;
piece02 = saying.substring(31, 35) ;
piece03 = saying.substr(31, 4) ;

console.log(piece01) ;
console.log(piece02) ;
console.log(piece03) ;
```

【相关说明】

```
let saying = 'Do one thing at a time, and do well.' ;
```

声明初始数据为字符串的变量 saying。

```
let piece01 = saying.slice(0, 12) ;
let piece02 = saying.substring(0, 12) ;
let piece03 = saying.substr(0, 12) ;
```

在这 3 个语句的等号右侧，均会被返回子字符串'Do one thing'，并赋给等号左侧的变量。saying.slice(0, 12)的含义，是在变量 saying 的字符串'Do one thing at a time, and do well.' 里，截取索引位置【0】到【12 - 1】的子字符串'Do one thing'。saying.substring(0, 12)的含义，是在变量 saying 的字符串里，截取索引位置【0】至【12 - 1】的子字符串'Do one thing'。saying.substr(0, 12)虽然也返回'Do one thing'，但是含义有所不同，是指在变量 saying 的字符串里，从索引位置【0】的位置，向后截取【12 个】字符的子字符串'Do one thing'。

```
console.log(piece01) ;
console.log(piece02) ;
console.log(piece03) ;
console.log('') ;

let start = saying.indexOf(',') ;
```

此语句声明了初始数值为【saying.indexOf(',')返回的索引值 22】的变量 start。

```
let end = saying.indexOf('.') ;
```

此语句声明了初始数值为【saying.indexOf('.')返回的索引值 35】的变量 end。

```
// piece01 = saying.slice(start + 2, end) ;
```

此语句若被开放之后，saying.slice(start + 2, end)当前等价于 saying.slice(24, 35)，会返回子字符串'and do well'。

```
piece01 = saying.slice(start + 2, -1) ;
```

saying.slice(start + 2, -1)当前等价于 saying.slice(24, -1)，也会返回子字符串'and do well'。其中，函数 slice()的第 2 个参数的数值-1，代表倒数第 2 个（-1 -1）字符的索引位置。若为-2，则代表倒数第 3 个（-2 -1）字符的索引位置，以此类推。

```
piece02 = saying.substring(start + 2, end) ;
```

saying.substring(start + 2, end)当前等价于 saying.substring(24, 35)，亦返回子字符串'and do well'。

```
piece03 = saying.substr(start + 2, end - start - 2) ;
```

saying.substr(start + 2, end-start-2)当前等价于 saying.substr(24, 11)，返回子字符串'and do well'。

```
console.log(piece01) ;
console.log(piece02) ;
console.log(piece03) ;
console.log('') ;

piece01 = saying.slice(start + 2) ;
```

saying.slice(start + 2)当前等价于 saying.slice(24, 36)，返回子字符串'and do well.'。

```
piece02 = saying.substring(start + 2) ;
```

saying.substring(start + 2)当前等价于 saying.substring(24, 36)，也返回子字符串'and do well.'。

```
piece03 = saying.substr(start + 2) ;
```

saying.substr(start + 2)当前等价于 saying.substr(24, 12)，亦返回子字符串'and do well.'。

```
console.log(piece01) ;
console.log(piece02) ;
console.log(piece03) ;
console.log('') ;

piece01 = saying.slice(-5, -1) ;
```

saying.slice(-5, -1)返回【从最后第 5 个开始，至倒数第 2 个（-1 - 1）字符】的子字符串'well'。

```
piece02 = saying.substring(31, 35) ;
```

saying.substring(31, 35)也返回子字符串'well'。

```
piece03 = saying.substr(31, 4) ;
```

saying.substr(31, 4)亦返回子字符串'well'。

7.1.4　子字符串的替换

若欲在特定字符串里，替换其中一段或者符合特定模式（pattern）的子字符串，成为新的子字符串，则可调用 String 对象实例所支持的函数 replace()。下面通过具体的示例，来说明子字符串的替换运用。

【7-1-4-sub-string-replacements.js】

```
let str01 = ' Dread   can  trap you, but optimism can   release  you. ' ;

let sentence = str01.replace(/\s{2,}/g, ' ') ;

console.log(sentence) ;

sentence = sentence.replace(/^\s|\s$/g, '') ;

console.log(sentence) ;
console.log('') ;

///
// let re01 = /(Dread)[a-z ]+(optimism)/ ;
let re01 = /(D)(read)([\w\s,]+)(o)(ptimism)/ ;
```

```
sentence = sentence.replace(re01, 'O$5$3d$2') ;

console.log(sentence) ;
```

【相关说明】

```
let str01 = '  Dread   can  trap you,  but  optimism can   release  you.  ' ;
```

声明初始数据为字符串的变量 str01。

```
let sentence = str01.replace(/\s{2,}/g, ' ') ;
```

此语句声明了变量 sentence，其初始数据为 str01.replace(/\s{2,}/g, ' ')所返回的字符串' Dread can trap you, but optimism can release you. '。str01.replace(/\s{2,}/g, ' ')在变量 str01 的数据字符串中，取代至少 2 个【连续】的空白字符，成为【单一】空格字符。正则表达式字面量/\s{2,}/g 代表【所有(g, global)】至少 2 个（{2,}）的空白字符（\s）。其中，空格（space）、制表符（tab）、换页符（form feed）、换行符（line feed）、回车符（carriage return）均是空白（whitespace）字符之一。

```
console.log(sentence) ;

sentence = sentence.replace(/^\s|\s$/g, '') ;
```

此语句将 sentence.replace(/^\s|\s$/g, ' ')所返回的字符串'Dread can trap you, but optimism can release you.'，赋给变量 sentence 本身。sentence.replace(/^\s|\s$/g, '')使得在变量 sentence 的数据字符串，被取代（replace）开头（^\s）与结尾（\s$）的空白字符。

```
console.log(sentence) ;
console.log('') ;

// let re01 = /(Dread)[a-z ]+(optimism)/ ;
let re01 = /(D)(read)([\w\s,]+)(o)(ptimism)/ ;
```

此语句声明了初始数据为正则表达式字面量/(D)(read)([\w\s,]+)(o)(ptimism)/的变量 re01。在正则表达式字面量/(D)(read)([\w\s,]+)(o)(ptimism)/中，存在 5 组小括号，可分别通过$1、$2、$3、$4与$5，来表示其【交换位置】的身份。

- $1 对应子字面量【(D)】的部分，用来匹配字符'D'。
- $2 对应子字面量【(read)】的部分，用来匹配字符串'read'。
- $3 对应子字面量【([\w\s,]+)】的部分，当前则匹配到字符串' can trap you, but '。
- $4 对应子字面量【(o)】的部分，用来匹配字符'o'。
- $5 对应子字面量【(ptimism)】的部分，用来匹配字符串'ptimism'。

其中，在子字面量【[\w\s,]+】里：

- 【\w】代表 1 个 a~z、A~Z、0~9、下画线字符。
- 【\s】代表 1 个**空白字符**。
- 【,】代表 1 个逗号字符。
- 【[\w\s,]+】代表**至少**为 1 个上述字符之一。

```
sentence = sentence.replace(re01, 'O$5$3d$2') ;
```

此语句将 sentence.replace(re01, 'O$5$3d$2')所返回的字符串'Optimism can trap you, but dread can release you.'，赋给变量 sentence 本身。值得注意的是，子字符串'Dread'和'optimism'在整体字符串所在的位置，已经被交换了。字面量'O$5$3d$2'暗示着要重新组合各部分的子字符串。其中：

- $5 代表子字符串'ptimism'。
- $3 代表子字符串' can trap you, but '。
- $2 代表子字符串'read'。

因此，字面量'O$5$3d$2'使得原来的字符串'Dread can trap you, but optimism'，被重新组合成为新的字符串'Optimism can trap you, but dread'。

7.1.5　字符串的大小写转换

欲将特定字符串中的所有字母，转换成大写字母，可调用 String 对象实例所支持的函数 toUpperCase()；反之，则调用函数 toLowerCase()。下面通过示例介绍字符串大小写转换的运用。

【7-1-5-String-toUpperCase-and-toLowerCase.js】

```javascript
let str01 = 'nelson' ;

let name01 = str01[0].toUpperCase() + str01.slice(1) ;

console.log(str01) ;
console.log(name01) ;
console.log('') ;

///
let str02 = 'ALEXANDER' ;

let name02 = str02[0] + str02.slice(1).toLowerCase() ;

console.log(str02) ;
console.log(name02) ;
```

【相关说明】

```javascript
let str01 = 'nelson' ;
```

声明初始数据为字符串的变量 str01。

```javascript
let name01 = str01[0].toUpperCase() + str01.slice(1) ;
```

此语句声明了变量 name01，其初始数据为【变量 str01 的字符串'nelson'，被设置第 1 个字母成为大写】之后的字符串'Nelson'。

```javascript
console.log(str01) ;
console.log(name01) ;
console.log('') ;

let str02 = 'ALEXANDER' ;
```

声明初始数据为字符串的变量 str02。

```
let name02 = str02[0] + str02.slice(1).toLowerCase() ;
```

此语句声明了变量 name02，其初始数据为【变量 str02 的字符串'ALEXANDER'，仅仅被保留第 1 个字母为大写】之后的字符串'Alexander'。

7.1.6 不同字符串的连接

连接多个字符串时，除可简单通过加号运算符【+】来串联 1 个以上的字符串之外，还可通过调用 String 对象支持的函数 concat()或者 Array 对象支持的函数 join()来完成。请看下面的示例。

【7-1-6-String-concat.js】

```
let prefix = 'Lucky numbers: ' ;

let numbers = Array.from(new Array(6), () => parseInt(100 * Math.random()) + ' ') ;

let message = prefix.concat(... numbers) ;

console.log(message) ;

message = prefix + numbers.join('') ;

console.log(message) ;
```

【相关说明】

```
let prefix = 'Lucky numbers: ' ;
```

声明初始数据为字符串的变量 prefix。

```
let numbers = Array.from(new Array(6), () => parseInt(100 * Math.random()) + ' ') ;
```

此语句声明了变量 numbers，其初始数据为数组实例，例如【Array.from(new Array(6), () => parseInt(100 * Math.random()) + ' ')】所产生的数组实例["47 ", "77 ", "43 ", "83 ", "85 ", "64 "]，并且内含 6 个带有 0～100 的整数值的字符串。

```
let message = prefix.concat(... numbers) ;
```

此语句声明了变量 message，其初始数据为 prefix.concat(... numbers)所返回的字符串。prefix.concat(... numbers)可串联（concate）变量 prefix 的数据字符串'Lucky numbers: '，以及从语法【... numbers】扩展出来的多个字符串。若变量 numbers 的数组实例为["47 ", "77 ", "43 ", "83 ", "85 ", "64 "]，则【prefix.concat(... numbers)】等价于【'Lucky numbers: '.concat("47 ", "77 ", "43 ", "83 ", "85 ", "64 ")】，会返回字符串'Lucky numbers: 47 77 43 83 85 64 '。

```
console.log(message) ;

message = prefix + numbers.join('') ;
```

prefix + numbers.join('')可返回字符串'Lucky numbers: 47 77 43 83 85 64 '。数组实例支持的函数 join('')，可将变量 numbers 的数组实例["47 ", "77 ", "43 ", "83 ", "85 ", "64 "]中的各字符串，不以任何字符加以分隔，并连接（join）成为新字符串'47 77 43 83 85 64 '。若是 numbers.join('/')，则会返

回字符串'47/77/43/83/85/64/'。

7.1.7 字符串的重复连接

如果要进行特定字符串的重复连接，可调用 String 对象实例所支持的函数 repeat()。请看下面的示例。

【7-1-7-string-repeatings.js】

```
let space = ' ' ;
let comma = ',' ;
let name = 'Alex' ;
let greeting = 'how are you today?' ;
let smile = '=^.^=' ;

let sentence = name + comma + space + greeting + space.repeat(5) + smile ;

console.log(sentence) ;
```

【相关说明】

```
let space = ' ' ;
```

声明初始数据为空格字符的变量 space。

```
let comma = ',' ;
```

声明初始数据为逗号字符的变量 comma。

```
let name = 'Alex' ;
```

声明初始数据为字符串'Alex'的变量 name。

```
let greeting = 'how are you today?' ;
```

声明初始数据为字符串'how are you today?'的变量 greeting。

```
let smile = '=^.^=' ;
```

声明初始数据为字符串'=^.^='的变量 smile。

```
let sentence = name + comma + space + greeting + space.repeat(5) + smile ;
```

此语句声明了初始数据为组合之后的字符串'Alex, how are you today?   =^.^='的变量 sentence。通过运算符【+】，可串联两个字符串。space.repeat(5)会使得变量 space 所代表的空格字符' '，被重复（repeat）5 次而返回'   '。

7.1.8 字符串的分割

如果要分割特定字符串，成为多个子字符串，并放入新数组实例中，则可调用 String 对象实例支持的函数 split()。可参看下面的例子。

【7-1-8-string-splitting.js】

```
let str01 = '10,20,50,40,70,90,80,30,60' ;
let str02 = ' 10, 20 , 50  ,40 , 70 , 90, 80 , 30  , 60  ' ;

let arr01 = str01.split(',') ;

console.log(arr01) ;

arr01 = str01.split(',', 5) ;

console.log(arr01) ;
console.log('') ;

///
let arr02 = str02.replace(/ /g, '').split(',') ;

console.log(arr02) ;
console.log('') ;

arr02 = str02.replace(/^\s*|\s*$/g, '').split(/\s*,\s*/) ;

console.log(arr02) ;
```

【相关说明】

```
let str01 = '10,20,50,40,70,90,80,30,60' ;
let str02 = ' 10, 20 , 50  ,40 , 70 , 90, 80 , 30  , 60  ' ;
```

这两个语句分别声明了初始数据为不同字符串的变量 str01 和 str02。

```
let arr01 = str01.split(',') ;
```

此语句声明了变量 arr01，其初始数据为 str01.split(',')所返回的数组实例。str01.split(',')可根据被传入的分隔符（seperator）','，加以分割（split）变量 str01 的数据字符串'10,20,50,40,70,90,80,30,60'，成为被返回的数组实例["10", "20", "50", "40", "70", "90", "80", "30", "60"]。

```
console.log(arr01) ;

arr01 = str01.split(',', 5) ;
```

此语句将 str01.split(',', 5)所返回的数组实例，赋给变量 arr01。str01.split(',', 5)可根据被传入的分隔符','，加以分割变量 str01 的数据字符串'10,20,50,40,70,90,80,30,60'的前 5 个项目，成为被返回的数组实例["10", "20", "50", "40", "70"]。

```
console.log(arr01) ;
console.log('') ;

let arr02 = str02.replace(/ /g, '').split(',') ;
```

此语句声明了初始数据为新数组实例的变量 arr02。其中，str02.replace(/ /g, '')会返回【变量 str02 的数据字符串' 10, 20 , 50 ,40 , 70 , 90, 80 , 30 , 60 '，被去除所有空格字符】之后的新字符串'10,20,50,40,70,90,80,30,60'。新字符串再经过 split(',')的处理，被分割成新数组实例["10", "20", "50", "40", "70", "90", "80", "30", "60"]。

```
console.log(arr02) ;
console.log('') ;

arr02 = str02.replace(/^\s*|\s*$/g, '').split(/\s*,\s*/) ;
```

此语句重新设置变量 arr02 的数据，成为另一新数组实例。str02.replace(/^\s*|\s*$/g, '')会返回【变量 str02 的数据字符串 ' 10, 20 , 50 ,40 , 70 , 90, 80 , 30   ,   60 '，被去除开头与结尾的空白字符】之后的新字符串'10, 20 , 50 ,40 , 70 , 90, 80 , 30   ,   60'。新字符串再经过 split(/\s*,\s*/)的处理，根据正则表达式字面量形式的分隔符【\s*,\s*/】，进一步被分割成为新数组实例["10", "20", "50", "40", "70", "90", "80", "30", "60"]。

replace(/^\s*|\s*$/g, '')用来取代（replace）开头或末尾的所有连续空白字符，成为空字符"。换言之，就是去除包括空格（space）字符在内的开头与末尾的所有空白（white-space）字符。其中：

- 【^】具有开头的含义。
- 【$】具有末尾的含义。
- 【g】具有全局（global）/所有的含义。
- 【|】具有逻辑或（logical or）的含义。
- 【*】具有**无和至少 1 个**，也就是可能存在至少 1 个的含义。
- 【^\s*】代表开头可能存在至少 1 个的连续空白字符。
- 【\s*$】代表末尾可能存在至少 1 个的连续空白字符。

split(/\s*,\s*/)根据正则表达式字面量【\s*,\s*/】所代表的【至少 1 个的连续空白字符,至少 1 个的连续空白字符】的复合式分隔符，例如',、', '、', ' , ',等，以分割（split）字符串'10, 20 , 50 ,40 , 70 , 90, 80 , 30   ,   60',成为新数组实例["10", "20", "50", "40", "70", "90", "80", "30", "60"]。

7.1.9 特定字符和 Unicode 数码的双向转换（ES6）

可被显示在计算机界面上的，无论是英文字母、数字或符号等字符（character）还是中文字，均可对应到 Unicode 字符集里的特定数码。JavaScript 语言支持【Unicode 字符集里的数码 ⇆ 可显示的字符】的双向转换。下面通过具体的示例来加以调试。

【7-1-9-codes-of-strings.js】

```
let name = '柯霖廷' ;
let str = 'abcdefg' ;

console.log(name.charCodeAt(0)) ;
console.log(name.codePointAt(0)) ;
console.log('') ;

console.log(name[0]) ;
console.log(name.charAt(0)) ;
console.log(String.fromCodePoint(26607)) ;
console.log(String.fromCharCode(26607)) ;
console.log('') ;

console.log(str.charCodeAt(1)) ;
```

```
console.log(str.codePointAt(1)) ;
console.log('') ;

console.log(str[1])
console.log(str.charAt(1)) ;
console.log(String.fromCodePoint(98)) ;
console.log(String.fromCharCode(98)) ;
console.log('') ;
```

【相关说明】

```
let name = '柯霖廷' ;
```

声明初始数据为字符串的变量 name。

```
let str = 'abcdefg' ;
```

声明初始数据为字符串的变量 str。

```
console.log(name.charCodeAt(0)) ;
```

因为变量 name 的数据为中文字符串'柯霖廷'，所以 name.charCodeAt(0)会返回中文字'柯'在【UTF-16】编码单元中的数码 26607。

在 UTF-16 编码单元中，其数码的范围仅介于 0 ~ 65536。

```
console.log(name.codePointAt(0)) ;
```

name.codePointAt(0)会返回中文字'柯'在【Unicode】编码单元中的数码 26607。在 Unicode 编码单元中，其数码可超过 65536，以表示更多文字。因为数码 26607 并未大于 65536，所以可看出中文字'柯'在【UTF-16】和【Unicode】编码单元的数码，均是相同的 26607。

```
console.log('') ;

console.log(name[0]) ;
console.log(name.charAt(0)) ;
```

name[0]和 name.charAt(0)具有相同的效果，均会返回变量 name 字符串的第 1 个中文字'柯'。

```
console.log(String.fromCodePoint(26607)) ;
```

String 对象支持的函数 fromCodePoint(26607)，会根据【Unicode】编码单元，返回被传入的数码 26607 所对应的文字'柯'。

```
console.log(String.fromCharCode(26607)) ;
```

String 对象支持的函数 fromCharCode(26607)，会根据【UTF-16】编码单元，返回被传入的数码 26607 所对应的文字'柯'。

```
console.log('') ;

console.log(str.charCodeAt(1)) ;
```

因为变量 str 的数据为字符串'abcdefg'，所以 str.charCodeAt(1)会返回字符'b'在【UTF-16】编码单元中的数码 98。

```
console.log(str.codePointAt(1)) ;
```

因为变量 str 的数据为字符串'abcdefg'，所以 str.CodePointAt(1)会返回字符'b'在【Unicode】编码单元中的数码 98。

```
console.log('') ;
console.log(str[1])
console.log(str.charAt(1)) ;
```

str[1]和 str.charAt(1)具有相同的效果，均会返回变量 str 字符串的第 2 个字符'b'。

```
console.log(String.fromCodePoint(98)) ;
```

String 对象支持的函数 fromCodePoint(98)，会根据【Unicode】编码单元，返回被传入的数码 98 所对应的文字'b'。

```
console.log(String.fromCharCode(98)) ;
```

String 对象支持的函数 fromCharCode(98)，会根据【UTF-16】编码单元返回被传入的数码 98 所对应的文字'b'。

7.1.10　重复填充子字符串于扩充后的字符串中（ES8）

如果欲在扩充之后的特定字符串里，重复填充新的子字符串，可调用 String 对象实例支持的函数 padStart()来实现。请看下面的示例。

【7-1-10-String-padStart.js】
```
let word01 = 'Happy ', word02 = ' Birthday' ;
let num = 567 ;

console.log(word01 + '^v^'.repeat(3) + word02.padStart(17, ' _ _')) ;
console.log('') ;
console.log(num.toString().padStart(16, '0')) ;
```

【相关说明】
```
let word01 = 'Happy ', word02 = ' Birthday' ;
```

此语句分别声明了初始数据为不同字符串的变量 word01 与 word02。

```
let num = 567 ;
```

声明初始数值为 567 的变量 num。

```
console.log(word01 + '^v^'.repeat(3) + word02.padStart(17, ' _ _')) ;
```

'^v^'.repeat(3)会产生字符串'^v^^v^^v^'。word02.padStart(17, ' _ _')会先根据变量 word02 的字符串' Birthday'的字符个数 9，以重复产生子字符串' _ _'，直到完成 17 - 9，也就是字符个数为 8 的字符串' _ _ _ _'，再衔接到字符串' Birthday'，成为字符个数为 17 的字符串' _ _ _ _ Birthday'。因此，word01 + '^v^'.repeat(3) + word02.padStart(17, ' _ _')，最终会返回字符串'Happy ^v^^v^^v^ _ _ _ _ Birthday'。

```
console.log('') ;
```

```
console.log(num.toString().padStart(16, '0')) ;
```

在此语句中，先由 num.toString()返回变量 num 的整数值 567 被转换之后的字符串'567'，再由 padStart(16, '0')，返回字符个数为 16 的字符串'0000000000000567'。

7.2　将冗长的字符串分割为多行（ES6）

在编程时，分割冗长的字符串到多行里，有助于舒缓编程人员的视觉负担。下面请看具体的示例。

【7-2---multiple-rows-of-strings.js】

```
let saying = 'Reading enriches our mind; conversation polishes it.' ;

console.log(saying) ;

saying = ['Reading', 'enriches', 'our', 'mind;', 'conversation', 'polishes',
  'it.'].join(' ') ;

console.log(saying) ;

saying = 'Reading enriches our mind; ' + 'conversation polishes it.' ;

console.log(saying) ;

saying = 'Reading enriches our mind; \
conversation polishes it.' ;

console.log(saying) ;

saying = 'Reading \
enriches \
our \
mind; \
conversation \
polishes \
it.' ;

console.log(saying) ;

saying = `Reading
enriches
our
mind;
conversation
polishes
it.`.split('\n').join(' ') ;

console.log(saying) ;
```

【相关说明】

```
let saying = 'Reading enriches our mind; conversation polishes it.' ;
```

声明初始数据为字符串的变量 saying。

```
console.log(saying) ;

saying = ['Reading', 'enriches', 'our', 'mind;', 'conversation', 'polishes',
  'it.'].join(' ') ;
```

此语句将数组实例内的多个字符串，连接（join）成为完整的字符串'Reading enriches our mind; conversation polishes it.'，并赋给变量 saying。

```
console.log(saying) ;

saying = 'Reading enriches our mind; ' + 'conversation polishes it.' ;
```

此语句将两个子字符串，串联成为新的字符串，再赋给变量 saying。

```
console.log(saying) ;

saying = 'Reading enriches our mind; \
conversation polishes it.' ;
```

通过反斜杠符号【\】，可跨多行地描述一个字符串。需特别注意的是，在【\】的右侧，不可再有包括任何空白字符在内的字符。此语句等价于将字符串'Reading enriches our mind; conversation polishes it.'，赋给变量 saying。

```
console.log(saying) ;

saying = 'Reading \
enriches \
our \
mind; \
conversation \
polishes \
it.' ;
```

此语句通过反斜杠符号【\】，衔接跨多行的字符串'Reading enriches our mind; conversation polishes it.'，并赋给变量 saying。

```
console.log(saying) ;

saying = `Reading
enriches
our
mind;
conversation
polishes
it.`.split('\n').join(' ') ;
```

在此语句中，其等号右侧的字符串，通过一对反引号（back quote）【`】，来保留其换行的效果。然后借助函数 split('\n')，以根据换行字符【\n】，来分割成为各个子字符串，并组成新数组实例["Reading", "enriches", "our", "mind;", "conversation", "polishes", "it."]，再交给函数 join(' ')来处理，重新将数组实例中的各字符串，组合成为最终的字符串"Reading enriches our mind; conversation polishes it."。

7.3 字符串的扩展运算（ES6）

通过扩展运算符（spread operator）【...】和中括号运算符[]，可打散特定字符串，成为在新数组实例中的**字符**（character）。

【7-3---string-spreadings.js】
```
let alphabet_string = 'abcdefghijklmnopqrstuvwxyz' ;

let letters = [... alphabet_string.toUpperCase()] ;

console.log(alphabet_string) ;
console.log(letters) ;
```

【相关说明】
```
let alphabet_string = 'abcdefghijklmnopqrstuvwxyz' ;
```
声明初始数据为字符串'abcdefghijklmnopqrstuvwxyz'的变量 alphabet_string。
```
let letters = [ ... alphabet_string.toUpperCase() ] ;
```
此语句声明了初始数据为数组实例的变量 letters。其中，alphabet_string.toUpperCase()可使得变量 alphabet_string 的数据字符串，被转换成为全部为大写字母的字符串"ABCDEFGHIJKLMNOPQRSTUVWXYZ"。通过扩展运算符【...】，让这个字符串被扩展在一组中括号[]当中，成为内含多个字符的数组实例["A", "B", "C", "D", "E", "F", "G", "H", "I", "J", "K", "L", "M", "N", "O", "P", "Q", "R", "S", "T", "U", "V", "W", "X", "Y", "Z"]。

7.4 字符串的插值格式化（ES6）

通过加号运算符【+】或函数 concat()，连接各字符串的任务，是相当烦琐的。不过，JavaScript 语言支持更为简单的操作方式：

- 在一对反引号【`】中，植入带有插值（interpolation）用途的各个子字符串【${变量名称或表达式}】，而成为模板（template）字符串。
- 再将上述模板字符串，赋给特定变量。

上述模板字符串，被转换完成之后，其变量的数据，自然会变成最终的字符串。关于字符串插值格式化的运用，请看下面的例子。

【7-4---string-interplation-formattings.js】

```
let subject = 'Jasper' ;
let object = 'Jessica' ;

let sentence = `${subject} lovingly loves lovely beloved of ${object}.` ;

console.log(sentence) ;
console.log('') ;

///
String.prototype.format = function ()
{
  let origin_string = this.valueOf() ;

  console.log(arguments) ;

  for (let prop in arguments)
  {
    origin_string = origin_string.replace(`{${prop}}`, arguments[prop]) ;
  }

  return origin_string ;
} ;

sentence = '{0} lovingly loves lovely beloved of {1}.'.format(subject, object) ;

console.log(sentence) ;
```

【相关说明】

```
let subject = 'Jasper' ;
let object = 'Jessica' ;
```

这两个语句分别声明了初始数据为不同字符串的变量 subject 与 object。

```
let sentence = `${subject} lovingly loves lovely beloved of ${object}.` ;
```

此语句声明了变量 sentence，其初始数据为字符串'Jasper lovingly loves lovely beloved of Jessica.'。插值是通过【${变量名称}】的形式，结合一般文本之后，被放置于一对反引号（back quote）【`】中的字符串里。模板字符串【`${subject} lovingly loves lovely beloved of ${object}.`】经过处理之后，会产生出最终的字符串'Jasper lovingly loves lovely beloved of Jessica.'。

```
console.log(sentence) ;
console.log('') ;

String.prototype.format = function ()
{
  let origin_string = this.valueOf() ;
```

内置的局部变量 this，被使用于自定义的函数 String.prototype.format()里。当'{0} lovingly loves lovely beloved of {1}.'.format(subject, object)被调用时，会间接让其内部的局部变量 this，指向内含字符串'{0} lovingly loves lovely beloved of {1}.'】的 String 对象实例，进而使得 this.valueOf()返回

String 对象实例的字符串'{0} lovingly loves lovely beloved of {1}.'。所以，此语句声明了初始数据为字符串'{0} lovingly loves lovely beloved of {1}.'的局部变量 origin_string。

```
console.log(arguments) ;
for (let prop in arguments)
```

在任意函数中的内置变量 arguments，是内含被传入的参数数据所构成的数组实例。举例来说，在语法【'{0} lovingly loves lovely beloved of {1}.'.format(subject, object)】中，调用了函数 format()，并传入了全局变量 subject 的数据字符串'Jasper'，以及变量 object 的数据字符串'Jessica'，使得内置变量 arguments 当前的数据，成为内含数组实例['Jasper', 'Jessica']的 Arguments 对象实例。

在 for ... in 的第 1 次迭代时，变量 prop 的数据为字符'0'，并不是数值 0；在其第 2 次迭代时，变量 prop 的数据则为字符'1'，而不是数值 1。

```
{
    origin_string = origin_string.replace(`{${prop}}`, arguments[prop]) ;
```

在语法【replace(`{${prop}}`, arguments[prop])】中，模板字符串【`{${prop}}`】在 for ... in 的第 1 次迭代时，被置换为字符串'{0}'，在其第 2 次迭代时，则会被置换为字符串'{1}'。换言之，在其第 1 次迭代时，【origin_string.replace(`{${prop}}`, arguments[prop])】会被置换为【'origin_string.'.replace('{0}', arguments[0])】，并等价于【'{0} lovingly loves lovely beloved of {1}.'.replace('{0}', 'Jasper')】，使得变量 orgin_string 的数据，进一步变成字符串'Jasper lovingly loves lovely beloved of {1}.'。

在其第 2 次迭代时，【origin_string.replace(`{${prop}}`, arguments[prop])】会被置换为【origin_string.'.replace('{1}', arguments[1])】，并等价于【'Jasper lovingly loves lovely beloved of {1}.'.replace('{1}', 'Jessica')】，使得变量 orgin_string 的数据，最后变成字符串'Jasper lovingly loves lovely beloved of Jessica.'。

```
}
return origin_string ;
```

此语句返回了变量 origin_string 的数据字符串'Jasper lovingly loves lovely beloved of Jessica.'。

```
} ;
```

通过语法【String.prototype.format = function () { ... } ;】，可动态定义任意 String 对象实例即将支持的函数 format()。

```
sentence = '{0} lovingly loves lovely beloved of {1}.'.format(subject, object) ;
```

在语法【'{0} lovingly loves lovely beloved of {1}.'.format(subject, object)】中，调用了函数 format(subject, object)，并返回了字符串'Jasper lovingly loves lovely beloved of Jessica.'。

7.5 原始字符串（ES6）

在所谓的原始（raw）字符串里，其控制字符（control character），例如'\f'、'\n'、'\t' 等，均不加以解析，直接视为普通字符！如果欲获取原始（raw）字符串，可按如下步骤操作：

步骤01 在一对反引号【`】中，植入带有插值（interpolation）用途的各个子字符串【${变量名称或表达式}】，以作为模板（template）字符串。

步骤02 在模板字符串的前方（左侧）冠上【String.raw】文本，代表以特殊方式，调用 String 对象中名称为 raw 的函数，例如【String.raw `▫▫▫ ${变量名称或表达式} ▫▫▫ ${变量名称或表达式} ▫▫▫`】。

步骤03 虽然 String.raw 被称为函数，但是衔接带有一对反引号【`】的模板字符串时，千万不可写成【String.raw(`▫▫▫`)】或是【String.raw(▫▫▫)】！

步骤04 将模板字符串，赋给特定变量。

步骤05 模板字符串被完成之后，特定变量的数据，就会变成最终的原始（raw）字符串。

关于原始字符串的运用，请看下面的示例。

【7-5---raw-strings.js】

```javascript
let amount = 7 ;
let str01 = `There are\n ${amount} categories.` ;
let str02 = String.raw `There are\n ${amount} categories.` ;

console.log(str01) ;
console.log(str02) ;

let str03 = String.raw({raw: 'abcd'}, '_', '_', '_') ;

console.log(str03) ;

str03 = String.raw({raw: 'abcd'}, ... '___') ;

console.log(str03) ;

let str04 = String.raw({raw: ['apple', 'banana', 'cherry', '']}, '= 15, ', '= 10, ', '= 18.') ;

console.log(str04) ;
```

【相关说明】

```
let amount = 7 ;
```

声明初始数值为 7 的变量 amount。

```
let str01 = `There are\n ${amount} categories.` ;
```

此语句声明了初始数据为【模板字符串`There are\n ${amount} categories.` 所产生的如下字符串】的变量 str01：

There are

7 categories.

```
let str02 = String.raw `There are\n ${amount} categories.` ;
```

此语句声明了初始数据为【模板字符串`There are\n ${amount} categories.`所产生的原始字符串'There are\n 7 categories.'】的变量 str02。

在原始字符串里，【\n】会被视为纯粹的字符【\】与【n】的组合，不再具有换行（line feed）的特征。

```
console.log(str01) ;
```

此语句显示出如下信息：

There are

　　7 categories.

```
console.log(str02) ;
```

此语句则仅仅显示出并无换行的【There are\n 7 categories.】的信息。

```
let str03 = String.raw({raw: 'abcd'}, '_', '_', '_') ;
```

此语句声明了初始数据为字符串'a_b_c_d'的变量 str03。调用 String 对象的函数 raw({raw: 'abcd'}, '_','_','_')，可使得字符串'abcd'和3个字符串'_'，被交错连接成为新的字符串'a_b_c_d'。

```
console.log(str03) ;

str03 = String.raw({raw: 'abcd'}, ... '___') ;
```

此语句将字符串'a_b_c_d'赋给变量 str03。调用 String 对象的函数 raw({raw: 'abcd'}, ... '___')，等价于调用 raw({raw: 'abcd'},'_','_','_')。换言之，单一参数【... '___'】，通过扩展运算符【...】，可被扩展成为3个参数的语法【'_','_','_'】。

```
console.log(str03) ;

let str04 = String.raw({raw: ['apple', 'banana', 'cherry', '']}, '= 15, ', '= 10, ', '= 18.') ;
```

此语句声明了初始数据为字符串'apple= 15, banana= 10, cherry= 18.'的变量 str04。调用 String 对象的函数 raw({raw: ['apple', 'banana', 'cherry', '']}, '= 15, ', '= 10, ', '= 18.')，可使得数组实例['apple', 'banana', 'cherry', '']中的各字符串元素，和3个字符串参数'= 15, '、'= 10, ' 和 '= 18.'，被交错连接成为新的字符串'apple= 15, banana= 10, cherry= 18.'。

7.6　正则表达式与黏性匹配（ES6）

在特定字符串中，寻找符合特定模式（pattern）的子字符串的操作，称为字符串的匹配（match）。正则表达式字面量（regular-expression literal）是用来进行字符串的匹配的模式字符串。

在 JavaScript 语言中，会自动将正则表达式字面量，转换成为 RegExp 对象实例，并支持内置的属性 lastIndex。

所谓的黏性匹配（sticky match），是指在特定字符串中，每次固定从属性 lastIndex 内含的索引值（index value）所对应的字符位置开始，进行仅仅一次的匹配】。换言之，在黏性匹配的过程中，前述索引值对应的字符位置，是【唯一】会被进行仅仅一次的匹配的开始位置。下面通过示例，说明正则表达式与黏性匹配的运用。

【7-6---regular-expression-sticky-matches.js】

```javascript
var user_input = 'z1 = x1 ^ 2 + y1 * 3 + 6' ;

// var pattern = new RegExp(/[a-zA-Z]\d/, 'g') ;
// var pattern = new RegExp(/[a-zA-Z]\d/g) ;
var pattern = /[a-zA-Z]\d/g ;

var matches = user_input.match(pattern) ;

console.log(matches) ;
console.log('') ;

///
matches = pattern.exec(user_input) ;

console.log(matches) ;

pattern.lastIndex = 1 ;

matches = pattern.exec(user_input) ;

console.log(matches) ;

pattern.lastIndex = 6 ;

matches = pattern.exec(user_input) ;

console.log(matches) ;
console.log('') ;

///
pattern = /[a-zA-Z]\d/y ;

pattern.lastIndex = 0 ;

matches = pattern.exec(user_input) ;

console.log(matches) ;

pattern.lastIndex = 1 ;

matches = pattern.exec(user_input) ;

console.log(matches) ;
```

【相关说明】

```javascript
var user_input = 'z1 = x1 ^ 2 + y1 * 3 + 6' ;
```

声明初始数据为字符串'z1 = x1 ^ 2 + y1 * 3 + 6'的变量 user_input。

```
// var pattern = new RegExp(/[a-zA-Z]\d/, 'g') ;
// var pattern = new RegExp(/[a-zA-Z]\d/g) ;
var pattern = /[a-zA-Z]\d/g ;
```

此语句声明了初始数据为正则表达式字面量/[a-zA-Z]\d/g 的变量 pattern。/[a-zA-Z]\d/g 用来匹配所有（g, global）1 个字母（[a-zA-Z]）衔接 1 个数字（\d）的组合，例如 x1、y1、z1 等。

```
var matches = user_input.match(pattern) ;
```

user_input.match(pattern)返回数组实例["z1", "x1", "y1"]，所以此语句声明了初始数据为数组实例["z1", "x1", "y1"]的变量 matches。其中，字符串"z1"、"x1"与"y1"，均是在变量 user_input 的数据字符串'z1 = x1 ^ 2 + y1 * 3 + 6'里，符合正则表达式字面量/[a-zA-Z]\d/g 的模式的子字符串。

```
console.log(matches) ;

matches = pattern.exec(user_input) ;
```

pattern.exec(user_input)返回数组实例["z1"]，所以此语句将内含单一元素的数组实例["z1"]，赋给变量 matches。其中，字符串"z1"是在变量 user_input 的数据字符串'z1 = x1 ^ 2 + y1 * 3 + 6'里，符合正则表达式字面量/[a-zA-Z]\d/g 的模式的第 1 个子字符串。

```
console.log(matches) ;

pattern.lastIndex = 1 ;
```

因为变量 pattern 当前内含正则表达式字面量，所以支持属性 lastIndex。此语句设置了变量 pattern 的属性 lastIndex 的数值为 1，代表在变量 user_input 的数据字符串里，从其索引值 1 的位置开始，向后进行匹配。

```
matches = pattern.exec(user_input) ;
```

pattern.exec(user_input)返回数组实例["x1"]，所以此语句将内含单一元素的数组实例["x1"]，赋给变量 matches。其中，字符串"x1"是在变量 user_input 的数据字符串'z1 = x1 ^ 2 + y1 * 3 + 6'里，符合正则表达式字面量/[a-zA-Z]\d/g 的模式的第 2 个子字符串。

```
console.log(matches) ;

pattern.lastIndex = 6 ;
```

此语句设置了变量 pattern 的属性 lastIndex 的数值为 6，代表在变量 user_input 的数据字符串里，从索引值 6 的位置开始，向后进行匹配。

```
matches = pattern.exec(user_input) ;
```

pattern.exec(user_input)返回数组实例["y1"]，所以此语句将内含单一元素的数组实例["y1"]，赋给变量 matches。其中，字符串"y1"是在变量 user_input 的数据字符串'z1 = x1 ^ 2 + y1 * 3 + 6'里，符合正则表达式字面量/[a-zA-Z]\d/g 的模式的第 3 个子字符串。

```
console.log('') ;

///
```

```
pattern = /[a-zA-Z]\d/y ;
```

此语句声明了初始数据为正则表达式字面量/[a-zA-Z]\d/y 的变量 pattern。上述字面量中的字母 y，意味着黏性匹配（y, sticky）的特征，代表在字符串'z1 = x1 ^ 2 + y1 * 3 + 6'中，每次均固定从属性 lastIndex 内含的索引值的位置开始，向后进行仅仅一次的正则表达式字面量/[a-zA-Z]\d/的匹配。

```
pattern.lastIndex = 0 ;
```

此语句设置了变量 pattern 的属性 lastIndex 的数值为 0，代表在变量 user_input 的数据字符串里，在索引值为 0 的位置，进行仅仅一次的匹配。

```
matches = pattern.exec(user_input) ;
```

pattern.exec(user_input) 返回数组实例["z1"]，所以此语句将内含单一元素的数组实例["z1"]，赋给变量 matches。其中，字符串"z1"是在变量 user_input 的数据字符串'z1 = x1 ^ 2 + y1 * 3 + 6'里，于索引值为 0 的位置上，刚好符合正则表达式字面量/[a-zA-Z]\d/y 的模式的子字符串。

```
console.log(matches) ;
pattern.lastIndex = 1 ;
```

此语句设置了变量 pattern 的属性 lastIndex 的数值为 1，代表在变量 user_input 的数据字符串里，于索引值为 1 的位置上，进行仅仅一次的匹配。

```
matches = pattern.exec(user_input) ;
```

patter.exec(user_input)在此返回空值 null，所以此语句将 null 赋给变量 matches，代表在变量 user_input 的数据字符串里，进行正则表达式字面量/[a-zA-Z]\d/y 的模式的匹配。并没有成功。

7.7 万国码字面量（ES6）

西方文字的字符种类较少，中文字符的种类则相当繁多，所以中文字在万国码（Unicode）中的编码长度，总是比西方文字，占用更多的比特位（bit）个数。其中，生僻字的编码长度，比起常用字，占用更多的比特位。中文字的 Unicode 数码，在 JavaScript 语言里，可通过万国码字面量'\u{特定中文字的数码}'来表示特定的中文字。关于万国码字面量的调试，可参考如下示例。

【7-7---unicode-literals.js】

```
var word01 = '柯' ;
var word02 = '𢒉' ;

console.log(word01.length) ;
console.log(word01.codePointAt(0).toString(16).toUpperCase()) ;
console.log(word01 === '\u{67EF}') ;
console.log(word01) ;
console.log('') ;

///
```

```
        console.log(word02.length) ;
        console.log(word02.codePointAt(0).toString(16)) ;
        console.log(word02.codePointAt(1).toString(16).toUpperCase()) ;
        console.log(word02 === '\u{26402}') ;
        console.log(word02) ;
        console.log('') ;

        console.log(word02.charAt(0)) ;
        console.log(word02.charAt(1)) ;
        console.log(word02.charAt(0) + word02.charAt(1)) ;
```

【相关说明】

```
        var word01 = '柯' ;
        var word02 = '𦐂' ;
```

此语句声明了初始数据为不同中文字的变量 word01 与 word02。

```
        console.log(word01.length) ;
```

因为变量 word01 内含的中文字'柯'，在【UTF-16】编码单元中，其数码为十进制数值 26607（十六进制数码 67EF），小于十进制数值 2^{16}=65536，所以仅被视为一个字符。

```
        console.log(word01.codePointAt(0).toString(16).toUpperCase()) ;
```

首先，word01.codePointAt(0)返回数值 26607；再由 toString(16)处理，而返回十六进制数码的字符串'67ef'；最后由 toUpperCase()转换，而返回其大写字母的字符串'67EF'。

```
        console.log(word01 === '\u{67EF}') ;
```

【word01 === '\u{67EF}'】的结果值为布尔值 true，代表变量 word01 内含的中文字'柯'与万国码字面量'\u{67EF}'，不仅其数据类型均为字符串（string），其所代表的文字，亦被视为相同。其中，\u 代表万国码（Unicode）的含义。

```
        console.log(word01) ;

        console.log(word02.length) ;
```

变量 word02 内含的中文字'𦐂'，并不存在于【UTF-16】编码单元里，仅存在于【Unicode】编码单元中，且其数码为十进制数值 156674（十六进制数码 26402），大于十进制数值 2^{16} = 65536，所以被视为两个字符。

```
        console.log(word02.codePointAt(0).toString(16)) ;
```

word02.codePointAt(0)在此返回数值 156674，再由 toString(16)处理，而返回其十六进制数码的字符串'26402'。

```
        console.log(word02.codePointAt(1).toString(16).toUpperCase()) ;
```

word02.codePointAt(1)在此则返回数值 56322，再由 toString(16)处理，而返回十六进制数码的字符串'dc02'；最后由 toUpperCase()转换，而返回其大写字母的字符串'DC02'。

```
        console.log(word02 === '\u{26402}') ;
```

【word02 === '\u{26402}'】的结果值为布尔值 true，代表变量 word02 内含的中文字'𦐂'与

'\u{26402}'，不仅其数据类型均为字符串（string），其所代表的文字，亦被视为相同。其中，\u 代表万国码（Unicode）的含义。

```
console.log(word02) ;
```

```
console.log(word02.charAt(0)) ;
```

word02.charAt(0)会返回字符'�'，代表此字符无法单独被完整呈现。

```
console.log(word02.charAt(1)) ;
```

word02.charAt(1)也会返回字符'�'，代表此字符无法单独被完整呈现。

```
console.log(word02.charAt(0) + word02.charAt(1)) ;
```

【word02.charAt(0) + word02.charAt(1)】会返回可被完整呈现的文字'𦐂'。

7.8 练 习 题

1. 已知变量 str01 的数据为字符串'今晨明曦暖春风，午后炙意似晴空，夕刻彩晖遍地羞，夜来寂阑愁思浓。'，试问 str01.indexOf('，')、str01.lastIndexOf('，') 和 str01.indexOf('晖')分别会返回什么？

2. 已知变量 str02 的数据为内含一首绝句或律诗的字符串，例如'建筑千层塔，通达万里跋，绽露盈满华，顶天立宏霸。'，并且已知简易辨识单一中文字的正则表达式【[\u4E00-\u9FA5]】。试编写 JavaScript 源代码，以判断变量 str02 的数据字符串为[1]绝句或律诗，以及[2]五言或七言的情况。

3. 已知变量 str03 的数据为【内含以逗点隔开的上底长度、下底长度和高度】的字符串，例如【'上底：15，下底：25，高度：35'】。除了其中的数值会变动之外，其余文本都是固定格式的。试编写 JavaScript 源代码，以计算并显示其梯形面积。

4. 已知变量 str04 的数据为【带有以逗号隔开的人名词汇】的字符串，例如【'zelda, peter, paul, jasper, alex, daisy, eric, john, stella, tommy, adam, betty, sean, frank, kevin, sammy, julia, yolanda, william'】。试编写 JavaScript 源代码，以重新输出并显示【按字母升序，且开头字母为大写】的每个人名词汇。

5. 试编写 JavaScript 源代码，以动态产生带有 1～1000 编号的如下字符串：

```
no. 0001
no. 0002
no. 0003
    ⋮
no. 0998
no. 0999
no. 1000
```

第 8 章

处理数组

在许多编程语言中,数组(array)一直是相当重要的数据类型。通过数据为特定数组实例(array instance)的变量,可灵活访问特定数组实例中**各元素**的数据。因此,处理特定数组实例和其各元素的任务,是相当常见而重要的。

8.1 Array 对象

在 Array 对象里,内置许多处理其各个元素的函数,以及少许的属性。

8.1.1 创建特定数组的副本(ES6)

欲复制特定数组实例,可通过 Array 对象支持的函数 of() 或 valueOf() 来实现。关于创建特定数组的副本的综合运用,可参看如下两个示例。

【8-1-1-e1-Array-of.js】

```
let array01 = Array.of(1, 'two', 3, 'four', 5, 'six') ;
let array02 = [1, 'two', 3, 'four', 5, 'six'] ;
let array03 = Array(1, 'two', 3, 'four', 5, 'six') ;
let array04 = new Array(1, 'two', 3, 'four', 5, 'six') ;

console.log(array01) ;
console.log(array02) ;
console.log(array03) ;
console.log(array04) ;
console.log('') ;
```

```
    let array05 = Array.of('January') ;
    let array06 = ['January'] ;
    let array07 = Array('January') ;
    let array08 = new Array('January') ;

    console.log(array05) ;
    console.log(array06) ;
    console.log(array07) ;
    console.log(array08) ;
    console.log('') ;

    ///
    let array09 = Array.of(5) ;
    let array10 = [5] ;

    let array11 = Array(5) ;
    let array12 = new Array(5) ;

    console.log(array09) ;
    console.log(array10) ;
    console.log(array11) ;
    console.log(array12) ;
```

【相关说明】

```
    let array01 = Array.of(1, 'two', 3, 'four', 5, 'six') ;
    let array02 = [1, 'two', 3, 'four', 5, 'six'] ;
    let array03 = Array(1, 'two', 3, 'four', 5, 'six') ;
    let array04 = new Array(1, 'two', 3, 'four', 5, 'six') ;
```

这 4 个语句声明了初始数据均为数组实例[1, "two", 3, "four", 5, "six"]的 4 个变量 array01 ~ array04。

```
    console.log(array01) ;
    console.log(array02) ;
    console.log(array03) ;
    console.log(array04) ;
    console.log('') ;

    let array05 = Array.of('January') ;
    let array06 = ['January'] ;
    let array07 = Array('January') ;
    let array08 = new Array('January') ;
```

这 4 个语句声明了初始数据均为数组实例["January"]的 4 个变量 array05 ~ array08。

```
    console.log(array05) ;
    console.log(array06) ;
    console.log(array07) ;
    console.log(array08) ;
    console.log('') ;

    let array09 = Array.of(5) ;
    let array10 = [5] ;
```

此两个语句声明了初始数据均为数组实例[5]的 2 个变量 array09 与 array10。

```
let array11 = Array(5) ;
let array12 = new Array(5) ;
```

这两个语句声明了初始数据均为内含 5 个空元素的数组实例的变量 array11 与 array12。

【8-1-1-e2-Array-valueOf.js】

```
let a01 = ['1st', '2nd', '3rd', '4th'] ;
let a02 = ['1st', '2nd', '3rd', '4th'] ;

console.log(a01 == a02) ;

let a03 = a01 ;

console.log(a01 == a03) ;
console.log('') ;

let a04 = a01.valueOf() ;

console.log(a01 == a01.valueOf()) ;
console.log(a01 == a04) ;
console.log('') ;

console.log(a01) ;
console.log(a04) ;
```

【相关说明】

```
let a01 = ['1st', '2nd', '3rd', '4th'] ;
let a02 = ['1st', '2nd', '3rd', '4th'] ;
```

这两个语句声明了初始数据均为数组实例['1st', '2nd', '3rd', '4th']的变量 a01 与 a02。

```
console.log(a01 == a02) ;
```

变量 a01 与 a02 的数据，虽然是其内容皆相同的数组实例；但是其所占用的内存地址并不相同，所以表达式【a01 == a02】的结果值为 false。

```
let a03 = a01 ;
```

此语句声明了初始数据为【变量 a01 的数组实例】的变量 a03。换言之，此语句使得变量 a03 与 a01 占用相同的内存地址。

```
console.log(a01 == a03) ;
```

因为变量 a01 与 a03 当前占用相同的内存地址，所以表达式【a01 == a03】的结果值为 true。

```
console.log('') ;

let a04 = a01.valueOf() ;
```

此语句声明了初始数据为【变量 a01 的数组实例】的变量 a04。在 JavaScript 引擎中，几乎所有内置的对象的实例，都支持函数 valueOf()，可用来取出特定对象实例中的原始数据（primitive data）。然而，对于 Array 对象的实例来说，其实例和【其实例经过函数 valueOf()处理之后所返回的新实例】，却占用相同的内存地址。

```
console.log(a01 == a01.valueOf()) ;
console.log(a01 == a04) ;
```

这两个语句均显示出布尔值 true，意味着当前变量 a01 的数组实例和变量 a04 的数组实例，以及函数 a01.valueOf()所返回的新实例，均是对应到相同内存地址的数组实例。

8.1.2　创建来自可迭代对象的新数组（ES6）

欲复制其他可迭代对象实例的当前数据，成为新数组实例的数据，可通过 Array 对象支持的函数 from()或 map()来实现。请看下面的两个例子。

【8-1-2-e1-Array-from.js】

```
let word = 'Happiness' ;
let a01 = Array.from(word) ;
let a02 = [... word] ;
let a03 = word.split('') ;

console.log(a01) ;
console.log(a02) ;
console.log(a03) ;
console.log('') ;

///
let r1 = 12, r2 = 15, r3 = 18 ;
let circle_area01 = r => (Math.PI * r ** 2).toFixed(2) ;
let a04 = Array.from([r1, r2, r3], circle_area01) ;
let a05 = Array.from([r1, r2, r3], r => (Math.PI * r ** 2).toFixed(2)) ;

console.log(a04) ;
console.log(a05) ;
console.log('') ;

function circle_area02(r)
{
  return (Math.PI * r ** 2).toFixed(2) ;
}

let a06 = Array.from([r1, r2, r3], circle_area02) ;

console.log(a06) ;
console.log('') ;

///
let a07 = Array.from({length: 5}, (value, index) => index ** 3) ;

console.log(a07) ;
```

【相关说明】

```
let word = 'Happiness' ;
```

声明初始数据为字符串的变量 word。

```
let a01 = Array.from(word) ;
```

此语句声明了初始数据为内含字符的数组实例["H", "a", "p", "p", "i", "n", "e", "s", "s"]的变量 a01。通过 Array 对象的函数 from(word)，可根据变量 word 中的字符串'Happiness'，创建由其各字符所构成的数组实例["H", "a", "p", "p", "i", "n", "e", "s", "s"]。

```
let a02 = [... word] ;
```

【... word】等价于【Array.from(word)】。此语句声明了初始数据为内含字符的数组实例["H", "a", "p", "p", "i", "n", "e", "s", "s"]的变量 a02。

```
let a03 = word.split('') ;
```

【word.split("")】也等价于【Array.from(word)】。此语句声明了初始数据为内含字符数组实例["H", "a", "p", "p", "i", "n", "e", "s", "s"]的变量 a03。

```
console.log(a01) ;
console.log(a02) ;
console.log(a03) ;
console.log('') ;

let r1 = 12, r2 = 15, r3 = 18 ;
```

此语句声明了初始数值为不同整数值的变量 r1、r2 与 r3。

```
let circle_area01 = r => (Math.PI * r ** 2).toFixed(2) ;
```

此语句声明了初始数据为箭头函数【r => (Math.PI * r ** 2).toFixed(2)】的定义的变量 circle_area01。此语句如同定义了函数 circle_area01(r)，用来计算出特定半径值的圆面积，并精确至小数第 2 位。

```
let a04 = Array.from([r1, r2, r3], circle_area01) ;
```

此语句声明了初始数据为数组实例["452.39", "706.86", "1017.88"]的变量 a04。其中，Array.from([r1, r2, r3], circle_area01)会使得变量 r1、r2 与 r3 内含的半径值，分别被传入函数 circle_area01()中，计算并返回内含其各个圆面积的数组实例["452.39", "706.86", "1017.88"]。

```
let a05 = Array.from([r1, r2, r3], r => (Math.PI * r ** 2).toFixed(2)) ;
```

此语句声明了初始数据为数组实例["452.39", "706.86", "1017.88"]的变量 a05。其中，Array.from([r1, r2, r3], r => (Math.PI * r ** 2).toFixed(2))会使得变量 r1、r2 与 r3 内含的半径值，分别被传入箭头函数【r => (Math.PI * r ** 2).toFixed(2)】中，并返回内含其各个圆面积的数组实例["452.39", "706.86", "1017.88"]。

```
console.log(a04) ;
console.log(a05) ;
console.log('') ;

function circle_area02(r)
{
  return (Math.PI * r ** 2).toFixed(2) ;
}
```

此语法定义了带有参数 r 的函数 circle_area02()，亦可用来计算出特定半径值的圆面积，并精确至小数第 2 位。

```
let a06 = Array.from([r1, r2, r3], circle_area02) ;
```

此语句声明了初始数据为数组实例["452.39", "706.86", "1017.88"]的变量 a06。其中，Array.from([r1, r2, r3], circle_area02)会使得变量 r1、r2 与 r3 内含的半径值，分别被传入函数 circle_area02()中，计算并返回内含其各个圆面积的数组实例["452.39", "706.86", 1017.88"]。

```
console.log(a06) ;
console.log('') ;

let a07 = Array.from({length: 5}, (value, index) => index ** 3) ;
```

此语句声明了初始数据为内含 5 个整数值的数组实例[0, 1, 8, 27, 64]的变量 a07。其中，【Array.from({length: 5}, (value, index) => index ** 3)】会先行创建存在 5 个空元素的新数组实例，再通过箭头函数【(value, index) => index ** 3)】，依次计算出当前参数 index 的数值的 3 次方值，作为各个空元素的新数值；最后返回内含其各个数值的 3 次方值的数组实例[0, 1, 8, 27, 64]。

【8-1-2-e2-Array-map.js】

```
let radius_list = [10, 18, 24, 50, 80] ;

let circle_areas = radius_list.map((r) => (Math.PI * r ** 2).toFixed(2)) ;

console.log(circle_areas) ;

circle_areas = Array.from(radius_list, (r) => (Math.PI * r ** 2).toFixed(2)) ;

console.log(circle_areas) ;
console.log('') ;

///
let bases = [1, 4, 9, 16, 25, 36, 49, 64, 81, 100] ;

let roots = bases.map(Math.sqrt) ;

console.log(roots) ;

roots = bases.map((b) => Math.sqrt(b)) ;

console.log(roots) ;

roots = bases.map((b) => b ** 0.5) ;

console.log(roots) ;
```

【相关说明】

```
let radius_list = [10, 18, 24, 50, 80] ;
```

此语句声明了初始数据为内含 5 个整数值的数组实例的变量 radius_list。其中，各整数值代表不同的半径值。

```
let circle_areas = radius_list.map((r) => (Math.PI * r ** 2).toFixed(2)) ;
```

此语句声明了变量 circle_areas，其初始数据为内含不同半径值的圆面积的数组实例["314.16", "1017.88", "1809.56", "7853.98", "20106.19"]。在此，被传入函数 map()的数据为匿名函数【(r) => (Math.PI * r ** 2).toFixed(2)】的定义，用来返回精确至小数第 2 位的特定圆面积的结果值。

因为变量 radius_list 的数据为数组实例，所以支持函数【map(匿名函数的定义或者特定函数名称)】，可将变量 radius_list 数组实例中的各整数值，间接通过函数 map()，传递给箭头函数【(r) => (Math.PI * r ** 2).toFixed(2)】。

```
console.log(circle_areas) ;

circle_areas = Array.from(radius_list, (r) => (Math.PI * r ** 2).toFixed(2)) ;
```

此语句将内含不同半径值的圆面积的数组实例["314.16", "1017.88", "1809.56", "7853.98", "20106.19"]，赋给变量 circle_areas。

```
console.log(circle_areas) ;
console.log('') ;

let bases = [1, 4, 9, 16, 25, 36, 49, 64, 81, 100] ;
```

此语句声明了变量 bases，其初始数据为内含 1～10 的平方值的数组实例。

```
let roots = bases.map(Math.sqrt) ;
```

在此，被传入函数 map()的数据为 Math 对象的内置函数 sqrt 的名称，用来返回特定整数值的平方根值。此语句声明了变量 roots，其初始数据为内含不同平方根值的数组实例[1, 2, 3, 4, 5, 6, 7, 8, 9, 10]。其中，可通过函数 map()，间接将变量 bases 的数组实例[1, 4, 9, 16, 25, 36, 49, 64, 81, 100]中的各整数值，传递给内置函数 Math.sqrt()。

```
console.log(roots) ;

roots = bases.map((b) => Math.sqrt(b)) ;
```

此语句将内含不同平方根值的数组实例[1, 2, 3, 4, 5, 6, 7, 8, 9, 10]，赋给变量 roots。
在此，【bases.map((b) => Math.sqrt(b))】和【bases.map(Math.sqrt)】具有相同的效果。

```
console.log(roots) ;

roots = bases.map((b) => b ** 0.5) ;
```

此语句也将内含不同平方根值的数组实例[1, 2, 3, 4, 5, 6, 7, 8, 9, 10]，赋给变量 roots。在此，【bases.map((b) => b ** 0.5)】和【bases.map(Math.sqrt)】也具有相同的效果。

8.1.3　数组元素数据所构成的字符串

欲将特定数组实例中各元素的数据，连接成为新的字符串，可通过 Array 对象支持的函数 toString()或 join()来实现。下面通过示例来加以调试。

【8-1-3-Array-toString-and-join.js】

```
let fruits = ['apricot', 'blueberry', 'cranberry', 'durian' ] ;
```

```
let str01 = fruits.toString() ;
let str02 = fruits.join(',') ;
let str03 = fruits.join() ;

console.log(str01) ;
console.log(str02) ;
console.log(str03) ;
console.log('') ;

///
let str04 = fruits.join('  ') ;

console.log(str04) ;
```

【相关说明】

```
let fruits = ['apricot', 'blueberry', 'cranberry', 'durian' ] ;
```

此语句声明了初始数据为内含多个字符串的数组实例的变量 fruits。

```
let str01 = fruits.toString() ;
let str02 = fruits.join(',') ;
let str03 = fruits.join() ;
```

这3个语句声明了初始数据为相同字符串'apricot,blueberry,cranberry,durian'的变量 str01、str02 与 str03。换言之，fruits.toString()、fruits.join(',') 与 fruits.join()在本示例中，具有相同的效果，均会返回【以逗号作为分隔符，并连接（join）变量 fruits 的数组实例中各字符串】的新字符串'apricot,blueberry,cranberry,durian'。

```
console.log(str01) ;
console.log(str02) ;
console.log(str03) ;
console.log('') ;

let str04 = fruits.join('  ') ;
```

此语句声明了初始数据为字符串'apricot blueberry cranberry durian'的变量 str04。fruits.join(' ')会返回【以2个空格字符' '，作为分隔符，并连接变量 fruits 的数组实例中各字符串】的新字符串'apricot blueberry cranberry durian'。

8.1.4 数组元素的放入和取出

欲放入新元素的数据，至特定数组实例中，可通过 Array 对象实例支持的函数 push()或 unshift()。欲取出特定数组实例中的元素，可通过 Array 对象实例支持的函数 pop()或 shift()来实现。下面通过示例进行说明。

【8-1-4-accessing-of-array-elements.js】

```
let balls = ['baseball', 'basketball'] ;

balls.push('marble') ;
```

```
    let new_length = balls.push('dodgeball', 'fireball', 'Earth' ) ;

    console.log(balls.length == new_length) ;

    console.log(balls) ;
    console.log('') ;

    ///
    let drawed_data = balls.pop() ;

    console.log(drawed_data) ;
    console.log(balls) ;
    console.log('') ;

    ///
    drawed_data = balls.shift() ;

    console.log(drawed_data) ;
    console.log(balls) ;
    console.log('') ;

    ///
    balls.unshift('Pluto') ;
    balls.unshift('Moon', 'Mars') ;

    console.log(balls) ;
```

【相关说明】

```
    let balls = ['baseball', 'basketball'] ;
```

此语句声明了初始数据为内含 2 个字符串的数组实例的变量 balls。

```
    balls.push('marble') ;
```

因为变量 balls 当前的数据为数组实例，所以支持函数 push()。语法 balls.push('marble')使得变量 balls 的数组实例在其尾端，被新增了 1 个数据为字符串'marble'的元素。

```
    let new_length = balls.push('dodgeball', 'fireball', 'Earth' ) ;
```

此语句声明了初值始为整数值 6 的变量 new_length。其中，balls.push('dodgeball', 'fireball', 'Earth')使得变量 balls 的数组实例，被新增其数据分别为字符串'dodgeball'、'fireball'与'Earth'的 3 个元素，而成为带有 6 个元素的数组实例["baseball", "basketball", "marble", "dodgeball", "fireball", "Earth"]，最后返回其元素个数 6。

```
    console.log(balls.length == new_length) ;
```

balls.length 也会返回当前变量 balls 的数组实例中的元素个数 6，所以其数值和变量 new_length 的数值是相等的。

```
    console.log(balls) ;
    console.log('') ;

    let drawed_data = balls.pop() ;
```

此语句声明了变量 drawed_data，其初始数据为【在变量 balls 的数组实例中，其尾端当前的元素】的数据字符串'Earth'。其中，语法 balls.pop()会取出并返回在变量 balls 的数组实例中，其尾端当前的元素的数据字符串'Earth'，使得变量 balls 的数组实例，变更为["baseball", "basketball", "marble", "dodgeball", "fireball"]。

```
console.log(drawed_data) ;
console.log(balls) ;
console.log('') ;

drawed_data = balls.shift() ;
```

此语句将【在变量 balls 的数组实例中，取出其**前端**当前的元素】的数据字符串'baseball'，并赋给变量 drawed_data。被取出其**前端**当前的元素之后，变量 balls 的数组实例变更为["basketball", "marble", "dodgeball", "fireball"]。

```
console.log(drawed_data) ;
console.log(balls) ;
console.log('') ;

balls.unshift('Pluto') ;
```

此语句在其**前端**，新增其数据为字符串'Pluto'的元素之后，变量 balls 的数组实例变更为["Pluto", "basketball", "marble", "dodgeball", "fireball"]。

```
balls.unshift('Moon', 'Mars') ;
```

此语句在其**前端**，新增其数据分别为字符串'Moon'与'Mars'的两个元素之后，变量 balls 的数组实例变更为["Moon", "Mars", "Pluto", "basketball", "marble", "dodgeball", "fireball"]。

8.1.5 新增或删除数组的多个元素

欲在特定数组实例中，新增或删除连续的多个元素，可通过 Array 对象实例支持的函数 splice()来实现。请看下面的示例。

【8-1-5-Array-splice.js】

```
let vegetables = ['Lettuce', 'broccoli', 'spinach', 'kale', 'cabbage', 'onion', 'green beans'] ;

console.log(vegetables) ;
console.log('') ;

let drawed_list = vegetables.splice(1, 2) ;

console.log(drawed_list) ;
console.log(vegetables) ;
console.log('') ;

vegetables.splice(1, 0, ... drawed_list) ;

console.log(vegetables) ;
console.log('') ;
```

```
vegetables.splice(3, 0, 'garlic') ;

console.log(vegetables) ;
```

【相关说明】

```
let vegetables = ['Lettuce', 'broccoli', 'spinach', 'kale', 'cabbage', 'onion', 'green beans'] ;
```

声明初始数据为内含多个字符串的数组实例的变量 vegetables。

```
console.log(vegetables) ;
console.log('') ;

let drawed_list = vegetables.splice(1, 2) ;
```

此语句声明了初始数据为数组实例["broccoli", "spinach"]的变量 drawed_list。其中，变量 vegetables 当前的数据为数组实例，所以支持函数 splice()，可在数组实例中，新增或删除连续的多个元素。vegetables.splice(1, 2)会取出【在变量 vegetables 的数组实例中，从其索引值为 1 开始的 2 个元素，并放入被返回的新数组实例["broccoli", "spinach"]中。因此，执行完此语句之后，变量 vegetables 的初始数据，成为了数组实例["Lettuce", "kale", "cabbage", "onion", "green beans"]。

```
console.log(drawed_list) ;
console.log(vegetables) ;
console.log('') ;

vegetables.splice(1, 0, ... drawed_list) ;
```

因为扩展运算符【...】的缘故，【vegetables.splice(1, 0, ... drawed_list)】等价于【vegetables.splice(1, 0, "broccoli", "spinach")】。其中，vegetables.splice(1, 0, ... drawed_list)会在变量 vegetables 的数组实例中，从索引值 1（第 2 个）的元素'kale'开始，在不删除任何元素（0）的情况下，插入来自【... drawed_list】扩展出来的数据字符串"broccoli"与"spinach"。执行完此语句之后，变量 vegetables 的数组实例，从["Lettuce", "kale", "cabbage", "onion", "green beans"]，变更为["Lettuce", "broccoli", "spinach", "kale", "cabbage", "onion", "green beans"]。

```
console.log(vegetables) ;
console.log('') ;

vegetables.splice(3, 0, 'garlic') ;
```

vegetables.splice(3, 0, 'garlic')会在变量 vegetables 的数组实例中，从索引值 3 的元素'kale'开始，在不删除任何元素（0）的情况下，插入其数据为字符串"garlic"的新元素。执行完此语句之后，变量 vegetables 的数组实例，从["Lettuce", "broccoli", "spinach", "kale", "cabbage", "onion", "green beans"]变更为["Lettuce", "broccoli", "spinach", "garlic", "kale", "cabbage", "onion", "green beans"]。

8.1.6 合并多个数组

欲将多个数组实例，合并成为新的数组实例，可通过 Array 对象实例支持的函数 concat()来实现。请看下面的示例。

【8-1-6-Array-concat.js】

```
let a01 = [1, 3, 5] ;
let a02 = [2, 4, 6] ;

let a03 = ['x', 'y', 'z'] ;

let combination = a01.concat(a02, a03) ;

console.log(combination) ;
console.log('') ;

///
combination = [... a01, ... a02, ... a03] ;

console.log(combination) ;
console.log('') ;
```

【相关说明】

```
let a01 = [1, 3, 5] ;
let a02 = [2, 4, 6] ;
```

这两个语句声明了初始数据为【分别内含 3 个整数值的数组实例】的变量 a01 与 a02。

```
let a03 = ['x', 'y', 'z'] ;
```

声明初始数据为内含 3 个字符的数组实例的变量 a03。

```
let combination = a01.concat(a02, a03) ;
```

声明变量 combination，其初始数据为合并 3 个变量 a01、a02 与 a03 的各个数组实例之后的新数组实例[1, 3, 5, 2, 4, 6, 'x', 'y', 'z']。

```
console.log(combination) ;
console.log('') ;
```

```
combination = [... a01, ... a02, ... a03] ;
```

在此语句中，其等号右侧的[... a01, ... a02, ... a03]等价于 a01.concat(a02, a03)。其中，通过扩展运算符【...】，使得变量 combination 的数据，再次成为数组实例[1, 3, 5, 2, 4, 6, 'x', 'y', 'z']。

8.1.7 切割数组

欲在特定数组实例中，复制数个连续的元素，放置于新的数组实例，可通过 Array 对象实例支持的函数 slice() 来达成。请看下面的示例。

【8-1-7-Array-slice.js】

```
let vegetables = ['Lettuce', 'broccoli', 'spinach', 'kale', 'cabbage', 'onion', 'green beans'] ;

let sliced_list = vegetables.slice(2) ;

console.log(sliced_list) ;
console.log('') ;
```

```
sliced_list = vegetables.slice(3, 6) ;

console.log(sliced_list) ;
console.log('') ;

sliced_list = vegetables.slice(-2) ;

console.log(sliced_list) ;
console.log('') ;
```

【相关说明】

```
let vegetables = ['Lettuce', 'broccoli', 'spinach', 'kale', 'cabbage', 'onion', 'green beans'] ;
```

声明初始数据为内含多个字符串的数组实例的变量 vegetables。

```
let sliced_list = vegetables.slice(2) ;
```

此语句声明了初始数据为数组实例["spinach", "kale", "cabbage", "onion", "green beans"]的变量 sliced_list。其中，vegetables.slice(2)会【在变量 vegetables 的数组实例中，分割（slice）而复制出从索引值为 2（第 3 个）的元素"spinach"开始，到最后一个元素"green beans"为止】的新数组实例 ["spinach", "kale", "cabbage", "onion", "green beans"]。

```
console.log(sliced_list) ;
console.log('') ;

sliced_list = vegetables.slice(3, 6) ;
```

此语句将数组实例["kale", "cabbage", "onion"]，赋给变量 sliced_list。其中，vegetables.slice(3, 6)会【在变量 vegetables 的数组实例中，分割而复制出从索引值为 3（第 4 个）的元素"kale"开始，到索引值为 5（第 6 个）的元素"onion"为止】的新数组实例["kale", "cabbage", "onion"]。

```
console.log(sliced_list) ;
console.log('') ;

sliced_list = vegetables.slice(-2) ;
```

此语句将数组实例["onion", "green beans"]，赋给变量 sliced_list。其中，vegetables.slice(-2)会【在变量 vegetables 的数组实例中，分割而复制出从索引值为-2（倒数第 2 个）的元素"onion"开始，到最后一个元素"green beans"为止】的新数组实例["onion", "green beans"]。

8.1.8　寻找符合特定条件的数组元素（ES6）

欲在特定数组实例中，找出符合特定条件的第 1 个元素的数据（data）或索引值（index value），可通过 Array 对象实例支持的函数 find()或 findIndex()来实现。请看下面的示例。

【8-1-8-Array-find-and-findIndex.js】

```
let numbers = [10, 30, 20, 55, 15, 70, 23] ;

numbers.sort() ;
```

```
console.log(numbers) ;

// greater than 30
let gt3 = n => n > 30 ;

console.log(numbers.find(gt3)) ;
console.log(numbers.findIndex(gt3)) ;
```

【相关说明】

```
let numbers = [10, 30, 20, 55, 15, 70, 23] ;
```

声明初始数据为内含多个整数值的数组实例的变量 numbers。

```
numbers.sort() ;
```

此语句将变量 numbers 的数组实例的各元素值,由小至大排序,使得其数组实例成为[10, 15, 20, 23, 30, 55, 70]。

```
console.log(numbers) ;

// greater than 30
let gt3 = n => n > 30 ;
```

此语句声明了初始数据为箭头函数【n => n > 30】的定义的变量 gt3。换言之,变量 gt3 就如同函数名称。表达式【n > 30】的结果值为布尔值 true 或者 false,会成为函数 gt3(n)的返回值。

```
console.log(numbers.find(gt3)) ;
```

numbers.find(gt3)会使得变量 numbers 的数组实例中各元素的整数值,间接被传递给名称为 gt3 的函数,成为其参数 n 的数值,并被判断是否为大于 30 的**第 1 个**整数值。若特定元素的整数值是大于 30 的**第 1 个**整数值,则会成为 numbers.find(gt3)的返回值;在此,其返回值为 55。

```
console.log(numbers.findIndex(gt3)) ;
```

numbers.findIndex(gt3)会使得变量 numbers 的数组实例中各元素的整数值,间接被传递给名称为 gt3 的函数,成为其参数 n 的数值,并被判断是否为大于 30 的**第 1 个**整数值。若特定元素的整数值是大于 30 的**第 1 个**整数值,则其元素所对应的索引值会成为 numbers.findIndex(gt3)的返回值;在此,其返回值为 5。

8.1.9 自我复制并覆盖数组的部分元素(ES6)

欲在特定数组实例中,先行自我复制连续的数个元素,并用来覆盖到特定索引位置开始的数个元素,则可通过 Array 对象实例支持的函数 copyWithin()来实现。请看下面的示例。

【8-1-9-Array-copyWithin.js】

```
let a00 = ['a', 'b', 'c', 'd', 'e', 'f'] ;
let a01 = ['a', 'b', 'c', 'd', 'e', 'f'] ;
let a02 = ['a', 'b', 'c', 'd', 'e', 'f'] ;
let a03 = ['a', 'b', 'c', 'd', 'e', 'f'] ;
let a04 = ['a', 'b', 'c', 'd', 'e', 'f'] ;
```

```
    // (target)
    a01.copyWithin(2) ;

    console.log(a00) ;
    console.log(a01) ;
    console.log('') ;

    // (target)
    a02.copyWithin(-2) ;

    console.log(a00) ;
    console.log(a02) ;
    console.log('') ;

    // (target, start)
    a03.copyWithin(0, 2) ;

    console.log(a00) ;
    console.log(a03) ;
    console.log('') ;

    // (target, start, end)
    a04.copyWithin(-3, 1, 3) ;

    console.log(a00) ;
    console.log(a04) ;
    console.log('') ;
```

【相关说明】

```
    let a00 = ['a', 'b', 'c', 'd', 'e', 'f'] ;
    let a01 = ['a', 'b', 'c', 'd', 'e', 'f'] ;
    let a02 = ['a', 'b', 'c', 'd', 'e', 'f'] ;
    let a03 = ['a', 'b', 'c', 'd', 'e', 'f'] ;
    let a04 = ['a', 'b', 'c', 'd', 'e', 'f'] ;
```

这 5 个语句分别声明了 5 个变量 a00～a04，其初始数据均为数组实例['a', 'b', 'c', 'd', 'e', 'f']。

```
    // (target)
    a01.copyWithin(2) ;
```

因为变量 a01 当前的数据为数组实例，所以支持函数 copyWithin()，以进行自我复制并覆盖本身数组实例的部分元素。a01.copyWithin(2)可使得变量 a01 的数组实例['a', 'b', 'c', 'd', 'e', 'f']，从索引值为 2（第 3 个）的元素'c'开始，被覆盖【从默认索引值 0（第 1 个）的元素'a'开始】的连续元素，进而使得变量 a01 的数组实例，变更为["a", "b", "a", "b", "c", "d"]。

```
    console.log(a00) ;
    console.log(a01) ;
    console.log('') ;

    // (target)
    a02.copyWithin(-2) ;
```

a02.copyWithin(-2)可使得变量 a02 的数组实例['a', 'b', 'c', 'd', 'e', 'f']，从索引值为-2（倒数第 2 个）的元素'e'开始，被连续覆盖【从默认索引值 0（第 1 个）的元素'a'开始】的连续元素，进而使得变量 a02 的数组实例，变更为["a", "b", "c", "d", "a", "b"]。

```
console.log(a00) ;
console.log(a02) ;
console.log('') ;

// (target, start)
a03.copyWithin(0, 2) ;
```

a03.copyWithin(0, 2)可使得变量 a03 的数组实例['a', 'b', 'c', 'd', 'e', 'f']，从索引值为 0（第 1 个）的元素'a'开始，被连续覆盖【从索引值 2（第 3 个）的元素'c'开始】的连续元素，进而使得变量 a03 的数组实例，变更为["c", "d", "e", "f", "e", "f"]。

```
console.log(a00) ;
console.log(a03) ;
console.log('') ;

// (target, start, end)
a04.copyWithin(-3, 1, 3) ;
```

a04.copyWithin(-3, 1, 3)可使得变量 a04 的数组实例['a', 'b', 'c', 'd', 'e', 'f']，从索引值为-3（倒数第 3 个）的元素'd'开始，被连续覆盖【从索引值 1（第 2 个）的元素'b'开始，到索引值为 3 - 1 = 2（第 3 个）的元素'c'为止】的两个元素，进而使得变量 a04 的数组实例，变更为["a", "b", "c", "b", "c", "f"]。

8.1.10 判断数组各元素是否符合特定条件

欲判断在特定数组实例中，其各元素的数据是否符合特定条件，可通过 Array 对象实例支持的函数 every()和 some()来实现。

【8-1-10-Array-every-and-some.js】

```
function comparing(value, index, array)
{
  return value > 0 && value < 20 ;
}

let a01 = [24, 36, 10, 5, -2] ;
let a02 = [1, 15, 8, 3, 19] ;

console.log(a01.every(comparing)) ;
console.log(a02.every(comparing)) ;
console.log('') ;

console.log(a01.some(comparing)) ;
console.log(a02.some(comparing)) ;
```

【相关说明】

```
function comparing(value, index, array)
{
  return value > 0 && value < 20 ;
}
```

此语句定义了带有 3 个参数 value、index 与 array 的函数 comparing()。其中，参数 value 的数值，是特定数组实例中特定元素的数值，而【return value > 0 && value < 20】最终会返回布尔值 true 或者 false。

```
let a01 = [24, 36, 10, 5, -2] ;
let a02 = [1, 15, 8, 3, 19] ;
```

这两个语句声明了初始数据为不同数组实例的变量 a01 和 a02。

```
console.log(a01.every(comparing)) ;
```

a01.every(comparing)将变量 a01 的数组实例[24, 36, 10, 5, -2]中的每个整数值，分别传入名称为 comparing 的函数，并判断其是否满足特定关系表达式所对应的条件。在此，a01.every(comparing)返回布尔值 false，意味着并不是所有的整数值，均满足【value > 0 && value < 20】的条件。

```
console.log(a02.every(comparing)) ;
```

a02.every(comparing)将变量 a02 的数组实例[1, 15, 8, 3, 19]中的每个整数值，分别传入名称为 comparing 的函数，并判断其是否满足特定关系表达式所对应的条件。在此，a02.every(comparing)返回布尔值 true，意味着所有的整数值，均满足【value > 0 && value < 20】的条件。

```
console.log('') ;

console.log(a01.some(comparing)) ;
```

a01.some(comparing)将变量 a01 的数组实例[24, 36, 10, 5, -2]中的每个整数值，分别传入名称为 comparing 的函数，并判断其中一些整数值，是否满足特定关系表达式所对应的条件。在此，a01.some(comparing)返回布尔值 true，意味着部分整数值满足了【value > 0 && value < 20】的条件。

```
console.log(a02.some(comparing)) ;
```

a02.some(comparing)将变量 a02 的数组实例[1, 15, 8, 3, 19]中的每个整数值，分别传入名称为 comparing 的函数，并判断其中一些整数值，是否满足特定关系表达式所对应的条件。在此，a02.some(comparing)返回布尔值 true，意味着部分整数值满足了【value > 0 && value < 20】的条件。

8.1.11 数组部分元素的填充（ES6）

欲在特定数组实例中，填充新元素，以覆盖部分的连续元素，可通过 Array 对象实例支持的函数 fill()来实现。请看下面的示例。

【8-1-11-Array-fill.js】

```
let template = ['a', 'b', 'c', 'd', 'e', 'f'] ;
let a01 = template.slice(0) ;

a01.fill('?') ;

console.log(a01) ;
console.log('') ;

a01 = template.slice(0) ;
```

```
a01.fill('$', 2) ;

console.log(a01) ;

a01 = template.slice(0) ;

a01.fill('*', 2, 5) ;

console.log(a01) ;

a01 = template.slice(0) ;

a01.fill('^', 1, -2) ;

console.log(a01) ;

a01 = Array(7).fill('@_@') ;

console.log(a01) ;
```

【相关说明】

```
let template = ['a', 'b', 'c', 'd', 'e', 'f'] ;
```

声明初始数据为内含多个字符的数组实例的变量 template。

```
let a01 = template.slice(0) ;
```

此语句声明了变量 a01，并且通过 Array 对象实例支持的函数 slice(0)，间接复制了变量 template 的数组实例['a', 'b', 'c', 'd', 'e', 'f']，成为变量 a01 的初始数据。

```
a01.fill('?') ;
```

此语句将变量 a01 的数组实例中各元素的数据，改写成为字符'?'，进而使得变量 a01 的数组实例，变更为["?", "?", "?", "?", "?", "?"]。

```
console.log(a01) ;
console.log('') ;

a01 = template.slice(0) ;
```

此语句间接将数组实例['a', 'b', 'c', 'd', 'e', 'f']，赋给变量 a01。

```
a01.fill('$', 2) ;
```

此语句使得在变量 a01 的数组实例中，从索引值为 2（第 3 个）的元素开始，到最后一个元素为止，其各个数据均被改写成为字符'$'，使得变量 a01 的数组实例，变更为["a", "b", "$", "$", "$", "$"]。

```
console.log(a01) ;

a01 = template.slice(0) ;
```

此语句再次间接将数组实例['a', 'b', 'c', 'd', 'e', 'f']，赋给变量 a01。

```
a01.fill('*', 2, 5) ;
```

在变量 a01 的数组实例中，从索引值为 2（第 3 个）的元素开始，到索引值为 5 - 1 = 4（第 5 个）的元素为止，其各个数据均被改写成为字符'*'，使得变量 a01 的数组实例，变更为["a", "b", "*", "*", "*", "f"]。

```
console.log(a01) ;

a01 = template.slice(0) ;
```

此语句再次将数组实例['a', 'b', 'c', 'd', 'e', 'f']，赋给变量 a01。

```
a01.fill('^', 1, -2) ;
```

在变量 a01 的数组实例中，从索引值为 1（第 2 个）的元素开始，到索引值为-2 - 1 = -3（倒数第 3 个）的元素为止，其各个数据均被改写成为字符'^'，使得变量 a01 的数组实例，变更为["a", "^", "^", "^", "e", "f"]。

```
console.log(a01) ;

a01 = Array(7).fill('@_@') ;
```

此语句使得变量 a01 的数组实例，变更为["@_@", "@_@", "@_@", "@_@", "@_@", "@_@", "@_@"]。其中，Array(7)等价于【new Array(7)】，可动态产生内含 7 个空元素的数组实例，进一步通过数组实例支持的函数 fill('@_@')，使得空元素的数据，被改写成为字符串'@_@'。

8.1.12 筛选出符合特定条件的数组元素（ES6）

欲在特定数组实例中，筛选出符合特定条件的所有元素，以组成新的数组实例，可使用 Array 对象实例支持的函数 filter()。请看如下示例。

【8-1-12-Array-filter.js】

```
let names = ['Eric', 'Sam', 'Jimmy', 'Paula', 'Daisy', 'Jennifer', 'Sean'] ;

function check(name)
{
  return name.length < 5 ;
}

let shorts = names.filter(check) ;

console.log(shorts) ;
console.log('') ;

///
let numbers = [123, 456, 789, 55, 22, 33] ;
let smalls = numbers.filter(num => num % 3 == 0) ;

console.log(smalls) ;
```

【相关说明】

```
let names = ['Eric', 'Sam', 'Jimmy', 'Paula', 'Daisy', 'Jennifer', 'Sean'] ;
```

声明初始数据为内含多个字符串的数组实例的变量 names。

```
function check(name)
{
  return name.length < 5 ;
}
```

若参数 name 的数据字符串小于 5 个字符，则此语句会返回布尔值 true；反之，则返回布尔值 false。

上述语法定义了带有参数 name 的函数 check()。

```
let shorts = names.filter(check) ;
```

此语句声明了初始数据为内含多个字符串的数组实例["Eric", "Sam", "Sean"]的变量 shorts。其中，names.filter(check)会返回【在变量 names 的数组实例['Eric', 'Sam', 'Jimmy', 'Paula', 'Daisy', 'Jennifer', 'Sean']中，其字符个数小于 5 的字符串】所构成的子数组实例["Eric", "Sam", "Sean"]。判断特定元素的数据字符串的字符个数，是否小于 5 的比较表达式，存在于名称为 check 的函数内部的语句【return name.length < 5 ;】里。

```
console.log(shorts) ;
console.log('') ;

let numbers = [123, 456, 789, 55, 22, 33] ;
```

声明初始数据为内含多个整数值的数组实例的变量 numbers。

```
let smalls = numbers.filter(num => num % 3 == 0) ;
```

此语句声明了初始数据为内含多个整数值的数组实例[123, 456, 789, 33]的变量 smalls。其中，numbers.filter(num => num % 3 == 0)会返回子数组实例[123, 456, 789, 33]，其各个元素的数值，来自于【在变量 numbers 的数组实例[123, 456, 789, 55, 22, 33]中，其除以 3 的余数为 0】的整数值。判断【除以 3 的余数为 0】的比较表达式，是在箭头函数内部的【num % 3 == 0】。在此，于箭头函数的箭头符号【=>】的右侧，其比较表达式【num % 3 == 0】等价于【{ return num % 3 == 0 ; }】。

8.1.13 循环访问数组各元素

欲循环访问特定数组实例中的各元素，除了通过循环语句的协助之外，亦可通过 Array 对象实例支持的函数 forEach()来达成。请看如下示例。

【8-1-13-Array-forEach.js】

```
let shapes = ['circle', 'triangle', 'rectangle', 'trapezoid', 'pentagons', 'hexagon',
  'heptagon', 'octagon', 'star'] ;

function display(value)
{
  console.log(value) ;
}
```

```
shapes.forEach(display) ;

console.log('') ;

shapes.forEach(value => console.log(value)) ;
```

【相关说明】

```
let shapes = ['circle', 'triangle', 'rectangle', 'trapezoid', 'pentagons', 'hexagon',
 'heptagon', 'octagon', 'star'] ;

function display(value)
{
  console.log(value) ;
}
```

此语法定义了带有参数 value 的函数 display()。

```
shapes.forEach(display) ;
```

变量 shapes 当前的数据为内含多个字符串的数组实例,所以支持函数 forEach()。此语句可使得变量 shapes 的数组实例的各个数据字符串,被传入名称为 display 的函数,并返回如下多行信息:

- circle
- triangle
- rectangle
- trapezoid
- pentagons
- hexagon
- heptagon
- octagon
- star

```
console.log('') ;

shapes.forEach(value => console.log(value)) ;
```

此语句等价于【shapes.forEach(display) ;】,并通过箭头函数的定义,来替换函数 display()的名称。

8.1.14 判断是否为数组的实例

欲判断特定变量当前的数据,是否为数组的实例,可通过 Array 对象支持的函数 isArray()来达成。请看如下示例。

【8-1-14-Array-isArray.js】

```
let instance ;

with(console)
{
```

```
    instance = [] ;
    log(Array.isArray(instance)) ;

    instance = Array(3) ;
    log(Array.isArray(instance)) ;

    instance = new Array(3) ;
    log(Array.isArray(instance)) ;
    log('') ;

    instance = {} ;
    log(Array.isArray(instance)) ;

    instance = parseInt(100 * Math.random()) ;
    log(Array.isArray(instance)) ;

    instance = instance < 50 ;
    log(Array.isArray(instance)) ;
}
```

【相关说明】

```
let instance ;
```

声明变量 instance。

```
with(console)
{
```

```
    instance = [] ;
```

将空数组实例，赋给变量 instance。

```
    log(Array.isArray(instance)) ;
```

通过 Array 对象支持的函数 isArray()，判断被传入的数据（变量 instance 当前的数据），是否为数组实例。此语句显示出布尔值 true，意味着变量 instance 当前的数据是数组实例。

```
    instance = Array(3) ;
```

此语句将 Array(3)所产生的内含 3 个空元素的数组实例，赋给变量 instance。

```
    log(Array.isArray(instance)) ;
```

此语句显示出布尔值 true，意味着变量 instance 当前的数据是数组实例。

```
    instance = new Array(3) ;
```

通过 JavaScript 引擎的解读，Array(3)和 new Array(3)均会返回内含 3 个空元素的数组实例。

```
    log(Array.isArray(instance)) ;
```

此语句显示出布尔值 true，意味着变量 instance 当前的数据是数组实例。

```
    log('') ;

    instance = {} ;
```

将空对象实例，赋给变量 instance。

```
log(Array.isArray(instance)) ;
```

此语句显示出布尔值 false，意味着变量 instance 当前的数据，**并非**数组实例。

```
instance = parseInt(100 * Math.random()) ;
```

将一个 0～100 的随机整数值，赋给变量 instance。

```
log(Array.isArray(instance)) ;
```

此语句显示出布尔值 false，意味着变量 instance 当前的数据，并非数组实例。

```
instance = instance < 50 ;
```

将比较表达式【instance < 50】返回的布尔值 true 或者 false，赋给变量 instance 本身。

```
log(Array.isArray(instance)) ;
```

此语句显示出布尔值 false，意味着变量 instance 当前的数据，并非数组实例。

8.1.15　访问并渐次处理数组各元素的数据

欲在特定数组实例中，访问并渐次处理各元素的数据，可通过 Array 对象实例支持的函数 reduce()或 reduceRight()来实现。请看如下**两个**示例。

【8-1-15-e1-Array-reduce.js】

```js
let numbers = [2, 3, 7, 11, 13, 17, 19, 23, 29] ;

let totals = numbers.reduce((pile, current) => pile + current) ;

console.log(totals) ;

totals = numbers.reduce((pile, current) => pile + current, 0) ;

console.log(totals) ;

totals = numbers.reduce((pile, current) => pile + current, 200) ;

console.log(totals) ;
console.log('') ;

///
let n01 = [3, 7, 2, 9] ;
let n02 = [10, 80, 50, 70] ;
let n03 = [200, 500, 100, 400] ;

let matrix = [n01, n02, n03] ;

function vertically_add(piles, current)
{
  for (let i = 0; i < piles.length; i++)
  {
```

```
    piles[i] += current[i] ;
  }

  return piles ;
}

let column_summations = matrix.reduce(vertically_add, Array(n01.length).fill(0)) ;

console.log(column_summations) ;
console.log('') ;

///
let box = ['apple', 'guava', 'durian', 'apple', 'guava', 'peach', 'apricot', 'peach', 'durian',
  'apple'] ;

function count_fruit(piles, current)
{
  console.log(current) ;

  if (! (current in piles))
    piles[current] = 1 ;
  else
    piles[current]++ ;

  return piles ;
}

let inventory = box.reduce(count_fruit, {}) ;

console.log(inventory) ;
```

【相关说明】

```
let numbers = [2, 3, 7, 11, 13, 17, 19, 23, 29] ;
```

声明初始数据为内含多个整数值的数组实例的变量 numbers。

```
let totals = numbers.reduce((pile, current) => pile + current) ;
```

此语句声明了变量 totals，其初始数值为变量 numbers 的数组实例中各整数的加总值 124。因为变量 numbers 当前的数据是数组实例，所以支持函数 reduce()，可将变量 numbers 数组实例中的各整数值，间接传递给特定函数。在此，箭头函数【(pile, current) => pile + current】具有两个参数 pile 与 current，并且会被调用 9 次，使得【pile + current】持续返回【在**前一次**被调用时，前几个整数值的加总值】。其中，参数 current 的数值就是变量 numbers 数组实例里的**其中一个整数值**。因此，参数 pile 的最终数值 124，是变量 numbers 的数组实例中各整数的加总值，最后成为 numbers.reduce() 的返回值。

```
console.log(totals) ;

totals = numbers.reduce((pile, current) => pile + current, 0) ;
```

此语句最后将将整数值 124，重新赋给变量 totals。其中，整数值 124，是 numbers.reduce() 的第 2 个参数数值 0，加上【变量 numbers 的数组实例中各整数】的加总值 124。

```
console.log(totals) ;

totals = numbers.reduce((pile, current) => pile + current, 200) ;
```

此语句将整数值 324，重新赋给变量 totals。其中，整数值 324，是 numbers.reduce()的第 2 个参数数值 200，加上【变量 numbers 的数组实例中各整数】的加总值 124。

```
console.log(totals) ;
console.log('') ;

let n01 = [3, 7, 2, 9] ;
let n02 = [10, 80, 50, 70] ;
let n03 = [200, 500, 100, 400] ;
```

这 3 个语句分别声明了初始数据为内含多个整数值的不同数组实例的变量 n01、n02 与 n03。

```
let matrix = [n01, n02, n03] ;
```

此语句声明了初始数据为【内含 3 个一维数组的二维数组实例】的变量 matrix。

```
function vertically_add(piles, current)
{
  for (let i = 0; i < piles.length; i++)
  {
    piles[i] += current[i] ;
  }

  return piles ;
}
```

此语法定义具有参数 piles 与 current 的函数 vertically_add()。每次函数 vertically_add()被调用时，这参数 piles 与 current 的数据均是**一维**数组实例。其中，函数 vertically_add()第 1 次被调用时，变量 current 的数据是数组实例[3, 7, 2, 9]，变量 piles 的数据为 数 组 实 例 [3, 7, 2, 9]。函数 vertically_add()**第 2 次**被调用时，变量 current 的数据是数组实例[10, 80, 50, 70]，变量 piles 的数据为数组实例[3 + 10, 7 + 80, 2 + 50, 9 + 70]，也就是[13, 87, 52, 79]。在第 3 回合完成时，变量 current 的数据是数组实例[200, 500, 100, 400]，，而变量 piles 的数据为数组实例[13 + 200, 87 + 500, 52 + 100, 79 + 400]，也就是[213, 587, 152, 479]。

```
let column_summations = matrix.reduce(vertically_add, Array(n01.length).fill(0)) ;
```

此语句声明了初始数据为内含多个整数值的数组实例[213, 587, 152, 479]的变量 column_summations。其中，matrix.reduce(vertically_add, Array(n01.length).fill(0))可将变量 matrix 的**二维**数组实例中的各**一维**数组实例，间接传递给名称为 vertically_add 的函数。

其中，Array(n01.length).fill(0)会返回 4 个元素值均为 0 的数组实例[0, 0, 0, 0]，所以 matrix.reduce(vertically_add, Array(n01.length).fill(0))当前等价于 matrix.reduce(vertically_add, [0, 0, 0, 0])。

```
console.log(column_summations) ;
console.log('') ;

let box = ['apple', 'guava', 'durian', 'apple', 'guava', 'peach', 'apricot', 'peach', 'durian',
  'apple'] ;
```

声明初始数据为内含多个字符串的数组实例的变量 box。

```
function count_fruit(piles, current)
{
  console.log(current) ;

  if (!(current in piles))
    piles[current] = 1 ;
```

在此,如果参数 current 的数据字符串所代表的属性名称,并不存在于【参数 piles 的对象实例】中,那么就在其对象实例中,新增前述字符串所代表的属性,并设置其属性值为 1。

```
  else
    piles[current]++ ;
```

在此,如果参数 current 的数据**字符串**所代表的属性名称,已经存在于【参数 piles 的对象实例】中,那么就在其对象实例中,【递增】前述字符串所代表的属性的数值。

```
  return piles ;
```

此语句返回参数 piles 的对象实例。

```
}
```

上述语法定义了具有参数 piles 与 current 的函数 count_fruit()。每次函数 count_fruit() 被调用时,参数 piles 的数据都是一个对象实例,参数 current 的数据则是变量 box 的数组实例中的一个字符串。

- 第 1 次被调用时,变量 current 的数据为字符串'apple',变量 piles 的数据为对象实例{apple: 1}。
- 第 2 次被调用时,变量 current 的数据为字符串'guava',变量 piles 的数据为对象实例{apple: 1, guava: 1}。
- 第 3 次被调用时,变量 current 的数据为字符串'durian',变量 piles 的数据为对象实例{apple: 1, guava: 1, durian: 1}。
- 第 4 次被调用时,变量 current 的数据为字符串'apple',变量 piles 的数据为对象实例{apple: 2, guava: 1, durian: 1}。
- 第 5 次被调用时,变量 current 的数据为字符串'guava',变量 piles 的数据为对象实例{apple: 2, guava: 2, durian: 1}。
- 第 6 次被调用时,变量 current 的数据为字符串'peach',变量 piles 的数据为对象实例{apple: 2, guava: 2, durian: 1, peach: 1}。
- 第 7 次被调用时,变量 current 的数据为字符串'apricot',变量 piles 的数据为对象实例{apple: 2, guava: 2, durian: 1, peach: 1, apricot: 1}。
- 第 8 次被调用时,变量 current 的数据为字符串'peach',变量 piles 的数据为对象实例{apple: 2, guava: 2, durian: 1, peach: 2, apricot: 1}。
- 第 9 次被调用时,变量 current 的数据为字符串'durian',变量 piles 的数据为对象实例{apple: 2, guava: 2, durian: 2, peach: 2, apricot: 1}。
- 第 10 次被调用时,变量 current 的数据为字符串'apple',变量 piles 的数据为对象实例{apple: 3, guava: 2, durian: 2, peach: 2, apricot: 1}。

```
let inventory = box.reduce(count_fruit, {}) ;
```

此语句声明了初始数据为对象实例{apple: 3, guava: 2, durian: 2, peach: 2, apricot: 1}的变量

inventory。其中，box.reduce(count_fruit, {})可将变量 box 的数组实例中的各字符串，间接传递给名称为 count_fruit 的函数。因为其第 2 个参数为空的对象实例{}，所以 box.reduce()最终返回的对象实例仅仅是{apple: 3, guava: 2, durian: 2, peach: 2, apricot: 1}。

【8-1-15-e2-Array-reduceRight.js】

```js
let numbers = [2, 3, 7, 11, 13, 17, 19, 23, 29] ;
let message = '' ;

function summation(pile, current)
{
  message += ' -> ' + current ;

  return pile + current ;
}

totals = numbers.reduce(summation, 0) ;

console.log(message) ;
console.log(totals) ;
console.log('') ;

message = '' ;
totals = numbers.reduceRight(summation, 0) ;

console.log(message) ;
console.log(totals) ;
```

【相关说明】

```js
let numbers = [2, 3, 7, 11, 13, 17, 19, 23, 29] ;
```

声明初始数据为内含多个整数值的数组实例的变量 numbers。

```js
let message = '' ;
```

声明初始数据为空字符串的变量 message。

```js
function summation(pile, current)
{
  message += ' -> ' + current ;
```

此语句用来串接变量 message 的字符串和其参数 current 的数值，成为新字符串，并赋给变量 message 本身。

```js
  return pile + current ;
```

此语句用来返回【参数 pile 的数值，加上 current 数值】的结果值。

```js
}
```

上述语法定义了具有参数 pile 与 current 的函数 summation()。

```js
totals = numbers.reduce(summation, 0) ;
```

在此语句中，调用函数 reduce()，会间接使得变量 numbers 的数组实例中的各整数值，【顺向被传递给名称为 summation 的函数。此语句最后将变量 numbers 的数组实例中各整数值的总和 124，赋给变量 totals。

```
console.log(message) ;
```

此语句显示出各整数值被**顺**向排列的信息【 -> 2 -> 3 -> 7 -> 11 -> 13 -> 17 -> 19 -> 23 -> 29】。

```
console.log(totals) ;
console.log('') ;

message = '' ;

totals = numbers.reduceRight(summation, 0) ;
```

在此语句中，调用函数 reduceRight()，会间接使得变量 numbers 的数组实例中的各整数值，【**反**】向被传递给名称为 summation 的函数。此语句最后将变量 numbers 的数组实例中各整数值的总和 124，赋给变量 totals。

```
console.log(message) ;
```

此语句显示出各整数值被**反**向排列的信息【 -> 29 -> 23 -> 19 -> 17 -> 13 -> 11 -> 7 -> 3 -> 2】。

8.1.16　反转数组各元素的顺序

欲在特定数组实例中，反转各元素的排列顺序，可通过 Array 对象实例支持的函数 reverse() 来实现。请参看如下的例子。

【8-1-16-Array-reverse.js】

```
let flowers = ['lily', 'jasmine', 'rose', 'daisy', 'daffodil', 'camellia'] ;

console.log(flowers) ;
console.log('') ;

flowers.reverse() ;

console.log(flowers) ;
```

【相关说明】

```
let flowers = ['lily', 'jasmine', 'rose', 'daisy', 'daffodil', 'camellia'] ;
```

声明初始数据为内含多个字符串的数组实例的变量 flowers。

```
console.log(flowers) ;
```

显示出["lily", "jasmine", "rose", "daisy", "daffodil", "camellia"]的信息。

```
console.log('') ;

flowers.reverse() ;
```

此语句返回了【变量 flowers 的数组实例["lily", "jasmine", "rose", "daisy", "daffodil", "camellia"]被**反**向排列之后的新数组实例["camellia", "daffodil", "daisy", "rose", "jasmine", "lily"]】。

8.1.17 数组各元素的重新排序

欲在特定数组实例中，根据各元素的数值或字符编码顺序，重新进行升序（ascending）或降序（descending）的排列，可通过 Array 对象实例支持的函数 sort() 来实现。关于数组各元素的重新排序的综合运用，可参考如下示例。

【8-1-17-Array-sort.js】

```
let a01 = Array.from({length: 7}, () => parseInt(100 * Math.random())) ;

console.log(a01) ;

let a02 = a01.sort() ;

console.log(a02) ;

function set_descending(previous, current)
{
  // Swap two adjacent numbers if positive.
  return current -previous ;
}

let a03 = a01.sort(set_descending) ;

console.log(a03) ;
```

【相关说明】

```
let a01 = Array.from({length: 7}, () => parseInt(100 * Math.random())) ;
```

此语句声明了初始数据为内含 7 个整数值的数组实例（例如 [69, 68, 80, 15, 4, 30, 11]）的变量 a01。在语法【Array.from({length: 7}, () => parseInt(100 * Math.random()))】中，{length: 7}使得被返回的新数组实例，具有 7 个元素。箭头函数【() => parseInt(100 * Math.random())】则是用来产生 0 ~ 100 的随机整数值。

```
console.log(a01) ;

let a02 = a01.sort() ;
```

此语句声明了初始数据为内含 7 个整数值的数组实例（例如 [11, 15, 30, 4, 68, 69, 80]）的变量 a02。其中，a01.sort()会返回【变量 a01 的数组实例中各元素被**顺向**排序】之后的新数组实例，例如 [11, 15, 30, 4, 68, 69, 80]。

```
console.log(a02) ;

function set_descending(previous, current)
{
  // Swap two adjacent numbers if positive.
  return current -previous ;
```

此语句返回了表达式【current-previous】的结果值。这个表达式还有一项重要意义，也就是被传入的当前数值（current），减掉被传入的上一个数值（previous）的结果值为正值，则交换前后两个数值。这就意味着，在特定数组实例中的各整数值，会被**由大至小**地排序。

```
}
```

上述语法定义了带有参数 previous 与 current 的函数 set_descending()。

```
let a03 = a01.sort(set_descending) ;
```

此语句声明了初始数据为**由大至小**地排序之后的新数组实例（例如[80, 69, 68, 30, 15, 11, 4]）的变量 a03。

8.1.18　判断特定元素的存在性（ES7）

欲判断特定元素，是否存在于特定数组实例中，可通过 Array 对象实例支持的函数 indexOf() 和 includes() 来实现。请看如下示例。

【8-1-18-Array-includes.js】

```javascript
let int_list = [1, 10, 18, 23, 29, 37, 56] ;
let str_list = ['keyboard', 'mouse', 'screen', 'speaker'] ;

let result ;

result = int_list.indexOf(23) ;

console.log(result) ;

result = int_list.includes(23) ;

console.log(result) ;
console.log('') ;

result = int_list.indexOf(45) ;

console.log(result) ;

result = int_list.includes(45) ;

console.log(result) ;
console.log('') ;

result = str_list.indexOf('speaker') ;

console.log(result) ;

result = str_list.includes('speaker') ;

console.log(result) ;
console.log('') ;

result = str_list.indexOf('joystick') ;

console.log(result) ;
```

```
result = str_list.includes('joystick') ;

console.log(result) ;
```

【相关说明】

```
let int_list = [1, 10, 18, 23, 29, 37, 56] ;
```

声明数据为数组实例[1, 10, 18, 23, 29, 37, 56]的变量 int_list。

```
let str_list = ['keyboard', 'mouse', 'screen', 'speaker'] ;
```

声明数据为数组实例['keyboard', 'mouse', 'screen', 'speaker']的变量 str_list。

```
let result ;
```

声明变量 result。

```
result = int_list.indexOf(23) ;
```

在此语句中，int_list.indexOf(23)会返回代表索引值的整数值 3，意味着数值为 23 的元素，是变量 int_list 的数组实例中的第 4 个（索引值为 3）元素。

```
console.log(result) ;

result = int_list.includes(23) ;
```

在此语句中，int_list.includes(23)会返回布尔值 true，意味着数值为 23 的元素，确实存在于变量 int_list 的数组实例中。

```
console.log(result) ;
console.log('') ;

result = int_list.indexOf(45) ;
```

在此语句中，int_list.indexOf(45)会返回代表索引值的整数值-1，意味着数值为 45 的元素，并不存在于变量 int_list 的数组实例中。

```
console.log(result) ;

result = int_list.includes(45) ;
```

在此语句中，int_list.includes(45)会返回布尔值 false，意味着数值为 45 的元素，并不存在于变量 int_list 的数组实例中。

```
console.log(result) ;
console.log('') ;

result = str_list.indexOf('speaker') ;
```

在此语句中，str_list.indexOf('speaker')会返回代表索引值的整数值 3，意味着数据为字符串 'speaker'的元素，是变量 str_list 的数组实例中的第 4 个（索引值为 3）元素。

```
console.log(result) ;

result = str_list.includes('speaker') ;
```

在此语句中，str_list.includes('speaker')会返回布尔值 true，意味着数据为字符串'speaker'的元素，存在于变量 str_list 的数组实例中。

```
console.log(result) ;
console.log('') ;

result = str_list.indexOf('joystick') ;
```

在此语句中，str_list.indexOf('joystick')会返回代表索引值的整数值-1，意味着数据为字符串'joystick'的元素，并不存在于变量 str_list 的数组实例中。

```
console.log(result) ;

result = str_list.includes('joystick') ;
```

在此语句中，str_list.indexOf('joystick')会返回布尔值 false，意味着数据为字符串'joystick'的元素，并不存在于变量 str_list 的数组实例中。

8.2 数组的扩展运算（ES6）

数组实例的扩展（spread）运算，是指通过扩展运算符【...】，衔接【其数据为数组实例】的变量的名称，例如【... a01】。其中，变量 a01 的数据为数组实例[1, 2, 3]。请看如下示例。

【8-2---array-spreadings.js】

```
let a01 = [1, 2, 3] ;
let a02 = [15, 25, 35] ;
let a03 = [137, 256, 391] ;

let result = [a01, a02, a03] ;

console.log(result[1]) ;
console.log(result[2][2]) ;
console.log('') ;

result = [... a01, ... a02, ... a03] ;

console.log(result) ;
console.log('') ;

console.log(a01) ;
console.log(... a01) ;
console.log(1, 2, 3) ;
```

【相关说明】

```
let a01 = [1, 2, 3] ;
let a02 = [15, 25, 35] ;
let a03 = [137, 256, 391] ;
```

这 3 个语句分别声明了初始数据均为内含 3 个不同整数值的数组实例的变量 a01、a02 与 a03。

```
let result = [a01, a02, a03] ;
```

此语句声明了变量 result，其初始数据为【变量 a01、a02 与 a03 的一维数组所构成的二维数组实例[[1, 2, 3], [15, 25, 35], [137, 256, 391]]。

```
console.log(result[1]) ;
```

result[1]会返回【在变量 result 当前的二维数组实例中，其索引值为 1（第 2 个）】的元素，也就是子数组实例[15, 25, 35]。

```
console.log(result[2][2]) ;
```

result[2][2]会返回【在变量 result 当前的二维数组实例中，其索引值为 2（第 3 个）】的元素，也就是在子数组实例[137, 256, 391]中，其索引值为 2（第 3 个）的元素的数值 391。

```
console.log('') ;

result = [... a01, ... a02, ... a03] ;
```

在此语句中，通过扩展运算符【...】，将变量 result 的数组实例[... [1, 2, 3], ... [15, 25, 35], ... [137, 256, 391]] 展开成为[1, 2, 3, 15, 25, 35, 137, 256, 391]。

```
console.log(result) ;
console.log('') ;

console.log(a01) ;
```

显示出变量 a01 的数组实例[1, 2, 3]。

```
console.log(... a01) ;
```

【... a01】会使得变量 a01 的数组实例[1, 2, 3]中的各元素，被展开成为【1, 2, 3】。所以 console.log(... a01)当前等价于 console.log(1, 2, 3)，会显示出【1 2 3】的信息。

```
console.log(1, 2, 3) ;
```

显示出【1 2 3】的信息。

8.3 数组元素的匹配（ES6）

在 JavaScript 语言的单一语句中，同时设置多个变量的数据时，例如【var a = 1, b = 2, c = 3 ;】，可借助数组实例中的元素的匹配语法，来实现同样的效果，例如【var [a, b, c] = [1, 2, 3] ;】。关于数组元素的匹配运用，可参考如下示例。

【8-3---array-matches.js】

```
var [a, b, c] = [1, 2, 3] ;

console.log(a, b, c) ;

[a, c] = [4, 6] ;
```

```
console.log(a, b, c);
```

【相关说明】

```
var [a, b, c] = [1, 2, 3];
```

此语句同时声明了初始数值为不同整数值的变量 a、b 与 c。

```
console.log(a, b, c);
```

显示出变量 a、b 和 c 的数值【1 2 3】。

```
[a, c] = [4, 6];
```

设置变量 a 的数值为 4，以及变量 c 的数值为 6。

```
console.log(a, b, c);
```

显示出变量 a、b 和 c 的最终数值【4 2 6】。

8.4 数据类型化的按位数组（ES6）

在 JavaScript 语言中，数据类型化（data-typed）的按位（bitwise）数组，会被 JavaScript 引擎视为：

- 每个元素占用特定数据类型（data type）所需的内存空间。
- 每个元素的数据均会被按位（bitwise）加以访问。

若需要处理音频、视频或是其他网络数据包（network packet）中的二进制数码，则可存储二进制数码，到特定数据类型化的按位数组里。请参看下面的例子。

【8-4---typed-arrays.js】

```
var buffer01 = new ArrayBuffer(32);

var item01 = {};

console.log(buffer01.byteLength);
console.log('');

item01.id = new Uint16Array(buffer01, 0, 1);
item01.name = new Uint8Array(buffer01, 2, 26);
item01.price = new Float32Array(buffer01, 28, 1);

item01.id = 13247;
item01.name = 'Sweet durian candy box';
item01.price = 100;

console.log(item01);
console.log(item01.id);
console.log('');

console.log(item01.name);
console.log('');
```

```
console.log(item01.name[0]) ;
console.log(item01.name[1]) ;
console.log(item01.name[2]) ;
console.log(item01.name[3]) ;
console.log(item01.name[4]) ;
console.log('') ;

console.log(item01.price) ;
```

【相关说明】

```
var buffer01 = new ArrayBuffer(32) ;
```

此语句声明了变量 buffer01，其初始数据为占用 **32 个**字节 (byte) 的 ArrayBuffer 对象实例，如同被存放于二进制的**数据缓冲区**（data buffer）里。

```
var item01 = {} ;
```

声明初始数据为空的对象实例{}的变量 item01。

```
console.log(buffer01.byteLength) ;
```

此语句显示出整数值 32，代表变量 buffer01 的数据缓冲区，占用了 32 个**字节长度**（byte length）的内存空间。

```
item01.id = new Uint16Array(buffer01, 0, 1) ;
```

此语句新增了变量 item01 的对象实例中的新属性 id，并设置其数据【在变量 buffer01 的数据缓冲区中，占用第 1 组（索引值为 0）的 **16 个比特位**（bit）】的空间。

在构造函数 Uint16Array(buffer01, 0, 1)中，【Uint16】意味着其每个元素的数据，如同是**无正负符号的 16 个比特位整数**（16-bit unsigned integer）；而【(buffer01, 0, 1)】代表在变量 buffer01 的数据缓冲区里，从其索引值为 0（第 1 个）的元素开始，仅仅占用 1 个元素（1 × 16 = 16 个比特位）的空间。

```
item01.name = new Uint8Array(buffer01, 2, 26) ;
```

此语句新增了变量 item01 的对象实例中的新属性 name，并设置其数据【在变量 buffer01 的数据缓冲区中，占用第 3 组（索引值为 2）的 **8 个比特位**】的空间。

在构造函数 Uint8Array(buffer01, 2, 26)中，【Uint8】意味着其每个元素的数据，如同是**无正负符号的 8 个比特位整数** (8-bit unsigned integer)；而【(buffer01, 2, 26)】代表在变量 buffer01 的数据缓冲区里，从其索引值为 2（第 3 个）的元素开始，连续占用 26 个元素（26 × 8 = 208 个比特位）的空间。

```
item01.price = new Float32Array(buffer01, 28, 1) ;
```

此语句新增了变量 item01 的对象实例中的新属性 price，并设置其数据【在变量 buffer01 的数据缓冲区中，占用第 29 组（索引值为 28）的 **32 个比特位**】的空间。

在构造函数 Float32Array(buffer01, 2, 27)中，【Float32】意味着其每个元素的数据，如同是 **32 个比特位浮点数**（32-bit floating-point number）；而【(buffer01, 28, 1)】代表在变量 buffer01 的数据缓冲区里，从其索引值为 28（第 29 个）的元素开始，仅仅占用 1 个元素（1 × 32 = 32 个比特位）的空间。

```
item01.id = 13247 ;
```

将整数值 13247，赋给变量 item01 的对象实例的属性 id。

```
item01.name = 'Sweet durian candy box' ;
```

将字符串'Sweet durian candy box'，赋给变量 item01 的对象实例的属性 name。

```
item01.price = 100 ;
```

将整数值 100，赋给变量 item01 的对象实例的属性 price。

```
console.log(item01) ;
```

显示出 {id: 13247, name: "Sweet durian candy box", price: 100} 的信息。

```
console.log(item01.id) ;
```

显示出整数值 13247 的信息。

```
console.log(item01.name) ;
```

显示出字符串'Sweet durian candy box'的信息。

```
console.log(item01.name[0]) ;
```

显示出字符'S'。

```
console.log(item01.name[1]) ;
```

显示出字符'w'。

```
console.log(item01.name[2]) ;
```

显示出字符'e'。

```
console.log(item01.name[3]) ;
```

显示出字符'e'。

```
console.log(item01.name[4]) ;
```

显示出字符't'。

```
console.log(item01.price) ;
```

显示出整数值 100。

8.5 练 习 题

1. 试编写**两种**版本的 JavaScript 源代码，均可转换特定字符串（例如'2013145205307752099'），成为【其各元素的数据是**单一字符**】的数组实例，例如["2", "0", "1", "3", "1", "4", "5", "2", "0", "5", "3", "0", "7", "7", "5", "2", "0", "9", "9"]。

2. 试编写 JavaScript 源代码，将如下数组实例，动态输出并显示出其元素的**排列顺序**，能**随机变化**的数组实例：

```
let juices = ['orange', 'apple', 'guava', 'pineapple', 'passion fruit', 'lemon', 'kiwi fruit',
    'watermelon', 'grapefruit', 'grape'] ;
```

第 9 章

处理自定义对象

在支持面向对象编程（OOP, object-oriented programming）的 JavaScript 语言里，其内置的对象，通常内含许多属性和函数/方法。

编程人员欲定义特定对象及其属性与函数，可以使用一对内含特定属性与函数的大括号，或者先行编写仅仅只有一对大括号{}的空对象实例（object instance），再通过后续的语法，逐一新增其属性和函数。欲处理自定义的对象实例，可通过 JavaScript 语言支持的 Object 或 Reflect 对象的内置函数，来加以实现。

9.1 自定义对象的属性名称、属性数据与副本

自定义对象实例的特定属性的名称，可被用来访问特定属性的数据。此外，在有些情况下，创建特定自定义对象实例的副本，可大幅简化编程的复杂度。

9.1.1 对象属性的名称或数据所构成的数组（ES8）

通过内置函数【Object.keys(内含自定义对象实例的变量名称)】或【Object.getOwnPropertyNames(内含自定义对象实例的变量名称)】，均可获取内含自定义对象实例的各属性名称的数组实例。

通过内置函数【Object.values(内置自定义对象实例的名称)】，可返回内含自定义对象实例的各属性数据的数组实例。请参看下面的示例。

【9-1-1-keys-and-values-of-objects.js】

```
let profile_array = ['Peter', 'male', 27, 'IT', 3.5] ;
let key_list01 = Object.keys(profile_array) ;
let key_list02 = Object.getOwnPropertyNames(profile_array) ;
let value_list = Object.values(profile_array) ;
```

```
console.log(key_list01) ;
console.log(key_list02) ;
console.log(value_list) ;
console.log('') ;

let profile_object = {name: 'Peter', gender: 'male', age: 27, department: 'IT',
 years_of_service: 3.5} ;

key_list01 = Object.keys(profile_object) ;

key_list02 = Object.getOwnPropertyNames(profile_object) ;

value_list = Object.values(profile_object) ;

console.log(key_list01) ;
console.log(key_list02) ;
console.log(value_list) ;
console.log('') ;

let all_entries = Object.entries(profile_object) ;

console.log(all_entries) ;
```

【相关说明】

```
let profile_array = ['Peter', 'male', 27, 'IT', 3.5] ;
```

声明初始数据为数组实例的变量 profile_array。

```
let key_list01 = Object.keys(profile_array) ;
```

此语句声明了变量 key_list01，其初始数据为【在变量 profile_array 的数组实例中，其各索引值被转换之后的字符串】所构成的新数组实例["0", "1", "2", "3", "4"]。

```
let key_list02 = Object.getOwnPropertyNames(profile_array) ;
```

此语句声明了变量 key_list02，其初始数据为【在变量 profile_array 的数组实例中，其各索引值被转换之后的字符串】，再包含属性 length，所共同构成的新数组实例["0", "1", "2", "3", "4", "length"]。

```
let value_list = Object.values(profile_array) ;
```

此语句声明了变量 value_list，其初始数据为【在变量 profile_array 的数组实例中，其各元素的数据】所构成的新数组实例["Peter", "male", 27, "IT", 3.5]。

```
console.log(key_list01) ;
console.log(key_list02) ;
console.log(value_list) ;
console.log('') ;

let profile_object = {name: 'Peter', gender: 'male', age: 27, department: 'IT',
 years_of_service: 3.5} ;
```

声明初始数据为对象实例的变量 profile_object。

```
key_list01 = Object.keys(profile_object) ;
```

此语句重新设置了变量 key_list01 的数据为【在变量 profile_object 的对象实例中，其各属性名称】所构成的新数组实例["name", "gender", "age", "department", "years_of_service"]。

```
key_list02 = Object.getOwnPropertyNames(profile_object) ;
```

此语句重新设置了变量 key_list02 的数据为【在变量 profile_object 的对象实例中，其各属性名称】所构成的新数组实例["name", "gender", "age", "department", "years_of_service"]。

由此可看出，当前 Object.keys(profile_object)与 Object.getOwnPropertyNames (profile_object)的效果相同。

```
value_list = Object.values(profile_object) ;
```

此语句重新设置了变量 value_list 的数据为【在变量 profile_object 的对象实例中，其各属性数据】所构成的新数组实例["Peter", "male", 27, "IT", 3.5]。

```
console.log(key_list01) ;
console.log(key_list02) ;
console.log(value_list) ;
console.log('') ;

let all_entries = Object.entries(profile_object) ;
```

Object.entries(profile_object) 会返回数组实例 [['name', 'Peter'], ['gender', 'male'], ['age', 27], ['department', 'IT'], ['years_of_services', 3.5]]，使得在变量 profile_object 的对象实例中，其各属性**名称**和**数据**的组合（例如 department: 'IT'），成为新数组实例中的子数组实例（例如：['department', 'IT']）。

9.1.2 精细创建特定对象实例的副本

欲更进一步精细创建特定自定义对象实例的副本，可通过内置函数 Object.create()来实现。以下举例说明。

【9-1-2-Object-create.js】

```
let person = {name: 'Jason', age: 37, married: true, occupation: 'finance manager', department:
  'Finance Team'} ;

let new_one = Object.create(person, {department: {value: 'Finance Team', writable: false}}) ;

console.log(new_one) ;
console.log('') ;

new_one.name = 'Alex' ;

console.log(new_one.name) ;
console.log(new_one.department) ;
console.log('') ;

console.log(new_one.age) ;
console.log(new_one.__proto__.age) ;
console.log('') ;
```

```
new_one.department = 'IT' ;

console.log(new_one.department) ;
///
let my_love = Object.create({name: 'Cindy', occupation: 'angel'}, {age: {value: 18, writable:
 false, enumerable: false, configurable: false}}) ;

console.log(my_love.age) ;
console.log('') ;

my_love.age = 27 ;

console.log(my_love.age) ;
console.log('') ;

for (let property in my_love)
{
  console.log(property) ;
}
```

【相关说明】

```
let person = {name: 'Jason', age: 37, married: true, occupation: 'finance manager', department:
 'Finance Team'} ;
```

声明初始数据为对象的实例（object instance）的变量 person。

```
let new_one = Object.create(person, {department: {value: 'Finance Team', writable: false}}) ;
```

此语句声明了初始数据为对象实例的变量 new_one，其中，这个对象实例继承自变量 person 的对象实例的原型（prototype）定义。在此，新属性 department 的初始数据为字符串'Finance Team'，而且被设置为不可再次被写入（writable: false）新数据的属性。

```
console.log(new_one) ;
```

此语句虽然在网页浏览器的调试工具【Console】面板里，仅仅显示出{department: "Finance Team"}的信息，但是此信息可被展开并显示出【可再度被展开】的内置属性【__proto__】。一旦点击内置属性【__proto__】的项目，便可看到继承自变量 person 的对象实例中的众属性 age、department、married、name 与 occupation。

```
console.log('') ;

new_one.name = 'Alex' ;
```

将字符串'Alex'赋给变量 new_one 对象实例的属性 name。

```
console.log(new_one.name) ;
```

显示出'Alex'的信息。

```
console.log(new_one.department) ;
```

显示出'Finance Team'的信息。

```
console.log('') ;

console.log(new_one.age) ;
console.log(new_one.__proto__.age) ;
```

这两个语句均显示出【在变量 new_one 的对象实例中，其属性 age】的数值 37。这个属性是继承自变量 person 的对象实例。

```
console.log('') ;

Finance Team
new_one.department = 'IT' ;
```

此语句试图将字符串'IT'，赋给变量 new_one 的对象实例中的属性 department，但是会失败！这是因为属性 department 在变量 new_one 被声明的阶段时，就已经被设置为不可再次被写入（writable: false）了。

```
console.log(new_one.department) ;
```

此语句仍然显示出 new_one.department 的初始数据'Finance Team'。

```
console.log('') ;

let my_love = Object.create({name: 'Cindy', occupation: 'angel'}, {age: {value: 18, writable: false, enumerable: false, configurable: false}}) ;
```

此语句声明了初始数据为对象实例的变量 my_love，其中，这个对象实例继承自对象实例{name: 'Cindy', occupation: 'angel'} 的原型定义。在此，其新属性 age 的初始数值为 18，但是被设置为不可再次被写入（writable: false）、不可被迭代 / 列举（enumerable: false）、不可再次被删除 / 修改 / 设置（configurable: false）。

```
console.log(my_love.age) ;
```

显示出整数值 18。

```
console.log('') ;

my_love.age = 27 ;
```

此语句**试图**将整数值 27 赋给变量 my_love 的对象实例中的属性 age，但是会失败！这是因为属性 age 在变量 my_love 被声明的阶段时，就已经被设置为不可再次被写入（writable: false）以及不可再次被删除 / 修改 / 设置（configurable: false）。其中，writable 或 configurable 被设置为 false 时，会使得属性 age 无法再次被写入新数据。

```
console.log(my_love.age) ;
```

此语句仍然显示出 my_love.age 的初始数值 18。

```
console.log('') ;

for (let property in my_love)
```

此 for 语句只会迭代两次。在第 1 次迭代时，变量 property 的数据为字符串'name'；在第 2 次迭代时，变量 property 的数据为字符串'occupation'。请留意，变量 property 的数据不会变成字符串

'age'，是因为属性 age 在变量 my_love 被声明的阶段时，就已经被设置为了不可迭代 / 列举（enumerable: false）了。

9.2 自定义对象实例之间的相同性与合并

本节即将说明，如何判断两个自定义对象实例之间的数据是否完全相同，或是将另一个对象实例的所有成员（属性和函数）合并至当前对象实例中。

9.2.1 判断两个对象实例的数据是否完全相同（ES6）

通过内置函数【Object.is(内含对象实例 1 的变量名称, 内含对象实例 2 的变量名称)】，可判断两个对象实例的数据，是否完全相同。请参看如下示例。

【9-2-1-Object-is.js】

```
console.log(Object.is('mars', 'Mars')) ;
console.log(Object.is('Mar' + 's', 'Mars')) ;
console.log('') ;

console.log(Object.is(document.body, document.body)) ;
console.log('') ;

console.log(Object.is([], [])) ;
console.log(Object.is(['good', 'day'], ['good', 'day'])) ;
console.log('') ;

var profile01 = {name: 'Jasper'} ;
var profile02 = {name: 'Jasper'} ;

console.log(Object.is(profile01, profile02)) ;
console.log('') ;

console.log(Object.is(null, null)) ;
console.log(Object.is(null, undefined)) ;
console.log(Object.is(undefined, undefined)) ;
console.log('') ;

console.log(Object.is(NaN, 0 % 0)) ;
console.log('') ;

console.log(Object.is(0, -0)) ;
console.log(Object.is(-0, -0)) ;
```

【相关说明】

```
console.log(Object.is('mars', 'Mars')) ;
```

Object.is(参数 1 的数据, 参数 2 的数据)若返回布尔值 true,则代表参数 1 与参数 2 内含相同的数据。在此，Object.is('mars', 'Mars')返回布尔值 false。

```
console.log(Object.is('Mar' + 's', 'Mars'));
```

Object.is('Mar' + 's', 'Mars')返回布尔值 true。

```
console.log('');
console.log(Object.is(document.body, document.body));
```

Object.is(document.body, document.body)返回布尔值 true。

```
console.log('');
console.log(Object.is([], []));
```

Object.is([], [])返回布尔值 false，代表两个空数组实例[]，占用不同的内存引址。

```
console.log(Object.is(['good', 'day'], ['good', 'day']));
```

Object.is(Object.is(['good', 'day'], ['good', 'day']))返回布尔值 false，代表两个数组实例['good', 'day']，也占用不同的内存引址。

```
console.log('');
var profile01 = {name: 'Jasper'};
var profile02 = {name: 'Jasper'};

console.log(Object.is(profile01, profile02));
```

Object.is(profile01, profile02)返回布尔值 false，意味着两个对象实例{name: 'Jasper'}占用不同的内存引址。

```
console.log('');
console.log(Object.is(null, null));
```

Object.is(null, null)返回布尔值 true。

```
console.log(Object.is(null, undefined));
```

Object.is(null, undefined)返回布尔值 false。

```
console.log(Object.is(undefined, undefined));
```

Object.is(undefined, undefined)返回布尔值 true。

```
console.log('');
console.log(Object.is(NaN, 0 % 0));
```

Object.is(NaN, 0 % 0)返回布尔值 true，间接意味着【0 % 0】无法被评估成为数值，而返回原始常量 NaN（not a number）。

```
console.log(Object.is(0, -0));
```

Object.is(0, -0)返回布尔值 false，意味着 0 和-0 仍然被视为不同的数据。

```
console.log(Object.is(-0, -0));
```

Object.is(-0, -0)返回布尔值 true。

9.2.2 合并多个对象实例的所有成员（ES6）

通过内置函数 Object.assign()，可将多个对象实例的所有成员（属性和函数），合并至特定对象实例中。

【9-2-2-Object-assign.js】

```
var profile = {name: 'Sean', gender: 'male', age: '25'} ;
var job = {department: 'IT', position: 'software engineer', monthly_salary: '10000' } ;

var info = {log_time: new Date().toLocaleString()} ;

Object.assign(info, profile, job) ;

console.log(info) ;
console.log(info.name) ;
console.log(info['position']) ;
```

【相关说明】

```
var profile = {name: 'Sean', gender: 'male', age: '25'} ;
var job = {department: 'IT', position: 'software engineer', monthly_salary: '10000' } ;
```

这两个语句声明了初始数据为不同对象实例的变量 profile 与 job。

```
var info = {log_time: new Date().toLocaleString()} ;
```

此语句声明了初始数据为仅具单一属性 log_time 的对象实例（例如{log_time: "2019/1/15 上午 10:24:21"}）的变量 info。其属性 log_time 的数据是由【new Date().toLocaleString()】动态产生的日期与时间的字符串，例如'2019/1/15 上午 10:24:21'。

```
Object.assign(info, profile, job) ;
```

此语句将变量 profile 与 job 的对象实例中的各成员（属性和函数），新增至变量 info 的对象实例中，最终变成{log_time: "2019/1/15 上午 10:24:21", name: "Sean", gender: "male", age: "25", department: "IT", position: 'software engineer', monthly_salary: '10000'}。

```
console.log(info) ;
console.log(info.name) ;
```

显示出字符串'Sean'的信息。

```
console.log(info['position']) ;
```

info['position']和 info.position 均可返回字符串'software engineer'。

9.3 对象实例的简短自定义语法（ES6）

JavaScript 语言可通过新对象实例的简短自定义语法，引用已经被声明的各变量名称，来创建**相同于各变量名称的各成员名称**。请参看如下示例。

【9-3---shorthand-syntax-of-object-definition.js】

```
var name = 'Jasper' ;
var age = 33 ;
var gender = 'male' ;

var person = {name, age, gender} ;

console.log(person) ;
console.log(person.name) ;
console.log(person['age']) ;
console.log('') ;

with (person) console.log(gender) ;
```

【相关说明】

```
var name = 'Jasper' ;
var age = 33 ;
var gender = 'male' ;
```

这 3 个语句分别声明了初始数据为字符串或整数值的变量 name、age 和 gender。

```
var person = {name, age, gender} ;
```

此语句声明了变量 person，其初始数据为变量 name、age 和 gender 的数据所共同构成的对象实例{name: "Jasper", age: 33, gender: "male"}。

```
console.log(person) ;
```

显示出{name: "Jasper", age: 33, gender: "male"}的信息。

```
console.log(person.name) ;
```

显示出字符串'Jasper'的信息。

```
console.log(person['age']) ;
```

person['age']和 person.age 均会返回整数值 33。

```
console.log('') ;
```

```
with (person) console.log(gender) ;
```

此语句等价于【console.log(person.gender)】，可显示出字符串'male'的信息。通过【with (内含对象实例的变量名称) { 访问其属性的多个语句 }】，可简化其大括号里的语法【内含对象实例的变量名称.属性名称】为简易的语法【属性名称】。例如，从语法 person.gender 简化为语法 gender。

9.4 自定义对象实例的动态成员名称（ES6）

JavaScript 语言支持将特定字符串表达式返回的字符串，作为自定义对象实例的特定成员名称。请参看以下示例。

【9-4---evaluated-member-name-of-object.js】
```
let id = 'CN24' ;

var one_item = {[id +'_name']: 'video set', [id + '_price']: 300} ;

console.log(one_item) ;
console.log(one_item.CN24_name) ;
console.log(one_item[id + '_price']) ;
```

【相关说明】
```
let id = 'CN24' ;
```
声明初始数据为字符串'CN24'的变量 id。
```
var one_item = {[id +'_name']: 'video set', [id + '_price']: 300} ;
```
此语句声明了初始数据为对象实例 {CN24_name: "video set", CN24_price: 300} 的变量 one_item。

请留意，通过在大括号里的中括号（例如[id +'_name']），可动态赋予特定对象实例的新成员的名称（例如 CN24_name）。
```
console.log(one_item) ;
```
显示出 {CN24_name: "video set", CN24_price: 300} 的信息。
```
console.log(one_item.CN24_name) ;
```
one_item.CN24_name、one_item['CN24_name']和 one_item[id +'_name']当前均可返回字符串 'video set'。
```
console.log(one_item[id + '_price']) ;
```
one_item[id +'_price']、one_item.CN24_price 和 one_item['CN24_price']当前均可返回数值 300。

9.5 对象实例的匹配（ES9）

JavaScript 语言可通过对象实例的匹配（match）语法，同时设置多个变量的数据。请参看如下示例。

【9-5---object-matches.js】

```
var person01 = {name: 'Jimmy', age: 20, gender: 'male'} ;
var person02 = {name: 'Erica', age: 40, gender: 'female'} ;
var person03 = {name: 'Frankie', age: 30, gender: 'male'} ;
var person04 = {name: 'Alexander', age: 28, gender: 'male'} ;

var {name, age, gender} = person01 ;

console.log(name, age, gender) ;
console.log('') ;

({name, age, gender} = person02) ;

console.log(name, age, gender) ;
console.log('') ;

var {name: value01, age: value02, gender: value03} = person03 ;

console.log(value01, value02, value03) ;

var {name, ... others} = person04 ;

console.log(name, others) ;
```

【相关说明】

```
var person01 = {name: 'Jimmy', age: 20, gender: 'male'} ;
var person02 = {name: 'Erica', age: 40, gender: 'female'} ;
var person03 = {name: 'Frankie', age: 30, gender: 'male'} ;
var person04 = {name: 'Alexander', age: 28, gender: 'male'} ;
```

这 4 个语句分别声明了变量 person01、person02、person03 与 person04，其初始数据分别为【其属性名称均相同、其属性数据可以不相同】的对象实例。

```
var { name, age, gender } = person01 ;
```

此对象实例的匹配语法，同时声明了初始数据为字符串'Jimmy'的变量 name、初始数值为 20 的变量 age，以及初始数据为字符串'male'的变量 gender。

```
console.log(name, age, gender) ;
```

显示出【Jimmy 20 male】的信息。

```
({name, age, gender} = person02) ;
```

因为变量 name、age 和 gender 已经被声明过了！若要再次通过对象实例的匹配语句，设置这 3 个变量的新数据，则必须将对象实例的匹配语法，放置于一对小括号之内才行。此语句使得变量 name 的数据成为字符串'Erica'、变量 age 的数值成为整数值 40，以及变量 gender 的数据成为字符串'female'。

```
console.log(name, age, gender) ;
```

显示出【Erica 40 female】的信息。

```
var {name: value01, age: value02, gender: value03} = person03 ;
```

此对象实例的匹配语法，同时声明了初始数据为字符串'Frankie'的变量 value01、初始数值为 30 的变量 value02，以及初始数据为字符串'male'的变量 value03。

请留意，在此语句中的 name、age 和 gender，并不是变量名称，而是用于匹配此对象实例的属性名称。

```
console.log(value01, value02, value03) ;
```

显示出【Frankie 30 male】的信息。

```
var {name, ... others} = person04 ;
```

此对象实例的匹配语法，同时声明了初始数据为字符串'Alexander'的变量 name，以及初始数据为对象实例{age: 28, gender: "male"}的变量 others。

```
console.log(name, others) ;
```

显示出【Alexander】与【{age: 28, gender: "male"}】的信息。

9.6 练 习 题

1. 已知如下 3 个对象实例：

```
let profile01 = {name: 'Alex Lee', age: 27, gender: 'male', married: false} ;
let profile02 = {name: 'Alex Lee', profession: 'software engineer', annual_salary: 15000, company: 'Weird Alien'} ;
let profile03 = {name: 'Alex Lee', country: 'China', place: 'Beijing', phone_number: 'secret'} ;
```

试编写 JavaScript 代码，以合并 3 个对象实例，成为全新的对象实例。

2. 已知存在数据为对象实例{name: 'computer', price: 5000}的变量 product。试编写 JavaScript 源代码，以替变量 product 的对象实例，额外设置【被写入新数据均无效】的属性 manufacturer 和其初始数据"~O_O~"，以及属性 model 和其初始数据"+^v^+"。

第 10 章

处理日期和时间

通过内置的 Date 对象实例的函数 / 方法，JavaScript 语言可以处理非常复杂的日期与时间的运算、转换和显示。

10.1 处理日期

各国的日期显示格式不尽相同，而且计算出不同日期中两个时间点之间的差距，是很复杂的；但是，却可通过 Date 对象实例支持的相关函数，轻松解决此类问题。

10.1.1 获取日期相关数据

通过【内含 Date 对象实例的变量名称】：

- 衔接【.getDate()】，可返回在特定日期中，代表特定月份第几日的整数值。
- 衔接【.getDay()】，可返回在特定日期中，代表星期几的整数值。
- 衔接【.getMonth()】，可返回在特定日期中，代表月份的整数值。
- 衔接【.getFullYear()】，可返回在特定日期中，代表公元年份的整数值。
- 衔接【.getUTCDate()】、【.getUTCDay()】、【.getUTCMonth()】和【.getUTCFullYear()】，可分别返回当前【协调世界时 (UTC, coordinated universal time)】的相关整数值。

关于获取日期相关数据的综合调试，可参看如下示例。

【10-1-1-date-gettings.js】

```
let special = new Date('2014/5/20') ;

console.log(special) ;
```

```
console.log('') ;

console.log(special.getDate()) ;

// return value 0 represents 'Sunday'.
console.log(special.getDay()) ;

// return value 0 represents 'January', and 11 represents 'December'.
console.log(special.getMonth()) ;

console.log(special.getFullYear()) ;
console.log('') ;

console.log(special.getUTCDate()) ;
console.log(special.getUTCDay()) ;
console.log(special.getUTCMonth()) ;
console.log(special.getUTCFullYear()) ;
console.log('') ;
```

【相关说明】

```
let special = new Date('2014/5/20') ;
```

此语句声明了变量 special，其初始数据为内含**日期 2014/5/20 相关数据**的 Date 对象实例。2014/5/20 在此被视为本地（中国）的日期。

```
console.log(special) ;
```

显示出类似【Tue May 20 2014 00:00:00 GMT+0800 (China Standard Time)】的信息。

```
console.log('') ;

console.log(special.getDate()) ;
```

special.getDate()会返回上述 Date 对象实例中的特定月份第几日的整数值 20。

```
// return value 0 represents 'Sunday'.
console.log(special.getDay()) ;
```

special.getDay()会返回上述 Date 对象实例中的星期几的数值 2，也就是代表星期二。若返回 0，则代表星期日；若返回 6，则代表星期六。

```
// return value 0 represents 'January', and 11 represents 'December'.
console.log(special.getMonth()) ;
```

special.getMonth()会返回上述 Date 对象实例中的月份的数值 4。若返回 0，代表 1 月；若返回 11，则代表 12 月。

```
console.log(special.getFullYear()) ;
```

special.getFullYear()会返回上述 Date 对象实例中的公元年份的数值 2014。

```
console.log('') ;

console.log(special.getUTCDate()) ;
```

special.getUTCDate()会先将变量 special 的 Date 对象实例中的本地日期 2014/5/20，转换成为协

调世界时（UTC, coordinated universal time）的日期 2014/5/19，再返回其特定月份第几日的数值 19。

```
console.log(special.getUTCDay()) ;
```

special.getUTCDay()会先将变量 special 的 Date 对象实例中的本地日期 2014/5/20，转换成为协调世界时的日期 2014/5/19，再返回其星期几的数值 1（星期一）。

```
console.log(special.getUTCMonth()) ;
```

special.getUTCMonth()会先将变量 special 的 Date 对象实例中的本地日期 2014/5/20，转换成为协调世界时的日期 2014/5/19，再返回其月份的数值 4（5 月）。若返回 0，则代表 1 月；若返回 11，则代表 12 月。

```
console.log(special.getUTCFullYear()) ;
```

special.getUTCFullYear()会先将变量 special 的 Date 对象实例中的本地日期 2014/5/20，转换成为协调世界时的日期 2014/5/19，再返回其年份的数值 2014。

10.1.2　设置日期相关数据

通过【内含 Date 对象实例的变量名称】：

- 衔接【.setDate()】，可设置在特定日期中，代表特定月份**第几日**的整数值。
- 衔接【.setMonth()】，可设置在特定日期中，代表**月份**的整数值。
- 衔接【.setFullYear()】，可设置在特定日期中，代表**公元年份**的整数值。
- 衔接【.setUTCDate()】、【.setUTCMonth()】和【.setUTCFullYear()】，可分别设置当前【协调世界时（UTC, coordinated universal time）】的相关整数值。

关于设置日期相关数据的综合调试，可参看如下示例。

【10-1-2-date-settings.js】

```
let special = new Date('2013/5/30') ;

console.log(special) ;
console.log('') ;

special.setFullYear(2014) ;
special.setMonth(11) ;
special.setDate(1) ;

console.log(special) ;
console.log('') ;

special.setUTCFullYear(2014) ;
special.setUTCMonth(11) ;
special.setUTCDate(1) ;

console.log(special) ;
```

【相关说明】

```
let special = new Date('2013/5/30') ;
```

此语句声明了变量 special，其初始数据为内含**日期 2013/5/30 相关数据**的 Date 对象实例。2013/5/30 在此被视为本地（中国）的日期。

```
console.log(special) ;
```

显示出类似【Thu May 30 2013 00:00:00 GMT+0800 (China Standard Time)】的信息。

```
console.log('') ;
special.setFullYear(2014) ;
```

重新设置变量 special 的 Date 对象实例中的**年份**数值为 2014。

```
special.setMonth(11) ;
```

重新设置变量 special 的 Date 对象实例中的**月份**数值为 11（12 月）。

```
special.setDate(1) ;
```

重新设置变量 special 的 Date 对象实例中的特定月份**第几日**的数值为 1。

```
console.log(special) ;
```

显示出类似【Mon Dec 01 2014 00:00:00 GMT+0800 (China Standard Time)】的信息。

```
console.log('') ;
special.setUTCFullYear(2014) ;
```

此语句依据协调世界时（UTC, coordinated universal time），重新设置变量 special 的 Date 对象实例中的年份数值为 2014。

```
special.setUTCMonth(11) ;
```

依据协调世界时，重新设置变量 special 的 Date 对象实例中的月份数值为 11（代表 12 月）。

```
special.setUTCDate(1) ;
```

依据协调世界时，重新设置在变量 special 的 Date 对象实例中的特定月份第几日的数值为 1。

```
console.log(special) ;
```

显示出类似【Tue Dec 02 2014 00:00:00 GMT+0800 (China Standard Time)】的信息。

10.1.3　带有日期的格式化字符串

通过【内含 Date 对象实例的变量名称】衔接【.toDateString()】、【.toISOString()】、【.toJSON()】和【.toLocaleDateString()】，可分别返回不同格式的日期字符串。请参看如下示例。

【10-1-3-date-formatting-strings.js】

```
let today = new Date() ;
```

```
with (console)
{
  log(today) ;
  log('') ;

  log(today.toDateString()) ;
  log(today.toISOString()) ;
  log(today.toJSON()) ;
  log('') ;

  log(today.toLocaleDateString()) ;
  log('') ;

  log(typeof today) ;
  log('') ;

  log(typeof today.toDateString()) ;
  log(typeof today.toISOString()) ;
  log(typeof today.toJSON()) ;
  log(typeof today.toLocaleDateString()) ;
}
```

【相关说明】

```
let today = new Date() ;
```

此语句声明了变量 today，其初始数据为内含**当前日期与时间**（例如：Thu Nov 29 2018 05:50:53 GMT+0800 (China Standard Time)）的 Date 对象实例。

```
with (console)
{

  log(today) ;
```

显示出类似【Thu Nov 29 2018 05:50:53 GMT+0800 (China Standard Time)】的信息。

```
  log('') ;

  log(today.toDateString()) ;
```

显示出类似【Thu Nov 29 2018】本地日期的信息。

```
  log(today.toISOString()) ;
  log(today.toJSON()) ;
```

这两个语句均会将本地的日期与时间，先转换成为协调世界时的日期与时间，再依据 ISO 8601 格式，显示出类似【2018-11-29T21:50:53.319Z】的信息。

```
  log('') ;

  log(today.toLocaleDateString()) ;
```

显示出类似【2018/11/29】的简略的本地日期。

```
  log('') ;

  log(typeof today) ;
```

变量 today 的数据类型为 object（对象），所以【typeof today】返回字符串'object'。

```
log('') ;

log(typeof today.toDateString()) ;
log(typeof today.toISOString()) ;
log(typeof today.toJSON()) ;
log(typeof today.toLocaleDateString()) ;
```

这 4 个语句均显示出'string'，意味着 today.toDateString()、today.toISOString()、today.toJSON() 与 today.toLocaleDateString()所返回的数据，均是字符串（string）。

10.2　处理时间

虽然各国的时间显示格式大致相同，但是计算出不同日期中两个时间点之间的差距，仍是过于复杂的问题；但是，却可通过 Date 对象实例支持的时间相关函数，轻松解决此类问题。

10.2.1　获取时间相关数据

通过【内含 Date 对象实例的变量名称】：
- 衔接【.getHours()】，可返回在特定时间点中，代表小时数的整数值。
- 衔接【.getMinutes()】，可返回在特定时间点中，代表分钟数的整数值。
- 衔接【.getSeconds()】，可返回在特定时间点中，代表秒数的整数值。
- 衔接【.getMilliseconds()】，可返回在特定时间点中，代表毫秒数的整数值。
- 衔接【.getTime()】，可返回从纪元时间（epoch time），也就是公元 1970 年 1 月 1 日 0 时 0 分 0 秒开始，至特定时间点为止的毫秒数。

此外，通过【内含 Date 对象实例的变量名称】衔接【.getUTCHours()】、【.getUTCMinutes()】、【.getUTCSeconds()】、【.getUTCMilliseconds()】和【.getTimezoneOffset()】，可返回【协调世界时（UTC, coordinated universal time）】的特定时间点的相关整数值。请参看如下示例。

【10-2-1-time-gettings.js】

```
let current = new Date() ;

with (console)
{
  log(current.getHours()) ;
  log(current.getMinutes()) ;
  log(current.getSeconds()) ;
  log(current.getMilliseconds()) ;
  log(current.getTime()) ;
  log('') ;

  log(current.getUTCHours()) ;
```

```
    log(current.getUTCMinutes()) ;
    log(current.getUTCSeconds()) ;
    log(current.getUTCMilliseconds()) ;
    log('') ;

    log(current.getTimezoneOffset()) ;
}
```

【相关说明】

```
let current = new Date() ;
```

此语句声明了变量 current，其初始数据为内含**当前日期与时间**的 Date 对象实例。

```
with (console)
{
```

```
    log(current.getHours()) ;
```

显示出变量 current 的 Date 对象实例中的本地【小时数】，例如 6。

```
    log(current.getMinutes()) ;
```

显示出变量 current 的 Date 对象实例中的本地【分钟数】，例如 13。

```
    log(current.getSeconds()) ;
```

显示出变量 current 的 Date 对象实例中的本地【秒数】，例如 30。

```
    log(current.getMilliseconds()) ;
```

显示出变量 current 的 Date 对象实例中的本地【毫秒数】，例如 883。

```
    log(current.getTime()) ;
```

current.getTime()返回【从 1970/1/1 00:00:00:000（0 时 0 分 0 秒 0 毫秒）开始，到变量 current 的 Date 对象实例中的本地**时间点**为止】的【毫秒数】，例如 1543465500880。

```
    log('') ;

    log(current.getUTCHours()) ;
```

依据协调世界时（UTC, coordinated universal time），显示出变量 current 的 Date 对象实例中的【小时数】，例如 22。

```
    log(current.getUTCMinutes()) ;
```

依据协调世界时，显示出变量 current 的 Date 对象实例中的【分钟数】，例如 13。

```
    log(current.getUTCSeconds()) ;
```

依据协调世界时，显示出变量 current 的 Date 对象实例中的【秒数】，例如 30。

```
    log(current.getUTCMilliseconds()) ;
```

依据协调世界时，显示出变量 current 的 Date 对象实例中的【毫秒数】，例如 883。

```
    log('') ;
```

```
log(current.getTimezoneOffset());
```

current.getTimezoneOffset()会先以分钟数作为测量单位，再返回其本地（中国）时间相对于协调世界时的时间差-480。

10.2.2　设置时间相关数据

通过【内含 Date 对象实例的变量名称】：

- 衔接【.setHours()】，可设置特定日期中，代表小时数的整数值。
- 衔接【.setMinutes()】，可设置特定日期中，代表分钟数的整数值。
- 衔接【.setSeconds()】，可设置特定日期中，代表秒数的整数值。
- 衔接【.setMilliseconds()】，可设置特定日期中，代表毫秒数的整数值。
- 衔接【.setTime()】，可设置从纪元时间（epoch time），也就是公元 1970 年 1 月 1 日 0 时 0 分 0 秒开始，至特定时间点为止的毫秒数。

此外，通过函数【内含 Date 对象实例的变量名称】衔接【.setUTCHours()】、【.setUTCMinutes()】、【.setUTCSeconds()】和【.setUTCMilliseconds()】，可分别设置【协调世界时（UTC, coordinated universal time）】的特定时间点的相关整数值。请参看如下示例。

【10-2-2-time-settings.js】

```
let special = new Date('2025/6/6 11:50:00');

console.log(special);
console.log('');

special.setHours(19);
special.setMinutes(45);
special.setSeconds(30);
special.setMilliseconds(777);

console.log(special);
console.log(special.getTime());
console.log('');

special.setUTCHours(19);
special.setUTCMinutes(45);
special.setUTCSeconds(30);
special.setUTCMilliseconds(777);

console.log(special);
console.log(special.getTime());
```

【相关说明】

```
let special = new Date('2025/6/6 11:50:00');
```

此语句声明了变量 special，其初始数据为【内含日期与时间 2025/6/6 11:50:00 (星期五)】的 Date 对象实例。

```
console.log(special);
```

显示出类似【Fri Jun 06 2025 11:50:00 GMT+0800 (China Standard Time)】的信息。

```
console.log('');
```

```
special.setHours(19);
```

重新设置变量 special 的 Date 对象实例中的【小时数】为 19。

```
special.setMinutes(45);
```

重新设置变量 special 的 Date 对象实例中的【分钟数】为 45。

```
special.setSeconds(30);
```

重新设置变量 special 的 Date 对象实例中的【秒数】为 30。

```
special.setMilliseconds(777);
```

重新设置变量 special 的 Date 对象实例中的【毫秒数】为 777。

```
console.log(special);
```

显示出类似【Fri Jun 06 2025 19:45:30 GMT+0800 (China Standard Time)】的信息。

```
console.log(special.getTime());
```

special.getTime()返回【从 1970/1/1 00:00:00:000（0 时 0 分 0 秒 0 毫秒）开始，到变量 special 的 Date 对象实例中的本地特定时间点为止】的毫秒数 1749210330777。

```
console.log('');
```

```
special.setUTCHours(19);
```

依据协调世界时（UTC, coordinated universal time），重新设置变量 special 的 Date 对象实例中的【小时数】为 19。

```
special.setUTCMinutes(45);
```

依据协调世界时，重新设置变量 special 的 Date 对象实例中的【分钟数】为 45。

```
special.setUTCSeconds(30);
```

依据协调世界时，重新设置变量 special 的 Date 对象实例中的【秒数】为 30。

```
special.setUTCMilliseconds(777);
```

依据协调世界时，重新设置变量 special 的 Date 对象实例中的【毫秒数】为 777。

```
console.log(special);
```

显示出类似【Sat Jun 07 2025 03:45:30 GMT+0800 (China Standard Time)】的信息。

```
console.log(special.getTime());
```

special.getTime()返回【从 1970/1/1 00:00:00:000（0 时 0 分 0 秒 0 毫秒）开始，到变量 special 的 Date 对象实例中的本地特定时间点为止】的毫秒数 1749239130777。

10.2.3　带有时间的格式化字符串

通过【内含 Date 对象实例的变量名称】衔接【.toLocaleTimeString()】、【.toLocaleString()】、【.toTimeString()】、【.toISOString()】和【.toString()】，可返回不同格式的时间字符串。请参看如下示例。

【10-2-3-time-formatting-strings.js】

```
let special = new Date('2100/8/8 10:30:00') ;

with (console)
{
  log(special) ;
  log('') ;

  log(special.toLocaleTimeString()) ;
  log(special.toLocaleString()) ;
  log(special.toTimeString()) ;
  log(special.toISOString()) ;
  log(special.toString()) ;
}
```

【相关说明】

```
let special = new Date('2100/8/8 10:30:00') ;
```

此语句声明了变量 special，其初始数据为【内含日期与时间 2100/8/8 10:30:00 (星期日)】的 Date 对象实例。

```
with (console)
{

  log(special) ;
```

显示出类似【Sun Aug 08 2100 10:30:00 GMT+0800 (China Standard Time)】的信息。

```
  log(special.toLocaleTimeString()) ;
```

显示出类似【上午 10:30:00】的简略本地时间。

```
  log(special.toLocaleString()) ;
```

显示出类似【2100/8/8 上午 10:30:00】的本地日期与时间。

```
  log(special.toTimeString()) ;
```

显示出类似【10:30:00 GMT+0800 (China Standard Time)】的本地时间。

```
  log(special.toISOString()) ;
```

special.toISOString()会先将本地的日期与时间,转换成为协调世界时的日期与时间,再依据 ISO 8601 格式，显示出类似【2100-08-08T02:30:00.000Z】的信息。

```
  log(special.toString()) ;
```

显示出类似【Sun Aug 08 2100 10:30:00 GMT+0800 (China Standard Time)】的本地日期与时间。

10.2.4　处理纪元时间至今的毫秒数（ES6）

纪元时间（epoch time）是指公元 1970 年 1 月 1 日 0 时 0 分 0 秒的时间点。通过函数【Date.now()】或【Date.parse(内含当前时间点的 Date 对象实例的变量名称)】，可返回纪元时间到当前时间点的毫秒数。请参看如下两个示例。

【10-2-4-e1-Date-now-and-parse.js】

```
let today = new Date() ;

let count = today.getTime() ;

console.log(count) ;
console.log(Date.now()) ;
console.log(Date.parse(today)) ;
console.log(Date.parse('2023/01/01 18:30:00')) ;
console.log('') ;

console.log(typeof count) ;
console.log(typeof Date.now()) ;
console.log(typeof Date.parse(today)) ;
```

【相关说明】

```
let today = new Date() ;
```

此语句声明了变量 today，其初始数据为内含当前日期与时间的 Date 对象实例。

```
let count = today.getTime() ;
```

today.getTime()会返回【从 1970/1/1 00:00:00:000（0 时 0 分 0 秒 0 毫秒）开始，到变量 today 的 Date 对象实例中的本地特定时间点为止】的毫秒数，例如 1543473928708。

```
console.log(count) ;

console.log(Date.now()) ;
```

Date.now()会返回【从 1970/1/1 00:00:00:000 开始，到执行此语句的当前时间点为止】的毫秒数，例如 1543473928709。

```
console.log(Date.parse(today)) ;
```

Date.parse(today)会返回【从 1970/1/1 00:00:00:000 开始，到执行此语句的当前时间点为止】较不精确的毫秒数，例如 1543473928000。

```
console.log(Date.parse('2023/01/01 18:30:00')) ;
```

Date.parse('2023/01/01 18:30:00')会返回【从 1970/1/1 00:00:00:000 开始，到 2023/01/01 18:30:00:000 为止】的毫秒数，例如 1672569000000。

```
console.log('') ;

console.log(typeof count) ;
console.log(typeof Date.now()) ;
console.log(typeof Date.parse(today)) ;
```

变量 count、Date.now()与 Date.parse(today)的返回数据，均是数值（number）类型的，所以这 3 个语句均显示出' number'的信息。

【10-2-4-e2-date-UTC-and-valueOf.js】

```
let ms_count = Date.UTC(2100, 10, 1, 19, 45, 30) ;

let special = new Date(ms_count) ;

with (console)
{
 log(Date.now()) ;
 log('') ;

 log(ms_count) ;
 log(special.getTime()) ;
 log(special.valueOf()) ;
 log(Date.parse(special)) ;
 log('') ;

 log(typeof ms_count) ;
 log(typeof special.getTime()) ;
 log(typeof special.valueOf()) ;
 log(typeof Date.parse(special)) ;
}
```

【相关说明】

```
let ms_count = Date.UTC(2100, 10, 1, 19, 45, 30) ;
```

此语句声明了变量 ms_count，其初始数据为【从 1970/1/1 00:00:00:000 开始，到协调世界时的日期与时间 2100/11/1 19:45:30 (星期一) 为止】的毫秒数 4128781530000。请留意，月份的数值 10 代表 11 月。

```
let special = new Date(ms_count) ;
```

此语句声明了变量 special，其初始数据为【内含协调世界时的日期与时间 2100/11/1 19:45:30（星期一），被转换成为本地（中国）日期与时间 2100/11/2 03:45:30（星期二）】之后的 Date 对象实例。

```
with (console)
{
 log(Date.now()) ;
```

Date.now()会返回【从 1970/1/1 00:00:00:000 开始，到执行此语句的当前时间点为止】的毫秒数，例如 1543474327408。

```
 log('') ;

 log(ms_count) ;
```

显示出毫秒数 4128781530000 的信息。

```
 log(special.getTime()) ;
```

```
log(special.valueOf()) ;
log(Date.parse(special)) ;
```

这 3 个语句均显示出【从 1970/1/1 00:00:00:000 开始，到变量 special 的 Date 对象实例中的本地特定时间点为止】的毫秒数 4128781530000。

```
log('') ;

log(typeof ms_count) ;
log(typeof special.getTime()) ;
log(typeof special.valueOf()) ;
log(typeof Date.parse(special)) ;
```

变量 ms_count、special.getTime()、special.valueOf()与 Date.parse(special)的返回数据，均是数值 (number) 类型的，所以这 4 个语句均显示出'number'的信息。

10.3　练　习　题

1. 试编写 JavaScript 源代码，以计算出从出生的那时刻起，至当前为止，您生活在地球上的天数。
2. 试编写 JavaScript 源代码，以显示出当前时间点的 100 天之后的日期和时间。

第11章

DOM 的事件处理（一）

文档对象模型（document object model，DOM）可将特定网页的内容，改以树状结构，来表示其内容里各节点的片段。编程人员可进一步通过 JavaScript 引擎支持的对象实例的相关属性和函数，在特定网页里，访问特定节点的片段；并可在特定事件（event）发生时，通过事先编写的源代码，和用户进行互动。本章内容主要介绍鼠标事件、键盘事件与表单事件的处理的说明。

11.1 鼠标事件

鼠标事件（mouse event）是指在特定网页里，用户借助指针设备，进行互动时所发生的事件。

11.1.1 单击和双击事件

在特定网页里，用户按下鼠标左键时，会发生单击（click）事件；用户快速按下鼠标的左键2次时，则会发生双击（double click）事件。请参看如下示例。

【11-1-1-mouse-onclick-and-ondblclick.js】

```
let element = null ;

for (let i = 1; i < 7; i++)
{
 element = document.createElement('button') ;

 element.id = 'btn0' + i ;
 element.innerHTML = 'TEST 0' + i ;

 element.style.margin = '5px' ;
 element.style.fontSize = '20px' ;
```

```
    element.style.borderRadius = '5px' ;
    element.style.backgroundColor = 'LightBlue' ;
    element.style.color = 'Green' ;

    document.body.appendChild(element) ;
}

btn01.onclick = function(event)
{
    console.log(`Button ${event.target.id} was clicked.`) ;
} ;
///
function actions(event)
{
    console.log(`Button ${event.target.id} was clicked.`) ;
}

btn02.onclick = actions ;
///
btn03.onclick = (event) => console.log(`Button ${event.target.id} was clicked.`) ;

btn04.addEventListener('click', (event) => console.log(`Button ${event.target.id} was
    clicked.`)) ;

btn05.ondblclick = (event) => console.log(`Button ${event.target.id} was doubly clicked.`) ;

btn06.addEventListener('dblclick', (event) => console.log(`Button ${event.target.id} was
    doubly clicked.`)) ;
console.log(`Button ${event.target.id} was doubly clicked.`)) ;
```

【相关说明】

```
let element = null ;
```

声明初始数据为 null（空值）的变量 element。

```
for (let i = 1; i < 7; i++)
```

这个 for 循环语句会迭代 6 次。

```
{
```

```
    element = document.createElement('button') ;
```

此语句将变量 element 的数据，设置为网页里代表按钮的 button 元素实例。

```
    element.id = 'btn0' + i ;
```

此语句设置了新 button 元素实例的属性 id 的数据，成为字符串'btn01'~'btn06'的其中之一，进而变成特定 button 元素实例的身份识别码。

```
    element.innerHTML = 'TEST 0' + i ;
```

此语句设置了新 button 元素实例的属性 innerHTML 的数据，成为字符串'TEST 01'~'TEST 06'，进而变成特定按钮上的文本。

```
element.style.margin = '5px' ;
```

此语句设置了新 button 元素实例的属性 margin 的数据，成为'5px'，使得每个按钮的边框之外的间隔距离，成为 5 像素，进而让按钮之间的距离，变成 5 + 5 = 10 像素。

```
element.style.fontSize = '20px' ;
```

此语句设置了新 button 元素实例的属性 fontSize 的数据，成为'20px'，使得每个按钮上单一文字的尺寸，变成 20 像素。

```
element.style.borderRadius = '5px' ;
```

此语句设置了新 button 元素实例的属性 borderRadius 的数据，成为'5px'，使得每个按钮的边框，变成半径为 5 像素的圆角边框。

```
element.style.backgroundColor = 'LightBlue' ;
```

此语句设置了新 button 元素实例的属性 backgroundColor 的数据，成为'LightBlue'，使得每个按钮的背景颜色，变成浅蓝色。

```
element.style.color = 'Green' ;
```

此语句设置了新 button 元素实例的属性 color，成为'Green'，使得每个按钮上的文本颜色，变成绿色。

```
document.body.appendChild(element) ;
```

此语句在网页里的 body 元素实例内，新增变量 element 所代表的新 button 元素实例。

```
}
btn01.onclick = function(event)
{
  console.log(`Button ${event.target.id} was clicked.`) ;
} ;
```

此语法在属性 id 的数据为'btn01'的按钮元素实例中，设置其属性 onclick 的数据，成为【显示特定按钮被用户单击之后的信息】的匿名函数的定义。在这个匿名函数所显示的信息里，包含特定按钮的属性 id 的数据字符串，例如【'btn01'、'btn02'、....、'btn06'】，使得属性 id 的数据为'btn01'的按钮元素实例，被监听着单击（click）动作。一旦此按钮元素实例被单击时，上述匿名函数就会被调用。

```
function actions(event)
{
  console.log(`Button ${event.target.id} was clicked.`) ;
}
```

此语法定义了带有参数 event 的函数 actions()，并且和上述的匿名函数，具有相同的效果。

```
btn02.onclick = actions ;
```

此语句在属性 id 的数据为'btn02'的按钮元素实例中，设置其属性 onclick 的数据，成为函数名称 actions。这意味着属性 id 的数据为'btn02'的按钮元素实例，被监听着单击（click）动作。一旦此按钮元素实例被单击时，函数 actions()就会被调用。

```
btn03.onclick = (event) => console.log(`Button ${event.target.id} was clicked.`) ;
```

此语句在属性 id 的数据为'btn03'的按钮元素实例中，设置其属性 onclick 的数据，成为箭头函数【(event) => console.log(`Button ${event.target.id} was clicked.`)】的定义，使得属性 id 的数据为'btn03'的按钮元素实例，被监听着单击（click）动作。一旦此按钮元素实例被单击时，上述箭头函数就会被调用。

```
btn04.addEventListener('click', (event) => console.log(`Button ${event.target.id} was
  clicked.`)) ;
```

此语句通过调用 addEventListener()函数，使得属性 id 的数据为'btn04'的按钮元素实例，被监听着单击动作。一旦属性 id 的数据为'btn04'的按钮元素实例被单击时，【显示特定按钮被用户单击之后的信息】的箭头函数【(event) => console.log(`Button ${event.target.id} was clicked.`)】就会被调用。

```
btn05.ondblclick = (event) => console.log(`Button ${event.target.id} was doubly clicked.`) ;
```

此语句在属性 id 的数据为'btn05'的按钮元素实例中，设置其属性 ondblclick 的数据，成为箭头函数【(event) => console.log(`Button ${event.target.id} was doubly clicked.`)】的定义，使得属性 id 的数据为'btn05'的按钮元素实例，被监听着双击（double click）动作。一旦此按钮元素实例被双击时，上述箭头函数就会被调用。

```
btn06.addEventListener('dblclick', (event) => console.log(`Button ${event.target.id} was
  doubly clicked.`)) ;
```

此语句通过调用 addEventListener()函数，使得属性 id 的数据为'btn06'的按钮元素实例，被监听双击（double click）动作。一旦属性 id 的数据为'btn06'对应的按钮元素实例被双击时，【显示特定按钮被用户双击之后的信息】的箭头函数【(event) => console.log(`Button ${event.target.id} was doubly clicked.`)】就会被调用。

11.1.2 上下文菜单事件

在特定网页里，用户按下鼠标右键或键盘上的菜单键（menu key）时，会发生右击（right click）/ 上下文菜单（context menu）事件。此时，在默认情况下，上下文菜单就会显示在特定的网页里。请参看如下示例。

【11-1-2-mouse-oncontextmenu.js】

```
let element = null ;

function display_message()
{
  event.preventDefault();

  console.log('Document was right clicked.') ;
}

// document.oncontextmenu = display_message ;
document.addEventListener('contextmenu', display_message) ;
```

【相关说明】

```
let element = null ;
```

此语句声明了初始数据为 null（空值）的变量 element。

```
function display_message()
{

  event.preventDefault();
```

此语句可使得鼠标右键被点击时，不会出现默认的上下文菜单。

```
  console.log('Document was right clicked.') ;
}
```

此语法定义了显示相关信息的函数 display_message()。

```
// document.oncontextmenu = display_message ;
document.addEventListener('contextmenu', display_message) ;
```

通过调用 addEventListener()函数，使得代表特定网页的 document 对象实例，被监听启动上下文菜单（context menu）的【鼠标右键被点击】动作。一旦在当前网页中，用户按下鼠标右键时，显示相关信息的函数 display_message()就会被调用。

11.1.3　鼠标按下与松开事件

在特定网页里，用户按下鼠标左键的瞬间，会发生鼠标按下（mouse down）事件；放开鼠标左键的瞬间，会发生鼠标松开（mouse up）事件。请参看如下示例。

【11-1-3-mouse-onmousedown-and-onmouseup.js】

```
let element = null ;

for (let i = 1; i < 5; i++)
{
  element = document.createElement('div') ;

  element.id = 'div0' + i ;
  element.innerHTML = 'Block 0' + i ;

  element.style.width = '150px' ;
  element.style.margin = '5px auto' ;
  element.style.textAlign = 'center' ;
  element.style.fontSize = '20px' ;
  element.style.borderRadius = '5px' ;
  element.style.backgroundColor = 'LightBlue' ;
  element.style.color = 'Green' ;

  document.body.appendChild(element) ;
}

function action01(event)
{
```

```
  console.log(`Block ${event.target.id} was pressed.`) ;
}

div01.onmousedown = action01 ;

div02.addEventListener('mousedown', action01) ;

///
function action02(event)
{
  console.log(`Block ${event.target.id} was released.`) ;
}

div03.onmouseup = action02 ;

div04.addEventListener('mouseup', action02) ;
```

【相关说明】

```
let element = null ;
```

声明初始数据为 null（空值）的变量 element。

```
for (let i = 1; i < 5; i++)
```

此 for 循环语句会迭代 4 次。

```
{
```

```
  element = document.createElement('div') ;
```

此语句使得变量 element 的数据，成为网页里的新的 div 元素实例。

```
  element.id = 'div0' + i ;
```

此语句设置了新 div 元素实例的属性 id 的数据，成为字符串'div01'~'div04'其中之一，进而变成这个 div 元素实例的身份识别码。

```
  element.innerHTML = 'Block 0' + i ;
```

此语句设置了新 div 元素实例的属性 innerHTML 的数据，成为字符串'Block 01'~'Block 04'其中之一，进而变成这个 div 元素实例上的文本。

```
  element.style.width = '150px' ;
  element.style.margin = '5px auto' ;
  element.style.textAlign = 'center' ;
  element.style.fontSize = '20px' ;
  element.style.borderRadius = '5px' ;
  element.style.backgroundColor = 'LightBlue' ;
  element.style.color = 'Green' ;
```

这 7 个语句分别设置了新 div 元素实例在网页里的不同部分的外观。

```
  document.body.appendChild(element) ;
```

此语句在网页 body 元素实例内，新增了变量 element 所代表的新 div 元素实例。

```
}
```

```
function action01(event)
{
  console.log(`Block ${event.target.id} was pressed.`) ;
```

此语句显示出特定 div 元素实例被按下鼠标左键的信息，并带有特定 div 元素实例的属性 id 的数据字符串'div01'、'div02'、'div03'或'div04'其中之一。

```
}
```

上述语法定义了带有参数 event 的函数 action01()。

```
div01.onmousedown = action01 ;
```

此语句在属性 id 的数据为'div01'的 div 元素实例中，设置其属性 onmousedown 的数据，成为函数名称 action01，使得属性 id 的数据为'div01'的 div 元素实例，被监听着被按下鼠标左键（mouse down / pressed）的动作。一旦在此 div 元素实例上，被按下鼠标左键，函数 action01()就会被调用。

```
div02.addEventListener('mousedown', action01) ;
```

此语句通过调用 addEventListener()函数，使得属性 id 的数据为'div02'的 div 元素实例，被监听着被按下鼠标左键（mouse down / pressed）的动作。一旦在属性 id 的数据为'div02'的 div 元素实例上，被按下鼠标左键时，函数 actions01()就会被调用。

```
function action02(event)
{
  console.log(`Block ${event.target.id} was released.`) ;
```

此语句显示出在特定 div 元素实例上，释放（released）鼠标左键的信息，并带有特定 div 元素实例的属性 id 的数据字符串'div01'、'div02'、'div03'或'div04'其中之一。

```
}
div03.onmouseup = action02 ;
```

此语句在属性 id 的数据为'div03'的 div 元素实例中，设置其属性 onmouseup 的数据，成为函数名称 action02，使得属性 id 的数据为'div03'的 div 元素实例，被监听着被释放鼠标左键（mouse up / released）的动作。一旦在此 div 元素实例上，被释放鼠标左键时，函数 action02()就会被调用。

```
div04.addEventListener('mouseup', action02) ;
```

此语句通过调用 addEventListener()函数，使得属性 id 的数据为'div04'的 div 元素实例，被监听着被释放鼠标左键（mouse up / released）的动作。一旦在属性 id 的数据为'div04'的 div 元素实例上，被释放鼠标左键时，函数 actions02()就会被调用。

11.1.4　鼠标指针相关进出事件

在特定网页里，鼠标指针被移入特定元素实例的范围时，会发生鼠标进入（mouse enter）事件和鼠标悬停（mouse over）事件；鼠标指针被移出特定元素实例的范围时，则会发生鼠标离开（mouse leave）事件和鼠标脱出（mouse out）事件。

鼠标进入事件和鼠标悬停事件的差异，在于鼠标进入事件仅会作用于当前元素实例**本身**，而不包括其**子**元素实例；鼠标悬停事件则会作用于当前元素实例，以及其所有的**子**元素实例！关于鼠标指针相关进出事件的运用，可参看如下示例。

【11-1-4-mouse-cursor-in-and-out.js】

```js
let element = null ;

element = document.createElement('div') ;

element.id = 'outer' ;
element.innerHTML = 'Outer' ;

element.style.width = '150px' ;
element.style.height = '100px' ;
element.style.margin = '5px auto' ;
element.style.fontSize = '1em' ;
element.style.borderRadius = '5px' ;
element.style.backgroundColor = 'LightBlue' ;
element.style.color = 'Green' ;

document.body.appendChild(element) ;

element = document.createElement('div') ;

element.id = 'inner' ;
element.innerHTML = 'Inner' ;

element.style.width = '100px' ;
element.style.height = '50px' ;
element.style.margin = '5px auto' ;
element.style.textAlign = 'center' ;
element.style.fontSize = '1em' ;
element.style.borderRadius = '5px' ;
element.style.backgroundColor = 'Teal' ;
element.style.color = 'GreenYellow' ;

outer.appendChild(element) ;

function action01(event)
{
  console.log(`Block ${event.target.id} was left.`) ;
}

// outer.onmouseenter = action01 ;
// outer.onmouseover = action01 ;
// outer.onmouseleave = action01 ;
outer.onmouseout = action01 ;
```

【相关说明】

```js
let element = null ;
```

声明了初始数据为 null（空值）的变量 element。

```
element = document.createElement('div') ;
```

此语句让变量 element 的数据,成为网页里的新的 div 元素实例。

```
element.id = 'outer' ;
```

设置新 div 元素实例的属性 id 的数据,成为字符串'outer',进而变成这个 div 元素实例的身份识别码。

```
element.innerHTML = 'Outer' ;
```

此语句设置了新 div 元素实例的属性 innerHTML 的数据,成为字符串'Outer',进而变成这个 div 元素实例上的文本。

```
element.style.width = '150px' ;
element.style.height = '100px' ;
element.style.margin = '5px auto' ;
element.style.fontSize = '1em' ;
element.style.borderRadius = '5px' ;
element.style.backgroundColor = 'LightBlue' ;
element.style.color = 'Green' ;
```

这些语句分别设置了新 div 元素实例在网页里的不同部分的特定外观。

```
document.body.appendChild(element) ;
```

此语句在网页里的 body 元素实例内,新增变量 element 所代表的新 div 元素实例。

```
element = document.createElement('div') ;
```

此语句设置了变量 element 的数据,成为在网页里的另一个新的 div 元素实例。

```
element.id = 'inner' ;
```

此语句设置了新 div 元素实例的属性 id 的数据,成为字符串'inner',进而变成这个 div 元素实例的身份识别码。

```
element.innerHTML = 'Inner' ;
```

此语句设置了新 div 元素实例的属性 innerHTML 的数据,成为字符串'Inner',进而变成这个 div 元素实例上的文本。

```
element.style.width = '100px' ;
element.style.height = '50px' ;
element.style.margin = '5px auto' ;
element.style.textAlign = 'center' ;
element.style.fontSize = '1em' ;
element.style.borderRadius = '5px' ;
element.style.backgroundColor = 'Teal' ;
element.style.color = 'GreenYellow' ;
```

这些语句分别设置了新 div 元素实例在网页里的不同部分的特定外观。

```
outer.appendChild(element) ;
```

此语句在其属性 id 的数据为'outer'的元素实例内,新增了变量 element 所代表的新 div 元素实例。

```
function action01(event)
{
  console.log(`Block ${event.target.id} was left.`) ;
```

此语句显示出特定 div 元素实例被移出（left）鼠标指针的信息，并带有此 div 元素实例的属性 id 的数据字符串'outer'或'inner'其中之一。

```
}
```

上述语法定义了带有参数 event 的函数 action01()。

```
// outer.onmouseenter = action01 ;
// outer.onmouseover = action01 ;
// outer.onmouseleave = action01 ;
outer.onmouseout = action01 ;
```

此语句在属性 id 的数据为'outer'的 div 元素实例中，设置其属性 onmouseout 的数据，成为函数名称 action01，使得属性 id 的数据为'outer'的 div 元素实例，被监听着移出鼠标指针（mouse out）的动作。一旦在这个 div 元素实例上，被移出鼠标指针时，函数 action01()就会被调用。

11.1.5 鼠标移动事件

在特定网页里，鼠标指针有任何被移动的动作时，都会发生鼠标移动（mouse move）事件。请参看如下示例。

【11-1-5-mouse-onmousemove.js】

```
let element = null ;

element = document.createElement('div') ;

element.id = 'div01' ;
element.innerHTML = 'Block 01' ;

element.style.width = '150px' ;
element.style.height = '100px' ;
element.style.margin = '5px auto' ;
element.style.fontSize = '1em' ;
element.style.borderRadius = '5px' ;
element.style.backgroundColor = 'LightBlue' ;
element.style.color = 'Green' ;

document.body.appendChild(element) ;

function action01(event)
{
  console.log(`Mouse moves on block ${event.target.id}.`) ;
}

div01.onmousemove = action01 ;
```

【相关说明】

```
let element = null ;
```

声明初始数据为 null（空值）的变量 element。

```
element = document.createElement('div') ;
```

此语句设置了变量 element 的数据，成为网页里的新的 div 元素实例。

```
element.id = 'div01' ;
```

此语句设置了新 div 元素实例的属性 id 的数据，成为字符串'div01'，进而变成这个 div 元素实例的身份识别码。

```
element.innerHTML = 'Block 01' ;
```

此语句设置了新 div 元素实例的属性 innerHTML 的数据，成为字符串'Block 01'，进而变成新 div 元素实例上的文本。

```
element.style.width = '150px' ;
element.style.height = '100px' ;
element.style.margin = '5px auto' ;
element.style.fontSize = '1em' ;
element.style.borderRadius = '5px' ;
element.style.backgroundColor = 'LightBlue' ;
element.style.color = 'Green' ;
```

这些语句分别设置了新 div 元素实例在网页里的不同部分的特定外观。

```
document.body.appendChild(element) ;
```

此语句在网页里的 body 元素实例内，新增了变量 element 所代表的新 div 元素实例。

```
function action01(event)
{
  console.log(`Mouse moves on block ${event.target.id}.`) ;
```

此语句显示出特定 div 元素实例被移动（move）鼠标的信息，并带有这个 div 元素实例的属性 id 的数据字符串'div01'。

```
}
```

上述语法定义了带有参数 event 的函数 action01()。

```
div01.onmousemove = action01 ;
```

此语句在属性 id 的数据为'div01'的 div 元素实例中，设置其属性 onmousemove 的数据，成为函数名称 action01，使得属性 id 的数据为'div01'的 div 元素实例，被监听着被移动鼠标（mouse move）的动作。一旦在这个 div 元素实例上，被移动鼠标时，函数 action01()就会被调用。

11.2 键盘事件

键盘事件（keyboard event）是指窗口（window）的焦点（focus）进入特定网页之后，用户按下或松开键盘（keyboard）上的任意按键（key）时，所发生的事件。

11.2.1 按压与按下按键事件

窗口（window）的焦点（focus）进入特定网页之后，用户按下特定按键（key）时，会发生按压按键（key pressed）事件和按下按键（key down）事件。

按压按键事件和按下按键事件的差异，在于按压按键事件仅来自于键盘上的字母键、数字键与符号键的动作；按下按键事件则来自于几乎键盘上的所有按键。关于按压与按下按键事件的运用，可参看如下示例。

【11-2-1-keyboard-onkeypress-and-onkeydown.js】

```
function action01(event)
{
 console.log('Any key is pushed.') ;
}

document.onkeypress = action01 ;
// document.onkeydown = action01 ;
```

【相关说明】

```
function action01(event)
{
 console.log('Any key is pushed.') ;
}
```

此语法定义了【可显示出按键被按下之后的信息】的函数 action01()。

```
document.onkeypress = action01 ;
```

此语句在代表特定网页的 document 对象实例中，设置了属性 onkeypress 的数据，成为函数名称 action01。一旦在特定网页里，键盘上的字母键、数字键、符号键被按下（key pressed）时，函数 action01() 就会被调用。然而，此语句并不支持上述按键【以外】的其他按键，被按下的动作。

```
// document.onkeydown = action01 ;
```

若改用属性【onkeydown】，则几乎在键盘上的所有按键，被按下的动作，都可被监听到。

11.2.2 松开按键事件

窗口（window）的焦点（focus）进入特定网页之后，用户松开特定按键（key）时，会发生松开按键（key up）事件。请参看如下示例。

【11-2-2-keyboard-onkeyup.js】

```
function action01(event)
{
  console.log('Any key is released.') ;
}

document.onkeyup = action01 ;
```

【相关说明】

```
function action01(event)
{
  console.log('Any key is released.') ;
}
```

此语法定义了【可显示出按键被松开之后的信息】的函数 action01()。

```
document.onkeyup = action01 ;
```

此语句在代表特定网页的 document 对象实例中，设置了其属性 onkeyup 的数据，成为函数名称 action01。一旦在特定网页里，键盘上的特定按键被释放（key up）时，函数 action01()就会被调用。

11.3　网页内容的装卸事件

特定网页的内容，因为被加载、卸载、隐藏、跳动或滚动时，均会发生装卸（load or unload）相关的事件。

11.3.1　出现错误事件

在特定网页里，若代表图像（image）的 img 元素实例，或内含 JavaScript 源代码的 script 元素实例，出现加载或执行上的问题时，则会发生**出现错误** (error) 事件。请参看如下示例。

【11-3-1-onerror.js】

```
let element = document.createElement('img') ;

element.src = 'http://www.xxx.cn/one_image.png' ;

element.style.display = 'block' ;
element.style.marginTop = '1em' ;

document.body.appendChild(element) ;

function display_message01(event)
{
```

```
  console.log('The file is not found...') ;
}

function display_message02(event)
{
  console.log('An exception is thrown...') ;
}

function expired()
{
  throw new Error('Time is out...') ;
}

element.onerror = display_message01 ;
window.onerror = display_message02 ;

setTimeout(expired, 2000) ;
```

【相关说明】

```
let element = document.createElement('img') ;
```

此语句声明了初始数据为【代表图像的 img 元素实例】的变量 element。

```
element.src = 'http://www.xxx.cn/one_image.png' ;
```

此语句将特定图像所在网址的字符串'http://www.xxx.cn/one_image.png'，赋给变量 element 的 img 元素实例中的属性 src。换言之，当 img 元素实例被新增于特定网页里时，浏览器就会尝试显示出其网址对应的图像。

```
element.style.display = 'block' ;
element.style.marginTop = '1em' ;
```

这两个语句分别设置了变量 element 的 img 元素实例的特定外观。

```
document.body.appendChild(element) ;
```

此语句在 body 元素实例中，新增了变量 element 所代表的 img 元素实例，进而使得 img 元素实例所代表的图像，正式显示于网页里。

```
function display_message01(event)
{
  console.log('The file is not found...') ;
}
```

此语法定义了带有参数 event 的函数 display_message01()。

```
function display_message02(event)
{
  console.log('An exception is thrown...') ;
}
```

此语法定义了带有参数 event 的函数 display_message02()。

```
function expired()
{
  throw new Error('Time is out...') ;
```

}
```

此语法定义了函数 expired()。

```
element.onerror = display_message01 ;
```

此语句在变量 element 所代表的 img 元素实例中，设置其属性 onerror 的数据，成为函数名称 display_message01，使得当浏览器开启网页时，若这个 img 元素实例所代表的图像，无法成功被加载时，函数 display_message01() 就会被调用，进而显示出【The file is not found...】的信息。

在显示出上述信息之前，浏览器会先行主动显示出默认的错误信息【GET http://www.xxx.cn/one_image.png net::ERR_NAME_NOT_RESOLVED】。

```
window.onerror = display_message02 ;
```

此语句在当前浏览器窗口（browser window）所代表的 window 对象实例中，设置其属性 onerror 的数据，成为函数名称 display_message02，使得浏览器执行特定网页里的 JavaScript 源代码时，若发生任何错误，则函数 display_message02() 会被调用，进而显示出【An exception is thrown...】的信息。在显示上述信息之后，浏览器会先行主动显示出默认的错误信息，例如【Uncaught Error: Time is out... at expired】。

```
setTimeout(expired, 2000) ;
```

此语句使得在 2000 毫秒（2 秒）之后，名称为 expired 的函数会被调用。

11.3.2 加载和页面显示事件

特定网页被加载完成时，会发生加载（load）事件；若特定网页正式被显示于浏览器窗口中，则会发生页面显示（page show）事件。请参看如下示例。

【11-3-2-onload-and-onpageshow.html】

```html
<!DOCTYPE html>
<html>
 <head>
  <title>整体页面的 pageshow 与 load 事件</title>
 </head>
 <body>
 </body>
 <script>
  // document.body.onload = () => console.log('This page has been loaded.') ;
  window.onload = () => console.log('This page has been loaded.') ;

  // document.body.onpageshow = () => console.log('This page shows up after being loaded.') ;
  window.onpageshow = () => console.log('This page shows up after being loaded.') ;

  ///
  function display()
  {
    console.log('An image is loaded.') ;
  }

  let element = document.createElement('img') ;
```
```

```
      element.src = 'images/thumbnail01.jpg' ;

      element.style.display = 'block' ;
      element.style.marginTop = '10px' ;

      document.body.appendChild(element) ;

      element.onload = display ;
    </script>
  </html>
```

【相关说明】

```
  <script>

    // document.body.onload = () => console.log('This page has been loaded.') ;
    window.onload = () => console.log('This page has been loaded.') ;
```

此语句在等号右侧，定义了用来显示信息【This page has been loaded.】的箭头函数，并赋给了 window 对象实例中的属性 onload，使得此网页文档，被浏览器加载完成时，上述箭头函数就会被调用。

```
    // document.body.onpageshow = () => console.log('This page shows up after being loaded.') ;
    window.onpageshow = () => console.log('This page shows up after being loaded.') ;
```

此语句将另一箭头函数的定义，赋给 window 对象实例中的属性 onpageshow，使得此网页文档，被浏览器显示完成时，上述箭头函数就会被调用。

```
    function display()
    {
      console.log('An image is loaded.') ;
    }
```

此语法定义了函数 display()，并用来显示【An image is loaded.】的信息。

```
      let element = document.createElement('img') ;
```

此语法声明了变量 element，其初始数据为代表图像的 img 元素实例。

```
      element.src = 'images/thumbnail01.jpg' ;
```

此语句将代表图像文档路径的字符串'images/thumbnail01.jpg'，赋给变量 element 的 img 元素实例中的属性 src，进而使得这个图像，被显示于 img 元素实例的范围内。

```
      element.style.display = 'block' ;
      element.style.marginTop = '10px' ;
```

这两个语句分别设置了变量 element 的 img 元素实例的特定外观。

```
      document.body.appendChild(element) ;
```

此语句将变量 element 所代表的 img 元素实例，新增至 body 元素实例内，进而使得 img 元素实例所代表的图像，被显示在网页里。

```
      element.onload = display ;
```

此语句将函数名称 display，赋给变量 element 所代表的 img 元素实例中的属性 onload，使得此 img 元素实例，被完整显示在网页里时，函数 display() 就会被调用。此网页的所有内容，被显示完成之前，浏览器会在其调试工具【Console】面板里，显示出如下 3 行信息：

- An image is loaded.
- This page has been loaded.
- This page shows up after being loaded.

从如上 3 行信息，可看出其执行顺序为【网页内的图像文档被加载完成→网页本身被加载完成→网页本身被完整显示于窗口中】。

11.3.3 卸载和页面隐藏事件

特定网页被卸载完成时，会发生卸载（unload）事件；若特定网页已经被隐藏起来了，则会发生页面隐藏（page hide）事件。请参看如下示例。

【11-3-3-onunload-and-onpagehide.html】

```html
<!DOCTYPE html>
<html>
  <head>
    <title>整体页面的 page hide 与 unload 事件</title>
  </head>
  <body>
    <a href="http://www.yahoo.com">按此到新页面</a>
  </body>
  <script>
    function display01(event)
    {
      console.log('This page is hidden.') ;
    }

    function display02(event)
    {
      console.log('This page is unloaded.') ;
    }

    window.onpagehide = display01 ;
    window.onunload = display02 ;
  </script>
</html>
```

【相关说明】

```
<script>
  function display01(event)
  {
    console.log('This page is hidden.') ;
  }
```

此语法定义了带有参数 event 的函数 display01()，并用来显示出【This page is hidden】的信息。

```
function display02(event)
{
  console.log('This page is unloaded.') ;
}
```

此语法定义了带有参数 event 的函数 display02()，并用来显示出【This page is unloaded.】。

```
window.onpagehide = display01 ;
```

此语句在 window 对象实例中，设置其属性 onpagehide 的数据，成为函数名称 display01，使得特定网页被隐藏完成时，函数 display01()就会被调用。

```
window.onunload = display02 ;
```

此语句在 window 对象实例中，设置其属性 onunload 的数据，成为函数名称 display02，使得此网页被卸载完成时，函数 display02()就会被调用。此网页被卸载完成之前，浏览器会在其调试工具【Console】面板里，显示出如下两行信息：

- This page is hidden.
- This page is unloaded.

从如上两行信息，可看出其执行顺序为【网页被隐藏完成→网页被卸载完成】。网页卸载的动作，并不容易被调试出来，建议采用如下操作：
- 让 Google Chrome 浏览器先行载入网页档。
- 按下快捷键 Ctrl + Shift + I，启动调试工具，并呈现默认的【Console】面板。
- 在网页里，每隔半秒左右，连续按下快捷键 Ctrl + R 或 F5，并在【Console】面板里，持续观察闪烁的信息。

11.3.4 先于卸载事件

特定网页【即将】被卸载完毕【之前】，会发生先于卸载（before unload）事件。换言之，先于卸载事件发生在卸载（unload）事件之前！请参看如下示例。

【11-3-4-onbeforeunload.js】

```
let element = document.createElement('a') ;

element.href = 'http://www.tup.tsinghua.edu.cn' ;

element.innerHTML = 'Click here to another page.' ;

element.style.display = 'block' ;
element.style.marginTop = '1em' ;

document.body.appendChild(element) ;

function goodbye(event)
{
  return 1 ;
}
```

```
// document.body.onbeforeunload = goodbye ;
window.onbeforeunload = goodbye ;
```

【相关说明】

```
let element = document.createElement('a') ;
```

此语句声明了初始数据为【在网页里，代表超链接的 a 元素实例】的变量 element。

```
element.href = 'http://www.tup.tsinghua.edu.cn' ;
```

此语句将代表特定网址的字符串'http://www.tup.tsinghua.edu.cn'，赋给变量 element 的 a 元素实例中的属性 href。换言之，当 a 元素实例所代表的超链接，被单击时，浏览器就会开启上述网址对应的网页。

```
element.innerHTML = 'Click here to another page.' ;
```

此语句将字符串'Click here to another page.'，赋给变量 element 的 a 元素实例中的属性 innerHTML，以便代表超链接上的文本。换言之，在网页里的 a 元素实例上，浏览器会显示出'Click here to another page.'，作为超链接的文本。

```
element.style.display = 'block' ;
element.style.marginTop = '1em' ;
```

这两个语句分别设置了变量 element 的 a 元素实例的特定外观。

```
document.body.appendChild(element) ;
```

此语句在 body 元素实例中，新增变量 element 所代表的 a 元素实例，进而正式显示在网页里。

```
function goodbye(event)
{
  return 1 ;
```

在本示例中的应用里，此语句也可以改写成为【return '' ;】、【return true ;】或【return false ;】。

```
}
```

上述语法定义了带有参数 event 的函数 goodbye()。

```
// document.body.onbeforeunload = goodbye ;
window.onbeforeunload = goodbye ;
```

此语句在代表浏览器窗口（window）的 window 对象实例中，设置其属性 onbeforeunload 的数据，成为函数名称 goodbye，使得当用户单击上述网页里的超链接时，浏览器会及时显示【要离开此网站吗？】的信息对话框。请留意，在函数 goodbye()的定义里，必须存在类似【return 1 ;】的语句。

11.3.5　网址散列变化事件

在浏览器的地址栏中，特定网址在符号【#】右侧的锚点（anchor）名称，被变更时，会发生网址散列变化（URL's hash changed）事件。网址中的散列（hash）发生变化，主要是因为在浏览器窗口中，其画面跳转至特定网页里的特定锚点（anchor）位置。请参看如下示例。

【11-3-5-onhashchange.js】

```
console.log(location.href) ;
console.log(location.hash) ;
console.log('') ;

function change_hash()
{
  // location.href += '#anchor01' ;
  location.hash = '#anchor01' ;

  console.log(location.href) ;
  console.log(location.hash) ;
}

function display_message()
{
  console.log('Hash part in URL is changed.') ;
}

setTimeout(change_hash, 2000) ;

// document.body.onhashchange = display_message ;
window.onhashchange = display_message ;
```

【相关说明】

```
console.log(location.href) ;
```

location.href 会返回特定网页文档在网站服务器上的【网址】字符串，或者返回【在硬盘中】的【文件路径】字符串，例如【file:///D:/works/books/publishers/示例文档/js_tester.html】。

```
console.log(location.hash) ;
```

location.hash 返回空字符串''，因为在文件路径的字符串里，尚未存在散列【#anchor01】。

```
console.log('') ;

function change_hash()
{
  // location.href += '#anchor01' ;
  location.hash = '#anchor01' ;
```

此语句等价于【location.href += '#anchor01';】，可将字符串'#anchor01',赋给属性 location.hash，使得当前网址的末尾，被加上作为散列的字符串【#anchor01】，例如【file:///D:/works/books/publishers/示例文档/js_tester.html**#anchor01**】。

```
  console.log(location.href) ;
```

location.href 在此会返回已经被加上字符串【#anchor01】的网址，例如【file:///D:/works/books/publishers/示例文档/js_tester.html**#anchor01**】。

```
  console.log(location.hash) ;
```

location.hash 在此会返回字符串"#anchor01"，因为在文件路径的字符串里，已经存在散列【#anchor01】的部分了。

```
}
```

上述语法定义了【用来动态修改网址里的散列】的函数 change_hash()。

```
function display_message()
{
  console.log('Hash part in URL is changed.') ;
}
```

此语法定义了函数 display_message()。

```
setTimeout(change_hash, 2000) ;
```

此语句设置了在 2000 毫秒（2 秒）之后，调用名称为 change_hash 的函数。

```
// document.body.onhashchange = display_message ;
window.onhashchange = display_message ;
```

此语句等价于【document.body.onhashchange = display_message ;】，可在浏览器窗口所代表的 window 对象实例中，设置其属性 onhashchange 的数据，成为函数名称 display_message，使得其网址里的散列被变更时，函数 display_message() 即会被调用，进而显示出【Hash part in URL is changed.】的信息。

11.3.6 滚动事件

在特定网页里，若 window 对象实例、body 元素实例或者其他子元素实例的内容，被滚动时，就会发生滚动（scroll）事件。请参看如下示例。

【11-3-6-onscroll.html】

```
<!DOCTYPE html>
<html>
  <head>
    <title>onscroll event</title>
    <style>
      body
      {
        width: 1500px ;
        height: 1500px ;
        background-image: url(images/sandy_texture.jpg) ;
      }
    </style>
  </head>
  <body>
  </body>
  <script>
    function action01()
    {
      console.log('This page is scrolling.') ;
    }
```

```
    window.onscroll = action01 ;
  </script>
</html>
```

【相关说明】

```
<script>

  function action01()
  {
    console.log('This page is scrolling.') ;
  }
```

此语法定义了函数 action01()，并用来显示出【This page is scrolling.】的信息。

```
  window.onscroll = action01 ;
```

此语句在 window 对象实例中，设置其属性 onscroll 的数据，成为函数名称 action01，使得其网页里的水平或垂直滚动条，被滚动时，函数 action01()就会被调用，并持续显示出【This page is scrolling.】的信息。

11.4 表单事件

在特定网页中，针对 form 元素实例本身或者其子元素实例，JavaScript 引擎可处理一系列的表单（form）事件。

11.4.1 内容变化事件

特定 input、select、textarea 元素实例的内容，被变更时，就会发生内容变化（change）事件。请参考如下示例。

【11-4-1-onchange.html】

```
<!DOCTYPE html>
<html>
  <head>
    <title>Form related events</title>
    <style>
      input, button, select
      {
        font-size: 1.2em ;
      }
    </style>
  </head>
  <body>
    <form style="text-align: center;">
      <h3>个人资料</h3>
      <input type="text" id="username" name="username" placeholder="username" size="16">
      <input type="password" id="password" name="password" placeholder="password" size="16">
      <p></p>
```

```html
        <select id="select01" name="select01">
          <option value="">choice of day-off</option>
          <option value="0">Sunday</option>
          <option value="1">Monday</option>
          <option value="2">Tuesday</option>
          <option value="3">Wednesday</option>
          <option value="4">Thursday</option>
          <option value="5">Friday</option>
          <option value="6">Saturday</option>
        </select>

        <p></p>
        <button type="submit">Login</button>
        <button type="reset">Reset</button>
    </form>
  </body>
  <script>
    select01.onchange = function (event)
    {
      console.log(event.target.value);
    };
  </script>
</html>
```

【相关说明】

```
<script>
  select01.onchange = function (event)
  {
    console.log(event.target.value);
```

在此语句中的 event.target，会对应到其属性 id 的数据为'select01'的 select 元素实例。在这个 select 元素实例中，欲访问被变更之后的选定项目的数据，可借助在 select 元素实例中，被内置的属性 value，来加以实现。

```
  };
```

上述语法将匿名函数的定义，赋给其属性 id 的数据为'select01'的 select 元素实例中的属性 onchange。一旦这个 select 元素实例的选定项目被变更，上述的匿名函数就会被调用，使得被变更之后的选定项目的数据，显示在浏览器的调试工具【Console】面板里。

11.4.2 获取和失去焦点相关事件

在特定 form 元素实例中，焦点移入（focus in）特定子元素实例（例如 input、button 元素实例等）时，会发生焦点移入（focus in）事件；焦点移出（focus out）特定子元素实例时，则会发生【失去焦点】的模糊（blur）事件。请参看如下示例。

【11-4-2-focus-in-and-out.html】

```
<!DOCTYPE html>
<html>
```

```html
<head>
  <title>Form related events</title>
  <style>
  input, button
  {
    font-size: 1.2em ;
  }
  </style>
</head>
<body>
  <form id="form01" name="form01" style="text-align: center;">
  <h3>个人资料</h3>
  <input type="text" id="username" name="username" placeholder="username" size="16">
  <input type="password" id="password" name="password" placeholder="password" size="16">

  <p></p>
  <button type="submit">Login</button>
  <button type="reset">Reset</button>
  </form>
</body>
<script>
  username.onblur = function (event)
  {
    event.target.style.outline = '1px solid Gray' ;
  } ;

  username.onfocus = function (event)
  {
    event.target.style.outline = '3px solid YellowGreen' ;
  } ;

  form01.addEventListener("focusin", () => console.log('Focus-in event occurs.')) ;

  form01.addEventListener("focusout", () => console.log('Focus-out event occurs.')) ;
</script>
</html>
```

【相关说明】

```
<script>
  username.onblur = function (event)
  {
    event.target.style.outline = '1px solid Gray' ;
```

在此语句中，event.target 会对应到刚才失去焦点（blur）而且【属性 id 的数据为'username'】的 input 元素实例。

```
  } ;
```

上述语法将带有参数 event 的匿名函数的定义，赋给了属性 id 的数据为'username'的 input 元素实例中的属性 onblur。一旦键盘光标离开 input 元素实例，也就是 input 元素实例失去焦点（blur）时，上述匿名函数就会被调用，使得 input 元素实例具有灰色轮廓（outline）。

```
username.onfocus = function (event)
{
  event.target.style.outline = '3px solid YellowGreen' ;
} ;
```

上述语法将带有参数 event 的匿名函数的定义,赋给属性 id 的数据为'username'的 input 元素实例的属性 onfocus。一旦键盘光标进入 input 元素实例,也就是 input 元素实例获取焦点(focus)时,上述的匿名函数就会被调用,使得 input 元素实例具有较粗的黄绿色轮廓。

```
form01.addEventListener("focusin", () => console.log('Focus-in event occurs.')) ;
```

此语句通过调用 addEventListener()函数,使得属性 id 的数据为'form01'的 form 元素实例,被监听着焦点进入(focus in)事件。一旦属性 id 的数据为'form01'的 form 元素实例,获取焦点,也就是键盘光标进入 form 元素实例时,显示【Focus-in event occurs.】信息的箭头函数就会被调用。

```
form01.addEventListener("focusout", () => console.log('Focus-out event occurs.')) ;
```

此语句通过调用 addEventListener()函数,使得属性 id 的数据为'form01'的 form 元素实例,被监听着焦点离开(focus out)事件。一旦属性 id 的数据为'form01'的 form 元素实例失去焦点,也就是键盘光标离开 form 元素实例时,显示【Focus-out event occurs.】信息的箭头函数就会被调用。

11.4.3 输入事件

当用户输入特定文本到特定 input 或 textarea 元素实例中时,会发生输入(input)事件。请参看如下示例。

【11-4-3-oninput.html】

```html
<!DOCTYPE html>
<html>
  <head>
    <title>Form related events</title>
    <style>
      input, button, select
      {
        font-size: 1.2em ;
      }
    </style>
  </head>
  <body>
    <form style="text-align: center;">
    <h3>个人资料</h3>
    <input type="text" id="username" name="username" placeholder="username" size="16">
    <input type="password" id="password" name="password" placeholder="password" size="16">
    <p></p>
    <select id="select01" name="select01">
      <option value="">choice of day-off</option>
      <option value="0">Sunday</option>
      <option value="1">Monday</option>
      <option value="2">Tuesday</option>
      <option value="3">Wednesday</option>
      <option value="4">Thursday</option>
```

```
        <option value="5">Friday</option>
        <option value="6">Saturday</option>
      </select>

      <p></p>
      <button type="submit">Login</button>
      <button type="reset">Reset</button>
    </form>
  </body>
  <script>
    username.oninput = function (event)
    {
      console.log(event.target.value) ;
    } ;
  </script>
</html>
```

【相关说明】

```
<script>

  username.oninput = function (event)
  {
    console.log(event.target.value) ;
  } ;
```

此语法将带有参数 event 的匿名函数的定义，赋给属性 id 的数据为'username'的 input 元素实例的属性 oninput。一旦 input 元素实例开始被输入（input）文本时，上述匿名函数就会被调用，使得 input 元素实例的数据（被输入的文本），被显示在浏览器的调试工具【Console】面板里。

11.4.4　无效事件

在网页里的 form 元素实例中，任何被设置为必填（required）或其文本必须符合特定模式（pattern）的 input 元素实例的数据，被提交（submitted）时，若未被填入任何文本或者不符合特定模式，则会发生无效（invalid）事件。请参看如下示例。

【11-4-4-oninvalid.html】

```
<!DOCTYPE html>
<html>
  <head>
    <title>Form related events</title>
    <style>
      input, button, select
      {
        font-size: 1.2em ;
      }
    </style>
  </head>
  <body>
    <form style="text-align: center;">
      <h3>个人资料</h3>
      <input type="text" id="username" name="username" placeholder="username" size="16"
```

```
            required>
            <input type="password" id="password" name="password" placeholder="password" size="16"
            required>
            <p></p>
            <select id="select01" name="select01">
              <option value="">choice of day-off</option>
              <option value="0">Sunday</option>
              <option value="1">Monday</option>
              <option value="2">Tuesday</option>
              <option value="3">Wednesday</option>
              <option value="4">Thursday</option>
              <option value="5">Friday</option>
              <option value="6">Saturday</option>
            </select>

            <p></p>
            <button type="submit">Login</button>
            <button type="reset">Reset</button>
        </form>
    </body>
    <script>
      username.oninvalid = function (event)
      {
        console.log('Need to type username!') ;
      } ;
    </script>
</html>
```

【相关说明】

```
<script>
  username.oninvalid = function (event)
  {
    console.log('Need to type username!') ;
  } ;
```

上述语法将带有参数 event 的匿名函数的定义，赋给了属性 id 的数据为'username'的 input 元素实例的属性 oninvalid。在 form 元素实例中，带有【type="submit"】的提交钮，被按下之后，一旦 input 元素实例被输入的文本，不符合要求时，上述匿名函数就会被调用，使得【Need to type username!】的信息，被显示在浏览器的调试工具【Console】面板里。

在本示例中，属性 id 的数据为'username'的 input 元素实例，被设置了属性 required（必填的），所以未输入任何文本，就被提交时，即是不符合要求的。

11.4.5 重置事件

在网页里的 form 元素实例中，输入的文本或选定的数据，被重置（reset）时，会发生重置（reset）事件。请参看如下示例。

【11-4-5-onreset.html】

```
<!DOCTYPE html>
```

```html
<html>
  <head>
    <title>Form related events</title>
    <style>
      input, button, select
      {
        font-size: 1.2em ;
      }
    </style>
  </head>
  <body>
    <form id="form01" name="form01" style="text-align: center;">
      <h3>个人资料</h3>
      <input type="text" id="username" name="username" placeholder="username" size="16"
        required>
      <input type="password" id="password" name="password" placeholder="password" size="16"
        required>
      <p></p>
      <select id="select01" name="select01">
        <option value="">choice of day-off</option>
        <option value="0">Sunday</option>
        <option value="1">Monday</option>
        <option value="2">Tuesday</option>
        <option value="3">Wednesday</option>
        <option value="4">Thursday</option>
        <option value="5">Friday</option>
        <option value="6">Saturday</option>
      </select>

      <p></p>
      <button type="submit">Login</button>
      <button type="reset">Reset</button>
    </form>
  </body>
  <script>
    form01.onreset = function (event)
    {
      console.log('Data of the form is reset!') ;
    } ;
  </script>
</html>
```

【相关说明】

```html
<script>

  form01.onreset = function (event)
  {
    console.log('Data of the form is reset!') ;
  } ;
```

上述语法将带有参数 event 的匿名函数的定义，赋给了属性 id 的数据为'form01'的 form 元素实例的属性 onreset。在 form 元素实例中，带有【type="reset"】的重置按钮，被按下时，匿名函数就会被调用，使得【Data of the form is reset!】的信息，被显示在浏览器的调试工具【Console】面板里。

所谓的重置（reset），就是指特定 form 元素实例内的默认数据，被复原成为浏览器载入时的初始状态。例如：若属性 id 的数据为'username'的 input 元素实例，被输入文本之后，带有【type="reset"】的重置按钮，才被按下时，则此 input 元素实例，会被复原成为**没有任何文本**的初始状态！

11.4.6 搜索事件

在网页里的 form 元素实例中，其属性 type 的数据为字符串'search'的 input 元素实例，被输入文本之后，被按下键盘上的 ENTER 键，或者单击 input 元素实例的右侧【X】形状的关闭按钮时，就会发生搜索（search）事件。请参看如下示例。

【11-4-6-onsearch.html】

```html
<!DOCTYPE html>
<html>
  <head>
    <title>Form related events</title>
    <style>
      input, button, select
      {
        font-size: 1.2em ;
        margin: 5px ;
      }
    </style>
  </head>
  <body>
    <form id="form01" name="form01" style="text-align: center;">
      <h3>个人资料</h3>
      <input type="text" id="username" name="username" placeholder="username" size="16"
        required>
      <input type="password" id="password" name="password" placeholder="password" size="16"
        required>
      <p></p>
      <select id="select01" name="select01">
        <option value="">choice of day-off</option>
        <option value="0">Sunday</option>
        <option value="1">Monday</option>
        <option value="2">Tuesday</option>
        <option value="3">Wednesday</option>
        <option value="4">Thursday</option>
        <option value="5">Friday</option>
        <option value="6">Saturday</option>
      </select>
      <input type="search" id="search" name="search" placeholder="job category.." size="9">

      <p></p>
      <button type="submit">Login</button>
      <button type="reset">Reset</button>
    </form>
  </body>
  <script>
    // IE and Firefox don't support this event.
    search.onsearch = function (event)
```

```
    {
      console.log('The query is searched.') ;
    } ;
  </script>
</html>
```

【相关说明】

```
<script>
  // IE and Firefox don't support this event.
  search.onsearch = function (event)
  {
    console.log('The query is searched.') ;
  } ;
```

上述语法将带有参数 event 的匿名函数的定义，赋给属性 id 的数据为'search'的 input 元素实例的属性 onsearch。在 input 元素实例内，被输入文本之后，被按下键盘上的 Enter 键，以进行搜索（search）时，上述匿名函数就会被调用，使得【The query is searched.】的信息，被显示在浏览器的调试工具【Console】面板里。请留意，Firefox 与 IE 浏览器并不支持 onsearch 事件的处理。

11.4.7 选定文本事件

在网页里的 input 或 textarea 元素实例中，其内容的任何文本，被选定（selected）时，都会发生选定（select）文本事件。请参看如下示例。

【11-4-7-onselect.html】

```
<!DOCTYPE html>
<html>
  <head>
    <title>Form related events</title>
    <style>
      input, button, select
      {
        font-size: 1.2em ;
        margin: 5px ;
      }
    </style>
  </head>
  <body>
    <form id="form01" name="form01" style="text-align: center;">
      <h3>个人资料</h3>
      <input type="text" id="username" name="username" placeholder="username" size="16" required>
      <input type="password" id="password" name="password" placeholder="password" size="16"
        required>
      <p></p>
      <select id="select01" name="select01">
        <option value="">choice of day-off</option>
        <option value="0">Sunday</option>
        <option value="1">Monday</option>
        <option value="2">Tuesday</option>
        <option value="3">Wednesday</option>
```

```
        <option value="4">Thursday</option>
        <option value="5">Friday</option>
        <option value="6">Saturday</option>
      </select>
      <input type="search" id="search" name="search" placeholder="job category.." size="9">

      <p></p>
      <button type="submit">Login</button>
      <button type="reset">Reset</button>
    </form>
  </body>
  <script>
    search.onselect = function (event)
    {
      console.log('The query text is selected.') ;
    } ;
  </script>
</html>
```

【相关说明】

```
<script>

  search.onselect = function (event)
  {
    console.log('The query text is selected.') ;
  } ;
```

上述语法将带有参数 event 的匿名函数的定义，赋给属性 id 的数据为'search'的 input 元素实例的属性 onselect。在 input 元素实例内，输入文本之后，而且其文本的一部分，被选定（selected）时，上述匿名函数就会被调用，使得【The query text is selected.】的信息，被显示在浏览器的调试工具【Console】面板里。

11.4.8　提交事件

在网页里的 form 元素实例中，其数据被提交（submitted）时，会发生提交（submit）事件。请参看如下示例。

【11-4-8-onsubmit.html】

```
<!DOCTYPE html>
<html>
  <head>
    <title>Form related events</title>
    <style>
      input, button, select
      {
        font-size: 1.2em ;
        margin: 5px ;
      }
    </style>
  </head>
  <body>
```

```html
    <form id="form01" name="form01" style="text-align: center;">
      <h3>个人资料</h3>
      <input type="text" id="username" name="username" placeholder="username" size="16"
        required>
      <input type="password" id="password" name="password" placeholder="password" size="16"
        required>
      <p></p>
      <select id="select01" name="select01">
        <option value="">choice of day-off</option>
        <option value="0">Sunday</option>
        <option value="1">Monday</option>
        <option value="2">Tuesday</option>
        <option value="3">Wednesday</option>
        <option value="4">Thursday</option>
        <option value="5">Friday</option>
        <option value="6">Saturday</option>
      </select>
      <input type="search" id="search" name="search" placeholder="job category.." size="9">

      <p></p>
      <button type="submit">Login</button>
      <button type="reset">Reset</button>
    </form>
  </body>
  <script>
    form01.onsubmit = function (event)
    {
      alert('Data of the form is submitted.') ;
    } ;
  </script>
</html>
```

【相关说明】

```
<script>

  form01.onsubmit = function (event)
  {
    alert('Data of the form is submitted.') ;
  } ;
```

此语法将带有参数 event 的匿名函数的定义，赋给属性 id 的数据为'form01'的 form 元素实例的属性 onsubmit。在 form 元素实例内，其所有必填的各元素实例（例如 input、select、textarea 元素实例等），被填妥之后，一旦带有【type="submit"】的提交按钮，被按下时，上述匿名函数就会被调用，使得带有信息【Data of the form is submitted.】的对话框，被显示在浏览器窗口当中。

11.5 练 习 题

1. 至少通过两种方式，改写如下 HTML 和 JavaScript 的源代码片段：

```
<button id="login_btn" onclick="display()">Login</button>
```

```
|
<script>
  function display()
  {
    |
  }
</script>
```

2. 已知当前的网址为 file:///.../html/6-3-6-onscroll.html#list01。试编写 JavaScript 源代码，以便当前网址被变更为如下网址时，可让显示特定信息的函数 display_message() 被调用：

```
file:///.../html/6-3-6-onscroll.html#list02
```

3. 已知在当前网页里，存在代表文本字段而且【属性 id 的数据为'username'】的 input 元素实例。至少编写两个版本的 JavaScript 源代码，以在当前网页被加载完成之后，自动将键盘光标，移入上述文本字段中。

第 12 章

DOM 的事件处理（二）

本章主要介绍多个与指针设备相关的拖动事件、与操作系统相关的剪贴板事件、与多媒体相关的视频和音频事件，以及与视觉体验相关的动画及过渡事件。

12.1 拖动事件

在特定网页里，任何元素实例，被持续拖动时，都会发生拖动（drag）相关的事件。

12.1.1 正在拖动事件

任何元素实例被拖动之后，一直到拖动结束之前，都会持续发生**正在拖动**（drag）事件。请参看如下示例

【12-1-1-ondrag.html】

```
<!DOCTYPE html>
<html>
  <head>
    <title>Drag related events</title>
    <style>
      #div01
      {
        padding: 5px ;
        width: 100px ;
        height: 100px ;
        background-color: RoyalBlue ;
      }

      [id^=span]
```

```
      {
        color: Teal ;
        width: 80px ;
        height: 30px ;
        display: block ;
        margin: 5px auto ;
        text-align: center ;
        background-color: GreenYellow ;
      }
    </style>
  </head>
  <body>
    <div id="div01">
      Block 1
      <span id="span01">
        box 1
      </span>
      <span id="span02">
        box 2
      </span>
    </div>
  </body>
  <script>
    for (let i = 1; i < 3; i++)
    {
      with (document.getElementById('span0' + i))
      {
        draggable = true ;

        ondrag = function (event)
        {
          console.log(`Box ${i} is dragged...`) ;
        } ;
      }
    }
  </script>
</html>
```

【相关说明】

```
<script>

  for (let i = 1; i < 3; i++)
```

此 for 循环语句会迭代 2 次循。

```
  {

    with (document.getElementById('span0' + i))
```

此语法在属性 id 的数据为'span01'与'span02'的 span 元素实例中,进一步简化了访问其特定属性的语法。例如:

- 【spna01.draggable = true ;】,在其大括号里,可被简化成【draggable = true ;】。
- 【span01.ondrag = function (event) { ... } ;】,在其大括号里,可被简化成【ondrag = function

(event) { ... } ;】。

```
{
  draggable = true ;
```

此语句将布尔值 true，赋给属性 id 的数据为'span01'和'span02'的 span 元素实例中的属性 draggable，使得这两个 span 元素实例，在网页里可被拖动。

```
ondrag = function (event)
{
  console.log(`Box ${i} is dragged...`) ;
} ;
```

上述语法将带有参数 event 的匿名函数的定义，赋给属性 id 的数据为'span01'与'span02'的 span 元素实例中的属性 ondrag。特定 span 元素实例被拖动时，上述匿名函数就会被调用，使得类似【Box X is dragged...】的信息，被显示在浏览器的调试工具【Console】面板里。

12.1.2 拖动结束事件

任何元素实例被拖动结束的瞬间，都会发生拖动结束（drag end）事件。请参看如下示例。

【12-1-2-ondragend.html】

```html
<!DOCTYPE html>
<html>
  <head>
    <title>Drag related events</title>
    <style>
      #div01
      {
        padding: 5px ;
        width: 100px ;
        height: 100px ;
        background-color: RoyalBlue ;
      }

      [id^=span]
      {
        color: Teal ;
        width: 80px ;
        height: 30px ;
        display: block ;
        margin: 5px auto ;
        text-align: center ;
        background-color: GreenYellow ;
      }
    </style>
  </head>
  <body>
    <div id="div01">
      Block 1
      <span id="span01">
```

```
      box 1
    </span>
    <span id="span02">
      box 2
    </span>
  </div>
</body>
<script>
  for (let i = 1; i < 3; i++)
  {
    with (document.getElementById('span0' + i))
    {
      draggable = true ;

      ondrag = function (event)
      {
        console.log(`Box ${i} is dragged...`) ;
      } ;

      ondragend = function (event)
      {
        console.log(`The dragging of Box ${i} is ended...`) ;
      } ;
    }
  }
</script>
</html>
```

【相关说明】

```
<script>

  for (let i = 1; i < 3; i++)
```

此 for 循环语句会迭代 2 次。

```
  {

    with (document.getElementById('span0' + i))
```

此语法在其属性 id 的数据为'span01'与'span02'的 span 元素实例中，简化了访问其特定属性的语法。例如：

- 【spna01.draggable = true ;】，在其大括号里，可被简化成【draggable = true ;】。
- 【span01.ondrag = function (event) { ... } ;】，在其大括号里，可被简化成【ondrag = function (event) { ... } ;】。

```
    {

      draggable = true ;
```

此语句将布尔值 true，赋给属性 id 的数据为'span01'和'span02'的 span 元素实例中的属性 draggable，使得这两个 span 元素实例在网页里，可被拖动。

```
      ondrag = function (event)
```

```
  {
    console.log(`Box ${i} is dragged...`) ;
  } ;
```

此语法将带有参数 event 的匿名函数的定义,赋给属性 id 的数据为'span01'与'span02'的 span 元素实例中的属性 ondrag。特定 span 元素实例被拖动时,上述匿名函数就会被调用,使得类似【Box X is dragged...】的信息,被显示在浏览器的调试工具【Console】面板里。

```
ondragend = function (event)
{
  console.log(`The dragging of Box ${i} is ended...`) ;
} ;
```

此语法将带有参数 event 的匿名函数的定义,赋给属性 id 的数据为'span01'与'span02'的 span 元素实例中的属性 ondragend。特定 span 元素实例被终止拖动时,上述匿名函数就会被调用,使得类似【The dragging of Box X is ended...】的信息,被显示在浏览器的调试工具【Console】面板里。

12.1.3　拖动进入事件

在任何元素实例被拖动当中,鼠标指针进入到另一元素实例的瞬间,会发生拖动进入(drag enter)事件。请参看如下示例。

【12-1-3-ondragenter.html】

```html
<!DOCTYPE html>
<html>
  <head>
    <title>Drag related events</title>
    <style>
      [id^=div]
      {
        padding: 5px ;
        width: 100px ;
        height: 100px ;
      }

      #div01
      {
        background-color: RoyalBlue ;
      }

      #div02
      {
        margin-top: 10px ;
        background-color: YellowGreen ;
      }

      [id^=span]
      {
        color: Teal ;
        width: 80px ;
        height: 30px ;
```

```
        display: block ;
        margin: 5px auto ;
        text-align: center ;
        background-color: GreenYellow ;
      }
    </style>
  </head>
  <body>
    <div id="div01">
      Block 1
      <span id="span01" draggable="true">
        box 1
      </span>
      <span id="span02" draggable="true">
        box 2
      </span>
    </div>

    <div id="div02"></div>
  </body>
  <script>
    div02.ondragenter = function (event)
    {
      console.log(`The dragged Box enters the target.`) ;
    } ;
  </script>
</html>
```

【相关说明】

```
<script>

  div02.ondragenter = function (event)
  {
    console.log(`The dragged Box enters the target.`) ;
  } ;
```

此语法将带有参数 event 的匿名函数的定义，赋给属性 id 的数据为'div02'的 div 元素实例中的属性 ondragenter。特定 span 元素实例被拖动，而进入属性 id 的数据为'div02'的 div 元素实例时，上述匿名函数就会被调用，使得【The dragged Box enters the target..】的信息，被显示在浏览器的调试工具【Console】面板里。

12.1.4 拖动离开事件

在任何元素实例被拖动当中，鼠标指针接触到上层元素实例的边缘的瞬间，会发生拖动离开（drag leave）事件。请参看如下示例。

【12-1-4-ondragleave.html】

```
<!DOCTYPE html>
<html>
  <head>
    <title>Drag related events</title>
```

```
    <style>
      [id^=div]
      {
        padding: 5px ;
        width: 100px ;
        height: 100px ;
      }

      #div01
      {
        background-color: RoyalBlue ;
      }

      #div02
      {
        margin-top: 10px ;
        background-color: YellowGreen ;
      }

      [id^=span]
      {
        color: Teal ;
        width: 80px ;
        height: 30px ;
        display: block ;
        margin: 5px auto ;
        text-align: center ;
        background-color: GreenYellow ;
      }
    </style>
  </head>
  <body>
    <div id="div01">
      Block 1
      <span id="span01" draggable="true">
        box 1
      </span>
      <span id="span02" draggable="true">
        box 2
      </span>
    </div>

    <div id="div02"></div>
  </body>
  <script>
    div01.ondragleave = function (event)
    {
      console.log(`The dragged Box leaves the source.`) ;
    } ;
  </script>
</html>
```

【相关说明】

```
<script>
```

```
div01.ondragleave = function (event)
{
  console.log(`The dragged Box leaves the source.`) ;
} ;
```

将带有参数 event 的匿名函数的定义赋给属性 id 的数据为'div01'的 div 元素实例的属性 ondragleave。特定 span 元素实例被拖动而离开属性 id 的数据为'div01'的 div 元素实例的瞬间，匿名函数就会被调用，使得【The dragged Box leaves the source.】的信息，显示在浏览器调试工具【Console】面板里。

12.1.5　拖动悬停事件

在任何元素实例被拖动当中，鼠标指针位于特定元素实例的范围内时，会发生拖动悬停（drag over）事件。请参看如下示例。

【12-1-5-ondragover.html】

```
<!DOCTYPE html>
<html>
  <head>
    <title>Drag related events</title>
    <style>
      [id^=div]
      {
        padding: 5px ;
        width: 100px ;
        height: 100px ;
      }

      #div01
      {
        background-color: RoyalBlue ;
      }

      #div02
      {
        margin-top: 10px ;
        background-color: YellowGreen ;
      }

      [id^=span]
      {
        color: Teal ;
        width: 80px ;
        height: 30px ;
        display: block ;
        margin: 5px auto ;
        text-align: center ;
        background-color: GreenYellow ;
      }
    </style>
  </head>
```

```
<body>
  <div id="div01">
    Block 1
    <span id="span01" draggable="true">
      box 1
    </span>
    <span id="span02" draggable="true">
      box 2
    </span>
  </div>

  <div id="div02"></div>
</body>
<script>
  div02.ondragover = function (event)
  {
    console.log(`The Box is being dragged over the target.`) ;
  } ;
</script>
</html>
```

【相关说明】

```
<script>

  div02.ondragover = function (event)
  {
    console.log(`The Box is being dragged over the target.`) ;
  } ;
```

此语法将带有参数 event 的匿名函数的定义，赋给属性 id 的数据为'div02'的 div 元素实例中的属性 ondragover。特定 span 元素实例仍然在属性 id 的数据为'div02'的 div 元素实例的范围内，而被拖动时，上述匿名函数就会被调用，使得【The Box is being dragged over the target.】的信息，会被显示在浏览器的调试工具【Console】面板里。

12.1.6 拖动开始事件

任何元素实例被开始拖动的瞬间，会发生拖动开始（drag start）事件。请参看如下示例。

【12-1-6-ondragstart.html】

```
<!DOCTYPE html>
<html>
  <head>
    <title>Drag related events</title>
    <style>
      [id^=div]
      {
        padding: 5px ;
        width: 100px ;
        height: 100px ;
      }
```

```
    #div01
    {
      background-color: RoyalBlue ;
    }

    #div02
    {
      margin-top: 10px ;
      background-color: YellowGreen ;
    }

    [id^=span]
    {
      color: Teal ;
      width: 80px ;
      height: 30px ;
      display: block ;
      margin: 5px auto ;
      text-align: center ;
      background-color: GreenYellow ;
    }
  </style>
</head>
<body>
  <div id="div01">
    Block 1
    <span id="span01" draggable="true">
      box 1
    </span>
    <span id="span02" draggable="true">
      box 2
    </span>
  </div>

  <div id="div02"></div>
</body>
<script>
  div01.ondragstart = function (event)
  {
    console.log(`The Box is being dragged over the source.`) ;
  } ;
</script>
</html>
```

【相关说明】

```
<script>

  div01.ondragstart = function (event)
  {
    console.log(`The Box is being dragged over the source.`) ;
  } ;
```

将带有参数 event 的匿名函数的定义赋给属性 id 的数据为'div01'的 div 元素实例的属性 ondragstart。特定 span 元素实例开始被拖动时，匿名函数就会被调用，使得【The Box is being dragged over the source.】的信息显示在浏览器调试工具【Console】面板里。

12.1.7　放下事件

拖动尚未结束之前，若鼠标指针仍然位于特定元素实例的范围里，则当鼠标左键被松开时，会发生放下（drop）事件。请参看如下示例。

【12-1-7-ondrop.html】

```html
<!DOCTYPE html>
<html>
  <head>
    <title>Drag related events</title>
    <style>
      [id^=div]
      {
        padding: 5px ;
        width: 100px ;
        height: 100px ;
      }

      #div01
      {
        background-color: RoyalBlue ;
      }

      #div02
      {
        margin-top: 10px ;
        background-color: YellowGreen ;
      }

      [id^=span]
      {
        color: Teal ;
        width: 80px ;
        height: 30px ;
        display: block ;
        margin: 5px auto ;
        text-align: center ;
        background-color: GreenYellow ;
      }
    </style>
  </head>
  <body>
    <div id="div01" ondragover="event.preventDefault();">
      Block 1
      <span id="span01" draggable="true">
        box 1
      </span>
      <span id="span02" draggable="true">
        box 2
      </span>
```

```
    </div>

    <div id="div02" ondragover="event.preventDefault();"></div>
  </body>
  <script>
    let ref = null ;

    for (let i = 1; i < 3; i++)
    {
      document.getElementById('span0' + i).ondragstart = function (event)
      {
        ref = event.target ;
      }

      document.getElementById('div0' + i).ondrop = function (event)
      {
        // event.preventDefault();
        event.target.appendChild(ref) ;

        console.log(`The Box is dropped in the target.`) ;
      }
    }
  </script>
</html>
```

【相关说明】

```
<body>

  <div id="div01" ondragover="event.preventDefault();">
    Block 1
    <span id="span01" draggable="true">
      box 1
    </span>
    <span id="span02" draggable="true">
      box 2
    </span>
  </div>

  <div id="div02" ondragover="event.preventDefault();"></div>
```

为了达成拖放机制，并简化读者们的理解，这里将【允许特定元素实例被拖动】的属性 draggable，以及【用来监听拖动悬停事件】的属性 ondragover，分别放置于属性 id 的数据为'span01'与'span02'的 span 元素实例本身，以及属性 id 的数据为'div01'与'div02'的 div 元素实例本身的 HTML 源代码里。

```
  </body>
  <script>

    let ref = null ;
```

此语句声明了初始数据为空值（null）的变量 ref。

```
    for (let i = 1; i < 3; i++)
```

此 for 循环语句会迭代 2 次。

```
    document.getElementById('span0' + i).ondragstart = function (event)
    {
      ref = event.target ;
    } ;
```

此语法将带有参数 event 的匿名函数的定义,赋给属性 id 的数据为'span01'与'span02'的 span 元素实例的属性 ondragstart。特定 span 元素实例开始被拖动时,上述匿名函数就会被调用,使得变量 ref 的数据,成为 event.target 对应的元素实例。在此,event.target 会对应到属性 id 的数据为'span01'或'span02'的元素实例。

```
    document.getElementById('div0' + i).ondrop = function (event)
    {
      // event.preventDefault();
      event.target.appendChild(ref) ;
```

此语句将属性 id 的数据为'span01'与'span02'的 span 元素实例,新增至属性 id 的数据为'div01'或'div02'的 div 元素实例中。

```
      console.log(`The Box is dropped in the target.`) ;
    } ;
```

上述语法将带有参数 event 的匿名函数的定义,赋给属性 id 的数据为'div01'和'div02'的 div 元素实例的属性 ondrop。特定 span 元素实例被放入特定 div 元素实例时,上述匿名函数就会被调用,以便允许在当前 event.target 所对应的 div 元素实例中,放入特定 span 元素实例。在此,event.target 会对应到属性 id 的数据为'div01'或'div02'的元素实例。

12.2　剪贴板事件

在特定网页里,特定元素实例内的文本,被复制、剪切或粘贴来自操作系统的剪贴板(clipboard)时,就会发生剪贴板(clipboard event)相关事件。请参看如下示例。

【12-2-^-clipboard-events.html】

```
<!DOCTYPE html>
<html>
  <head>
    <title>Form related events</title>
    <style>
      input, button, select
      {
        font-size: 1.2em ;
        margin: 5px ;
      }
    </style>
  </head>
```

```html
    <body>
        <form id="form01" name="form01" style="text-align: center;">
            <h3>个人资料</h3>
            <input type="text" id="username" name="username" placeholder="username" size="16" required>
            <input type="password" id="password" name="password" placeholder="password" size="16"
              required>
            <p></p>
            <select id="select01" name="select01">
                <option value="">choice of day-off</option>
                <option value="0">Sunday</option>
                <option value="1">Monday</option>
                <option value="2">Tuesday</option>
                <option value="3">Wednesday</option>
                <option value="4">Thursday</option>
                <option value="5">Friday</option>
                <option value="6">Saturday</option>
            </select>
            <input type="search" id="search" name="search" placeholder="job category.." size="9">

            <p></p>
            <button type="submit">Login</button>
            <button type="reset">Reset</button>
        </form>
    </body>
    <script>
      search.oncopy = function (event)
      {
        console.log('Data of the search input is copied.') ;
      } ;

      search.oncut = function (event)
      {
        console.log('Data of the search input is cut.') ;
      } ;

      search.onpaste = function (event)
      {
        console.log('Data is pasted into the search input.') ;
      } ;
    </script>
</html>
```

【相关说明】

请进一步参看本节中各小节的说明。

12.2.1 复制事件

在特定网页里，代表文本字段的特定元素实例内的文本，被复制到操作系统的剪贴板时，会发生剪贴板的复制（copy）事件。请参看如下示例。

【相关说明】

```html
<script>
```

```
search.oncopy = function (event)
{
  console.log('Data of the search input is copied.') ;
} ;
```

此语法将带有参数 event 的匿名函数的定义，赋给属性 id 的数据为'search'的 input 元素实例的属性 oncopy。input 元素实例内的文本，被复制（copy）时，上述匿名函数就会被调用，使得【Data of the search input is copied.】，被显示在浏览器的调试工具【Console】面板里。

12.2.2 剪切事件

在特定网页里，特定元素实例内的文本被剪切到操作系统的剪贴板时，会发生剪贴板的剪切（cut）事件。

【相关说明】

```
search.oncut = function (event)
{
  console.log('Data of the search input is cut.') ;
} ;
```

将带有参数 event 的匿名函数的定义赋给属性 id 的数据为'search'的 input 元素实例的属性 oncut。input 元素实例内的文本被剪切（cut）时，匿名函数就会被调用，使得【Data of the search input is cut.】显示在浏览器调试工具【Console】面板里。

12.2.3 粘贴事件

在特定网页里，于代表文本字段的特定元素实例内，被**粘贴**来自操作系统的剪贴板的文本时，会发生剪贴板的粘贴（paste）事件。请参看如下示例。

【相关说明】

```
search.onpaste = function (event)
{
  console.log('Data of the search input is pasted.') ;
} ;
```

此语法将带有参数 event 的匿名函数的定义，赋给属性 id 的数据为'search'的 input 元素实例的属性 onpaste。在代表文本字段的特定 input 元素实例内，粘贴（paste）特定文本时，上述匿名函数就会被调用，使得【Data is pasted into the search input.】被显示在浏览器的调试工具【Console】面板里。

12.3 视频和音频事件

在特定网页里，浏览器处理视频（video）和音频（audio）的过程中，会发生视频和音频的相关事件。

12.3.1 加载相关事件

浏览器加载特定视频或音频，至特定网页里的 video 或 audio 元素实例时，就会发生加载（load）相关事件。请参看如下示例。

【12-3-1-loading-events.html】

```html
<!DOCTYPE html>
<html>
  <head>
    <title>Multimedia related events</title>
  </head>
  <body>
    <video id="video" src="videos/toystory.mp4" type="video/mp4" width="400" controls>
    </video>
  </body>
  <script>
    video.onloadstart = function (event)
    {
      console.log('The video is being loaded.') ;
    } ;

    video.ondurationchange = function (event)
    {
      console.log('The video\'s duration has changed.') ;
    } ;

    video.onloadedmetadata = function (event)
    {
      console.log('The video\'s meta data has been loaded.') ;
    } ;

    video.onprogress = function (event)
    {
      console.log('The video is being downloaded.') ;
    } ;

    video.onloadeddata = function (event)
    {
      console.log('The video\'s first frame has been loaded.') ;
    } ;

    video.oncanplay = function (event)
    {
      console.log('The video has been loaded enough to play now.') ;
    } ;

    video.oncanplaythrough = function (event)
    {
      console.log('The video has been loaded "completely", and can play now.') ;
    } ;
  </script>
</html>
```

【相关说明】

```
<script>
 video.onloadstart = function (event)
 {
   console.log('The video is being loaded.') ;
 } ;
```

此语法将带有参数 event 的匿名函数的定义，赋给属性 id 的数据为'video'的 video 元素实例的属性 onloadstart。特定 video 元素实例的视频（video）已经开始被加载时，上述匿名函数就会被调用，使得【The video is loading.】，被显示在浏览器的调试工具【Console】面板里。

```
 video.ondurationchange = function (event)
 {
   console.log('The video\'s duration has changed.') ;
 } ;
```

此语法将带有参数 event 的匿名函数的定义，赋给属性 id 的数据为'video'的 video 元素实例的属性 ondurationchange。对于特定 video 元素实例的视频（video），其持续时间已经从 NaN (not a number)，变更为实际的时间长度时，上述匿名函数就会被调用，使得【The video's duration has changed.】，被显示在浏览器的调试工具【Console】面板里。

```
 video.onloadedmetadata = function (event)
 {
   console.log('The video\'s meta data has been loaded.') ;
 } ;
```

此语法将带有参数 event 的匿名函数的定义，赋给属性 id 的数据为'video'的 video 元素实例的属性 onloadedmetadata。特定 video 元素实例的视频（video）的元数据（metadata），已经被加载完成时，上述匿名函数就会被调用，使得【The video's meta data has been loaded.】，被显示在浏览器的调试工具【Console】面板里。

```
 video.onprogress = function (event)
 {
   console.log('The video is being downloaded.') ;
 } ;
```

此语法将带有参数 event 的匿名函数的定义，赋给属性 id 的数据为'video'的 video 元素实例的属性 onprogress。特定 video 元素实例的视频（video），处于被下载的状态时，上述匿名函数就会被调用，使得【The video is being downloaded.】，被显示在浏览器的调试工具【Console】面板里。

```
 video.onloadeddata = function (event)
 {
   console.log('The video\'s first frame has been loaded.') ;
 } ;
```

此语法将带有参数 event 的匿名函数的定义，赋给属性 id 的数据为'video'的 video 元素实例的属性 onloadeddata。特定 video 元素实例的视频（video）的第 1 帧（first frame），已经被加载完成时，上述匿名函数就会被调用，使得【The video's first frame has been loaded.】，被显示在浏览器的调试工具【Console】面板里。

```
video.oncanplay = function (event)
{
  console.log('The video has been loaded enough to play now.') ;
} ;
```

此语法将带有参数 event 的匿名函数的定义，赋给属性 id 的数据为'video'的 video 元素实例的属性 oncanplay。特定 video 元素实例的视频（video），处于可被播放（can be played）的状态时，上述匿名函数就会被调用，使得【The video has been loaded enough to play now.】，被显示在浏览器的调试工具【Console】面板里。

```
video.oncanplaythrough = function (event)
{
  console.log('The video has been loaded "completely", and can play now.') ;
} ;
```

此语法将带有参数 event 的匿名函数的定义，赋给属性 id 的数据为'video'的 video 元素实例的属性 oncanplaythrough。特定 video 元素实例的视频（video）已经被加载完成，并处于**从头到尾可被播放**的状态时，上述匿名函数就会被调用，使得【The video has been loaded "completely", and can play now.】，被显示在浏览器的调试工具【Console】面板里。

本示例会使得浏览器在其调试工具【Console】面板里，分别显示出如下的信息：

- The video is being loaded.
- The video's duration has changed.
- The video's meta data has been loaded.
- The video is being downloaded.
- The video's first frame has been loaded.
- The video has been loaded enough to play now.
- The video has been loaded "completely", and can play now.

由此可知，特定视频从远程服务器，被加载到客户端的浏览器之前，会经过【开始进行加载→其持续时间发生变化→其元数据被加载→其主要数据开始被下载→其第 1 帧被加载完成→其数据充分被加载并可被播放→其数据被加载完成】等阶段。

12.3.2 清空事件

在特定网页里，，特定 video 或 audio 元素实例所对应的视频或音频被清空时，会发生清空（emptied）事件。请参看如下示例。

【12-3-2-onemptied.html】

```
<!DOCTYPE html>
<html>
  <head>
    <title>Multimedia related events</title>
  </head>
  <body>
    <video id="video" src="videos/toystory.mp4" type="video/mp4" width="400" controls>
    </video>
```

```
  </body>
  <script>
    video.onemptied = function (event)
    {
      console.log('The video has been removed.') ;
    } ;

    setTimeout(() => video.src = '', 2000) ;
  </script>
</html>
```

【相关说明】

```
<script>
  video.onemptied = function (event)
  {
    console.log('The video\'s playlist has been removed.') ;
  } ;
```

此语法将带有参数 event 的匿名函数的定义，赋给属性 id 的数据为'video'的 video 元素实例的属性 onemptied。特定 video 元素实例的视频（video）被清空（emptied）时，上述匿名函数就会被调用，使得【The video's playlist has been removed.】，被显示在浏览器的调试工具【Console】面板里。

```
setTimeout(() => video.src = '', 2000) ;
```

此语法设置了在 2000 毫秒（2 秒）之后，调用箭头函数【() => video.src = ''】，使得在属性 id 的数据为'video'的 video 元素实例中，其属性 src 的数据，被设置为空字符''，进而等同于这个 video 元素实例所对应的视频，被清空而触发清空（emptied）事件。

12.3.3 播放结束事件

在特定网页里，特定 video 或 audio 元素实例对应的视频或音频，已经被播放完成时，会发生播放结束（ended）事件。请参看如下示例。

【12-3-3-onended.html】

```
<!DOCTYPE html>
<html>
  <head>
    <title>Multimedia related events</title>
  </head>
  <body>
    <video id="video" src="videos/toystory.mp4" type="video/mp4" width="400" controls>
    </video>
  </body>
  <script>
    video.onended = function (event)
    {
      console.log('The video has fully played.') ;
    } ;
```

```
    </script>
</html>
```

【相关说明】

```
<script>
  video.onended = function (event)
  {
    console.log('The video has fully played.') ;
  } ;
```

此语法将带有参数 event 的匿名函数的定义，赋给属性 id 的数据为'video'的 video 元素实例的属性 onended。特定 video 元素实例的视频（video），被播放到结束（ended）的位置时，上述匿名函数就会被调用，使得【The video has fully played.】，被显示在浏览器的调试工具【Console】面板里。

12.3.4 异常相关事件

在特定网页里，特定 video 或 audio 元素实例对应的视频或音频，于加载（load）阶段，遇到异常状况时，会发生异常（exception）相关事件。请参看如下示例。

【12-3-4-exception-events.html】

```
<!DOCTYPE html>
<html>
  <head>
    <title>Multimedia related events</title>
  </head>
  <body>
    <video id="video" src="videos/_toystory.mp4" type="video/mp4" width="400" controls>
    </video>
  </body>
  <script>
    video.onerror = function (event)
    {
      console.log('Error occurs.') ;
    } ;

    video.onstalled = function (event)
    {
      console.log('The data of video is not available.') ;
    } ;

    video.onsuspend = function (event)
    {
      console.log('The data of video is suspended.') ;
    } ;
  </script>
</html>
```

【相关说明】

```
<script>
```

```
video.onerror = function (event)
{
  console.log('Error occurs.') ;
} ;
```

此语法将带有参数 event 的匿名函数的定义，赋给属性 id 的数据为'video'的 video 元素实例的属性 onerror。特定 video 元素实例的视频（video），在加载阶段中，发生任何错误（error）时，上述匿名函数就会被调用，使得【Error occurs.】被显示在浏览器的调试工具【Console】面板里。

```
video.onstalled = function (event)
{
  console.log('The data of video is not available.') ;
} ;
```

此语法将带有参数 event 的匿名函数的定义，赋给属性 id 的数据为'video'的 video 元素实例的属性 onstalled。特定 video 元素实例的视频（video），在加载阶段中，突然停滞（stalled）时，上述匿名函数就会被调用，使得【The data of video is not available.】，被显示在浏览器的调试工具【Console】面板里。

```
video.onsuspend = function (event)
{
  console.log('The data of video is suspended.') ;
} ;
```

此语法将带有参数 event 的匿名函数的定义，赋给属性 id 的数据为'video'的 video 元素实例的属性 onsuspend。特定 video 元素实例的视频（video），在加载阶段中，突然被停止（suspended）加载时，上述匿名函数就会被调用，使得【The data of video is suspended.】，被显示在浏览器的调试工具【Console】面板里。

在本示例中，因为难以模拟【停滞（stalled）】或【突然被停止（suspended）】的状态，所以只能调试出【发生错误（error）】的状态。

12.3.5　播放与暂停相关事件

在特定网页里，特定 video 或 audio 元素实例对应的视频或音频，被播放与暂停时，会发生播放（play）与暂停（pause）相关事件。请参看如下示例。

【12-3-5-play-and-pause-events.html】

```
<!DOCTYPE html>
<html>
  <head>
    <title>Multimedia related events</title>
  </head>
  <body>
    <video id="video" src="videos/toystory.mp4" type="video/mp4" width="400" controls>
    </video>
  </body>
  <script>
    video.onpause = function (event)
```

```
      {
        console.log('The video is paused.') ;
      } ;

      video.onplay = function (event)
      {
        console.log('The video is starting.') ;
      } ;

      video.onplaying = function (event)
      {
        console.log('The video is restarting after paused.') ;
      } ;
    </script>
</html>
```

【相关说明】

```
<script>
  video.onpause = function (event)
  {
    console.log('The video is paused.') ;
  } ;
```

此语法将带有参数 event 的匿名函数的定义，赋给属性 id 的数据为'video'的 video 元素实例的属性 onpause。特定 video 元素实例的视频（video），被暂停（pause）时，上述匿名函数就会被调用，使得【The video is paused.】，被显示在浏览器的调试工具【Console】面板里。

```
  video.onplay = function (event)
  {
    console.log('The video is starting.') ;
  } ;
```

此语法将带有参数 event 的匿名函数的定义，赋给属性 id 的数据为'video'的 video 元素实例的属性 onplay。特定 video 元素实例的视频（video），被播放（play）时，上述匿名函数就会被调用，使得【The video is starting.】，被显示在浏览器的调试工具【Console】面板里。

```
  video.onplaying = function (event)
  {
    console.log('The video is restarting after paused.') ;
  } ;
```

此语法将带有参数 event 的匿名函数的定义，赋给属性 id 的数据为'video'的 video 元素实例的属性 onplaying。特定 video 元素实例的视频（video），从暂停状态，重新被播放（playing）时，上述匿名函数就会被调用，使得【The video is restarting after paused.】，被显示在浏览器的调试工具【Console】面板里。

12.3.6　播放速率变化事件

在特定网页里，特定 video 或 audio 元素实例对应的视频或音频的播放速率（playback rate），被变更时,会发生播放速率变化(rate change)事件。请参看如下示例。

【12-3-6-onratechange.html】

```
<!DOCTYPE html>
<html>
  <head>
    <title>Multimedia related events</title>
  </head>
  <body>
    <video id="video" src="videos/toystory.mp4" type="video/mp4" width="400" controls autoplay>
    </video>
  </body>
<script>
  video.onratechange = function (event)
  {
    console.log('The video\'s rate has been changed.') ;
  } ;

  setTimeout(() => video.playbackRate = 2, 2000) ;
</script>
</html>
```

【相关说明】

```
<script>

  video.onratechange = function (event)
  {
    console.log('The video\'s rate has been changed.') ;
  } ;
```

此语法将带有参数 event 的匿名函数的定义,赋给属性 id 的数据为'video'的 video 元素实例的属性 onratechange。特定 video 元素实例的视频(video)的播放速率(rate),被变更(change)时,上述匿名函数就会被调用,使得【The video's rate has been changed.】,被显示在浏览器的调试工具【Console】面板里。

```
  setTimeout(() => video.playbackRate = 2, 2000) ;
```

此语句设置了 2000 毫秒(2 秒)之后,调用箭头函数【() => video.playbackRate = 2】,进而在属性 id 的数据为'video'的 video 元素实例中,将数值 2,赋给其属性 playbackRate,意味着其播放速率变更为 200%。

12.3.7 播放位置变化相关事件

在特定网页里,特定 video 或 audio 元素实例对应的视频或音频的当前播放位置(playback position),被变更时,就会发生播放位置变化(playback position changed)相关事件。请参看如下示例。

【12-3-7-playback-position-changed-events.html】

```
<!DOCTYPE html>
```

```html
<html>
  <head>
    <title>Multimedia related events</title>
  </head>
  <body>
    <video id="video" src="videos/toystory.mp4" type="video/mp4" width="400" controls>
    </video>
  </body>
  <script>
    video.onseeked = function (event)
    {
      console.log('Another position in the video has been seeked.') ;
    } ;

    video.onseeking = function (event)
    {
      console.log('Another position in the video is seeking.') ;
    } ;

    video.ontimeupdate = function (event)
    {
      console.log('The playing position in the video has been changed.') ;
    } ;
  </script>
</html>
```

【相关说明】

```html
<script>
  video.onseeked = function (event)
  {
    console.log('Another position in the video has been seeked.') ;
  } ;
```

此语法将带有参数 event 的匿名函数的定义，赋给属性 id 的数据为'video'的 video 元素实例的属性 onseeked。特定 video 元素实例的视频（video），被用户搜索（seeked）至另一时间点的瞬间，上述匿名函数就会被调用，使得【Another position in the video has been seeked.】，被显示在浏览器的调试工具【Console】面板里。

```html
  video.onseeking = function (event)
  {
    console.log('Another position in the video is seeking.') ;
  } ;
```

此语法将带有参数 event 的匿名函数的定义，赋给属性 id 的数据为'video'的 video 元素实例的属性 onseeking。特定 video 元素实例的视频（video），被用户搜索（seeking）的瞬间，也就是用户单击视频**播放栏**（playback bar）的任意时间位置的瞬间，上述匿名函数就会被调用，使得【Another position in the video is seeking.】，被显示在浏览器的调试工具【Console】面板里。

```html
  video.ontimeupdate = function (event)
  {
    console.log('The playing position in the video has been changed.') ;
  } ;
```

此语法将带有参数 event 的匿名函数的定义，赋给属性 id 的数据为'video'的 video 元素实例的属性 ontimeupdate。特定 video 元素实例的视频（video）的播放时间点，持续被播放或者被手动变更时，上述匿名函数就会被调用，使得【The playing position in the video has been changed.】，被显示在浏览器的调试工具【Console】面板里。

12.3.8　音量变化事件

在特定网页里，特定 video 或 audio 元素实例对应的视频或音频的音量（volume），出现变化时，会发生音量变化（volume change）事件。请参看如下示例。

【12-3-8-onvolumechange.html】
```
<!DOCTYPE html>
<html>
  <head>
    <title>Multimedia related events</title>
  </head>
  <body>
    <video id="video" src="videos/toystory.mp4" type="video/mp4" width="400" controls>
    </video>
  </body>
  <script>
    video.onvolumechange = function (event)
    {
      console.log('The volume of video has been changed.') ;
    } ;
  </script>
</html>
```

【相关说明】

```
<script>
  video.onvolumechange = function (event)
  {
    console.log('The volume of video has been changed.') ;
  } ;
```

此语法将带有参数 event 的匿名函数的定义，赋给属性 id 的数据为'video'的 video 元素实例的属性 onvolumechange。特定 video 元素实例的视频（video）的音量（volume），被调整时，上述匿名函数就会被调用，使得【The volume of video has been changed.】，被显示在浏览器的调试工具【Console】面板里。

12.3.9　缓冲等待事件

在特定网页里，特定 video 或 audio 元素实例对应的视频或音频，加载不完全而被等待下一帧（next frame）的二进制数据时，会发生缓冲等待（waiting）事件。请参看如下示例。

【12-3-9-onwaiting.html】

```html
<!DOCTYPE html>
<html>
  <head>
    <title>Multimedia related events</title>
  </head>
  <body>
    <video id="video" src="videos/toystory.mp4" type="video/mp4" width="400" controls>
    </video>
  </body>
  <script>
    video.onwaiting = function (event)
    {
      console.log('The video is waiting for buffering its next frame.') ;
    } ;
  </script>
</html>
```

【相关说明】

```html
<script>
  video.onwaiting = function (event)
  {
    console.log('The video is waiting for buffering its next frame.') ;
  } ;
```

此语法将带有参数 event 的匿名函数的定义，赋给属性 id 的数据为'video'的 video 元素实例的属性 onwaiting。当特定 video 元素实例的视频（video），处于被等待其下一帧（next frame）被加载的状态时，上述匿名函数就会被调用，使得【The video is waiting for buffering its next frame.】，被显示在浏览器的调试工具【Console】面板里。

在本示例中，难以模拟【因为网络连接问题，而导致处于等待下一帧】的状态，所以理解本示例的工作原理即可。

12.4　动画及过渡事件

在特定网页里的特定元素实例中，被设置的动画（animation）或过渡（transition）效果，在播放进程中，前后会发生动画或过渡效果的相关事件。

12.4.1　动画相关事件

特定元素实例被设置的动画，在其播放进程中，前后会发生动画开始（animation start）、动画结束（animation end）和动画迭代（animation iteration）的事件。请参看如下示例。

【12-4-1-animation-events.html】

```html
<!DOCTYPE html>
```

```html
<html>
  <head>
    <title>Multimedia related events</title>
    <style>
      @keyframes text_color
      {
        from {color: Gold ;}
        to {color: RoyalBlue ;}
      }

      #div01
      {
        font-size: 2em ;
        animation: text_color 2s 3 ;
      }
    </style>
  </head>
  <body>
    <div id="div01">Hello</div>
  </body>
  <script>
    div01.addEventListener('animationstart', () => console.log('The animation is starting.')) ;

    div01.addEventListener('animationend', () => console.log('The animation is over.')) ;

    div01.addEventListener('animationiteration', () => console.log('The animation is iterated.')) ;
  </script>
</html>
```

【相关说明】

```
<style>
  @keyframes text_color
  {

    from {color: Gold ;}
    to {color: RoyalBlue ;}
```

这两行 CSS 语法，设置了【从**金色**，变成**宝蓝色**的文本颜色】的变换动画。

```
  }
  #div01
  {
    font-size: 2em ;

    animation: text_color 2s 3 ;
```

这行 CSS 语法，则设置了以 2 秒的时间，重复 **3 次**，播放名称为 text_color 的上述动画。

```
<script>
  div01.addEventListener('animationstart', () => console.log('The animation is starting.')) ;
```

此语句通过调用 addEventListener()函数，使得属性 id 的数据为'div01'的 div 元素实例，被监听着动画启动（animation start）事件。一旦作用于属性 id 的数据为'div01'的 div 元素实例的动画（animation），被启动（start）时，显示【The animation is starting.】信息的箭头函数，就会被调用。

```
div01.addEventListener('animationend', () => console.log('The animation is over.')) ;
```

此语句通过调用 addEventListener()函数，使得属性 id 的数据为'div01'的 div 元素实例，被监听着动画结束（animation end）事件。一旦作用于属性 id 的数据为'div01'的 div 元素实例的动画（animation），已结束（end）时，显示【The animation is over.】信息的箭头函数，就会被调用。

```
div01.addEventListener('animationiteration', () => console.log('The animation is iterated.')) ;
```

此语句通过调用 addEventListener()函数，使得属性 id 的数据为'div01'的 div 元素实例，被监听着动画迭代（animation iteration）事件。一旦作用于属性 id 的数据为'div01'的 div 元素实例的动画（animation），再次被播放／迭代（iterate）时，显示【The animation is iterated.】信息的箭头函数，就会被调用。

12.4.2　过渡结束事件

特定元素实例被设置的过渡效果，被播放到结束时，会发生过渡结束（transition end）事件。请参看如下示例。

【12-4-2-transition-events.html】

```html
<!DOCTYPE html>
<html>
  <head>
    <title>Multimedia related events</title>
    <style>
      #div01
      {
        font-size: 2em ;
        width: 100px ;
        height: 100px ;
        border-radius: 5px ;
        background-color: GoldenRod ;
        transition: transform 2s ;
        position: relative ;
        left: 50px ;
        top: 50px ;
      }

      #div01:hover
      {
        transform: rotate(360deg) ;
      }
    </style>
  </head>
  <body>
    <div id="div01"></div>
```

```
    </body>
    <script>
      div01.addEventListener('transitionend', () => console.log('The transition is over.')) ;
    </script>
</html>
```

【相关说明】

```
    <style>
      #div01
      {
        font-size: 2em ;
        width: 100px ;
        height: 100px ;
        border-radius: 5px ;
        background-color: GoldenRod ;

        transition: transform 2s ;
```

这行 CSS 语法，设置了 2 秒的过渡（transition）特效。

```
        position: relative ;
        left: 50px ;
        top: 50px ;
      }

      #div01:hover
      {

        transform: rotate(360deg) ;
```

这行 CSS 语法，设置了过渡特效为【当鼠标指针被移入（hover）其属性 id 的数据为'div01'的元素实例的范围内时，将这个元素实例，旋转 360 度】。

```
      }
    <script>
      div01.addEventListener('transitionend', () => console.log('The transition is over.')) ;
```

此语句通过调用 addEventListener()函数，使得属性 id 的数据为'div01'的 div 元素实例，被监听着过渡结束（transition end）事件。一旦作用于属性 id 的数据为'div01'的 div 元素实例的过渡特效已结束（end）时，显示【The transition is over.】信息的箭头函数，就会被调用。

12.5　其 他 事 件

本节将提及 details 元素实例的切换事件、鼠标滚轮（mouse wheel）事件、触摸（touch）相关事件，以及接收服务器数据（receiving server data）相关事件。

12.5.1　details 元素实例的切换事件

在特定网页里，特定 details 元素实例，被展开或折叠时，会发生**切换**（toggle）事件。请参看如下示例。

【12-5-1-ontoggle.html】

```html
<!DOCTYPE html>
<html>
  <head>
    <title>Touch related events</title>
    <style>
      #details01
      {
        width: 400px ;
        height: 400px ;
        color: RoyalBlue ;
        text-align: center ;
        border-radius: 5px ;
      }
    </style>
  </head>
  <body>
    <details id="details01">
      <summary style="color: Tomato">
        About Lycopene
      </summary>
      Absorption of lycopene requires that it be combined with bile salts and fat to form
        micelles.<p></p>
      Intestinal absorption of lycopene is enhanced by the presence of fat and by cooking.<p></p>
      Lycopene dietary supplements (in oil) may be more efficiently absorbed than lycopene from
        food.<p></p>
      Lycopene is not an essential nutrient for humans, but is commonly found in the diet mainly
        from dishes prepared from tomatoes.
    </details>
  </body>
  <script>
    details01.ontoggle = function (event)
    {
      console.log('Block of the article is toggled to open/close.') ;
    } ;
  </script>
</html>
```

【相关说明】

```
<script>

  details01.ontoggle = function (event)
  {
    console.log('Block of the article is toggled to open/close.') ;
  } ;
```

此语法将带有参数 event 的匿名函数的定义，赋给属性 id 的数据为'details01'的 details 元素实例的属性 ontoggle。特定 details 元素实例的内容，被**展开**或**折叠**时，上述匿名函数就会被调用，使得【Block of the article is toggled to open/close.】，被显示在浏览器的调试工具【Console】面板里。

12.5.2 鼠标滚轮事件

在特定网页里，通过鼠标滚轮（mouse wheel），使得特定元素实例中的内容，被滚动时，会发生鼠标滚轮事件。请参看如下示例。

【12-5-2-onwheel.html】

```
<!DOCTYPE html>
<html>
  <head>
    <title>Touch related events</title>
    <style>
      #div01
      {
        /*width: 400px ;*/
        /*height: 400px ;*/
        color: RoyalBlue ;
        border-radius: 5px ;
        font-size: 2em ;
      }
    </style>
  </head>
  <body>
    <div id="div01">
      <p>Curcumin is a bright yellow chemical produced by some plants. It is the principal curcuminoid
        of turmeric (Curcuma longa), a member of the ginger family, Zingiberaceae.It is sold
        as an herbal supplement, cosmetics ingredient, food flavoring, and food coloring.</p>
      <p>Chemically, curcumin is a diarylheptanoid, belonging to the group of curcuminoids, which
        are natural phenols responsible for turmeric's yellow color. It is a tautomeric compound
        existing in enolic form in organic solvents, and as a keto form in water.</p>
      <p>Although thoroughly studied in laboratory and clinical studies, curcumin has no confirmed
        medical uses, and has proved frustrating to scientists who state that it is unstable, not
        bioavailable, and unlikely to produce useful leads for drug development.</p>
    </div>
  </body>
  <script>
    // div01.onmousewheel = function (event)
    div01.onwheel = function (event)
    {
      console.log('Block of the article is scrolled by mouse wheel.') ;
    } ;
  </script>
</html>
```

【相关说明】

```
<script>
  // div01.onmousewheel = function (event)
  div01.onwheel = function (event)
  {
```

```
      console.log('Block of the article is scrolled by mouse wheel.') ;
    } ;
```

此语句将带有参数 event 的匿名函数的定义，赋给属性 id 的数据为'div01'的 div 元素实例的属性 onwheel。特定 div 元素实例的内容，通过鼠标滚轮，被加以滚动时，上述匿名函数就会被调用，使得【Block of the article is scrolled by mouse wheel.】，被显示在浏览器的调试工具【Console】面板里。

12.5.3　触摸相关事件

在触摸屏（touch screen）中的特定网页里，特定元素实例被触摸时，会发生触摸（touch）相关事件，例如触摸开始（touch start）事件、触摸结束（touch end）事件、触摸移动（touch move）事件，以及触摸撤销（touch cancel）事件。请参看如下示例。

【12-5-3-touch-events.html】

```
<!DOCTYPE html>
<html>
  <head>
    <title>Touch related events</title>
    <style>
      #div01
      {
        width: 400px ;
        height: 400px ;
        color: RoyalBlue ;
        text-align: center ;
        border-radius: 5px ;
        background-color: Gold ;
      }
    </style>
  </head>
  <body>
    <div id="div01">Block 1</div>
  </body>
  <script>
    div01.ontouchstart = function (event)
    {
      console.log('Block is pressed.') ;
    } ;

    div01.ontouchend = function (event)
    {
      console.log('Block is released.') ;
    } ;

    div01.ontouchmove = function (event)
    {
      console.log('Touch point is moving.') ;
    } ;
```

```
    div01.ontouchcancel = function (event)
    {
      console.log('Touch point is changed.') ;
    } ;
  </script>
</html>
```

【相关说明】

```
<script>

  div01.ontouchstart = function (event)
  {
    console.log('Block is pressed.') ;
  } ;
```

此语法将带有参数 event 的匿名函数的定义，赋给属性 id 的数据为'div01'的 div 元素实例的属性 ontouchstart。特定 div 元素实例被触摸 (touch start) 时，上述匿名函数就会被调用，使得【Block is pressed.】，被显示在浏览器的调试工具【Console】面板里。

若读者欲在台式电脑和笔记本电脑，通过鼠标的点击，模拟被触摸的动作，则可借助浏览器的调试工具来完成。以 Google Chrome 浏览器为例，可执行如下步骤：

- 在窗口中，浏览此示例文档。
- 按下快捷键 Ctrl + Shift + I，以启动调试工具的【Console】面板。
- 务必单击【Console】面板一次，再按下快捷键 Ctrl + Shift + M，以切换至【移动设备】的模拟状态。
- 此时，必须按快捷键 F5 或者 Ctrl + R，重新加载此示例文档！
- 接着，通过鼠标单击的动作，即可在移动设备里，模拟出上述 div 元素实例被触摸（touch）的动作。

```
  div01.ontouchend = function (event)
  {
    console.log('Block is released.') ;
  } ;
```

此语法将带有参数 event 的匿名函数的定义，赋给属性 id 的数据为'div01'的 div 元素实例的属性 ontouchend。特定 div 元素实例，被触摸已结束 (touch end) 时，上述匿名函数就会被调用，使得【Block is released.】，被显示在浏览器的调试工具【Console】面板里。其中，所谓的被触摸已结束，即是指用户的手指离开触摸屏。

```
  div01.ontouchmove = function (event)
  {
    console.log('Touch point is moving.') ;
  } ;
```

此语法将带有参数 event 的匿名函数的定义，赋给属性 id 的数据为'div01'的 div 元素实例的属性 ontouchmove。特定 div 元素实例正在被触摸，并且其触摸点仍然在移动时，就会持续调用上述匿名函数，使得【Touch point is moving.】，被持续显示在浏览器的调试工具【Console】面板里。其中，所谓的触摸点仍然在移动，即是指用户的手指，持续在触摸屏上滑动。

```
    div01.ontouchcancel = function (event)
```

```
      {
        console.log('Touch point is changed.') ;
      } ;
```

此语句将带有参数 event 的匿名函数的定义，赋给属性 id 的数据为'div01'的 div 元素实例的属性 ontouchcancel。在特定 div 元素实例中，若已经存在一个触摸点，却又出现另一个新的触摸点时，上述匿名函数就会被调用，使得【Touch point is changed.】，被显示在浏览器的调试工具【Console】面板里。

所谓的已经存在一个触摸点，却又出现另一个新的触摸点，就是指用户在 div 元素实例的范围内，前后不止放入一根手指。也因此，这个动作的调试，无法在触摸屏以外的屏幕上进行。

12.5.4　接收服务器数据相关事件

在特定网页里的 script 元素实例中，可编写访问 EventSource 对象实例的源代码，让浏览器通过 EventSource 对象实例，创建【至特定网站服务器的特定网址】的连接，以便接收来自特定网站服务器的信息（message）数据。

每次接收到来自特定网站服务器的信息数据时，都会发生接收服务器数据（receiving server data）相关事件，例如错误（error）事件、接收信息（message）事件和连接敞开（open）事件。请参看如下带有两个源代码文档的示例。

【12-5-4-receiving-server-data-events.html】

```
<!DOCTYPE html>
<html>
  <head>
    <title>Server-sent events</title>
  </head>
  <body>
  </body>
  <script>
    let source = new EventSource('6-9-4-receiving-server-data-testing.php') ;

    source.onopen = function (event)
    {
      console.log('Connection to the server is established.') ;
    }

    source.onmessage = function (event)
    {
      console.log('A signal from the server is received.') ;
      console.log(event.data) ;
    } ;

    source.onerror = function (event)
    {
      console.log('An error occurs.') ;
    } ;
  </script>
</html>
```

第 12 章 DOM 的事件处理（二） | 345

【相关说明】

```
<script>
  let source = new EventSource('6-9-4-receiving-server-data-testing.php');
```

此语句声明了变量 source，其初始数据为可访问特定网站服务器端的 PHP 源代码文档 12-5-4-receiving-server-data-testing.php 的 EventSource 对象实例。变量 source 的 EventSource 对象实例，可通过 HTTP 协议来，接收特定网站服务器端所发送的数据。

请留意，本示例只能正常运作于支持 HTTP 协议的网站服务器网址，例如 http://localhost/12-5-4-receiving-server-data-testing.php。直接通过浏览器打开 file:///D:/works/books/publishers/示例文档/html/12-5-4-receiving-server-data-testing.php，是无法使得本示例正常运作的。

```
  source.onopen = function (event)
  {
    console.log('Connection to the server is established.');
  }
```

此语法将带有参数 event 的匿名函数的定义，赋给变量 source 的 EventSource 对象实例的属性 onopen。当 EventSource 对象实例被浏览器启动之后，以连接到网站服务器的 PHP 源代码 12-5-4-receiving-server-data-testing.php 时，上述匿名函数就会被调用，使得【Connection to the server is established.】，被显示在浏览器的调试工具【Console】面板里。

```
  source.onmessage = function (event)
  {
    console.log('A signal from the server is received.');
    console.log(event.data);
  };
```

此语法将带有参数 event 的匿名函数的定义，赋给变量 source 的 EventSource 对象实例的属性 onmessage。当 EventSource 对象实例接收到【网站服务器**解析** PHP 源代码 12-5-4-receiving-server-data-testing.php】之后的返回数据时，上述匿名函数就会被调用，使得【A signal from the server is received.】和【current number = XX】的信息，分别被显示在浏览器的调试工具【Console】面板里。其中，【current number = XX】是来自于网站服务器端，并且由 event.data 所返回的字符串信息。

```
  source.onerror = function (event)
  {
    console.log('An error occurs.');
  };
```

此语法将带有参数 event 的匿名函数的定义，赋给变量 source 的 EventSource 对象实例的属性 onerror。当浏览器对网站服务器的 PHP 源代码文档 12-5-4-receiving-server-data-testing.php 的连接，被中断时，上述匿名函数就会被调用，使得【An error occurs.】，被显示在浏览器的调试工具【Console】面板里。

【12-5-4-receiving-server-data-testing.php】

```
<?php
  header('Content-Type: text/event-stream');

  $number = rand(1, 100);
```

```
    echo "data: current number = $number\n\n" ;

    flush();
?>
```

【相关说明】

```
<?php

    header('Content-Type: text/event-stream') ;
```

通过此语句，网站服务器可发送事件流 (event stream)。所谓的事件流，是指定时回复至客户端的信息数据。

```
    $number = rand(1, 100) ;
```

此语句会动态产生 1～100 之间的随机整数值。

```
    echo "data: current number = $number\n\n" ;
```

此语句的返回数据，是被用来定时回复至客户端的信息数据，例如"data: current number = 24\n\n"。在此，用来表示回复用途的开头字符串"data:" 及其末尾的字符串"\n\n"，都是必要的！

```
    flush();
```

此语句将上述信息数据，尽量即时地传送到客户端，以尽快显示到其浏览器窗口中。

12.6 练习题

1. 试编写 JavaScript 源代码，以显示出用户在特定网页上，从拖动特定元素实例开始，至拖动结束为止，以像素（pixel）为计量单位的拖动距离。（提示：DragEvent.clientX 和 DragEvent.clientY。）

2. 试编写 JavaScript 源代码，使得在当前网页中，其属性 id 的数据为'video'的影片实例，被暂停播放时，显示出带有特定信息的 div 元素实例；而影片被继续播放时，则隐藏这个 div 元素实例。

3. 试编写 JavaScript 源代码，以便在当前网页中，显示出用户在 5 秒钟之内，点击鼠标左键的次数、输入字符的个数，以及滚动鼠标滚轮的总刻度。

第 13 章

Reflect 对象

本章内容主要介绍 Reflect 对象的多个函数的运用，以利于编程人员在特定**自定义**的**对象实例**中，更加细腻地对其各个属性，进行创建、定义、删除、获取、判断和列举等访问的动作。

13.1　Reflect 对象介绍（ES6）

Reflect 对象实例的内置函数，可用来替代或简化【访问 Object 对象实例的属性（property）】的相关语法。请参看如下示例。

【13-1-^-Reflect-object.js】

```
let candy_amount = {durian: 30, strawberry: 55, cranberry: 10, blueberry: 13, cherry: 60, orange: 18, lemon: 10} ;

Object.defineProperty(candy_amount,'apple', {value: 17, writable: true, enumerable: true, configurable: true}) ;

candy_amount.watermelon = 33 ;

let symbol01 = Symbol('mixed') ;

candy_amount[symbol01] = 5 ;

console.log(Reflect.ownKeys(candy_amount)) ;
console.log('') ;

console.log(candy_amount.orange) ;
console.log(candy_amount.watermelon) ;
console.log('') ;

console.log(candy_amount.mixed) ;
```

```
console.log(candy_amount['mixed']) ;
console.log('') ;

console.log(candy_amount[symbol01]) ;
```

【相关说明】

```
let candy_amount = {durian: 30, strawberry: 55, cranberry: 10, blueberry: 13, cherry: 60, orange:
    18, lemon: 10} ;
```

声明初始数据为对象实例的变量 candy_amount。

```
Object.defineProperty(candy_amount,'apple', {value: 17, writable: true, enumerable: true,
    configurable: true}) ;
```

此语句通过内置函数 Object.definePreperty()，在变量 candy_amount 的对象实例中，新增初始数值为 17 的属性 apple，并设置其可再次被写入（writable）、可被迭代／列举（enumerable）和可再次被配置（configurable）的特征。

```
candy_amount.watermelon = 33 ;
```

此语句将整数值 33，赋给变量 candy_amount 对象实例的属性 watermelon。

```
let symbol01 = Symbol('mixed') ;
```

此语句声明了变量 symbol01，其初始数据为内含字符串'mixed'的 Symbol 对象实例，使得变量名称 symbol01 等价于 Symbol('mixed')，可作为特定对象实例的**新属性**的标识符（identifier）。

```
candy_amount[symbol01] = 5 ;
```

此语句将整数值 5，赋给变量 candy_amount 对象实例中的标识符为【Symbol('mixed')】的新属性。在此，通过变量名称 symbol01，来代表新属性的标识符【Symbol('mixed')】。换言之，【candy_amount[symbol01] = 5 ;】等价于【candy_amount[Symbol('mixed')] = 5 ;】。

```
console.log(Reflect.ownKeys(candy_amount)) ;
```

通过内置函数 Reflect.ownKeys()，可使得变量 candy_amount 的对象实例中的各个新属性，组成新数组实例["durian", "strawberry", "cranberry", "blueberry", "cherry", "orange", "lemon", "apple", "watermelon", Symbol(mixed)]。

```
console.log(candy_amount.orange) ;
```

candy_amount.orange 当前会返回整数值 18。

```
console.log('') ;

console.log(candy_amount.watermelon) ;
```

candy_amount.watermelon 当前会返回整数值 33。

```
console.log('') ;

console.log(candy_amount.mixed) ;
console.log(candy_amount['mixed']) ;
```

candy_amount.mixed 或 candy_amount['mixed'] 当前均会返回 undefined，代表在变量

candy_amount 的对象实例中，并不存在属性 mixed。

```
console.log('') ;

console.log(candy_amount[symbol01]) ;
```

candy_amount[symbol01]当前会返回整数值 5，代表在变量 candy_amount 的对象实例中，存在**标识符**为 Symbol('mixed')的属性。

13.2　间接应用特定函数（ES6）

通过函数 Reflect.apply()，可传入多个数据，至即将被间接应用的其他函数。请参看如下示例。

【13-2-^-Reflect-apply.js】

```
let code_list = [108, 111, 118, 101, 32, 121, 111, 117] ;

result = String.fromCharCode(... code_list) ;
console.log(result) ;

result = Reflect.apply(String.fromCharCode, undefined, code_list) ;
console.log(result) ;

let pointer = null ;
let range_list = [10, 10, 10, 50, 50, 50, 100, 100, 100] ;

function generate(... list)
{
  let number_list = [] ;

  console.log(this.valueOf()) ;

  for (let i = 0; i < list.length; i++)
  {
    number_list.push(parseInt(list[i]* Math.random())) ;
  }

  return number_list ;
}

result = Reflect.apply(generate,'test message', range_list) ;
console.log(result) ;
```

【相关说明】

```
let code_list = [108, 111, 118, 101, 32, 121, 111, 117] ;
```

声明初始数据为内含多个整数值的数组实例的变量 code_list。

```
result = String.fromCharCode(... code_list) ;
```

此语句将字符串'love you'，赋给变量 result。其中，语法【String.fromCharCode(... code_list)】等价

于语法【String.fromCharCode(... [108, 111, 118, 101, 32, 121, 111, 117])】，最后也等同于语法
【String.fromCharCode(108, 111, 118, 101, 32, 121, 111, 117)】，均会返回各数码所对应的字符'l'
（108）、'o'（111）、'v'（118）、'e'（101）、''（32）、'y'（121）、'o'（111）与'u'（117），并组
成新的字符串'love you'。

```
console.log(result) ;

result = Reflect.apply(String.fromCharCode, undefined, code_list) ;
```

【Reflect.apply(String.fromCharCode, undefined, code_list)】等同于【String.fromCharCode(... code_list)】。若是【String.fromCharCode(... code_list)】可被改写成为【其数据为字符串的变量名称.fromCharCode(... code_list)】，则【Reflect.apply(String.fromCharCode, **undefined**, code_list)】可被改写成为【Reflect.apply(fromCharCode, **其数据为字符串的变量名称**, code_list)】。

可惜的是，函数 String.fromCharCode() 并不能被改成为【其数据为字符串的变量名称.fromCharCode()】，所以【Reflect.apply(String.fromCharCode, **undefined**, code_list)】中的第 2 个参数的数据必须是 undefined。

```
console.log(result) ;

let pointer = null ;
```

声明初始数据为空值（null）的变量 pointer。

```
let range_list = [10, 10, 10, 50, 50, 50, 100, 100, 100] ;
```

声明初始数据为内含多个整数值的数组实例的变量 range_list。

```
function generate(... list)
{
  let number_list = [] ;
```

声明初始数据为空数组实例的局部变量 number_list。

```
  console.log(this.valueOf()) ;
```

this.valueOf()当前会返回字符串'test message'。

```
  for (let i = 0; i < list.length; i++)
```

依据参数 list 的数组实例中的元素个数 9，此 for 循环语句会迭代 9 次。

```
  {
    number_list.push(parseInt(list[i]* Math.random())) ;
```

此语句分别将参数 list 的数组实例中的各整数值，作为不同的上限值，以新增【0 ~ 上限值】的随机整数值，至变量 number_list 的数组实例中。

```
  }
  return number_list ;
```

此语句会返回变量 number_list 所代表的【被新增 9 个整数值】的数组实例，例如 [0, 4, 6, 49, 24,

13, 62, 54, 75]。

}

上述语法定义了函数 generate()，并用来返回内含 9 个整数值的新数组实例。其中，每个整数值均大于或等于 0；而且，小于 10、小于 50 和小于 100 的整数值，各有 3 个，例如[0, 4, 6, 49, 24, 13, 62, 54, 75]。在上述函数内部，变量 range_list 的数组实例[10, 10, 10, 50, 50, 50, 100, 100, 100]中的各整数值，在被返回的新数组实例（例如 [0, 4, 6, 49, 24, 13, 62, 54, 75]）中，被视为不同的上限值。

```
result = Reflect.apply(generate,'test message', range_list) ;
```

此语句将【间接被调用的函数 generate()所返回】的新数组实例（例如 [0, 4, 6, 49, 24, 13, 62, 54, 75]），赋给变量 result。此语句可使得字符串'test message'，成为在名称为 generate 的函数内部，其内置的局部变量 this 的数据，并可将变量 range_list 的数组实例[10, 10, 10, 50, 50, 50, 100, 100, 100]中的各整数值，作为传入至名称为 generate 的函数的各参数数据，如同【generate(... range_list)】被调用一样。

```
result = generate(... range_list) ;
```

除了作为函数 Reflect.apply()的第 2 个参数数据'test message'之外，此语句几乎等价于【result = Reflect.apply(generate,'test message', range_list) ;】。

13.3　创建特定对象的实例（ES6）

通过函数 Reflect.construct()，可传入多个数据，至即将被间接调用的其他对象的构造函数里，进而创建特定对象的新实例。请参看如下示例。

【13-3-^-Reflect-construct.js】

```
let a01 = new Array(7) ;
let a02 = Reflect.construct(Array, [7]) ;

console.log(a01) ;
console.log(a02) ;
console.log('') ;

///
let d01 = new Date() ;
let d02 = Reflect.construct(Date, []) ;

console.log(d01) ;
console.log(d02) ;
console.log('') ;

///
let n01 = new Number(2591.8) ;
let n02 = Reflect.construct(Number, [2591.8]) ;
```

```
console.log(n01) ;
console.log(n02) ;
console.log('') ;
```

【相关说明】

```
let a01 = new Array(7) ;
let a02 = Reflect.construct(Array, [7]) ;
```

在这两个语句里，【new Array(7)】和【Reflect.construct(Array, [7])】是等价的，会使得变量 a01 与 a02 的初始数据，均成为可容纳 7 个元素的数组实例。其中，语法【[7]】中的整数值 7，被视为传入 Array 对象的构造函数 Array(7)中的参数数据。

```
console.log(a01) ;
console.log(a02) ;
```

这两个语句均显示出【(7)[empty × 7]】的信息。

```
console.log('') ;

let d01 = new Date() ;
let d02 = Reflect.construct(Date, []) ;
```

在这两个语句中，【new Date()】和【Reflect.construct(Date, [])】是等价的，会使得变量 d01 与 d02 的初始数据，均成为内含当前日期与时间的 Date 对象实例。其中，[]意味着没有任何被传入 Date 对象的构造函数 Date()中的参数数据。

```
console.log(d01) ;
console.log(d02) ;
```

这两个语句均显示出当前的日期与时间，例如【Sat Dec 01 2018 23:25:37 GMT+0800 (China Standard Time)】。

```
console.log('') ;

let n01 = new Number(2591.8) ;
let n02 = Reflect.construct(Number, [2591.8]) ;
```

在这两个语句中，【new Number(2591.8)】和【Reflect.construct(Number, [2591.8])】是等价的，会使得变量 n01 与 n02 的初始数据，均成为内含数值 2591.8 的 Number 对象实例。其中，[2591.8]中的数值 2591.8，被视为传入 Date 对象的构造函数 Number(2591.8)的参数数据。

```
console.log(n01) ;
console.log(n02) ;
```

这两个语句均显示出【Number{2591.8}】的信息。

13.4 精细定义新属性（ES6）

通过函数 Reflect.defineProperty()，可间接对特定对象的实例，**精细定义**其新属性及其特征。请参看如下示例。

【13-4-^-Reflect-defineProperty.js】

```
let obj01 = {}, obj02 = {} ;
let result01 = null , result02 = null ;

result01 = Object.defineProperty(obj01,'name', {value: 'Alex', writable: true, enumerable:
  true, configurable: true}) ;
result02 = Reflect.defineProperty(obj02,'name', {value: 'Alex', writable: true, enumerable:
  true, configurable: true}) ;

console.log(obj01) ;
console.log(obj02) ;
console.log('') ;

console.log(result01) ;
console.log(result02) ;
```

【相关说明】

```
let obj01 = {}, obj02 = {} ;
```

此语句分别声明了初始数据均为空对象实例{}的变量 obj01 与 obj02。

```
let result01 = null , result02 = null ;
```

此语句分别声明了初始数据均为空值（null）的变量 result01 与 result02。

```
result01 = Object.defineProperty(obj01, 'name', {value: 'Alex', writable: true, enumerable:
  true, configurable: true}) ;
result02 = Reflect.defineProperty(obj02, 'name', {value: 'Alex', writable: true, enumerable:
  true, configurable: true}) ;
```

这两个语句均在变量 obj01 与 obj02 的空对象实例中，将字符串'Alex'，赋给新属性 name。不同的是，在成功定义新属性的情况下，函数 Object.defineProperty()会返回新属性 name 的对象实例{name: "Alex"}，而函数 Reflect.defineProperty()则返回布尔值 true。

```
console.log(obj01) ;
console.log(obj02) ;
```

这两个语句均显示出{name: "Alex"}的信息。

```
console.log('') ;

console.log(result01) ;
```

显示出{name: "Alex"}的信息。

```
console.log(result02) ;
```

显示出布尔值 true 的信息。

13.5 删除特定属性（ES6）

通过函数 Reflect.deleteProperty()，可间接删除特定对象实例的特定属性。请参看如下示例。

【13-5-^-Reflect-deleteProperty.js】
```
let inventory = {apple: 10, banana: 18, durian: 8, grapefruit: 30} ;
let result = null ;

result = delete inventory.apple ;

console.log(result) ;
console.log(inventory) ;
console.log('') ;

result = Reflect.deleteProperty(inventory,'grapefruit') ;

console.log(result) ;
console.log(inventory) ;
```

【相关说明】

```
let inventory = {apple: 10, banana: 18, durian: 8, grapefruit: 30} ;
```

声明初始数据为对象实例的变量 inventory。

```
let result = null ;
```

声明初始数据为空值（null）的变量 result。

```
result = delete inventory.apple ;
```

此语句将变量 inventory 的对象实例中的属性 apple，被试图删除。若成功删除，则返回布尔值 true，并赋给变量 result。

```
console.log(result) ;
```

显示出布尔值 true 的信息。

```
console.log(inventory) ;
```

此语句显示出 {banana: 18, durian: 8, grapefruit: 30} 的信息。从中可以看出，属性 apple 及其数值 10，已被成功删除。

```
console.log('') ;

result = Reflect.deleteProperty(inventory,'grapefruit') ;
```

此语句通过函数 Reflect.deleteProperty()，试图删除变量 inventory 的对象实例中的属性

grapefruit。若成功删除，则返回布尔值 true，并赋给变量 result。

```
console.log(result) ;
```

显示出布尔值 true 的信息。

```
console.log(inventory) ;
```

此语句显示出{banana: 18, durian: 8}的信息。从中可以看出，属性 grapefruit 及其数值 30，已被成功删除。

13.6　获取特定属性的数据（ES6）

通过函数 Reflect.get()，可间接获取特定对象实例中的特定属性的数据。请参看如下示例。

【13-6-^-Reflect-get.js】

```
let inventory = {apple: 10, banana: 18, durian: 8, grapefruit: 30} ;

let result = inventory.banana ;

console.log(result) ;
console.log('') ;

result = Reflect.get(inventory,'grapefruit') ;
console.log(result) ;
```

【相关说明】

```
let inventory = {apple: 10, banana: 18, durian: 8, grapefruit: 30} ;
```

声明初始数据为对象实例的变量 inventory。

```
let result = null ;
```

声明初始数据为空值（null）的变量 result。

```
result = inventory.banana ;
```

inventory.banana 会返回在变量 inventory 的对象实例中，其属性 banana 的数值 18，并赋给变量 result。

```
console.log(result) ;
console.log('') ;

result = Reflect.get(inventory,'grapefruit') ;
```

通过函数 Reflect.get()，返回在变量 inventory 的对象实例中，其属性 grapefruit 的数值 30，并赋给变量 result。

13.7 返回特定属性的描述器（ES8）

通过函数 Reflect.getOwnPropertyDescriptor()，可间接返回特定对象实例的特定属性的描述器（descriptor）。请参看如下示例。

【13-7-Reflect-getOwnPropertyDescriptor.js】

```javascript
let profile =
{
 firstname: 'Jason',
 lastname: 'Alex',
 gender: 'male',
 age: 28,
 position: 'Product Manager',

 fullname01()
 {
   return this.firstname + ' ' + this.lastname ;
 },

 get fullname02()
 {
   return this.firstname + ' ' + this.lastname
 }
} ;

let d01 = Object.getOwnPropertyDescriptor(profile,'position') ;
let d02 = Object.getOwnPropertyDescriptor(profile,'fullname01') ;
let d03 = Object.getOwnPropertyDescriptor(profile,'fullname02') ;

console.log(d01) ;
console.log(d02) ;
console.log(d03) ;
console.log('') ;

console.log(profile.fullname01()) ;
console.log(profile.fullname02) ;
console.log('') ;

///
d01 = Reflect.getOwnPropertyDescriptor(profile,'position') ;
d02 = Reflect.getOwnPropertyDescriptor(profile,'fullname01') ;
d03 = Reflect.getOwnPropertyDescriptor(profile,'fullname02') ;

console.log(d01) ;
console.log(d02) ;
console.log(d03) ;
console.log('') ;

let d04 = Object.getOwnPropertyDescriptors(profile) ;

console.log(d04) ;
```

第 13 章 Reflect 对象 | 357

【相关说明】

```
let profile =
{
  firstname: 'Jason',
  lastname: 'Alex',
  gender: 'male',
  age: 28,
  position: 'Product Manager',
```

这 5 个语句分别定义了变量 profile 的对象实例中的 5 个新属性。

```
  fullname01()
  {
    return this.firstname + ' ' + this.lastname ;
```

此语句返回了【在变量 profile 的对象实例中，其属性 firstname 的字符串'Jason'与属性 lastname 的字符串'Alex'，被合并】之后的字符串'Jason Alex'。其中，内置的局部变量 this，指向变量 profile 的对象实例本身。

```
  },
```

上述语法定义了变量 profile 的对象实例中的新函数 fullname01()，可用来返回字符串'Jason Alex'。

```
  get fullname02()
  {
    return this.firstname + ' ' + this.lastname
  }
```

此语法通过关键字 get，将函数 fullname02()，设置为 getter。getter 被访问（调用）时，必须省略其小括号，例如【profile.fullname02】并不带小括号。

```
} ;
```

上述语法声明了初始数据为内含多个成员的对象实例的变量 profile。

```
let d01 = Object.getOwnPropertyDescriptor(profile,'position') ;
```

此语句通过函数 Object.getOwnPropertyDescriptor()，返回在变量 profile 的对象实例中，其属性 position 的相关描述{value: "Product Manager", writable: true, enumerable: true, configurable: true}，并赋给变量 d01。

```
let d02 = Object.getOwnPropertyDescriptor(profile,'fullname01') ;
```

此语句通过函数 Object.getOwnPropertyDescriptor()，返回在变量 profile 的对象实例中，其函数 fullname01()的相关描述{value: ƒ, writable: true, enumerable: true, configurable: true}，并赋给变量 d02。

```
let d03 = Object.getOwnPropertyDescriptor(profile,'fullname02') ;
```

此语句通过函数 Object.getOwnPropertyDescriptor()，返回在变量 profile 的对象实例中，作为 getter 的 fullname02()的相关描述{get: ƒ, set: undefined, enumerable: true, configurable: true}，并赋给

变量 d03。

```
console.log(d01) ;
console.log(d02) ;
console.log(d03) ;
console.log('') ;

console.log(profile.fullname01()) ;
console.log(profile.fullname02) ;
```

【profile.fullname01()】与【profile.fullname02】均会返回字符串'Jason Alex'。因为 fullname02 被定义成为变量 profile 对象实例中的 getter，所以在使用时，不可加上一对小括号。

```
console.log('') ;

d01 = Reflect.getOwnPropertyDescriptor(profile,'position') ;
```

【Reflect.getOwnPropertyDescriptor(profile,'position')】等价于【Object.getOwnPropertyDescriptor(profile,'position')】，会返回在变量 profile 的对象实例中，其属性 position 的相关描述{value: "Product Manager", writable: true, enumerable: true, configurable: true}。

```
d02 = Reflect.getOwnPropertyDescriptor(profile,'fullname01') ;
```

【Reflect.getOwnPropertyDescriptor(profile,'fullname01')】等价于【Object.getOwnPropertyDescriptor(profile,'fullname01')】，会返回在变量 profile 的对象实例中，其函数 fullname01()的相关描述{value: ƒ, writable: true, enumerable: true, configurable: true}。

```
d03 = Reflect.getOwnPropertyDescriptor(profile,'fullname02') ;
```

【Reflect.getOwnPropertyDescriptor(profile,'fullname02')】等价于【Object.getOwnPropertyDescriptor(profile,'fullname02')】，会返回在变量 profile 的对象实例中，其作为 getter 的 fullname02()的相关描述{get: ƒ, set: undefined, enumerable: true, configurable: true}

```
console.log(d01) ;
console.log(d02) ;
console.log(d03) ;

console.log('') ;

let d04 = Object.getOwnPropertyDescriptors(profile) ;
```

Object.getOwnPropertyDescriptors(profile)可用来返回在变量 profile 的对象实例中，其所有自定义成员（属性、函数、getter 和 setter）的如下相关描述：

- age: {value: 28, writable: true, enumerable: true, configurable: true}
- firstname: {value: "Jason", writable: true, enumerable: true, configurable: true}
- fullname01: {value: ƒ, writable: true, enumerable: true, configurable: true}
- fullname02: {get: ƒ, set: undefined, enumerable: true, configurable: true}
- gender: {value: "male", writable: true, enumerable: true, configurable: true}
- lastname: {value: "Alex", writable: true, enumerable: true, configurable: true}
- position: {value: "Product Manager", writable: true, enumerable: true, configurable: true}

13.8 返回特定对象的原型（ES6）

通过函数 Reflect.getPrototypeOf()，可返回特定对象的原型（prototype）。请参看如下示例。

【13-8-^-Reflect-getPrototypeOf.js】

```
let profile =
{
  firstname: 'Jason',
  lastname: 'Alex',
  gender: 'male',
  age: 28,
  position: 'Product Manager',

  fullname01()
  {
    return this.firstname + ' ' + this.lastname ;
  },

  get fullname02()
  {
    return this.firstname + ' ' + this.lastname
  }
} ;

let result01 = Object.getPrototypeOf(profile) ;
let result02 = Reflect.getPrototypeOf(profile) ;

console.log(result01) ;
console.log(result02) ;
console.log('') ;

console.log(result01 === result02) ;
```

【相关说明】

```
let profile =
{
  firstname: 'Jason',
  lastname: 'Alex',
  gender: 'male',
  age: 28,
  position: 'Product Manager',

  fullname01()
  {
    return this.firstname + ' ' + this.lastname ;
  },

  get fullname02()
  {
    return this.firstname + ' ' + this.lastname
  }
```

上述语法声明了初始数据为内含多个成员的对象实例的变量 profile。

```
let result01 = Object.getPrototypeOf(profile) ;
let result02 = Reflect.getPrototypeOf(profile) ;
```

【Object.getPrototypeOf(profile)】与【Reflect.getPrototypeOf(profile)】均会返回变量 profile 的对象实例的原型（prototype）{constructor: ƒ, __defineGetter__: ƒ, __defineSetter__: ƒ, hasOwnProperty: ƒ, __lookupGetter__: ƒ, …}。

```
console.log(result01) ;
console.log(result02) ;
console.log('') ;

console.log(result01 === result02) ;
```

在此，表达式【result01 === result02】被评估为布尔值 true，意味着变量 result01 与 result02 的数据，被视为占用相同内存引址的原型。

13.9　判断特定属性的存在性（ES6）

通过函数 Reflect.has()，可判断特定对象实例，是否存在特定名称的成员（属性、函数、getter 和 setter）。请参看如下示例。

【13-9-^-Reflect-has.js】

```
let profile =
{
 firstname: 'Jason',
 lastname: 'Alex',
 gender: 'male',
 age: 28,
 position: 'Product Manager',

 fullname01()
 {
   return this.firstname + ' ' + this.lastname ;
 },

 get fullname02()
 {
   return this.firstname + ' ' + this.lastname
 }
} ;

let result = Reflect.has(profile,'age') ;
console.log(result) ;

result = Reflect.has(profile,'fullname02') ;
console.log(result) ;
```

```
result = Reflect.has(profile,'fullname03') ;
console.log(result) ;
```

【相关说明】

```
let profile =
{
 firstname: 'Jason',
 lastname: 'Alex',
 gender: 'male',
 age: 28,
 position: 'Product Manager',

 fullname01()
 {
  return this.firstname + ' ' + this.lastname ;
 },

 get fullname02()
 {
  return this.firstname + ' ' + this.lastname
 }
} ;
```

上述语法声明了初始数据为内含多个成员的对象实例的变量 profile。

```
let result = Reflect.has(profile,'age') ;
```

此语句声明了初始数据为布尔值 true 的变量 result。其中，Reflect.has(profile,'age')判断出在变量 profile 的对象实例中，的确存在属性 age，并返回布尔值 true。

```
console.log(result) ;

result = Reflect.has(profile,'fullname02') ;
```

此语句将布尔值 true，赋给变量 result。其中，Reflect.has(profile,'fullname02')判断出在变量 profile 对象实例中，的确存在名称为 fullname02 的 getter，并返回布尔值 true。

```
console.log(result) ;

result = Reflect.has(profile,'fullname03') ;
```

此语句将布尔值 false，赋给变量 result。其中，Reflect.has(profile,'fullname03')判断出变量 profile 的对象实例中，并不存在名称为 fullname03 的成员（属性、函数、getter 或 setter），并返回布尔值 false。

13.10 判断与设置特定对象的扩展性（ES6）

通过函数 Reflect.isExtensible()，可判断特定对象实例，是否可被扩展新的属性；通过函数 Reflect.preventExtensions()，可制止特定对象实例，被扩展新的属性。请参看如下示例。

【13-10-^-Reflect-isExtensible-and-preventExtensions.js】

```
let profile =
{
  firstname: 'Jason',
  lastname: 'Alex',
  gender: 'male',
  age: 28,
  position: 'Product Manager',

  fullname01()
  {
    return this.firstname + ' ' + this.lastname ;
  },

  get fullname02()
  {
    return this.firstname + ' ' + this.lastname
  }
} ;

// let result = Object.isExtensible(profile) ;
let result = Reflect.isExtensible(profile) ;

console.log(result) ;
console.log('') ;

///
// Object.seal(profile) ;
// Object.freeze(profile) ;
// Object.preventExtensions(profile) ;
Reflect.preventExtensions(profile) ;

result = Object.isExtensible(profile) ;
console.log(result) ;

profile.department = 'Production' ;
console.log(profile) ;
```

【相关说明】

```
let profile =
{
  firstname: 'Jason',
  lastname: 'Alex',
  gender: 'male',
  age: 28,
  position: 'Product Manager',

  fullname01()
  {
    return this.firstname + ' ' + this.lastname ;
  },

  get fullname02()
  {
```

```
        return this.firstname + ' ' + this.lastname
    }
} ;
```

上述语法声明了初始数据为内含多个成员的对象实例的变量 profile。

```
// let result = Object.isExtensible(profile) ;
let result = Reflect.isExtensible(profile) ;
```

【Object.isExtensible(profile)】与【Reflect.isExtensible(profile)】均会返回布尔值 true，并赋给变量 result，意味着变量 profile 的对象实例，是可被扩充新成员（属性、函数、getter 与 setter）的。

```
console.log(result) ;
console.log('') ;

// Object.seal(profile) ;
// Object.freeze(profile) ;
// Object.preventExtensions(profile) ;
Reflect.preventExtensions(profile) ;
```

这 4 个语句具有相同的效果，均可使得变量 profile 的对象实例，禁用扩充机制。

```
result = Object.isExtensible(profile) ;
```

Object.isExtensible(profile)当前返回布尔值 false，并赋给变量 result，意味着变量 profile 的对象实例，当前无法被扩充新成员（属性、函数、getter 与 setter）。

```
console.log(result) ;

profile.department = 'Production' ;
```

此语句试图新增变量 profile 的对象实例中的新属性 department 和其数据字符串'Production'。然而，变量 profile 的对象实例的扩充机制，已经被禁用，所以上述新增的动作会失败。

```
console.log(profile) ;
```

此语句显示出{firstname: "Jason", lastname: "Alex", gender: "male", age: 28, position: "Product Manager", …}的信息。可看出并不存在属性 department。

13.11　简易定义新属性（ES6）

通过函数 Reflect.set()，可间接在特定对象实例中，简易定义其新属性。请参看如下示例。

【13-11-^-Reflect-set.js】

```
let shapes = {} ;

shapes['diamond'] = 4 ;
shapes.parallelogram = 4 ;

Reflect.defineProperty(shapes,'trapezoid', {value: 4,writable: false, enumerable: true,
 configurable: false}) ;
```

```
    Reflect.set(shapes,'square', 4) ;
    Reflect.set(shapes,'triangle', 3) ;

    console.log(shapes) ;
    console.log('') ;

    ///
    let metals = ['Zinc'] ;

    metals.push('Silver') ;
    metals.unshift('Gold') ;
    metals[3] = 'Copper' ;
    metals.splice(1, 2,'Iron','Tin') ;

    Reflect.set(metals, 4,'Aluminum') ;
    console.log(metals) ;

    Reflect.set(metals,'length', 10) ;
    console.log(metals) ;
```

【相关说明】

```
let shapes = {} ;
```

声明了初始数据为空对象实例{}的变量 shapes。

```
shapes['diamond'] = 4 ;
shapes.parallelogram = 4 ;
```

这两个语句均可在变量 shapes 的对象实例中，添加新的属性及其数据，例如属性 diamond 与 parallelogram 及其相同的数值 4。

```
Reflect.defineProperty(shapes,'trapezoid', {value: 4,writable: false, enumerable: true,
  configurable: false}) ;
```

此语句在变量 shapes 的对象实例中，精细添加新的属性、其数值，以及其特征（writable、enumerable 与 configurable）。例如属性 trapezoid 和其数值 4，以及其各项特征。

```
Reflect.set(shapes,'square', 4) ;
Reflect.set(shapes,'triangle', 3) ;
```

这两个语句均通过了函数 Reflect.set()，在变量 shapes 的对象实例中，简易添加新的属性 square 和其数值 4，以及属性 triangle 和其数值 3。

```
console.log(shapes) ;
```

显示出变量 shapes 的对象实例{diamond: 4, parallelogram: 4, square: 4, triangle: 3, trapezoid: 4}。

```
console.log('') ;

let metals = ['Zinc'] ;
```

声明初始数据为数组实例['Zinc']的变量 metals。

```
metals.push('Silver') ;
```

此语句使得变量 metals 的数组实例，在其**尾**端，被新增字符串'Silver'，成为数组实例['Zinc', 'Silver']。

```
metals.unshift('Gold') ;
```

此语句使得变量metals的数组实例，在其**前**端，被新增字符串'Gold'，成为数组实例['Gold', 'Zinc', 'Silver']。

```
metals[3] = 'Copper' ;
```

此语句将字符串'Cooper'，赋给在变量 metals 的数组实例中，其索引值为 3（第 4 个）的元素，成为数组实例['Gold', 'Zinc', 'Silver', 'Copper']。

```
metals.splice(1, 2,'Iron','Tin') ;
```

此语句在变量 metals 的数组实例['Gold', 'Zinc', 'Silver', 'Copper']中，从索引值 1（第 2 个）的元素开始，取出两个元素的数据字符串'Zinc'和'Silver'，成为数组实例['Gold', 'Copper']，然后一样在索引值 1（第 2 个）的元素开始，插入数据字符串为'Iron'与'Tin'的两个元素，成为['Gold', 'Iron', 'Tin', 'Copper']。

```
Reflect.set(metals, 4,'Aluminum') ;
```

此语句将字符串'Aluminum'，赋给在变量 metals 的数组实例中，其索引值为 4（第 5 个）的元素，成为数组实例['Gold','Iron','Tin','Copper','Aluminum']。

```
console.log(metals) ;

Reflect.set(metals,'length', 10) ;
```

执行此语句之前，变量 metals 的数组实例中的元素个数为 5。此语句将整数值 10 赋给变量 metals 的数组实例中的属性 length，进而设置了变量 metals 的数组实例，可容纳的元素个素为 10。换言之，此语句使得变量 metals 的数组实例['Gold', 'Iron', 'Tin', 'Copper', 'Aluminum', empty × 5]，成为可容纳 10 个元素的数组实例，其中最后 5 个是**空元素**。

13.12　重新设置特定对象的原型（ES6）

通过函数 Reflect.setPrototypeOf()，可重新将特定对象实例的原型（prototype），设置为另一对象实例。请参看如下示例。

【13-12-^-Reflect-setPrototypeOf.js】

```
let book01 = {title: 'New Thoughts', price: 150, author: 'Brat Terminator', Publisher:
 'TsingHua', date: '2018/06/06'} ;

let book02 = {title: 'New Thinkings', price: 180, author: 'Brat Terminator', Publisher:
 'TsingHua', date: '2018/07/07', version: 3} ;

let result = Object.setPrototypeOf(book01, String) ;
```

```
console.log(result);
console.log(book01);
console.log('');

result = Reflect.setPrototypeOf(book02, Number);

console.log(result);
console.log(book02);
console.log(book02.EPSILON);
```

【相关说明】

```
let book01 = {title: 'New Thoughts', price: 150, author: 'Brat Terminator', Publisher:
 'TsingHua', date: '2018/06/06'};
```

此语句声明了初始数据为内含多个属性的对象实例的变量 book01。

```
let book02 = {title: 'New Thinkings', price: 180, author: 'Brat Terminator', Publisher:
 'TsingHua', date: '2018/07/07', version: 3};
```

此语句声明了变量 book02，其初始数据为内含包括属性 version 在内的多个属性的对象实例。

```
let result = Object.setPrototypeOf(book01, String);
```

此语句声明了变量 result，其初始数据为内含 String 对象的原型（prototype）的实例{title: "New Thoughts", price: 150, author: "Brat Terminator", Publisher: "TsingHua", date: "2018/06/06", __proto__:f String()}。其中，从【__proto__:f String()】的描述中可看出，变量 book01 对象实例的原型被置换成为 String 对象的原型。

值得留意的是，Object.setPrototypeOf(book01, String)会返回变量 book01 的对象实例本身。

```
console.log(result);
console.log(book01);
```

这两个语句均显示出{title: "New Thoughts", price: 150, author: "Brat Terminator", Publisher: "TsingHua", date: "2018/06/06", __proto__:f String()}的信息。

```
console.log('');

result = Reflect.setPrototypeOf(book02, Number);
```

重新设置变量 result 的数据为内含 Number 对象原型（prototype）的实例{title: "New Thinkings", price: 180, author: "Brat Terminator", Publisher: "TsingHua", date: "2018/07/07", version: 3, __proto__:f Number()}。其中，从【__proto__:f Number()】的描述中可以看出，变量 book02 的对象实例的原型，被置换成为 Number 对象的原型。请留意，Reflect.setPrototypeOf(book02, Number)返回布尔值 true，意味着成功设置了变量 book02 的对象实例的原型。

```
console.log(result);
```

显示出布尔值 true。

```
console.log(book02);
```

显示出{title: "New Thinkings", price: 180, author: "Brat Terminator", Publisher: "TsingHua", date: "2018/07/07", version: 3, __proto__:f Number()}的信息。

```
console.log(book02.EPSILON) ;
```

因为变量 book02 当前的原型，已被置换成为 Number 对象的原型，所以变量 book02 就被内置了常量属性 EPSILON，使得 book02.EPSILON 可返回数值 2.220446049250313e-16。

13.13　返回与列举特定对象的自定义属性（ES6）

通过函数 Reflect.ownKeys()，可返回特定对象实例的自定义属性所构成的数组实例，进而搭配循环语句，以列举出特定对象实例的自定义属性。请参看如下**两个**示例。

【13-13-^-e1-Reflect-ownKeys.js】

```
let profile =
{
 firstname: 'Jason',
 lastname: 'Alex',
 gender: 'male',
 age: 28,
 position: 'Product Manager',

 fullname01()
 {
  return this.firstname + ' ' + this.lastname ;
 },

 get fullname02()
 {
  return this.firstname + ' ' + this.lastname
 }
} ;

// let result = Object.keys(profile) ;
let result = Reflect.ownKeys(profile) ;

console.log(result) ;
```

【相关说明】

```
let profile =
{
 firstname: 'Jason',
 lastname: 'Alex',
 gender: 'male',
 age: 28,
 position: 'Product Manager',

 fullname01()
 {
  return this.firstname + ' ' + this.lastname ;
 },
```

```
    get fullname02()
    {
      return this.firstname + ' ' + this.lastname
    }
} ;
```

上述语法声明了初始数据为内含多个成员的对象实例的变量 profile。

```
// let result = Object.keys(profile) ;
let result = Reflect.ownKeys(profile) ;
```

【Object.keys(profile)】与【Reflect.ownKeys(profile)】均可返回【在变量 profile 对象实例中，其各成员（属性、函数、getter 与 setter）的名称所构成】的新数组实例["firstname", "lastname", "gender", "age", "position", "fullname01", "fullname02"]。

【13-13-^-e2-enumerations-of-object-properties.js】

```
let book01 = {title: 'New Thoughts', price: 150, author: 'Brat Terminator', Publisher:
 'TsingHua', date: '2018/06/06'} ;

let book02 = {title: 'New Thinkings', price: 180, author: 'Brat Terminator', Publisher:
 'TsingHua', date: '2018/07/07', version: 3} ;

console.log(Object.keys(book01)) ;
console.log(Reflect.ownKeys(book02)) ;

let p = '' ;

for (p of Object.keys(book01))
{
  console.log(`The value of property'${p}'= ${book01[p]}.`) ;
}

console.log('\n\n') ;

for (p of Reflect.ownKeys(book02))
{
  console.log(`The value of property'${p}'= ${book01[p]}.`) ;
}

console.log('') ;
```

【相关说明】

```
let book01 = {title: 'New Thoughts', price: 150, author: 'Brat Terminator', Publisher:
 'TsingHua', date: '2018/06/06'} ;
```

此语句声明了初始数据为内含多个属性的对象实例的变量 book01。

```
let book02 = {title: 'New Thinkings', price: 180, author: 'Brat Terminator', Publisher:
 'TsingHua', date: '2018/07/07', version: 3} ;
```

此语句声明了变量 book02，其初始数据为内含包括属性 version 在内的多个属性的对象实例。

```
console.log(Object.keys(book01)) ;
```

【Object.keys(book01)】与【Reflect.ownKeys(book01)】均可返回【在变量 book01 对象实例中，

其各成员（属性、函数、getter 与 setter）的名称所构成】的新数组实例["title", "price", "author", "Publisher", "date"]。

```
console.log(Reflect.ownKeys(book02)) ;
```

【Object.keys(book02)】与【Reflect.ownKeys(book02)】均可返回【在变量 book02 对象实例中，其各成员的名称所构成】的新数组实例["title", "price", "author", "Publisher", "date", "version"]。

```
let p = '' ;
```

声明初始数据为空字符串"的变量 p。

```
for (p of Object.keys(book01))
```

因为 Object.keys(book01)会返回元素个数为 5 的新数组实例["title", "price", "author", "Publisher", "date"]，所以此 for...of 循环语句，会迭代 5 次，使得变量 p 均会被赋予上述数组实例里的其中一个字符串。

```
{
  console.log(`The value of property'${p}'= ${book01[p]}.`) ;
```

此语句在迭代 5 次中，分别显示出如下信息：

- The value of property 'title' = New Thoughts.
- The value of property 'price' = 150.
- The value of property 'author' = Brat Terminator.
- The value of property 'Publisher' = TsingHua.
- The value of property 'date' = 2018/06/06.

```
}
console.log('\n\n') ;
for (p of Reflect.ownKeys(book02))
```

因为 Reflect.ownKeys(book02)会返回元素个数为 6 的新数组实例["title", "price", "author", "Publisher", "date", "version"]，所以此 for...of 循环语句，会迭代 6 次，使得变量 p 均会被赋予数组实例里的其中一个字符串。

```
{
  console.log(`The value of property'${p}'= ${book01[p]}.`) ;
```

此语句在迭代 6 次中，分别显示出如下信息：

- The value of property 'title' = New Thoughts.
- The value of property 'price' = 150.
- The value of property 'author' = Brat Terminator.
- The value of property 'Publisher' = TsingHua.
- The value of property 'date' = 2018/06/06.
- The value of property 'version' = undefined.

13.14 练 习 题

1. 已知如下 JavaScript 源代码片段:

```
let profile = {name: 'Andy', age: 33, country: 'China'} ;

Object.defineProperty(profile, 'department', {value: 'IT', writable: true, enumerable: true,
  configurable: false}) ;
```

试改写【Object.defineProperty(...)】的部分，成为等价语法，使得属性 department 可被新增和设置。

2. 已知如下 JavaScript 源代码的语句:

```
let result = Reflect.has(profile, 'department') ;
```

试改写上述语句，成为等价的语句。

3. 已知如下 JavaScript 源代码的语句:

```
let result = Reflect.ownKeys(profile) ;
```

试改写上述语句，成为等价的语句。

第 14 章

Proxy 对象、Intl 对象和 navigator 对象实例

本章内容主要聚焦于【用来改造特定对象实例的访问机制】的 Proxy 对象、【用来格式化各国语言相关的字符串】的 Intl 对象，以及【提供浏览器相关信息和当前地理位置】的 navigator 对象实例。

14.1 Proxy 对象

Proxy 对象主要用来改造特定对象实例的访问机制，成为新的访问代理（proxy）机制，以便不存在的属性，被访问到时，或者超出范围的数据，被赋予特定属性时；被改造之后的访问机制，可返回动态产生的数据。请参看如下示例。

【14-1-^-Proxy-object.js】

```js
function get_value(obj, attr)
{
  let choice ;

  choice = attr in obj ? obj[attr]: 'N/A' ;

  return choice ;
}

function set_value(obj, attr, value)
{
  let new_value ;

  switch (attr)
  {
    case 'age':
      if (value > 150)
```

```
        new_value = 'Impossibly old...' ;
      break ;

    case 'gender':
      if (! (value in ['male','female']))
        new_value = 'Impossible gender...' ;
      break ;

    default:
      new_value = value ;
  }
  obj[attr] = new_value ;
}
var cur_obj = {name: 'Daisy', gender: 'female', age: 30} ;

// var proxy_handler = {get: get_value} ;
var proxy_handler = {get: get_value , set: set_value } ;
var p01 = new Proxy(cur_obj, proxy_handler) ;

p01.department = 'finance' ;
p01.position = 'manager' ;

console.log(p01) ;
console.log(p01.name) ;
console.log(p01.position) ;
console.log(p01.company) ;
console.log(p01.salary) ;
console.log('') ;

var p02 = new Proxy(cur_obj, proxy_handler) ;

p02.age = 152 ;
p02.gender = 'unknown' ;

console.log(p02) ;
```

【相关说明】

请参看本节中各小节的相关说明。

14.1.1　创建访问代理机制的构造函数（ES6）

欲创建特定对象实例的访问代理机制，需要通过 Proxy 对象的构造函数来达成。

```
var p01 = new Proxy(cur_obj, proxy_handler) ;
```

此语句声明了初始数据为 Proxy 对象实例的变量 p01。在其等号右侧，调用了 Proxy 对象的构造函数 Proxy(cur_obj, proxy_handler)，并将变量 cur_obj 的对象实例、变量 proxy_handler 的对象实例，作为参数的数据，传递至 Proxy 对象的构造函数内部。因此，在变量 p01 的 Proxy 对象实例中，同时带有在变量 cur_obj 的对象实例中，可被访问的属性成员 name、gender 和 age，以及在变量

proxy_handler 的对象实例中，不可被访问的 getter 成员 get_value 和 setter 成员 set_value。

```
var p02 = new Proxy(cur_obj, proxy_handler) ;
```

此语句声明了初始数据为 Proxy 对象实例的变量 p02。在其等号右侧，调用了 Proxy 对象的构造函数 Proxy(cur_obj, proxy_handler)，并将变量 cur_obj 的对象实例、变量 proxy_handler 的对象实例，作为参数数据，传递至 Proxy 对象的构造函数内部。因此，在变量 p02 的 Proxy 对象实例中，同时带有在变量 cur_obj 的对象实例中，可被访问的属性成员 name、gender 和 age，以及在变量 proxy_handler 的对象实例中，不可被访问的 getter 成员 get_value 和 setter 成员 set_value。

14.1.2　确定被代理的特定对象（ES6）

运用 Proxy 对象的构造函数之前，必须确定欲被代理的特定对象实例。

```
var cur_obj = {name: 'Daisy', gender: 'female', age: 30} ;
```

此语句声明了初始数据为内含 3 个属性成员的对象实例的变量 cur_obj。

14.1.3　自定义代理函数（ES6）

运用 Proxy 对象的构造函数之前，必须先定义达成访问代理机制的代理函数。

```
function get_value(obj, attr)
{
  let choice ;
```

声明局部变量 choice。

```
  choice = attr in obj ? obj[attr]: 'N/A' ;
```

此语句用来判断参数 attr 的数据字符串，是否为参数 obj 的对象实例中的属性名称。若是，则将其属性的数据，赋给局部变量 choice；若否，则将字符串'N/A'，赋给局部变量 choice。

```
  return choice ;
```

返回局部变量 choice 的数据。

```
}
```

上述语法定义了带有参数 obj 与 attr 的函数 get_value()，并作为特定 Proxy 对象实例的 getter 成员。

```
function set_value(obj, attr, value)
{
  let new_value ;
```

声明局部变量 new_value。

```
  switch (attr)
```

此语法用来进一步判断参数 attr 的数据，是否为字符串'age'或'gender' 等。

```
{
  case 'age':
```

参数 attr 的数据，若是字符串'age'，则执行下方到最接近 break 语句为止的源代码。

```
    if (value > 150)
      new_value = 'Impossibly old...' ;
```

在此，参数 value 的数值，会被设置为属性 age 的数值。若参数 value 的数值大于 150，则将字符串'Impossibly old...'，赋给局部变量 new_value。

```
    break ;
  case 'gender':
```

参数 attr 的数据，若是字符串'gender'，则执行下方到最接近 break 语句为止的源代码。

```
    if (! (value in ['male','female']))
      new_value = 'Impossible gender...' ;
```

在此，参数 value 的数据，会被设置为属性 gender 的数据字符串。若参数 value 的数据字符串，**并不是**'male'或'female'，则将字符串'Impossibly gender...'，赋给局部变量 new_value。

```
    break ;
  default:
```

参数 attr 的数据，若是其他字符串，则执行下方至末尾处为止的源代码。

```
    new_value = value ;
```

此语句将参数 value 的数据，赋给局部变量 new_value。

```
}
obj[attr] = new_value ;
```

此语句将局部变量 new_value 最终的数据，赋给【在参数 obj 的对象实例中，其名称为参数 attr 的数据字符串所代表】的属性。

```
}
```

上述语法定义了带有参数 obj、attr 与 value 的函数 set_value()，并作为特定 Proxy 对象实例的 setter 成员。

```
// var proxy_handler = {get: get_value} ;
var proxy_handler = {get: get_value , set: set_value } ;
```

此语句声明了初始数据为内含 getter 与 setter 成员的对象实例的变量 proxy_handler。

14.1.4 调试访问代理机制（ES6）

设置好实现访问代理机制的 Proxy 对象实例之后，可通过访问 Proxy 对象实例的方式，调试其代理机制是否正常工作。

```
p01.department = 'finance' ;
p01.position = 'manager' ;
```

这两个语句分别新增了变量 p01 的 Proxy 对象实例中的新属性 department 和 manager。其中，属性 department 的初始数据为字符串'finance'，属性 position 的初始数据则为字符串'manager'。

```
console.log(p01) ;
```

此语句显示出 Proxy{name: "Daisy", gender: "female", age: 30, department: "finance", position: "manager"}的信息。从中可以看出，在变量 p01 的对象实例中，带有【在变量 cur_obj 的对象实例中，可被访问】的属性成员 name、gender 与 age。

```
console.log(p01.name) ;
```

显示出字符串'Daisy'的信息。

```
console.log(p01.position) ;
```

显示出字符串'manager'的信息。

```
console.log(p01.company) ;
console.log(p01.salary) ;
```

这两个语句均显示出字符串'N/A'的信息。这是因为，p01.company 与 p01.salary 用于个别取出属性 company 与 salary 的数据，并会访问变量 p01 的 Proxy 对象实例中的 getter 成员 get_value，进而间接调用名称为 get_value 的函数，使得在函数 get_value()内，语句【choice = attr in obj？obj[attr]: 'N/A' ;】被执行，并判断出名称为 company 及 salary 的成员，并不存在于变量 cur_obj 的 Proxy 对象实例中。所以，函数 get_value() 最终会返回字符串'N/A'，并间接成为 p01.company 与 p01.salary 的数据。

```
p02.age = 152 ;
```

此语句试图在变量 p02 的 Proxy 对象实例中，重新设置其成员属性 age 的数据，成为整数值 152，但是并没有成功；反而使得 p02.age 的数据，变成字符串'Impossibly old...'。这是因为，语句【p02.age = 152 ;】被用来设置其属性 age 的数据，并会访问变量 p02 的 Proxy 对象实例中的 setter 成员 set_value，进而间接调用了名称为 set_value 的函数，并判断出名称为 age 的属性成员的数值，是否试图被设置为大于 150 的数值，使得在函数 set_value()内，语句【new_value = 'Impossibly old...';】和【obj[attr] = new_value ;】先后被执行，最终导致 p02.age 的数据，间接成为字符串'Impossibly old...'。

```
p02.gender = 'unknown' ;
```

此语句试图在变量 p02 的 Proxy 对象实例中，重新设置其成员属性 gender 的数据，成为字符串'unknown'，但是并没有成功；反而使得 p02.gender 的数据，变成字符串'Impossible gender...'。这是因为，【p02.gender = 'unknown' ;】被用来设置其属性 gender 的数据，并会访问变量 p02 的 Proxy 对象实例中的 setter 成员 set_value，进而间接调用了名称为 set_value 的函数，并判断出名称为 gender

的属性成员的数据，是否试图被设置为'male'和'female'**以外**的字符串，使得在函数 set_value()内，语句【new_value = 'Impossible gender...' ;】和【obj[attr] = new_value ;】先后被执行，最终导致 p02.gender 的数据，间接成为字符串'Impossibly gender...'。

```
console.log(p02) ;
```

此语句显示出 Proxy{name: "Daisy", gender: "Impossible gender...", age: "Impossibly old...", department: "finance", position: "manager"}的信息。从中可以看出，在变量 p02 的 Proxy 对象实例中，带有【在变量 cur_obj 的对象实例中，可被访问】的属性成员 name、gender 与 age。其中，原本试图被设置属性 age 的数值 152 和属性 gender 的数据'unknown'，因为不符合【小于 150】和【只能是字符串'male'或'female'】的规定，导致其属性 age 的数据，间接变成'Impossibly old...'，而其属性 gender 的数据，间接变成'Impossible gender...'。

14.2　Intl 对象

Intl 对象实例的函数和属性，均有关于语言敏感（language sensitive）的处理机制，包括字符串的比较、日期与时间的格式以及数字的格式。

14.2.1　精确比较语言敏感的字符串（ECMA-402）

通过构造函数 Intl.Collator()，可进一步达成**精确比较**语言敏感的字符串的任务。请参看如下示例。

【14-2-1-Intl-Collator.js】

```
var list = ['z','ɑ','a','å','ó','ä'] ;

var collator = new Intl.Collator() ;
var collator_de = new Intl.Collator("de") ;
var collator_sv = new Intl.Collator("sv") ;

// in German, "ä" sorts with "a"
// return -1 means list[5]< list[0]
console.log(collator_de.compare(list[5], list[0])) ;

// in Swedish, "ä" sorts after "z"
// return 1 means list[5]> list[0]
console.log(collator_sv.compare(list[5], list[0])) ;

console.log(collator.compare(list[2],'a')) ;
console.log('') ;

console.log(list.sort(collator.compare)) ;
console.log(list.sort(collator_de.compare)) ;
console.log(list.sort(collator_sv.compare)) ;
```

第 14 章　Proxy 对象、Intl 对象和 navigator 对象实例

【相关说明】

```
var list = ['z','α','a','å','ά','ä'] ;
```

此语句声明了初始数据为内含特殊字母的数组实例的变量 list。

```
var collator = new Intl.Collator() ;
```

此语句声明了初始数据为【Intl 对象中的 Collator 子对象实例】的变量 collator。因为 Intl.Collator() 并未被传入参数的数据，所以变量 collator 的对象实例，可用来比较本地语言（中文）相关的字符串。

```
var collator_de = new Intl.Collator("de") ;
```

此语句声明了初始数据为【Intl 对象中的 Collator 子对象实例】的变量 collator_de。Intl.Collator("de")被传入的参数数据为"de"，所以变量 collator_de 的对象实例，可用来比较【德文】相关的字符串。

```
var collator_sv = new Intl.Collator("sv") ;
```

此语句声明了初始数据为【Intl 对象中的 Collator 子对象实例】的变量 collator_sv。Intl.Collator("sv")被传入的参数数据为"sv"，所以变量 collator_sv 的对象实例，可用来比较【瑞典文】相关的字符串。

```
// in German, "ä" sorts with "a"
// return -1 means list[5]< list[0]
console.log(collator_de.compare(list[5], list[0])) ;
```

在此语句中，list[5]当前在变量 list 的数组实例中，代表其索引值为 5（第 6 个）的元素的字符'ä'，而 list[0]则代表在变量 list 的数组实例中，其索引值为 0（第 1 个）的元素的字符'z'。因此，【collator_de.compare(list[5], list[0])】等同于【collator_de.compare('ä','z')】，返回-1，意味着在【德文】中，字符'ä'的数码值，小于'z'的数码值。换言之，字符'ä'被编排在字符'z'【之前】。

```
// in Swedish, "ä" sorts after "z"
// return 1 means list[5]> list[0]
console.log(collator_sv.compare(list[5], list[0])) ;
```

collator_sv.compare(list[5], list[0])等价于 collator_sv.compare('ä','z')，会返回整数值 1，意味着在【瑞典文】中，字符'ä'的数码值，大于'z'的数码值。换言之，字符'ä'被编排在字符'z'【之后】。

```
console.log(collator.compare(list[2],'a')) ;
```

在此语句中，list[2]当前在变量 list 的数组实例中，代表其索引值为 2（第 3 个）的元素的字符'a'，因此 collator_de.compare(list[2],'a')等价于 collator_de.compare('a','a')，会返回整数值 0，意味着在【中文】里，字符'a'的数码值，等同于'a'的数码值。

```
console.log(list.sort(collator.compare)) ;
```

此语句显示出["ά", "a", "å", "ä", "z", "α"]的信息，代表在变量 list 的数组实例中，各字符在【中文】里的编码顺序。

```
console.log(list.sort(collator_de.compare)) ;
```

此语句显示出["a", "á", "å", "ä", "z", " α "]的信息，代表在变量 list 的数组实例中，各字符在【德文】里的编码顺序。

```
console.log(list.sort(collator_sv.compare)) ;
```

显示出["a", "á", "z", "å", "ä", " α "]的信息，代表在变量 list 的数组实例中，各字符在【瑞典文】里的编码顺序。从中可以看出，字符'z'在瑞典文，明显和在德文与中文里的编码顺序，有所不同。

14.2.2 语言敏感的日期与时间格式（ECMA-402）

通过构造函数 Intl.DateTimeFormat()，可进一步获取语言敏感的特定格式的日期与时间。请参看如下示例。

【14-2-2-Intl-DateTimeFormat.js】

```
let special = new Date(2100, 11, 12, 14, 30, 0) ;
let result = new Intl.DateTimeFormat('en').format(special) ;

console.log(result) ;

let options =
{
  year: 'numeric',
  month: 'numeric',
  day: 'numeric',
  hour: 'numeric',
  minute: 'numeric',
  second: 'numeric',
  hour12: true,
  timeZone: 'Asia/Shanghai'
} ;

result = new Intl.DateTimeFormat('zh', options).format(special) ;

console.log(result) ;
```

【相关说明】

```
let special = new Date(2100, 11, 12, 14, 30, 0) ;
```

此语句声明了初始数据为【内含日期与时间 2100/12/12 14:30:00】相关数据的 Date 对象实例的变量 special。其中，第 2 个参数的数值 11，代表 12 月份。

```
let result = new Intl.DateTimeFormat('en').format(special) ;
```

此语句声明了变量 result，其初始数据为英文（en）格式的日期与时间的字符串'12/12/2100'。其中，Intl.DateTimeFormat('en').format(special)会返回【在变量 special 的 Date 对象实例中，其日期与时间 2100/12/12 14:30:00 相关数据，被转换】之后的英文格式的日期与时间字符串'12/12/2100'。

```
console.log(result) ;

let options =
{
```

```
    year: 'numeric',
```
设置其年份为公元年份的 4 位整数值。
```
    month: 'numeric',
```
设置其月份为整数值。
```
    day: 'numeric',
```
设置其【日】为整数值。
```
    hour: 'numeric',
```
设置其【时】为整数值。
```
    minute: 'numeric',
```
设置其【分】为整数值。
```
    second: 'numeric',
```
设置其【秒】为整数值。
```
    hour12: true,
```
设置其【时】为 12 小时制。
```
    timeZone: 'Asia/Shanghai'
```
设置其【时区】为【亚洲/上海市】。
```
};
```
上述语法声明了变量 options,其初始数据为内含多个成员属性的对象实例。
```
result = new Intl.DateTimeFormat('zh', options).format(special) ;
```
此语句声明了变量 result,其初始数据为中文(zh)格式的日期与时间的字符串'2100/12/12 下午 2:30:00'。其中,Intl.DateTimeFormat('zh', options).format(special)会返回【在变量 special 的 Date 对象实例中,其日期与时间 2100/12/12 14:30:00 相关数据,被转换】之后的中文格式的日期与时间字符串'2100/12/12 下午 2:30:00'。而变量 options 的对象实例中的属性 year、month、day、hour、minute、second、hour12 与 timeZone,会影响被返回的字符串的日期与时间格式。

14.2.3 语言敏感的数值格式(ECMA-402)

通过构造函数 Intl.NumberFormat(),可进一步获取语言敏感的特定数值格式的字符串。请参看如下示例。

【14-2-3-Intl-NumberFormat.js】

```
let number = 352591.8 ;
let result = new Intl.NumberFormat('en').format(number) ;

console.log(result) ;
```

```
result = new Intl.NumberFormat('cn', {style: 'currency', currency: 'cny'}).format(number) ;
console.log(result) ;
```

【相关说明】

```
let number = 352591.8 ;
```

声明初始值为 352591.8 的变量 number。

```
let result = new Intl.NumberFormat('en').format(number) ;
```

此语句声明了变量 result，其初始数据为英文（en）数值格式的字符串'352,591.8'。其中，Intl.NumberFormat('en'). format(number)会返回【在变量 number 中，其数值 352591.8】的英文数值格式的字符串'352,591.8'。

```
console.log(result) ;
```

```
result = new Intl.NumberFormat('cn', {style: 'currency', currency: 'cny'}).format(number) ;
```

此语句声明了变量 result，其初始数据为中文（cn）货币数值格式的字符串'CN￥352,591.80'。Intl.NumberFormat('cn', {style: 'currency', currency: 'cny'}).format(number)会返回【在变量 number 中，其数值 352591.8】的中文货币数值格式的字符串'CN￥352,591.80'。其中，子对象实例{style: 'currency', currency: 'cny'}中的属性 style 与 currency，会影响被返回的字符串的货币数值格式。

14.2.4　返回规范化语言环境名称（ECMA-402）

通过函数 Intl.getCanonicalLocales()，可进一步获取特定规范化语言环境名称（canonical locale names）。请参看如下示例。

【14-2-4-Intl-getCanonicalLocales.js】

```
let locale = Intl.getCanonicalLocales('zh-cn') ;
console.log(locale) ;

locale = Intl.getCanonicalLocales('en-us') ;
console.log(locale) ;

locale = Intl.getCanonicalLocales('en-uk') ;
console.log(locale) ;

locale = Intl.getCanonicalLocales('en_kk') ;
console.log(locale) ;
```

【相关说明】

```
let locale = Intl.getCanonicalLocales('zh-cn') ;
```

Intl.getCanonicalLocales('en-cn')会返回["en-CN"]。所以，此语句声明了变量 locale，其初始数据为数组实例["zh-CN"]，并内含规范化语言环境名称（canonical locale name）的字符串"zh-CN"。

```
console.log(locale) ;
```

```
locale = Intl.getCanonicalLocales('en-us') ;
```

Intl.getCanonicalLocales('en-us')会返回["en-US"]。所以，此语句将内含规范化语言环境名称的字符串"en-US"的数组实例["en-US"]，赋给变量 locale。

```
console.log(locale) ;

locale = Intl.getCanonicalLocales('en-uk') ;
```

Intl.getCanonicalLocales('en-uk')会返回["en-UK"]。所以，此语句将内含规范化语言环境名称的字符串"en-UK"的数组实例["en-UK"]，赋给变量 locale。

```
console.log(locale) ;

locale = Intl.getCanonicalLocales('en_kk') ;
```

字符串'en-kk'并非语言环境名称，所以 Intl.getCanonicalLocales('en-kk')会产生【Uncaught RangeError: Invalid language tag: en_kk at Object.getCanonicalLocales (native)】的错误信息。

14.3　window.navigator 对象实例

window.navigator 对象实例，可简写为 navigator 对象实例，其函数和属性，可用来返回浏览器的相关情报或者当前地理位置的相关数据。

14.3.1　获取浏览器相关信息

在 windows.navigator 对象实例中，其所有属性和一些函数，均会返回浏览器的相关信息。请参看如下示例。

【14-3-1-browser-information.js】

```
// console.log(window.navigator.cookieEnabled) ;
console.log(navigator.cookieEnabled) ;
console.log(navigator.onLine) ;
console.log(navigator.javaEnabled()) ;
console.log('') ;

// console.log(window.navigator.appCodeName) ;
console.log(navigator.appCodeName) ;
console.log(navigator.appName) ;
console.log('') ;

console.log(navigator.appVersion) ;
console.log(navigator.userAgent) ;
console.log('') ;

console.log(navigator.platform) ;
console.log(navigator.product) ;
```

```
console.log('');
```

```
console.log(navigator.language);
```

【相关说明】

```
// console.log(window.navigator.cookieEnabled);
console.log(navigator.cookieEnabled);
```

在 navigator 对象实例中，其属性 cookieEnabled 的数据，若为布尔值 true，则意味着浏览器支持的 cookie 机制，当前处于开启状态。

```
console.log(navigator.onLine);
```

在 navigator 对象实例中，其属性 onLine 的数据，若为布尔值 true，则意味着浏览器当前处于在线（on-line）状态；反之，则处于离线（off-line）状态。

```
console.log(navigator.javaEnabled());
```

在 navigator 对象实例中，其属性 javaEnabled 的数据，若为布尔值 true，则代表浏览器支持的 Java applet 处理机制，当前处于开启状态。

```
console.log('');
```

```
// console.log(window.navigator.appCodeName);
console.log(navigator.appCodeName);
```

在 navigator 对象实例中，其属性 appCodeName 会返回浏览器的代码名称（code name），例如 Mozilla。

```
console.log(navigator.appName);
```

在 navigator 对象实例中，其属性 appName 会返回浏览器的应用名称（application name），例如 Netscape。

```
console.log('');
```

```
console.log(navigator.appVersion);
```

在 navigator 对象实例中，其属性 appVersion 会返回浏览器的应用版本（application version），例如【5.0 (Windows NT 10.0; Win64; x64) AppleWebKit/537.36 (KHTML, like Gecko) Chrome/70.0.3538.110 Safari/537.36】。

```
console.log(navigator.userAgent);
```

在 navigator 对象实例中，其属性 userAgent 会返回浏览器的用户代理（user agent）的相关数据，例如【Mozilla/5.0 (Windows NT 10.0; Win64; x64) AppleWebKit/537.36 (KHTML, like Gecko) Chrome/70.0.3538.110 Safari/537.36】。

```
console.log('');
```

```
console.log(navigator.platform);
```

在 navigator 对象实例中，其属性 platform 会返回浏览器所在的操作平台（platform）的版本，例如【Win32】。

```
console.log(navigator.product) ;
```

在 navigator 对象实例中，其属性 product 会返回浏览器的引擎的产品（product）名称，例如【Gecko】。

```
console.log('') ;
```

```
console.log(navigator.language) ;
```

在 navigator 对象实例中，其属性 language 会返回浏览器当前被设置的界面语言（language）的版本，例如代表简体中文的【zh-CN】。

根据官方的建议，navigator 对象实例中的属性 appCodeName、appName、appVersion、product 已经不合时宜了，所以应避免使用。

14.3.2 获取当前地理定位相关数据

window.navigator.geolocation 对象实例，可被简写为 navigator.geolocation 对象实例，其属性和函数可返回当前地理位置的相关数据。请参看如下示例。

【14-3-2-navigator-geolocation.js】

```
let options = {maximumAge: 0, timeout: 7000, enableHighAccuracy: true} ;

function succeed(position)
{
  with (position.coords)
  {
    let message = `Current location:\n\tLatitude = ${latitude}\n\tLongitude =
      ${longitude}\n\tdistance accuracy = ${accuracy}meters` ;
    ///Latitude = ???
    ///Longitude = ???
    ///distance accuracy = ??? meters
    console.log(message) ;
  }
} ;

function fail(error)
{
  console.warn(`${error.code}: ${error.message}`) ;
} ;

navigator.geolocation.getCurrentPosition(succeed, fail, options) ;
```

【相关说明】

```
let options = {maximumAge: 0, timeout: 7000, enableHighAccuracy: true} ;
```

此语句声明了变量 options，其初始数据为内含 3 个属性的对象实例。其中，属性 maximumAge 用来设置可允许沿用前一个地理位置的暂存数据的间隔毫秒数。若其间隔毫秒数为 0，则意味着浏览器应该立即返回当前地理位置的数据，而不是沿用前一个暂存数据。属性 timeout 用来设置可允许访问当前地理位置的逾时毫秒数，在此被设置为 7000 毫秒（7 秒）内，若用户仍然没有回复，则浏览器**无权**访问当前的地理位置。

在浏览器所在的计算机或者移动设备，支持高准确率（high accuracy）的定位机制的前提下，若其属性 enableHighAccuracy 的数据为布尔值 true，则启动高准确率（high accuracy）的定位机制。

```
function succeed(position)
{
  with (position.coords)
```

在此，为了简化源代码，通过语法【with (position.coords)】，使得 position.coords.latitude、position.coords.longitude 与 position.coords.accuracy，可在其下方的大括号内，被精简成为 latitude、longitude 和 accuracy。

```
  {
    let message = `Current location:\n\tLatitude = ${latitude}\n\tLongitude = ${longitude}\n\tdistance accuracy = ${accuracy}meters` ;
```

此语句声明了局部变量 message，其初始数据为类似如下被格式化的字符串：

Current location:

Latitude = XX.XXXXXXXXXXXXXXX

Longitude = XXX.XXXXXXX

distance accuracy = XXXX meters

其中，'\n'代表换行（line feed）字符，'\t'代表制表（tab）字符。

```
    console.log(message) ;
  }
} ;
```

上述语法定义了带有参数 position 的函数 succeed()。

```
function fail(error)
{
  console.warn(`${error.code}: ${error.message}`) ;
```

此语句显示出内含错误代码与原因的错误信息，例如【1: User denied Geolocation】或【3: Timeout expired】。

```
} ;
```

上述语法定义了带有参数 error 的函数 fail()。

```
navigator.geolocation.getCurrentPosition(succeed, fail, options) ;
```

此语句启动了浏览器支持的定位机制。启动时，会出现询问用户是否允许的信息对话框，例如带有信息【http://XXX.XXX.XXX 想要获取您的位置，允许或禁止？】的对话框。此时，若用户单击【允许】按钮，则名称为 succeed 的函数会被调用；若用户单击【禁止】按钮，或逾时未按下任何按钮，则名称为 fail 的函数会被调用。

在此语句中，变量名称 options，成为了内置函数 navigator.geolocation.getCurrentPosition()的第 3 个参数数据。对于浏览器处理当前地理位置的机制而言，变量 options 的对象实例，可用来设置其暂存数据的间隔毫秒数、允许访问当前地理位置的逾时毫秒数和高准确率的定位机制等特征。

14.4 练 习 题

1. 已知存在初始数据为对象实例{type: 'english', weekday: 0}的变量 one_day，试通过 Proxy 对象的访问代理机制，编写 JavaScript 源代码，以实现如下功能：

当 one_day.weekday 的数值为 1 至 7 的整数值时，让 console.log(one_day.weekday)显示出对应的'星期一'至'星期日'其中一个字符串。

2. 承上题，通过 Proxy 对象的访问代理机制，编写 JavaScript 源代码，以实现如下功能：

当 one_day.weekday 的数值为 1 至 7 的整数值，并且 one_day.type 的数据字符串为'english'时，让 console.log(one_day.weekday)显示出对应的'Monday'至'Sunday'其中一个字符串。

第 15 章

window.document 对象实例

window.document 对象实例,可被简写为 document 对象实例,即代表当前网页中的**根节点**实例,并且被内置了许多【处理各个**后代节点**实例】的属性和函数。本章通过许多示例的讲解,谈及对象实例的使用。

15.1 返回焦点所在的元素

属性 document.activeElement 可返回当前焦点(focus)所在的元素实例。请参看如下示例。

【15-1-^-document-activeElement.html】

```
<!DOCTYPE html>
<html>
  <head>
    <title>document activeElement</title>
    <style>
      input, button, select
      {
        font-size: 1.2em ;
        margin: 5px ;
      }
    </style>
  </head>
  <body>
    <form id="form01" name="form01" style="text-align: center;">
      <h3>个人资料</h3>
      <input type="text" id="username" name="username" placeholder="username" size="16" required>
      <input type="password" id="password" name="password" placeholder="password" size="16" required>
      <p></p>
      <select id="select01" name="select01">
```

```html
        <option value="">choice of day-off</option>
        <option value="0">Sunday</option>
        <option value="1">Monday</option>
        <option value="2">Tuesday</option>
        <option value="3">Wednesday</option>
        <option value="4">Thursday</option>
        <option value="5">Friday</option>
        <option value="6">Saturday</option>
      </select>
      <input type="search" id="search" name="search" placeholder="job category.." size="9">

      <p></p>
      <button type="submit" id="Login_btn">Login</button>
      <button type="reset" id="reset_btn">Reset</button>
    </form>
  </body>
  <script>
    form01.onclick = function (event)
    {
      with (document.activeElement)
      {
        console.log(`Tag name = ${tagName}, id = ${id}`) ;
      }
    } ;
  </script>
</html>
```

【相关说明】

```
<script>

  form01.onclick = function (event)
  {

    with (document.activeElement)
    {
      console.log(`Tag name = ${tagName}, id = ${id}`) ;
    }
```

通过此语法，使得代表特定子元素实例的标签名称（tag name）/元素名称的 document.activeElement.tagName，在其大括号里，可被简化成为 tagName；而代表特定子元素实例的属性 id 的 document.activeElement.id，在其大括号里，则可被简化成为 id。其中，document.activeElement 会返回【在当前网页中，其获取焦点（focus）】的元素实例。在此，被单击的元素实例，即会获得焦点。

```
  } ;
```

上述语法将带有参数 event 的匿名函数的定义，赋给属性 id 的数据为'form01'的 form 元素实例的属性 onclick。在 form 元素实例的范围内，任何子元素实例被单击时，上述匿名函数就会被调用，使得类似【Tag name = SELECT, id = select01】的信息，被显示在浏览器的调试工具【Console】面板里。

15.2 附加事件处理器至特定元素

通过内置的全局函数 addEventListener()，可在特定元素实例上，附加作为事件处理器（event handler）的自定义函数。请参看如下示例。

【15-2-^-DOM-addEventListener.html】

```html
<!DOCTYPE html>
<html>
  <head>
    <title>DOM addEventListener()</title>
    <style>
      input, button, select
      {
        font-size: 1.2em ;
        margin: 5px ;
      }
    </style>
  </head>
  <body>
    <form id="form01" name="form01" style="text-align: center;">
      <h3>个人资料</h3>
      <input type="text" id="username" name="username" placeholder="username" size="16" required>
      <input type="password" id="password" name="password" placeholder="password" size="16" required>
      <p></p>
      <select id="select01" name="select01">
        <option value="">choice of day-off</option>
        <option value="0">Sunday</option>
        <option value="1">Monday</option>
        <option value="2">Tuesday</option>
        <option value="3">Wednesday</option>
        <option value="4">Thursday</option>
        <option value="5">Friday</option>
        <option value="6">Saturday</option>
      </select>
      <input type="search" id="search" name="search" placeholder="job category.." size="9">

      <p></p>
      <button type="submit" id="Login_btn">Login</button>
      <button type="reset" id="reset_btn">Reset</button>
    </form>
  </body>
  <script>
    function display(event)
    {
      with (document.activeElement)
      {
        console.log(`Tag name = ${tagName}, id = ${id}`) ;
      }
    }
```

```
    // form01.addEventListener('click', display, true) ;
    form01.addEventListener('click', display, {capture: true}) ;

    function warning(event)
    {
      console.log('Username field is clicked.') ;
    }

    username.addEventListener('click', warning) ;

    ///
    function only_once_message(event)
    {
      console.log('Password field is clicked.') ;
    }

    password.addEventListener('click', only_once_message, {once: true}) ;
  </script>
</html>
```

【相关说明】

```
<script>

  function display(event)
  {
    with (document.activeElement)
    {
      console.log(`Tag name = ${tagName}, id = ${id}`) ;
    }
  }
```

上述语法定义了带有参数 event 的函数 display()。

```
    // form01.addEventListener('click', display, true) ;
    form01.addEventListener('click', display, {capture: true}) ;
```

此语句通过调用函数 addEventListener()，使得属性 id 的数据为'form01'的 form 元素实例，被监听着单击（click）动作。一旦属性 id 的数据为'form01'的 form 元素实例被单击时，名称为 display 的函数就会被调用，进而显示出类似【Tag name = INPUT, id = username】的信息。

在此，form01.addEventListener('click', display, {capture: true})和 form01.addEventListener('click', display, true)是等价的语法。

传入函数 addEventListener()的第 3 个参数的数据{capture: true}或 true，是用来设置特定事件的捕获（capture）流程，使得进行属性 id 的数据为'form01'的 form 元素实例本身的事件处理之后，会再被进行其子元素实例的事件处理。

若属性 id 的数据为'username'的 input 元素实例被单击时，则会先进行属性 id 的数据为'form01'的 form 元素实例的事件处理；然后在显示出【Tag name = INPUT, id = username】的信息之后，接着才会进行属性 id 的数据为'username'的 input 元素实例的事件处理，并显示出【Username field is clicked.】的信息。

```
    function warning(event)
    {
```

```
    console.log('Username field is clicked.') ;
  }
```

上述语法定义了带有参数 event 的函数 warning()。

```
username.addEventListener('click', warning) ;
```

此语句通过调用函数 addEventListener()，使得属性 id 的数据为'username'的 input 元素实例，会被监听着单击（click）动作。一旦属性 id 的数据为'username'的 input 元素实例被单击时，名称为 warning 的函数就会被调用，进而显示出类似【Username field is clicked.】的信息。

```
function only_once_message(event)
{
  console.log('Password field is clicked.') ;
}
```

上述语法定义了带有参数 event 的函数 only_once_message()。

```
password.addEventListener('click', only_once_message, {once: true}) ;
```

此语句通过调用函数 addEventListener()，使得属性 id 的数据为'password'的 input 元素实例，会监听着单击（click）动作。一旦属性 id 的数据为'password'的 input 元素实例被单击时，名称为 only_once_message 的函数就会被调用，进而显示出类似【Password field is clicked.】的信息。其中，其第 3 个参数的数据【{once: true}】，是用来限制其单击事件，仅会被处理 1 次而已！

15.3　收养特定节点

通过函数 document.adpotNode()，使得在当前网页中，特定节点实例会被收养到其他节点实例中，或者代表子网页的 iframe 元素实例里的节点实例中。请参看如下示例。

【15-3-^-document-adoptNode.html】

```
<!DOCTYPE html>
<html>
  <head>
    <title>documentadoptNode</title>
    <style>
      [id^=ul]
      {
        border-radius: 5px ;
        color: RoyalBlue ;
        width: 100px ;
        text-align: center ;
        padding: 5px ;
        list-style-type: none ;
        background-color: Gold ;
        margin: 20px ;
        display: inline-block ;
      }

      #ul02
```

```
    {
      background-color: GreenYellow ;
    }
  </style>
</head>
<body>
  <ul id="ul01">
    <li>apple</li>
    <li>blueberry</li>
    <li>cherry</li>
    <li>durian</li>
    <li>grape</li>
  </ul>

  <ul id="ul02">
    <li>banana</li>
  </ul>
</body>
<script>
  let temp = null ;

  function select_item(event)
  {
    temp = document.adoptNode(event.target) ;

    // ul02.appendChild(event.target) ;
    ul02.appendChild(temp) ;
  }

  ul01.addEventListener('click', select_item) ;
</script>
</html>
```

【相关说明】

```
<script>

  let temp = null ;
```

声明初始数据为空值（null）的变量 temp。

```
  function select_item(event)
  {
    temp = document.adoptNode(event.target) ;
```

此语句将当前 event.target 所代表的 **li 元素实例**，赋给变量 temp。

```
    // ul02.appendChild(event.target) ;
    ul02.appendChild(temp) ;
```

此语句将变量 temp 的 li 元素实例，新增至属性 id 的数据为'ul02'的 ul 元素实例中。

```
  }
```

上述语法定义了带有参数 event 的函数 select_item()。

```
ul01.addEventListener('click', select_item) ;
```

此语句通过调用函数 addEventListener()，使得属性 id 的数据为'ul01'的 ul 元素实例，会监听着单击（click）动作。一旦属性 id 的数据为'ul01'的 ul 元素实例被单击时，名称为 select_item 的函数就会被调用，进而让特定 li 元素实例，从属性 id 的数据为'ul01'的元素实例内，被移植到属性 id 的数据为'ul02'的 ul 元素实例中。

15.4　返回所有锚点元素构成的集合

通过属性 document.anchors，可返回【在当前网页中，**带有属性 name** 而代表锚点（anchor）】的所有 a 元素实例的集合。请参看如下示例。

【15-4-^-document-anchors.html】

```html
<!DOCTYPE html>
<html>
  <head>
    <title>documentanchors</title>
    <style>
      [id^=div]
      {
        display: block ;
        height: 500px ;
      }

      #div01{ background-color: Gold ; }
      #div02{ background-color: GreenYellow ; }
      #div03{ background-color: YellowGreen ; }
      #div04{ background-color: RoyalBlue ; }
      #div05{ background-color: Chocolate ; }
    </style>
  </head>
  <body>
    <a href="#a01">To a01</a>
    <a href="#a02">To a02</a>
    <a href="#a03">To a03</a>
    <a href="#a04">To a04</a>
    <a href="#a05">To a05</a>
    <p></p>

    <a name="a01">a01</a>
    <div id="div01">
      ...
    </div>

    <a name="a02">a02</a>
    <div id="div02">
      ...
    </div>
```

```
        <a name="a03">a03</a>
        <div id="div03">
            ...
        </div>

        <a name="a04">a04</a>
        <div id="div04">
            ...
        </div>

        <a name="a05">a05</a>
        <div id="div05">
            ...
        </div>
    </body>
    <script>
        let anchors = document.anchors ;

        console.log(anchors) ;
    </script>
</html>
```

【相关说明】

```
<script>

    let anchors = document.anchors ;
```

此语句声明了变量 anchors，其初始数据为【在当前网页中，带有属性 name 的 a 元素实例】的集合，例如【<a name="a01">a01</a>】~【<a name="a05">a05</a>】。在网页上，这种带有属性 name 及其数据的 a 元素实例，可被称为锚点（anchor）。

```
    console.log(anchors) ;
```

此语句显示出一组可展开的信息，代表当前网页中的 a 元素实例的集合。举例来说，当【<a href="#a03">To a03</a>】所代表的超链接，被单击时，浏览器并不会跳转至其他网页，而是滚动至当前网页的【<a name="a03">a03</a>】所代表的锚点位置！

请留意，代表超链接的 a 元素实例，带有属性 href 及其数据字符串"#特定锚点名称"或者"特定网址"，例如【href="#a03"】或者【href="http://www.tup.tsinghua.edu.cn"】；而代表锚点的 a 元素实例，则带有属性 name 及其数据"锚点名称"，例如【name="a03"】。

15.5　返回当前网址的相关属性

在 window.document 对象实例中，存在多个可返回当前网址的多个属性，例如属性 documentURI、baseURI 与 URL。此外，window.location.href 亦可返回当前网址。请参看如下示例。

【15-5-^-web-addresses.js】

```
console.log(document.documentURI) ;
```

```
console.log(document.baseURI) ;
console.log(document.URL) ;

// readable and writable
console.log(location.href) ;
```

【相关说明】

```
console.log(document.documentURI) ;
console.log(document.baseURI) ;
console.log(document.URL) ;
```

这 3 个语句中的 document.documentURI、document.baseURI 与 document.URL，均为**只读**（read only）属性，可返回当前网页所在的路径或网址，例如【file:///D:/examples/html/js_tester.html】或者【http://www.tup.tsinghua.edu.cn】。

```
// readable and writable
console.log(location.href) ;
```

location.href 是可读与可写（readable and writable）的属性，也可返回当前网页所在的路径或网址。

15.6 在当前网页中动态生成 HTML 源代码

通过函数 document.open()、document.write()、document.writeln()和 document.close()，可动态生成浏览器可执行的 HTML 源代码，成为当前网页里的源代码的一部分。请参看如下示例。

【15-6-^-dynamically-writing-HTML-codes-on-document-body.js】

```
document.open() ;

document.writeln('<hr color="Green">') ;
document.write('<br>') ;

document.write('<div align="center">Time is money.</div>') ;
document.write('<br>') ;

document.writeln('<hr color="Cyan">') ;

document.close() ;

document.body.style.backgroundColor = 'YellowGreen' ;

document.body.innerHTML += '<h1 style="color: RoyalBlue; text-align: center">
World Peace...</h1>' ;

console.log(document.body) ;
///   <hr color="Green">
///   <br>
///   <div align="center">Time is money.</div>
///   <br>
```

```
///    <hr color="Cyan">
///    <h1 style="color: RoyalBlue; text-align: center">World Peace...</h1>
///  </body>
console.log('') ;
```

【相关说明】

```
document.open() ;
```

此语句用来开启一个输出流，以便收集后续通过调用 document.write(代表特定 HTML 源代码片段的字符串)或 document.writeln(代表特定 HTML 源代码片段的字符串)，而被输出的字符串。

```
document.writeln('<hr color="Green">') ;
```

此语句通过调用 document.writeln(代表特定 HTML 源代码片段的字符串)，可将特定 HTML 源代码片段，动态输出至网页里，成为源代码的一部分。浏览器会在此语句之后的位置，动态生成 HTML 源代码片段【<hr color="Green">】，进而在当前网页上，呈现一条绿色（green）的水平线（horizon）。

```
document.write('<br>') ;
```

此语句使得浏览器在此语句之后的位置，动态生成 HTML 源代码片段【
】，进而在网页上，完成在视觉上的断行（break）/ 换行（line feed）。

```
document.write('<div align="center">Time is money.</div>') ;
```

此语句使得浏览器在此语句之后的位置，动态生成 HTML 源代码片段【<div align="center">Time is money.</div>】，进而在当前网页上，呈现带有居中对齐（center align）的文本【Time is money.】的 div 元素实例。

```
document.write('<br>') ;
```

此语句使得浏览器在此语句之后的位置，动态生成 HTML 源代码片段【
】，进而在当前网页上，完成在视觉上的断行。

```
document.writeln('<hr color="Cyan">') ;
```

此语句使得浏览器在此语句之后的位置，动态生成 HTML 源代码片段【<hr color="Cyan">】，进而在当前网页上，呈现一条青色（cyan）的水平线（horizon）。

```
document.close() ;
```

此语句用来**关闭**输出流，并正式呈现【document.write()和 document.writeln()所动态生成】的各个 HTML 源代码片段。

```
document.body.style.backgroundColor = 'YellowGreen' ;
```

此语句设置了当前网页的背景颜色为黄绿色（yellow green）。

```
document.body.innerHTML += '<h1 style="color: RoyalBlue; text-align: center">
World Peace...</h1>' ;
```

此语句在当前网页的 body 元素实例中，新增带有宝蓝色（RoyalBlue）、居中对齐（text-align: center）的文本【World Peace...】的 h1 元素实例。

```
console.log(document.body) ;
```

此语句显示出【可被展开】的 body 元素实例本身的源代码片段。展开之后，可进一步看到如下 body 元素实例中的源代码片段：

```
<body style="background-color: yellowgreen;">
  <hr color="Green">
  <br>
  <div align="center">Time is money.</div>
  <br>
  <hr color="Cyan">
  <h1 style="color: RoyalBlue; text-align: center">World Peace...</h1>
</body>
```

15.7　内含特定服务器相关数据的 cookie

所谓的 cookie，并不是真的指饼干，而是指网站服务器借助客户端的浏览器，存储于客户端的相关数据的小型文档，以便加速客户端日后和网站服务器的连接。关于内含相关数据的 cookie 的调试，可参看如下示例。

【15-7-^-document-cookie.html】

```
<!DOCTYPE html>
<html>
  <head>
    <title>document cookie</title>
  </head>
  <body>
  </body>
  <script>
    console.log(document.cookie) ;
    console.log('') ;

    let future_date = new Date() ;

    future_date.setTime(future_date.getTime() + 1 * 60 * 60 * 1000 ) ;

    document.cookie = 'username=jasper337' ;
    document.cookie = 'ip_address=111.222.333.777'+' ; expires= '+ future_date.toUTCString() ;

    console.log(document.cookie) ;
    console.log('') ;

    document.cookie = 'username=; expires=Thu, 01 Jan 1970 00:00:00 GMT' ;
    document.cookie = 'ip_address=; expires=Thu, 01 Jan 1970 00:00:00 GMT' ;

    console.log(document.cookie) ;
```

```
    </script>
</html>
```

【相关说明】

```
<script>
    console.log(document.cookie) ;
```

通过 HTTP 或 HTTPS 协议，浏览器连接至特定网站服务器的当前网页的文档时，当前网页中的 document.cookie，会返回当前网页的相关 cookie 的数据字符串（例如'username=jasper337; ip_address=111.222.333.777'）。若是让浏览器直接打开**本地**的当前网页，则当前网页中的 document.cookie 会返回空字符串"。

```
    console.log('') ;

    let future_date = new Date() ;
```

此语句声明了变量 future_date，其初始数据为内含当前日期与时间的 Date 对象实例。

```
    future_date.setTime(future_date.getTime() + 1 * 60 * 60 * 1000 ) ;
```

此语句将变量 future_date 的 Date 对象实例中的时间，往后推移 1 小时（1 × 60 × 60 × 1000 = 3600000 毫秒）。

```
    document.cookie = 'username=jasper337' ;
    document.cookie = 'ip_address=111.222.333.777'+' ; expires= '+ future_date.toUTCString() ;
```

这两个语句在当前网页中的 cookie 里，新增了数据为字符串'jasper337'的参数 username，以及其数据为字符串'111.222.333.777'的参数 ip_address。

举例来说，像是子字符串【; expires=Sun, 2Dec 2018 19:27:17 GMT"】的部分，用来让浏览器知道，参数 ip_address 的数据字符串，在协调世界时的时间点【2018/12/2 19:27:17】即会到期，并会被删除。

```
    console.log(document.cookie) ;
```

document.cookie 当前返回字符串'username=jasper337; ip_address=111.222.333.777'。

```
    console.log('') ;

    document.cookie = 'username=; expires=Thu, 01 Jan 1970 00:00:00 GMT' ;
    document.cookie = 'ip_address=; expires=Thu, 01 Jan 1970 00:00:00 GMT' ;
```

这两个语句会使得在当前网页相关的 cookie 里，其参数 username 和 ip_address 的数据被清空。

```
    console.log(document.cookie) ;
```

于上述两个语句中，其 cookie 的两个参数的数据，均已经被清空，所以 document.cookie 当前会返回空字符串"。

15.8 返回当前网页编码字符集的名称

通过属性 document.characterSet，可返回当前网页的编码字符集（encoding character set）的名称。请参看如下简短的示例。

【15-8-^-document-characterSet.js】
```
console.log(document.characterSet) ;
```

document.characterSet 会返回当前网页的编码字符集（encoding character set）的名称，例如 UTF-8 (Unicode Transformation Format - 8 bits)，意味着包括字母、数字、符号、中文和其他国家的文字，均通过 1 到 4 组【8 个比特位（bit）】，被加以编码。换句话说，在网页上的单一文字所占用的内存空间，是介于 8 个比特位（1 组）到 32 个比特位（4 组）之间。

15.9 创建代表新属性的节点

通过 document.createAttribute()，可创建并返回【在特定元素实例中，代表新属性】的节点实例。请参看如下示例。

【15-9-^-document-createAttribute.js】
```
let e01 = document.createElement('div') ;

e01.id = 'div01' ;
e01.innerHTML = 'A person who never made a mistake never tried anything new.' ;

e01.style.color = 'Gold' ;
e01.style.backgroundColor = 'RoyalBlue' ;

// e01.title = 'word count = 11' ;
let a01 = document.createAttribute('title') ;

a01.value = 'word count = 11' ;

e01.setAttributeNode(a01) ;

document.body.appendChild(e01) ;
```

【相关说明】
```
let e01 = document.createElement('div') ;
```
此语句声明了变量 e01，其初始数据为新的 div 元素实例。

```
e01.id = 'div01' ;
```
此语句将字符串'div01'，赋给变量 e01 的 div 元素实例中的属性 id。

```
e01.innerHTML = 'A person who never made a mistake never tried anything new.' ;
```

此语句将字符串'A person who never made a mistake never tried anything new.'，赋给变量 e01 的 div 元素实例中的属性 innerHTML，进而使得上述字符串，呈现在 div 元素实例的范围内。

```
e01.style.color = 'Gold' ;
```

将字符串'Gold'，赋给变量 e01 所代表的 div 元素实例的属性 style.color，使得此 div 元素实例中的文本在网页上呈现出金色（gold）字体。

```
e01.style.backgroundColor = 'RoyalBlue' ;
```

将字符串'RoyalBlue'，赋给变量 e01 所代表的 div 元素的属性 style.backgroundColor，使得此 div 元素实例的背景颜色在网页上呈现为宝蓝色（royal blue）。

```
// e01.title = 'word count = 11' ;
let a01 = document.createAttribute('title') ;
```

此语句声明了变量 a01，其初始数据是【在特定元素实例中，其名称为 title】的属性（attribute）的节点实例。

```
a01.value = 'word count = 11' ;
```

此语句将字符串'word count = 11'，赋给【在变量 a01 所代表的名称为 title 的属性节点实例中，其名称为 value】的属性。

请留意，在 HTML 编程语言里，特定元素实例的属性（attribute）在英文中，称为 attribute；在 JavaScript 编程语言里，特定对象实例的属性（property）在英文中，称为 property。

```
e01.setAttributeNode(a01) ;
```

此语句将变量 a01 所代表的名称为 title 的属性（attribute）节点实例，设置成为变量 e01 所代表的 div 元素实例的属性 title。执行完此语句之后，浏览器会在当前网页里，动态生成上述 div 元素实例的如下源代码：

```
<div id="div01" style="color: gold; background-color: royalblue;" title="word count = 11">
 A person who never made a mistake never tried anything new.</div>
```

由此可看出，在上述 div 元素实例中，带有属性 title 及其数据字符串'word count = 11'，使用户将鼠标指针，移入 div 元素实例的范围内时，便会出现提示信息【word count = 11】。

```
document.body.appendChild(e01) ;
```

此语句将变量 e01 所代表的 div 元素实例，新增至 body 元素实例中。执行此语句之后，浏览器会将变量 e01 所代表 div 元素实例，正式显示在网页上。

15.10 创建代表新注释的节点

通过函数 document.createComment()，可创建代表新注释（comment）的节点实例。请参看如下示例。

【15-10-^-document-createComment.js】

```
let c01 = document.createComment('This is the body for the main content.') ;

document.body.appendChild(c01) ;

console.log(document.body) ;
```

【相关说明】

```
let c01 = document.createComment('This is the body for the main content.') ;
```

此语句声明了变量 c01，其初始数据为【内含字符串'This is the body for the main content'的新**注释**（comment）】的节点实例。

```
document.body.appendChild(c01) ;
```

此语句将变量 c01 所代表的新注释的节点实例，新增至 body 元素实例中。

```
console.log(document.body) ;
```

此语句显示出如下的信息：

<body>
<input type="file" name="file_selector" id="file_selector" accept=".js">
<script>...</script>
<!--This is the body for the main content.-->
</body>

从上述信息可看出，注释文本【<!--This is the body for the main content.-->】，存在于如上的 HTML 源代码片段里。其中，HTML 源代码里的注释文本，必须放置于【<!--】与【-->】之间。

15.11 创建代表新片段或新元素的节点

通过函数 document.createDocumentFragment()，可创建【随后可再被新增其他子节点】的片段（fragment）的节点实例；而通过函数 document.createElement()，则可创建新的元素实例。请参看如下示例。

【15-11-^-document-createDocumentFragment-and-createElement.js】

```
let e01 = document.createElement('select') ;
let e02 = null ;

e01.style.fontSize = '1.2em' ;
e01.style.marginTop = '10px' ;

let fruits = ['apple','banana','cherry','durian','guava'] ;

let f01 = document.createDocumentFragment() ;
```

```
    for (let i = 0; i < fruits.length; i++)
    {
      e02 = document.createElement('option') ;

      e02.value = fruits[i] ;

      // e02.innerHTML = fruits[i] ;
      e02.textContent = fruits[i] ;

      f01.appendChild(e02) ;
    }

    e01.appendChild(f01) ;

    document.body.appendChild(e01) ;
```

【相关说明】

```
let e01 = document.createElement('select') ;
```

此语句声明了变量 e01，其初始数据为新的 select 元素实例。document.createElement('select') 用来创建并返回在网页中的新的 select 元素实例。

```
let e02 = null ;
```

声明初始数据为空值（null）的变量 e02。

```
e01.style.fontSize = '1.2em' ;
e01.style.marginTop = '10px' ;
```

这两个语句分别为变量 e01 代表的 select 元素实例，设置了字体大小（font size）和顶部边距（margin on the top）的外观。

```
let fruits = ['apple', 'banana', 'cherry', 'durian', 'guava'] ;
```

此语句声明了初始数据为内含多个字符串的数组实例的变量 fruits。

```
let f01 = document.createDocumentFragment() ;
```

此语句声明了变量 f01，其初始数据为【可内含多个元素实例的源代码片段 / 文档片段 (document fragment)】的节点实例。

```
for (let i = 0; i < fruits.length; i++)
```

因为变量 fruits 的数组实例的元素个数为 5，所以此 for 语句会迭代 5 次。

```
{
```

```
  e02 = document.createElement('option') ;
```

此语句将新的 option 元素实例，赋给变量 e02。

```
  e02.value = fruits[i] ;
```

此语句将【在变量 fruits 的数组实例中，其第 i + 1 个（其索引值为变量 i 的数值）元素】的数据字符串，赋给变量 e02 所代表的 option 元素实例的新属性 value。

```
  // e02.innerHTML = fruits[i] ;
```

```
e02.textContent = fruits[i] ;
```

此语句将【在变量 fruits 数组实例中，其第 i+1 个（索引值为变量 i 的数值）元素】的数据字符串，赋给变量 e02 所代表的 option 元素实例的新属性 textContent。

```
f01.appendChild(e02) ;
```

此语句将变量 e02 所代表的 option 元素实例，新增至变量 f01 所代表的节点实例中。

```
}
e01.appendChild(f01) ;
```

此语句将变量 f01 所代表的内含多个 option 元素实例的节点实例，新增至变量 e01 所代表的 select 元素实例中。所以，变量 e01 当前代表如下源代码片段的 select 元素实例：

```
<select style="font-size: 1.2em; margin-top: 10px;">
  <option value="apple">apple</option>
  <option value="banana">banana</option>
  <option value="cherry">cherry</option>
  <option value="durian">durian</option>
  <option value="guava">guava</option>
</select>
```

```
document.body.appendChild(e01) ;
```

此语句将变量 e01 所代表的 select 元素实例，新增至 body 元素实例中。执行此语句之后，浏览器会将变量 e01 所代表的 select 元素实例，正式显示在网页上。

15.12　创建代表新文本的节点

通过函数 document.createTextNode()，可创建代表新文本的节点实例。请参看如下示例。

【15-12-^-document-createTextNode.js】

```
let e01 = document.createElement('div') ;

e01.id = 'div01' ;

let t01 = document.createTextNode('A person who never made a mistake never tried anything new.') ;

e01.appendChild(t01) ;

e01.style.color = 'Gold' ;
e01.style.backgroundColor = 'RoyalBlue' ;

// e01.title = 'word count = 11' ;
let a01 = document.createAttribute('title') ;
```

```
a01.value = 'word count = 11' ;
e01.setAttributeNode(a01) ;

document.body.appendChild(e01) ;
```

【相关说明】

```
let e01 = document.createElement('div') ;
```

此语句声明了变量 e01，其初始数据为新的 div 元素实例。

```
e01.id = 'div01' ;
```

此语句将字符串'div01'，赋给变量 e01 所代表的 div 元素实例的属性 id，使得 div01 成为这个 div 元素实例的身份识别码。

```
let t01 = document.createTextNode('A person who never made a mistake never tried anything
    new.') ;
e01.appendChild(t01) ;
```

document.createTextNode('A person who never made a mistake never tried anything new.')会返回内含字符串'A person who never made a mistake never tried anything new.'的文本节点（text node）实例。

执行这两个语句的效果，等同于执行了单一语句【e01.innerHTML = 'A person who never made a mistake never tried anything new.' ;】，可将字符串'A person who never made a mistake never tried anything new.'，赋给变量 e01 所代表的 div 元素实例的属性 innerHTML，进而使得带有上述字符串文本的 div 元素实例，呈现于当前网页上。

```
e01.style.color = 'Gold' ;
e01.style.backgroundColor = 'RoyalBlue' ;
```

这两个语句分别使得变量 e01 所代表的 div 元素实例，以金色（gold）的字体颜色和宝蓝色（royal blue）的背景颜色，呈现于网页上。

```
// e01.title = 'word count = 11' ;
let a01 = document.createAttribute('title') ;
```

此语句声明了变量 a01，其初始数据为【在特定元素实例中，其名称为 title】的属性（attribute）节点实例。

```
a01.value = 'word count = 11' ;
```

此语句将字符串'word count = 11'，赋给【在变量 a01 所代表的名称为 title 的属性节点实例中，其名称为 value】的属性。

```
e01.setAttributeNode(a01) ;
```

此语句将【变量 a01 所代表的名称为 title 的属性节点实例】，设置成为【变量 e01 所代表的 div 元素实例】的属性 title。执行完此语句之后，浏览器会在当前网页里，动态生成上述 div 元素实例的如下源代码：

```
<div id="div01" style="color: gold; background-color: royalblue;" title="word count = 11">
    A person who never made a mistake never tried anything new.</div>
```

由此可看出，在上述 div 元素实例中，存在属性 title 及其数据字符串'word count = 11'，使用户将鼠标指针，移入 div 元素实例的范围内时，便会出现提示信息【word count = 11】。

```
document.body.appendChild(e01) ;
```

此语句将变量 e01 所代表的 div 元素实例，新增至 body 元素实例中。执行此语句之后，浏览器会将变量 e01 所代表 div 元素实例，正式显示在网页上。

15.13　返回当前网页的根元素

通过属性 document.documentElement，可返回当前网页的根元素（root）实例。当前网页的根元素实例，即是指当前网页中的 html 元素实例【<html>...</html>】。请参看如下示例。

【15-13-^-document-documentElement.js】

```
let root = document.documentElement ;
let children = root.children ;

console.log(children) ;

children[1].style.backgroundColor = 'GreenYellow' ;

console.log(children[1]) ;
```

【相关说明】

```
let root = document.documentElement ;
```

document.documentElement 会返回当前网页（document）的根元素（root）实例。此语句声明了变量 root，其初始数据为当前网页的根元素实例。

若当前网页的大致架构如下，则根元素实例即是 html 元素实例，并内含 head 和 body 子元素实例。

<html>
　<head>...</head>
　<body style="background-color: GreenYellow;">...</body>
</html>

```
let children = root.children ;
```

root.children 会返回根元素实例中的子元素实例所构成的特殊集合[head, body]，内含 head 与 body 元素实例。此语句声明了变量 children，其初始数据为【根元素实例中的所有子元素实例】构成的特殊集合。上述特殊集合的数据类型是 HTMLCollection 对象实例，相似于数组实例，可内含网页上的各种元素实例。

```
console.log(children) ;
```

显示出【HTMLCollection(2) [head, body]】的信息。

```
children[1].style.backgroundColor = 'GreenYellow' ;
```

在此语句中，children[1]对应到 body 元素实例，所以 children[1]当前等价于 document.body，使得此语句等同于【document.body.style.backgroundColor = 'GreenYellow' ;】，可设置网页的背景颜色（background color）为绿黄色（green yellow）。

```
console.log(children[1]) ;
```

显示出【<body style="background-color: GreenYellow;">...</body>】的信息。

15.14　访问当前网址的域名

通过属性 document.domain，可返回或者动态变更当前网页所在网站服务器（server）的域名（domain name）。请参看如下示例。

【15-14-^-document-domain.html】

```
<!DOCTYPE html>
<html>
  <head>
    <title>document domain</title>
  </head>
  <body>
  </body>
  <script>
    // Assume the domain is www.local.com
    console.log(document.domain) ;

    document.domain = 'local.com' ;

    console.log(document.domain) ;
  </script>
</html>
```

【相关说明】

```
<script>

  // Assume the domain is www.local.com
  console.log(document.domain) ;
```

document.domain 会返回当前网址的域名 (domain name)，例如 www.local.com。

```
  document.domain = 'local.com' ;
```

若当前网址的域名为 www.local.com，则最多只能将【上一层】域名 local.com，赋给 document.domain；否则浏览器会显示出类似【Uncaught DOMException: Failed to set the 'domain' property on 'Document': 'global.com' is not a suffix of 'www.local.com'.】的错误信息。

15.15 返回所有 embed 元素构成的集合

通过属性 document.embeds，可返回当前网页中的所有 embed 元素实例构成的集合。请参看如下示例。

【15-15-^-document-embeds.html】

```
<!DOCTYPE html>
<html>
  <head>
    <title>document embeds</title>
  </head>
  <body>
    <embed src="animations/1st.gif" width="150" height="100">
    <embed src="animations/2nd.gif" width="150" height="100">
    <embed src="animations/3rd.gif" width="150" height="100">
    <embed src="animations/4th.gif" width="150" height="100">
    <embed src="animations/5th.gif" width="150" height="100">
  </body>
  <script>
    let embeds = document.embeds ;

    console.log(embeds) ;
    console.log(embeds.length) ;
    console.log(embeds[1].src) ;
  </script>
</html>
```

【相关说明】

```
<script>

  let embeds = document.embeds ;
```

此语句声明了变量 embeds，其初始数据为当前网页中的所有 embed 元素实例构成的集合。

```
  console.log(embeds) ;
```

显示出可展开的【HTMLCollection(5)】信息。

```
  console.log(embeds.length) ;
```

embeds.length 会返回整数值 5，代表变量 embeds 的集合中的 embed 元素实例个数。

```
  console.log(embeds[1].src) ;
```

embeds[1].src 会返回【在变量 embeds 的集合里，其第 2 个（索引值为 1）embed 元素实例中的属性 src】的数据字符串，例如'file:///D:/examples/html/animations/2nd.gif'。

15.16 返回所有 form 元素实例构成的集合

通过属性 document.forms，可返回当前网页中的所有 form 元素实例构成的集合。请参看如下示例。

【15-16-^-document-forms.html】

```
<!DOCTYPE html>
<html>
  <head>
    <title>document forms</title>
  </head>
  <body>
    <form id="form01">
      <input type="text" id="username">
      <input type="password" id="password">
      <br>
      ...
    </form>

    <form id="form02">
      <p>...</p>
    </form>

    <form id="form03">
      <p>...</p>
    </form>
  </body>
  <script>
    let forms = document.forms ;

    console.log(forms) ;
    console.log('') ;

    console.log(forms[0].id) ;
    console.log(forms[0].elements[0].id) ;
    console.log('') ;

    // username.style.backgroundColor = 'Gold' ;
    forms[0].elements[0].style.backgroundColor = 'Gold' ;

    console.log(forms[1].id) ;
  </script>
</html>
```

【相关说明】

```
<script>

  let forms = document.forms ;
```

此语句声明了变量 forms，其初始数据为【在当前网页中，被 document.forms 返回的所有 form

元素实例】构成的集合。

```
console.log(forms) ;
```

显示出可展开的【HTMLCollection(3)】的信息。

```
console.log('') ;
console.log(forms[0].id) ;
```

forms[0].id 会返回【在变量 forms 的集合里,其第 1 个（索引值为 0）form 元素实例的属性 id】的数据字符串'form01'。

```
console.log(forms[0].elements[0].id) ;
```

forms[0].elements[0].id 会返回【在变量 forms 的集合里的第 1 个（索引值为 0）form 元素实例中,其第 1 个（索引值为 0）子元素实例的属性 id】的数据字符串'username'。

由此可知,可通过 forms[m].elements[n],快速访问第（m + 1）个 form 元素实例中的第（n + 1）个子元素实例。

```
console.log('') ;
// username.style.backgroundColor = 'Gold' ;
forms[0].elements[0].style.backgroundColor = 'Gold' ;
```

在本示例中,语法【forms[0].elements[0].style.backgroundColor = 'Gold' ;】当前等价于【username.style.backgroundColor = 'Gold' ;】,可将属性 id 的数据为'username'的 input 元素实例的背景颜色,设置为金色 (gold)。

```
console.log(forms[1].id) ;
```

forms[1].id 会返回【在变量 forms 的集合里,其第 2 个（索引值为 1）form 元素实例中的属性 id】的数据字符串'form02'。

15.17 返回特定身份识别码的元素实例

通过函数 document.getElementById(),可返回属性 id 的数据为'特定名称'的元素实例。请参看如下示例。

【15-17-^-document-getElementById.html】

```html
<!DOCTYPE html>
<html>
  <head>
    <title>document getElementById()</title>
  </head>
  <body>
    <h1 id="header">
      Hello, Earth!
    </h1>
```

```
    <form>
      <input type="text" id="username" name="username">
      <input type="password" id="password" name="password">
      <p></p>
      <button type="submit" id="submit_btn">Login</button>
      <button type="reset" id="reset_btn">reset</button>
    </form>
  </body>
  <script>
    document.body.style.textAlign = 'center' ;

    let title = document.getElementById('header') ;

    title.style.color = 'YellowGreen' ;

    username.style.color = 'RoyalBlue' ;
    username.style.fontSize = '20px' ;
    username.style.backgroundColor = 'Gold' ;
    username.style.width = '120px' ;
    username.placeholder = 'username...' ;
    username.style.padding = '2px' ;
    username.style.margin = '5px' ;

    username.focus() ;

    password.placeholder = 'password...' ;

    with (password.style)
    {
      color = 'Teal' ;
      fontSize = '1.2em' ;
      padding = '2px' ;
      margin = '5px' ;
      backgroundColor = 'Yellow' ;
      width = '120px' ;
    }

    with (submit_btn.style)
    {
      fontSize = '1.2em' ;
      margin = '0 5px' ;
    }

    with (reset_btn.style)
    {
      fontSize = '1.2em' ;
      margin = '0 5px' ;
    }
  </script>
</html>
```

【相关说明】

```
<script>
```

```
document.body.style.textAlign = 'center' ;
```

此语句使得在 body 元素实例中的文本（text）和子元素实例，均呈现出居中对齐（align center）的外观。欲在网页上呈现出来的子元素实例，均应放置在 document.body 元素实例中。JavaScript 源代码中的 document.body，其实就是对应到 HTML 源代码的【<body>...</body>】元素实例。

```
let title = document.getElementById('header') ;
```

此语句将属性 id 的数据为'header'的 h1 元素实例，赋给变量 title。此语句等价于简短的语句【let title = header ;】。

```
title.style.color = 'YellowGreen' ;
```

此语句将字符串'YellowGreen'，赋给变量 title 的 h1 元素实例的属性 style.color，使得 h1 元素实例内的字体颜色，成为黄绿色（yellow green）。

```
username.style.color = 'RoyalBlue' ;
```

此语句将字符串'RoyalBlue'，赋给属性 id 的数据为'username'的 input 元素实例的属性 style.color，使得 input 元素实例内的字体颜色，成为宝蓝色（royal blue）。

```
username.style.fontSize = '20px' ;
```

此语句将字符串'20px'，赋给属性 id 的数据为'username'的 input 元素实例的子属性 style.fontSize，使得 input 元素实例的字体尺寸（font size），成为 20 像素。

```
username.style.backgroundColor = 'Gold' ;
```

此语句将字符串'Gold'，赋给属性 id 的数据为'username'的 input 元素实例的子属性 style.backgroundColor，使得 input 元素实例的背景颜色（background color），成为金色（gold）。

```
username.style.width = '120px' ;
```

此语句将字符串'120px'，赋给属性 id 的数据为'username'的 input 元素实例的子属性 style.width，使得 input 元素实例的宽度（width），成为 120 像素。

```
username.placeholder = 'username...' ;
```

此语句将字符串'username...'，赋给属性 id 的数据为'username'的 input 元素实例的属性 placeholder，使得 input 元素实例内的提示用途的占位（place holder）文本，成为【username...】。

```
username.style.padding = '2px' ;
```

此语句将字符串'2px'，赋给属性 id 的数据为'username'的 input 元素实例的子属性 style.padding，使得 input 元素实例的内侧边缘，到文本的内边距／填充间隔（padding），成为 2 像素。

```
username.style.margin = '5px' ;
```

此语句将字符串'5px'，赋给属性 id 的数据为'username'的 input 元素实例的子属性 style.margin，使得 input 元素实例的边框，到其他元素实例或者浏览器窗口边缘的外边距／外部间隔（margin），成为 5 像素。

```
username.focus() ;
```

此语句使得属性 id 的数据为'username'的 input 元素实例，获取焦点（focus），进而让键盘光标呈现在 input 元素实例的内部，以提示用户输入文本。

```
password.placeholder = 'password...' ;
```

此语句将字符串'password...'，赋给属性 id 的数据为'password'的 input 元素实例的属性 placeholder，使得 input 元素实例内的提示用途的占位（place holder）文本，成为【password...】。

```
with (password.style)
```

此语法可简化【属性 id 的数据为'password'的 input 元素实例】的访问语句。例如：简化【password.style.color】，成为在其大括号里的【color】，或者简化【password.style.backgroundColor】，成为在其大括号里的【backgroundColor】。

```
{
  color = 'Teal' ;
  fontSize = '1.2em' ;
  padding = '2px' ;
  margin = '5px' ;
  backgroundColor = 'Yellow' ;
  width = '120px' ;
}

with (submit_btn.style)
```

此语法可简化【属性 id 的数据为'submit_btn'的 button 元素实例】的访问语句。例如：简化【submit_btn.style.fontSize】，成为在其大括号里的【fontSize】，或者简化【submit_btn.style.margin】，成为在其大括号里的【margin】。

```
{
  fontSize = '1.2em' ;

  margin = '0 5px' ;
```

此语句设置 button 元素实例的【上下】外边距为 0、【左右】外边距为 5 像素。

```
}

with (reset_btn.style)
```

此语法可简化属性 id 的数据为'reset_btn'的 button 元素实例的访问语句。例如：简化【reset_btn.style.fontSize】成为在其大括号里的【fontSize】，或简化【reset_btn.style.margin】，成为在其大括号里的【margin】。

15.18 返回被设置带有特定 CSS 类名的所有元素实例的集合

通过函数 document.getElementsByClassName()，可返回被设置带有特定 CSS 类名（class name）的所有元素实例的集合。请参看如下示例。

【15-18-^-document-getElementsByClassName.html】

```html
<!DOCTYPE html>
<html>
  <head>
    <title>document getElementsByClassName()</title>
    <style>
      div
      {
        text-align: center ;
        width: 100px ;
        height: 100px ;
        border-radius: 20px ;
        height: 50px ;
        display: table-cell ;
        vertical-align: middle ;
      }

      .color01
      {
        color: RoyalBlue ;
        background-color: Gold ;
      }

      .color02
      {
        color: ForestGreen ;
        background-color: Pink ;
      }

      .border01
      {
        border: 2px Teal solid ;
      }

      .border02
      {
        border: 2px SkyBlue dashed ;
      }
    </style>
  </head>
  <body>
    <div class="color01 border01">Hello</div>
    <div class="color01 border02">World</div>
    <div class="color02 border01">and</div>
    <div class="color02 border02">Solar System.</div>
  </body>
  <script>
    let elist01 = document.getElementsByClassName('color01') ;

    elist01[0].style.opacity = 0.2 ;
    elist01[1].style.opacity = 0.2 ;

    let elist02 = document.getElementsByClassName('color02 border02') ;
```

```
    elist02[0].style.textDecoration = 'underline' ;
  </script>
</html>
```

【相关说明】

```
<script>

  let elist01 = document.getElementsByClassName('color01') ;
```

此语句声明了变量 elist01，其初始数据为【CSS 类名（class name）被设置带有 color01】的元素实例的集合。

在【\<div class="color01 border01"\>Hello\</div\>】与【\<div class="color01 border02"\>World\</div\>】两个 div 元素实例中，其属性 class 的数据，均带有子字符串'color01'，这就意味着带有 color01 的 CSS 类名。

其 CSS 类名带有 color01 的元素实例，会被应用选择器名称（selector name）为【.color01】的如下 CSS 样式，从而使得其文本颜色成为宝蓝色（royal blue），其背景颜色成为金色（gold）：

.color01

{

　　color: RoyalBlue ;

　　background-color: Gold ;

}

```
    elist01[0].style.opacity = 0.2 ;
    elist01[1].style.opacity = 0.2 ;
```

elist01[0]当前代表着 div 元素实例【\<div class="color01 border01"\>Hello\</div\>】；elist02[1]当前代表着 div 元素实例【\<div class="color01 border02"\>World\</div\>】。

这两个语句分别在上述两个 div 元素实例其中之一，设置其子属性 style.opacity 的数值为 0.2 (20%)，进而使得上述两个 div 元素实例，均呈现出不透明度（opacity）为 20%的外观。

```
    let elist02 = document.getElementsByClassName('color02 border02') ;
```

此语句声明了变量 elist02，其初始数据为【CSS 类名被设置同时带有 color02 与 border02】的元素实例的集合。

在 div 元素实例【\<div class="color02 border02"\>Solar System.\</div\>】中，其属性 class 的数据，同时带有子字符串'color02'与'border02'，这就意味着【其 CSS 类名同时带有 color02 与 border02】的含义。

其 CSS 类名带有 color02 的元素实例，会被应用选择器名称为【.color02】的如下 CSS 样式，从而使得其文本颜色成为森林绿色（forest green），背景颜色成为粉红色（pink）：

.color02

{

　　color: ForestGreen ;

　　background-color: Pink ;

}

CSS 类名带有 border02 的元素实例，则会被应用选择器名称为【.border02】的如下 CSS 样式，从而使得其边框成为宽度，成为 2 像素（2px）的破折号型（dashed）虚线的天蓝色（sky blue）边框（border）：

.border02
{
  border: 2px SkyBlue dashed ;
}

```
elist02[0].style.textDecoration = 'underline' ;
```

elist02[0]当前代表 div 元素实例【<div class="color02 border02">Solar System.</div>】。其子属性 style.textDecoration 代表文本装饰（text decoration）为下画线（underline），进而使得 div 元素实例的文本装饰（text decoration）具有下画线（underline）的外观。

15.19　返回特定标签名称的所有元素实例的集合

通过函数 document.getElementsByTagName()，可返回具有特定标签名称（tag name）的所有元素实例的集合。请参看如下示例。

【15-19-^-document-getElementsByTagName.html】

```html
<!DOCTYPE html>
<html>
  <head>
    <title>documentgetElementsByTagName()</title>
    <style>
     div
     {
       text-align: center ;
       width: 100px ;
       height: 100px ;
       border-radius: 20px ;
       height: 50px ;
       display: table-cell ;
       vertical-align: middle ;
     }

     .color01
     {
       color: RoyalBlue ;
       background-color: Gold ;
     }

     .color02
     {
       color: ForestGreen ;
       background-color: Pink ;
     }
```

```
    .border01
    {
      border: 2px Teal solid ;
    }

    .border02
    {
      border: 2px SkyBlue dashed ;
    }
  </style>
</head>
<body>
  <div class="color01 border01">Hello</div>
  <div class="color01 border02">World</div>
  <div class="color02 border01">and</div>
  <div class="color02 border02">Solar System.</div>
</body>
<script>
  let elist = document.getElementsByTagName('div') ;

  for (let i = 0; i < elist.length; i++)
  {
    elist[i].style.opacity = 0.2 * (1 + i) ;
  }
</script>
</html>
```

【相关说明】

```
<script>

  let elist = document.getElementsByTagName('div') ;
```

此语句声明了变量 elist，其初始数据为【在当前网页上，其元素名称 / 标签名称（tag name）为 div】的所有元素实例的集合。

```
  for (let i = 0; i < elist.length; i++)
  {

    elist[i].style.opacity = 0.2 * (1 + i) ;
```

elist[i]代表【在变量 elist 中，其第 i + 1 个（索引值为变量 i 的整数）】的 div 元素实例的子属性 style.opacity，使得各 div 元素实例，呈现在网页上时，具有不同的不透明度（opacity）的外观，例如不透明度为 20%（0.2）、40%（0.4）、60%（0.6）或 80%（0.8）。

```
  }
```

上述 for 语句会迭代 4 次，进而访问不同的 div 元素实例。

15.20 判断当前网页是否存在焦点

通过函数 document.hasFocus()，可判断焦点（focus）是否已进入当前网页中。请参看如下示例。

【15-20-^-document-hasFocus.html】

```html
<!DOCTYPE html>
<html>
  <head>
    <title>document hasFocus()</title>
  </head>
  <body>
  </body>
  <script>
    setInterval(check_focus, 500) ;

    function check_focus()
    {
      let focused = document.hasFocus() ;

      if (focused)
      {
        document.body.style.backgroundColor = 'GreenYellow' ;
      }
      else
      {
        document.body.style.backgroundColor = 'Pink' ;
      }
    }
  </script>
</html>
```

【相关说明】

```
<script>
  setInterval(check_focus, 500) ;
```

此语句使得每隔 500 毫秒（半秒），即调用名称为 check_focus 的函数。

```
function check_focus()
{
  let focused = document.hasFocus() ;
```

若当前网页获得焦点（focus），document.hasFocus()会返回布尔值 true；若当前网页失去焦点，则 document.hasFocus()会返回 false。单击当前网页内的任何位置，都可使得当前网页获得焦点。所以，此语句声明了变量 focused，其初始数据为布尔值 false 或 true。

```
if (focused)
{
```

```
        document.body.style.backgroundColor = 'GreenYellow' ;
```
若当前网页获得焦点，则执行此语句，进而使得网页的背景颜色，成为绿黄色（green yellow）。
```
      }
      else
      {
        document.body.style.backgroundColor = 'Pink' ;
```
若当前网页失去焦点，则执行此语句，进而使得网页的背景颜色，成为粉红色（pink）。
```
      }
    }
```

上述语法定义了函数 check_focus()。

15.21 返回当前网页的 head 元素实例

通过属性 document.head，可返回当前网页的 head 元素实例。请参看如下示例。

【15-21-^-document-head.html】
```
<!DOCTYPE html>
<html>
  <head>
    <title>document head</title>
    <style>
      h1
      {
        color: RoyalBlue ;
        padding: 5px ;
        border-radius: 10px ;
        background-color: Gold ;
      }
    </style>
  </head>
  <body>
    <h1>World Peace</h1>
  </body>
  <script>
    console.log(document.head) ;

    let my_title = document.head.getElementsByTagName('title')[0] ;

    my_title.innerHTML = my_title.innerHTML.toUpperCase() ;
  </script>
</html>
```

【相关说明】
```
  <script>
```

```
console.log(document.head) ;
```

此语句显示出可展开的【<head>...</head>】的信息。加以展开之后,可进一步看到 head 元素实例内的子元素实例:

  <title>USING DOCUMENT HEAD OBJECT</title>

  <style>...</style>

```
let my_title = document.head.getElementsByTagName('title')[0] ;
```

此语句声明了变量 my_title,其初始数据为 title 元素实例【<title>document head</title>】。document.head.getElementsByTagName('title')会返回其元素名称 / 标签名称(tag name)为 title 的所有元素实例的集合。在此,元素名称为 title 的元素实例仅有 1 个,所以在其集合内,只有 1 个 title 元素实例,也就是索引值为 0 的 title 元素实例。因此,document.head.getElementsByTagName ('title')[0] 会返回其元素名称为 title 的元素实例。

```
my_title.innerHTML = my_title.innerHTML.toUpperCase() ;
```

此语句可使得变量 my_title 的 title 元素实例内的文本【document head】,变成所有字母均大写的【DOCUMENT HEAD】。在此语句的等号右侧,my_title.innerHTML 可返回 title 元素实例内的文本字符串 'document head'。因为是字符串,所以支持函数 toUpperCase(),使得 my_title.innerHTML.toUpperCase()返回字符串'DOCUMENT HEAD'。

15.22 返回当前网页所有 image 元素实例的集合

通过属性 document.images,可返回当前网页的所有 image 元素实例的集合。请参看如下示例。

【15-22-^-document-images.html】

```
<!DOCTYPE html>
<html>
  <head>
    <title>document images</title>
    <style>
      img
      {
        width: 100px ;
        height: 70px ;
      }
    </style>
  </head>
  <body>
  </body>
  <script>
    let count = 1, suffix = '' ;

    for (let i = 1; i < 21; i++)
    {
      suffix = i.toString().padStart(2,'0') ;
```

```
      document.writeln(`<img id="image${suffix}" src="images/w${suffix}.jpg">`) ;
    }

    let image_list = document.images ;
    let color = '' ;

    for (let j = 0; j < image_list.length; j++)
    {
      color = '#'+ parseInt(0xffffff * Math.random()).toString(16).padStart(6,'0') ;
      image_list[j].style.border = `5px ${color}double` ;
    }
  </script>
</html>
```

【相关说明】

```
<script>

  let count = 1, suffix = '' ;
```

此语句声明了初始值为 1 的变量 count，以及初始数据为空字符串"的变量 suffix。

```
    for (let i = 1; i < 21; i++)
    {
      suffix = i.toString().padStart(2,'0') ;
```

此语句将变量 i 被迭代的整数值 1～20，分别转换成为字符串'01'、'02'、'03'、…、'20'，并赋给变量 suffix。其中，i.toString()会返回【转换变量 i 的整数值】之后的字符串'1'、'2'、'3'、…、'20'，再经过 padStart(2,'0')的【左侧补 0】的处理之后，最终返回'01'、'02'、'03'、…、'20'。

```
      document.writeln(`<img id="image${suffix}" src="images/w${suffix}.jpg">`) ;
```

此语句可在网页里，动态写入（write）一个新的 img 元素实例的源代码片段（例如<img id="image05" src="images/w05.jpg">）之后，再写入换行（ln, line）字符，使得各 img 元素实例所代表的影像，呈现在网页上时，分别被换行而纵向排列。

```
    }
```

上述 for 语句会迭代 20 次。

```
    let image_list = document.images ;
```

此语句声明了变量 image_list，其初始数据为当前网页中的所有 img 元素实例的集合。

```
    let color = '' ;
```

声明初始数据为空字符串"的变量 color。

```
    for (let j = 0; j < image_list.length; j++)
    {
      color = '#' + parseInt(0xffffff * Math.random()).toString(16).padStart(6,'0') ;
```

此语句将字符'#'，衔接至【被随机产生而介于 0～16^6 - 1（0xffffff）】之间的十六进制数码，例如 #a9e2f8 或 #04028a 等。

因为 parseInt(0xffffff * Math.random()).toString(16)有可能会返回未满 6 位而类似【4028a】的十六进制数码，所以必须通过 padStart(6,'0')，以进行【左侧补 0】的处理，进而转换成为类似【04028a】的十六进制数码。最后，变量 color 的数据字符串，才能成为表达特定颜色的颜色数码，例如【#04028a】。

```
        image_list[j].style.border = `5px ${color}double`;
```

此语句使得在变量 image_list 的集合中的第 j + 1 个 img 元素实例，被设置了宽度为 5 像素（5px）的特定颜色的双线（double）边框，使得 img 元素实例的源代码片段，变成【<img id="image05" src="images/w05.jpg" style="border: 5px double #04028a;">】。若通过浏览器，加以检查其源代码片段，则会看到被自动转换成为**颜色语法**不同的源代码片段，例如【<img id="image05" src="images/w05.jpg" style="border: 5px double **rgb(4, 2, 138);**">】。

```
    }
```

上述 for 语句会迭代 20 次。

15.23 创建当前网页或者子网页里的特定节点实例的副本

通过函数 document.importNode()或【特定元素实例的 id】衔接函数【.cloneNode()】的调用，可在当前网页或者 iframe 元素实例中的子网页里，创建特定节点实例的副本。请参看如下**两个示**例。

【15-23-^-e1-document-importNode.html】

```html
<!DOCTYPE html>
<html>
  <head>
    <title>document importNode()</title>
    <style>
      [id^=div]
      {
        width: 120px ;
        padding: 5px ;
        margin: 5px ;
        border-radius: 5px ;
        border: 1px Cyan solid ;
      }
      img
      {
        width: 100px ;
        height: 60px ;
      }
    </style>
  </head>
```

```
    <body>
      <div id="div01" align="center">
        <img src="images/w01.jpg">
        <hr>
        <span>Natural waterfall</span>
      </div>
    </body>
    <script>
      let ref = null ;

      ref = document.importNode(div01, true) ;

      ref.id = 'div02' ;

      ref.getElementsByTagName('img')[0].src = 'images/w18.jpg' ;

      ref.getElementsByTagName('span')[0].innerHTML = 'Flowers and the bees' ;

      document.body.appendChild(ref) ;
    </script>
</html>
```

【相关说明】

```
    <script>

      let ref = null ;
```

声明初始数据为空值（null）的变量 ref。

```
      ref = document.importNode(div01, true) ;
```

此语句将 document.importNode(div01, true)返回的元素实例，赋给了变量 ref。其中，document.importNode(div01, true)可返回【属性 id 的数据为'div01'的 div 元素实例】的副本。函数 importNode()的第 2 个参数的数据为布尔值 true，意味着对属性 id 的数据为'div01'的 div 元素实例，进行复制时，必须一并复制其内部的所有子元素实例。

```
      ref.id = 'div02' ;
```

此语句设置了【在变量 ref 的 div 元素实例中，其属性 id 的数据】成为'div01'。

```
      ref.getElementsByTagName('img')[0].src = 'images/w18.jpg' ;
```

此语句将特定影像文档路径的字符串'images/w18.jpg'，赋给了【在变量 ref 的新元素实例中，其第 1 个 img 子元素实例】的属性 src。

```
      ref.getElementsByTagName('span')[0].innerHTML = 'Flowers and the bees' ;
```

此语句将字符串'Flowers and the bee'，赋给了【在变量 ref 的新元素实例中，其第 1 个 span 子元素实例】的属性 innerHTML。

```
      document.body.appendChild(ref) ;
```

此语句将变量 ref 的新元素实例，新增至 body 元素实例中。在本示例的网页上，除了呈现出影像文档和说明文本的原始区块之外，还动态呈现出其源代码架构大致相同，但内含不同的影像文档和说明文本的新区块。

【15-23-^-e2-DOM-cloneNode.html】

```html
<!DOCTYPE html>
<html>
  <head>
    <title>DOM cloneNode()</title>
    <style>
      div
      {
        text-align: center ;
        color: IndianRed ;
        background-color: Pink ;
        width: 80px ;
        height: 80px ;
        margin: 10px ;
      }

      h3
      {
        line-height: 1em ;
      }

      a
      {
        text-decoration: none ;
        color: Yellow ;
      }
    </style>
  </head>
  <body>
    <div id="div01">
      <p>Super</p>
      <a href="#">happy time</a>
    </div>
  </body>
  <script>
    let element = div01.cloneNode(true) ;

    element.id = 'div02' ;

    element.innerHTML = '<p href="#">Nice</p>' ;

    element.innerHTML += '<a href="#">amazing hour</a>' ;

    document.body.appendChild(element) ;
  </script>
</html>
```

【相关说明】

```
<script>

  let element = div01.cloneNode(true) ;
```

此语句声明了变量 element，其初始数据为【属性 id 的数据为'div01'的 div 元素实例】的副本。div01.cloneNode()会返回【属性 id 的数据为'div01'的 div 元素实例】的副本，但不包含其内部的各个子元素实例；而 div01.cloneNode(**true**)则会返回包含其所有子元素实例在内的副本。

```
element.id = 'div02' ;
```

此语句将字符串'div02'，赋给变量 element 所代表的 div 元素实例副本的属性 id。

```
element.innerHTML = '<p href="#">Nice</p>' ;
```

此语句将字符串'<p href="#">Nice</p>'，赋给变量 element 所代表的 div 元素实例副本的属性 innerHTML。换言之，此语句间接在 div 元素实例副本中，新增了 p 元素实例【<p href="#">Nice</p>】。

```
element.innerHTML += '<a href="#">amazing hour</a>' ;
```

此语句将字符串'<a href="#">amazing hour</a>'，通过运算符【+=】，串接至【在变量 element 所代表的 div 元素实例副本中，其属性 innerHTML】的数据字符串里。换言之，此语句间接在 div 元素实例副本中，新增了 a 元素实例【<a href="#">amazing hour</a>】。

```
document.body.appendChild(element) ;
```

此语句将变量 element 所代表的 div 元素实例副本，新增至 body 元素实例中。因此，在网页上，会看到两个布局相同，但文本不同的 div 元素实例。

15.24 获取当前网页的最近被修改的日期和时间

通过属性 document.lastModified，可返回当前网页的最近被修改的日期与时间。请参看如下示例。

【15-24-^-document-lastModified.html】

```
<!DOCTYPE html>
<html>
  <head>
    <title>document lastModified</title>
  </head>
  <body>
  </body>
  <script>
    console.log(document.lastModified) ;
    console.log('') ;

    let mdate = new Date(document.lastModified) ;

    console.log(mdate) ;
    console.log('') ;

    console.log(mdate.getFullYear()) ;
    console.log(mdate.getMonth()) ;
    console.log(mdate.getDate()) ;
```

```
    console.log('') ;

    console.log(mdate.getDay()) ;
    console.log('') ;

    console.log(mdate.getHours()) ;
    console.log(mdate.getMinutes()) ;
    console.log(mdate.getSeconds()) ;
  </script>
</html>
```

【相关说明】

```
<script>

    console.log(document.lastModified) ;
```

dcoument.lastModified 返回当前网页的最近被修改的【日期与时间】字符串，例如 12/03/2018 20:17:40。

```
    console.log('') ;

    let mdate = new Date(document.lastModified) ;
```

此语句声明了变量 mdate，其初始数据为【内含当前网页的最近被修改的日期与时间】的 Date 对象实例，例如日期与时间为【Mon Dec 03 2018 20:17:40 GMT+0800】的 Date 对象实例。

```
    console.log(mdate) ;
    console.log('') ;

    console.log(mdate.getFullYear()) ;
```

mdate.getFullYear()返回【在当前网页的最近被修改的日期中，其代表**公元年份**】的数值。

```
    console.log(mdate.getMonth()) ;
```

mdate.getMonth()返回【在当前网页的最近被修改的日期中，其代表**月份**】的数值。其中，数值 0 代表 1 月份，数值 11 代表 12 月份。

```
    console.log(mdate.getDate()) ;
```

mdate.getDate()返回【在当前网页的最近被修改的日期中，其代表**日**】的数值。

```
    console.log('') ;

    console.log(mdate.getDay()) ;
```

mdate.getDay()返回【在当前网页的最近被修改的日期中，其代表**星期几**】的数值。其中，数值 0 代表星期日，数值 6 代表星期六。

```
    console.log('') ;

    console.log(mdate.getHours()) ;
```

mdate.getHours()返回【在当前网页的最近被修改的时间中，其代表**时数**】的数值。

```
    console.log(mdate.getMinutes()) ;
```

mdate.getMinutes()返回【在当前网页的最近被修改的时间中，其代表**分钟**数】的整数值。

```
console.log(mdate.getSeconds()) ;
```

mdate.getSeconds()返回【在当前网页的最近被修改的时间中，其代表**秒**数】的整数值。

15.25　返回当前网页中的所有超链接元素实例的集合

通过属性 document.links，可返回当前网页中的所有超链接（hyper link）元素实例构成的集合。其中，带有属性 href 及其数据的 a 元素实例与 area 元素实例，均被视为**超链接**元素实例。请参看如下示例。

【15-25-^-document-links.html】

```
<!DOCTYPE html>
<html>
  <head>
    <title>document links</title>
  </head>
  <body>
    <a href="#" id="a01">Test 1</a>
    <a href="#" id="a02">Test 2</a>
    <hr color="ForestGreen">
    <a href="#" id="a03">Test 3</a>
    <a href="#" id="a04">Test 4</a>
    <hr color="RoyalBlue">
    <a href="#" id="a05">Test 5</a>
    <a href="#" id="a06">Test 6</a>
    <a href="#" id="a07">Test 7</a>
  </body>
  <script>
    let refs = document.links ;

    console.log(refs) ;

    for (let i = 0; i < refs.length; i++)
    {
      refs[i].style.color = '#'+ parseInt(0xffffff * Math.random()).toString(16).
        padStart(6,'0') ;
    }
  </script>
</html>
```

【相关说明】

```
<script>

  let refs = document.links ;
```

此语句声明了变量 refs，其初始数据为当前网页中的所有超链接 a 元素实例构成的集合。

```
console.log(refs) ;
```

显示出可展开的【HTMLCollection(7)】的信息。

```
for (let i = 0; i < refs.length; i++)
{

  refs[i].style.color = '#' + parseInt(0xffffff * Math.random()).toString(16).
    padStart(6,'0') ;
```

在上述 for 循环语句的每次迭代中，refs[i]代表其中一个超链接 a 元素实例，使得其属性 style.color，随机被设置成为特定颜色数码，例如'#a9e2f8'。此语句将字符'#'衔接到【被随机产生而介于 0～16^6-1（0xffffff）】之间的十六进制数码，例如 #a9e2f8 或 #04028a 等。因为 parseInt(0xffffff * Math.random()).toString(16)有可能返回未满 6 位而类似【4028a】的十六进制数码，所以必须通过 padStart(6,'0')，进行【左侧补 0】的处理，进而转换成为类似【04028a】的十六进制数码。最后，变量 color 的数据字符串，才能成为表达特定颜色的颜色数码，例如【#04028a】。

读者在浏览器的调试工具【Console】面板里，输入【refs[0].style.color】时，会看到已经被浏览器转换之后的类似'rgb(4, 2, 138)'的字符串，而不是类似'#04028a'的字符串。

```
}
```

上述 for 循环语句会迭代 7 次，这是因为 refs.length 当前返回整数值 7，意味着变量 refs 的集合内，内含 7 个元素实例。

15.26 返回特定 CSS 选择器名称对应的元素实例或集合

通过函数 document.querySelector()，可返回特定 CSS 选择器名称（selector name）对应的第 1 个元素实例；通过 document.querySelectorAll()，可返回特定 CSS 选择器名称对应的**所有**元素实例构成的**集合**。请参看如下示例。

【15-26-^-document-querySelector-and-querySelectorAll.html】

```
<!DOCTYPE html>
<html>
  <head>
    <title>documentquerySelector() and querySelectorAll()</title>
    <style>
      [id^=div]
      {
        width: 120px ;
        padding: 5px ;
        margin: 5px ;
        border-radius: 5px ;
        border: 1px Cyan solid ;
      }

      img
```

```
      {
        width: 100px ;
        height: 60px ;
      }
    </style>
  </head>
  <body>
    <div id="div01" align="center">
      <img src="images/w01.jpg">
      <br>
      <span>Natural waterfall</span>
    </div>

    <div id="div02" align="center">
      <img src="images/w12.jpg">
      <br>
      <span>Beautiful flower</span>
    </div>

    <div id="div03" align="center">
      <img src="images/w18.jpg">
      <br>
      <span>flowers and the bee</span>
    </div>
  </body>
  <script>
    let ref01 = document.querySelector('div') ;
    let ref02 = document.querySelector('#div01') ;

    console.log(ref01 == ref02) ;

    ref01.style.border = '5px Gold dotted' ;
    ref02.style.outline = '5px Pink dashed' ;

    ///
    let ref03 = document.querySelectorAll('div img') ;

    console.log('') ;
    console.log(ref03) ;

    ref03[1].style.opacity = 0.3 ;

    ///
    let ref04 = document.querySelectorAll('div>span') ;

    console.log('') ;
    console.log(ref04) ;

    ref04[0].style.color = 'SkyBlue' ;
  </script>
</html>
```

【相关说明】

```
<script>
```

```
        let ref01 = document.querySelector('div') ;
```

此语句声明了变量 ref01，其初始数据为 document.querySelector('div')所返回的 div 元素实例。

```
        let ref02 = document.querySelector('#div01') ;
```

此语句声明了变量 ref02，其初始数据为 document.querySelector('#div01')所返回的 div 元素实例。被传入函数 document.querySelector('#div01')的参数数据，是 CSS 选择器名称（selector name）的字符串'#div01'。举例来说，'div'代表选择【在当前网页中，其元素名称 / 标签名称为 div】的元素实例，而'#div01'则代表选择【在当前网页中，其属性 id 的数据为'div01'】的元素实例。

```
        console.log(ref01 == ref02) ;
```

在此，【ref01 == ref02】返回布尔值 true，意味着变量 ref01 与 ref02，均对应到带有相同内存引址的 div 元素实例。

```
        ref01.style.border = '5px Gold dotted' ;
        ref02.style.outline = '5px Pink dashed' ;
```

因为变量 ref01 与 ref02 当前均对应到同一个 div 元素实例，所以这两个语句，均在同一个 div 元素实例上，分别设置了边框（border）和外框（outline）。

```
        let ref03 = document.querySelectorAll('div img') ;
```

此语句声明了变量 ref03，其初始数据为 document.querySelectorAll('div img')所返回的多个元素实例构成的集合。在此，被传入函数 document.querySelectorAll('div img')的参数数据字符串'div img'，是 CSS 选择器名称（selector name）的字符串，意味着选择【在当前网页中，其**外层**的**其中之一**，是 div 元素实例】的所有 img 元素实例。

```
        console.log('') ;

        console.log(ref03) ;
```

显示出可展开的【NodeList(3)】的信息。

```
        ref03[1].style.opacity = 0.3 ;
```

ref03[1]当前对应到网页中的第 2 个（索引值为 1）img 元素实例。此语句设置了上述 img 元素实例的子属性 style.opacity 的数值为 0.3，意味着上述 img 元素实例，以不透明度 30%（0.3）的外观，呈现在网页上。

```
        let ref04 = document.querySelectorAll('div>span') ;
```

此语句声明了变量 ref04，其初始数据为 document.querySelectorAll('div>span')所返回的多个元素实例构成的集合。被传入函数 document.querySelectorAll('div>span')的参数数据字符串'div>span'，是 CSS 选择器名称（selector name）的字符串，意味着选择【在当前网页中，其上一层**刚好**是 div 元素实例】的所有 span 元素实例。

```
        console.log('') ;

        console.log(ref04) ;
```

显示出可展开的【NodeList(3)】的信息。

```
ref04[0].style.color = 'SkyBlue' ;
```

ref04[0]当前对应到网页中第 1 个（索引值为 0）span 元素实例。此语句设置了上述 span 元素实例的属性 style.color 的数据为字符串'SkyBlue'，意味着上述 span 元素实例，以天蓝色（sky blue）文本颜色的外观，呈现在网页上。

15.27　返回和处置当前网页的加载状态

通过属性 document.readyState，可返回当前网页的加载（loading）/ 就绪（ready）状态；通过属性 document.onreadystatechange，可设置一旦当前网页的就绪状态（ready state）发生任何变化（change）时，就调用【作为事件处理器（event handler）】的函数。请参看如下示例。

【15-27-^-document-readyState-and-onreadystatechange.html】

```
<!DOCTYPE html>
<html>
 <head>
  <title>document readyState and onreadystatechange</title>
 </head>
 <body>
 </body>
 <script>
  console.log(document.readyState) ;

  document.onreadystatechange = function (event)
  {
   switch (document.readyState)
   {
    case 'loading':
     console.log('The document is being loaded.\n\n') ;
     break ;
    case 'interactive':
     console.log('The DOM is ready to be accessed.\n\n') ;
     break ;
    case 'complete':
     console.log('The document has been completely loaded.\n\n') ;
     break ;
   }
  } ;
 </script>
</html>
```

【相关说明】

```
<script>

  console.log(document.readyState) ;
```

document.readyState 在此返回代表网页当前加载状态的字符串'loading'。

```
        document.onreadystatechange = function (event)
        {
            switch (document.readyState)
```

此语法通过 document.readyState 所返回的网页当前加载状态的字符串，加以选择特定的执行分支。

```
            {
                case 'loading':
                    console.log('The document is being loaded.\n\n') ;
                    break ;
                case 'interactive':
                    console.log('The DOM is ready to be accessed.\n\n') ;
                    break ;
                case 'complete':
                    console.log('The document has been completely loaded.\n\n') ;
                    break ;
```

从上述各 case 语句来看，可将其加载状态，大致分为网页加载中（loading）、子资源（影像、框架源代码、样式表源代码等）加载中（interactive），以及加载完成（complete）。

```
            }
        } ;
```

在上述语法中，将其等号右侧的匿名函数的定义，赋给了代表网页本身的 document 对象实例的属性 onreadystatechange。一旦网页当前的加载状态，发生变化时，上述匿名函数就会被调用。

上述语法前后会使得 3 个信息，按照如下顺序，呈现在浏览器的调试工具【Console】面板里：

- The document is being loaded.
- The DOM is ready to be accessed.
- The document has been completely loaded.

15.28 返回跳转前的网址

通过属性 document.referrer，可返回【跳转至当前网页的前一网页】的网址。请参看如下带有两个源代码文档的示例。请留意，本示例的源代码文档，必须通过 HTTP 协议，才能调试出其正确的结果。

【15-28-^-document-referrer.html】

```html
<!DOCTYPE html>
<html>
  <head>
    <title>document referrer</title>
  </head>
  <body>
```

```
    </body>
    <script>
      console.log(document.referrer) ;
    </script>
</html>
```

【相关说明】

```
<script>

  console.log(document.referrer) ;
```

若浏览器加载这个网页文档,是因为在类似【**http**://主机名/文件夹名称/15-28-^-prior-to-document-referrer.html】所代表的网页中,用户单击了特定**超链接**,则在浏览器的调试工具【Console】面板里,类似【**http**://主机名/文件夹名称/15-28-^-prior-to-document-referrer.html】的信息,会被显示出来。这也就意味着【http://主机名/文件夹名称/15-28-^-prior-to-document-referrer.html】所代表的网页,即是本网页的引用来源(referrer)。若让浏览器直接加载本网页,则document.referrer只会返回【http://主机名/文件夹名称/】的不完整信息。

【15-28-^-prior-to-document-referrer.html】

```
<!DOCTYPE html>
<html>
  <head>
    <title>prior to document referrer</title>
  </head>
  <body>
    <a href="15-28-^-document-referrer.html">To Main Page</a>
    <p></p>
    <iframe src="15-28-^-document-referrer.html"></iframe>
  </body>
</html>
```

【相关说明】

```
      <iframe src="15-28-^-document-referrer.html"></iframe>
```

在本网页的 iframe 元素实例中,加载了其属性 src 的数据字符串所代表的网页【15-28-^-document-referrer.html】。而在网页【15-28-^-document-referrer.html】中,语句【console.log(document.referrer) ;】会被执行。所以,在浏览器的调试工具【Console】面板中,亦会显示出类似【http://主机名/文件夹名称/15-28-^-prior-to-document-referrer.html】的信息。这就意味着在浏览器的窗口里,之所以间接呈现出网页【15-28-^-document-referrer.html】的画面,是因为本网页的缘故!

因此,若用户单击了超链接【<a href="15-28-^-document-referrer.html">To Main Page</a>】,则浏览器会加载网页【15-28-^-document-referrer.html】,并间接执行【console.log(document.referrer) ;】,从而显示出类似【http://主机名/文件夹名称/15-28-^-prior-to-document-referrer.html】的信息。

15.29 解除已被附加的事件处理器

通过函数 document.removeEventListener()，可解除已被附加至特定元素实例的事件处理器（event handler）。请参看如下示例。

【15-29-^-document-removeEventListener.html】

```
<!DOCTYPE html>
<html>
  <head>
    <title>document removeEventListener()</title>
  </head>
  <body>
  </body>
  <script>
    let xy_list = [], count = 0 ;

    document.addEventListener('click', save_xy) ;

    function save_xy(event)
    {
      xy_list.push(`(${event.clientX}, ${event.clientY})`) ;

      count++ ;

      if (count == 4)
      {
        document.removeEventListener('click', save_xy) ;

        console.log(xy_list) ;
      }
    }
  </script>
</html>
```

【相关说明】

```
<script>

    let xy_list = [], count = 0 ;
```

此语句声明了初始数据为空数组实例[]的变量 xy_list，以及初始值为 0 的变量 count。

```
    document.addEventListener('click', save_xy) ;
```

此语句通过调用函数 addEventListener()，使得 document 对象实例所代表的当前网页，会被监听着单击（click）动作。当前网页被单击时，名称为 save_xy 的函数就会被调用。

```
    function save_xy(event)
    {
      xy_list.push(`(${event.clientX}, ${event.clientY})`) ;
```

此语句将类似'(36, 47)'的字符串，新增至变量 xy_list 的数组实例中。其中，event.clientX 与 event.clientY 分别返回了【用户在当前网页内，单击时】的坐标位置的 *x* 值与 *y* 值。

```
    count++ ;
```

变量 count 的数值被递增。

```
    if (count == 4)
```

若变量 count 的数值等于 4，则意味着用户已经在当前网页上，单击了 4 次，因此存在 4 次单击时的坐标位置。

```
    {
        document.removeEventListener('click', save_xy) ;
```

此语句通过调用函数 removeEventListener()，使得 document 对象实例所代表的当前网页，被移除【监听着单击 (click) 动作】的处理机制。换言之，记录了 4 个坐标位置之后，就会立即停止再继续记录。

```
        console.log(xy_list) ;
```

显示出类似["(36, 47)", "(144, 32)", "(37, 191)", "(142, 191)"]的信息，可看出变量 xy_list 当前代表着内含 4 个坐标位置的数组实例。

```
    }
}
```

上述语法定义了带有参数 event 的函数 save_xy()。

15.30　返回当前网页中的所有 script 元素实例构成的集合

通过属性 document.scripts，可返回当前网页中的所有 script 元素实例构成的集合。请参看如下示例。

【15-30-^-document-scripts.html】

```
<!DOCTYPE html>
<html>
  <head>
    <title>document scripts</title>
    <script src= "https://code.jquery.com/jquery-3.2.1.slim.min.js"
      integrity="sha384-KJ3o2DKtIkvYIK3UENzmM7KCkRr/rE9/Qpg6aAZGJwFDMVNA/GpGFF93hXpG5KkN"
      crossorigin="anonymous"></script>

    <script src= "https://cdnjs.cloudflare.com/ajax/libs/popper.js/1.12.3/umd/popper.min.js"
      integrity="sha384-vFJXuSJphROIrBnz7yo7oB41mKfc8JzQZiCq4NCceLEaO4IHwicKwpJf9c9IpFgh"
      crossorigin="anonymous"></script>
```

```html
    <script src="https://maxcdn.bootstrapcdn.com/bootstrap/4.0.0-beta.2/js/
      bootstrap.min.js" integrity="sha384-alpBpkh1PFOepccYVYDB4do5UnbKysX5WZXm3XxPqe5iKT
      fUKjNkCk9SaVuEZflJ" crossorigin="anonymous"></script>
  </head>
  <body>
  </body>
  <script>
    let scripts = document.scripts ;

    console.log(scripts.length) ;
    console.log(scripts) ;
    console.log('') ;

    console.log(scripts[3].innerHTML) ;
  </script>
</html>
```

【相关说明】

```
<script>

    let scripts = document.scripts ;
```

此语句声明了变量 scripts，其初始数据为 document.scripts 返回的所有 script 元素实例构成的集合。

```
    console.log(scripts.length) ;
```

scripts.length 在此返回了整数值 4，意味着在当前网页里，存在 4 个 script 元素实例。

```
    console.log(scripts) ;
```

显示出可展开的【HTMLCollection(4)】的信息。

```
    console.log('') ;

    console.log(scripts[3].innerHTML) ;
```

scripts[3]返回了当前网页中的第 4 个（索引值为 3）script 元素实例；scripts[3].innerHTML 进一步返回上述 script 元素实例中的所有 JavaScript 源代码片段。

15.31　访问当前网页的标题文本

通过属性 document.title，可设置或者获取当前网页的标题（title）文本。请参看如下示例。

【15-31-^-document-title.html】

```
<!DOCTYPE html>
<html>
  <head>
    <title>document title</title>
  </head>
```

```
<body>
</body>
<script>
 console.log(document.title) ;
 console.log('') ;

 document.title = 'Funny Place' ;

 console.log(document.title) ;
</script>
</html>
```

【相关说明】

```
<script>

 console.log(document.title) ;
```

document.title 返回当前网页的标题文本。

```
 console.log('') ;

 document.title = 'Funny Place' ;
```

此语句将字符串'Funny Place'，赋给代表当前网页的 document 对象实例的属性 title，使得当前网页的标题文本，更改为'Funny Place'。

15.32 练 习 题

1. 已知如下 JavaScript 源代码片段：

```
function display(event)
{
  |
}

form01.addEventListener('click', display) ;
```

试将上述源代码片段，改写成为不调用函数 addEventListener()的等价语句，使得 form01 代表的 form 元素实例，被单击时，函数 display()即会被调用。

2. 已知如下源代码语句：

```
current_ref = document.getElementById('div01') ;
```

在不修改变量名称 current_ref 和任何符号的情况下，试改写上述语句，成为更简短的等价语句。

3. 已知如下源代码片段：

```
username.style.color = 'RoyalBlue' ;
username.style.backgroundColor = 'GoldenRod' ;
username.style.padding = '5px' ;
```

```
username.style.margin = '10px' ;
username.style.fontSize = '30px' ;
```

在不修改属性 id 的数据为字符串'username'，以及不修改任何符号的情况下，试改写上述源代码片段，成为更简短的等价语法。

第 16 章

DOM 的元素实例

在特定网页中的任意元素实例,均可被视为在文档对象模型(DOM, document object model)中的元素实例(element instance)。通过被内置在各元素实例中的属性和函数,可访问各元素实例的相关数据。

16.1 设置焦点跳转至特定元素实例上的快捷键

现在的浏览器均支持【在特定网页上,将焦点(focus)移至特定元素实例上】的快捷键(shortcut / hot key),以代替费时的鼠标操作。请参看如下示例。

【16-1-^-DOM-accessKey.html】

```
<!DOCTYPE html>
<html>
  <head>
    <title>DOM accessKey</title>
    <style>
      a
      {
        display: inner-block ;
        margin: 5px ;
      }

      [id^=div]
      {
        display: block ;
        padding: 10px ;
        height: 500px ;
      }
```

```html
      #div01{ background-color: Gold ; }
      #div02{ background-color: GreenYellow ; }
      #div03{ background-color: YellowGreen ; }
      #div04{ background-color: RoyalBlue ; }
      #div05{ background-color: Chocolate ; }
    </style>
  </head>
  <body>
    <a href="#a01" accesskey="1">To a01</a>
    <a href="#a02" accesskey="2">To a02</a>
    <a href="#a03">To a03</a>
    <a href="#a04">To a04</a>
    <a href="#a05">To a05</a>
    <br>
    <a href="http://www.tup.tsinghua.edu.cn" target="new" accesskey="6">To Official Website</a>
    <p></p>

    <a name="a01">a01</a>
    <div id="div01">
      <input type="text" id="username" size="8" placeholder="username...">
      <input type="password" id="password" size="8" placeholder="password...">
      <br>
      ...
    </div>

    <a name="a02">a02</a>
    <div id="div02">
      ...
    </div>

    <a name="a03">a03</a>
    <div id="div03">
      ...
    </div>

    <a name="a04">a04</a>
    <div id="div04">
      ...
    </div>

    <a name="a05">a05</a>
    <div id="div05">
      ...
    </div>
  </body>
  <script>
    let links = document.links ;

    console.log(links.length) ;

    links[2].accessKey = '3' ;
    links[3].accessKey = '4' ;
    links[4].accessKey = '5' ;

    username.accessKey = 'u' ;
```

```
        password.accessKey = 'p' ;
   </script>
</html>
```

【相关说明】

```
<script>

   let links = document.links ;
```

此语句声明了变量 links，其初始数据为 document.links 返回的所有超链接（hyperlink）元素实例所构成的集合。

【<a href="#a03">To a03</a>】才算是代表超链接的 a 元素实例；然而，【<a name="a03">a03</a>】并不代表超链接，而是代表锚点（anchor）的 a 元素实例。

```
   console.log(links.length) ;
```

links.length 在此返回整数值 6，意味着在当前网页中，总共有 6 个超链接元素实例。

```
   links[2].accessKey = '3' ;
```

links[2]在此会返回【在变量 links 的集合中，其第 3 个（索引值为 2）】的超链接元素实例。此语句将字符'3'，赋给上述超链接元素实例的属性 accessKey，使得按下快捷键【Alt + 键盘上方的数字键 3】的效果，等价于单击在网页上带有【To a03】字样的上述超链接元素实例。

```
   links[3].accessKey = '4' ;
```

links[3]在此会返回【在变量 links 的集合中，其第 4 个（索引值为 3）】的超链接元素实例。此语句将字符'4'，赋给上述超链接元素实例的属性 accessKey，使得按下快捷键【Alt + 键盘上方的数字键 4】的效果，等价于单击在网页上带有【To a04】字样的上述超链接元素实例。

```
   links[4].accessKey = '5' ;
```

links[4]在此会返回【在变量 links 的集合中，其第 5 个（索引值为 4）】的超链接元素实例。此语句将字符'5'，赋给上述超链接元素实例的属性 accessKey，使得按下快捷键【Alt + **键盘上方**的数字键 5】的效果，等价于单击在网页上带有【To a05】字样的上述超链接元素实例。

```
   username.accessKey = 'u' ;
```

此语句将字符'u'，赋给属性 id 的数据为'username'的 input 元素实例的属性 accessKey，使得按下快捷键【Alt + u】的效果，等价于单击在网页上其属性 id 的数据为'username'的 input 元素实例。

```
   password.accessKey = 'p' ;
```

将字符'p'，赋给属性 id 的数据为'password 的 input 元素实例的属性 accessKey，使得按下快捷键【Alt + p】的效果，等价于单击在网页上其属性 id 的数据为'password'的 input 元素实例。

16.2　创建特定元素实例的动画效果（Web Animations）

通过【特定元素实例的 id】衔接函数【.animate()】的调用，可创建特定元素实例的动画效果。

请参看如下示例。

【16-2-^-DOM-animate.html】

```
<!DOCTYPE html>
<html>
  <head>
    <title>DOM animate()</title>
    <style>
      @keyframes text_color
      {
        from{color: Gold ;}
        to{color: RoyalBlue ;}
      }

      #div01
      {
        font-size: 2em ;
        animation: text_color 2s 3 ;
      }
    </style>
  </head>
  <body>
    <div id="div01">Hello</div>
  </body>
  <script>
    let frame01 = {transform: 'translateY(0px)'} ;
    let frame02 = {transform: 'translateY(20px)'} ;
    let keyframes = [frame01, frame02] ;
    let options = {duration: 1000, iterations: 4, direction: 'alternate', easing: 'ease-in-out'} ;

    div01.animate(keyframes, options) ;
  </script>
</html>
```

【相关说明】

```
<script>

  let frame01 = {transform: 'translateY(0px)'} ;
```

此语句声明变量 frame01，其初始数据为内含属性 transform 及其数据字符串'translateY(0px)' 的对象实例。在此，对象实例的属性 transform 是对应到 CSS 样式里的属性 transform，并代表着特定转变（transform）动画；而其属性 transform 的数据字符串'translateY(0px)'，代表 y 值为 0 像素的坐标位置，并且意味着特定元素实例的位移（translate）**起点**。

```
  let frame02 = {transform: 'translateY(20px)'} ;
```

此语句声明了变量 frame02，其初始数据为内含属性 transform 及其数据字符串'translateY(20px)' 的对象实例。在此，属性 transform 的数据字符串'translateY(20px)'，代表 y 值为 20 像素的坐标位置，并且意味着特定元素实例的位移**终点**。

```
  let keyframes = [frame01, frame02] ;
```

此语句声明了变量 keyframes，其初始数据为内含变量 frame01 和 frame02 分别代表的对象实例的数组实例。

```
let options = {duration: 1000, iterations: 4, direction: 'alternate', easing: 'ease-in-out'} ;
```

此语句声明了变量 options，其初始数据为对象实例{duration: 1000, iterations: 4, direction: 'alternate', easing: 'ease-in-out'}。其中：

- 属性 duration 的数值为 1000，用来设置动画的时间长度（duration）为 1000 毫秒 (1 秒)。
- 属性 iterations 的数值为 4，用来设置动画迭代（iteration）/ 播放的次数为 4。
- 属性 direction 的数据字符串为'alternate'，用来设置动画的方向（direction）为【从头到尾，再从尾逆向到头】的来回交替（alternate）。

```
div01.animate(keyframes, options) ;
```

此语句在属性 id 的数据为'div01'的 div 元素实例中，正式设置其动画成为【变量 keyframes 的数组实例和 options 的对象实例】所定义的模样。

16.3　添加新元素实例和访问特定元素实例的所有属性

通过【document.body】或【数据为特定元素实例的变量名称】，衔接函数【.appendChild()】的调用，可添加新元素实例，至当前网页中的 body 元素实例或者特定元素实例中，进而成为其内部的子元素实例。通过【数据为特定元素实例的变量名称】衔接属性【.attributes】，则可访问特定元素实例的各属性。请参看如下示例。

【16-3-^-DOM-appendChild-and-attributes.html】

```
<!DOCTYPE html>
<html>
 <head>
  <title>DOM appendChild() and attributes</title>
 </head>
 <body>
 </body>
 <script>
  let ref = document.createElement('h1') ;

  ref.innerHTML = 'Box World' ;

  with (ref.style)
  {
   color = 'RoyalBlue' ;
   textDecoration = 'underline' ;
  }

  document.body.appendChild(ref) ;
```

```
      ///
      ref = document.createElement('div') ;

      ref.innerHTML = 'Be diligent' ;
      ref.id = 'big_block' ;

      with (ref.style)
      {
        color = 'Gold' ;
        backgroundColor = 'RoyalBlue' ;
        width = height = '110px' ;
        textAlign = 'center' ;
        margin = 'auto' ;
        padding = '5px' ;
      }

      document.body.appendChild(ref) ;

      ///
      ref = document.createElement('h3') ;

      ref.innerHTML = 'Great' ;
      ref.id = 'small_block' ;

      with (ref.style)
      {
        color = 'GreenYellow' ;
        background = 'ForestGreen' ;
        width = height = '60px' ;
        textAlign = 'center' ;
        margin = '10px auto' ;
        borderRadius = '10px' ;
        padding = '5px' ;
      }

      big_block.appendChild(ref) ;

      ///
      console.log(small_block.attributes) ;
      console.log('') ;

      console.log(small_block.attributes.length) ;
      console.log('') ;

      console.log(small_block.attributes[0]) ;
      console.log(small_block.attributes[0].nodeValue) ;
      console.log('') ;

      console.log(small_block.attributes[1]) ;
      console.log(small_block.attributes[1].nodeValue) ;
    </script>
  </html>
```

【相关说明】

```
<script>
  let ref = document.createElement('h1') ;
```

此语句声明了变量 ref，其初始数据为 document.createElement('h1')返回的新的 h1 元素实例。

```
  ref.innerHTML = 'Box World' ;
```

此语句将字符串'Box World'，赋给变量 ref 当前代表的 h1 元素实例的属性 innerHTML，使得 h1 元素实例在网页上，呈现出文本'Box World'。

```
  with (ref.style)
  {
    color = 'RoyalBlue' ;
    textDecoration = 'underline' ;
  }
```

此语法等同于执行了【ref.style.color = 'RoyalBlue' ;】和【ref.style.textDecoration = 'underline' ;】。

```
  document.body.appendChild(ref) ;
```

此语句将变量 ref 当前代表的 h1 元素实例，新增至 body 元素实例中，正式呈现在网页上。

```
  ref = document.createElement('div') ;
```

此语句将 document.createElement('div')所返回的新的 div 元素实例，赋给变量 ref。

```
  ref.innerHTML = 'Be diligent' ;
```

此语句将字符串'Be diligent'，赋给变量 ref 当前代表的 div 元素实例的属性 innerHTML，使得 div 元素实例在网页上，呈现出文本'Be diligent'。

```
  ref.id = 'big_block' ;
```

此语句将字符串'big_block'，赋给变量 ref 当前代表的 div 元素实例的属性 id，使得此 div 元素实例的属性 id 的数据，成为'big_block'。

```
  with (ref.style)
  {
    color = 'Gold' ;
    backgroundColor = 'RoyalBlue' ;
    width = height = '110px' ;
    textAlign = 'center' ;
    margin = 'auto' ;
    padding = '5px' ;
  }
```

此语法简化了类似【ref.style.color = 'Gold' ;】等语句，成为在大括号里的【color = 'Gold' ;】等语句，以设置变量 ref 代表的 div 元素实例，呈现在网页上的各个外观。

```
  document.body.appendChild(ref) ;
```

此语句将变量 ref 当前代表的 div 元素实例，新增至 body 元素实例中，以正式呈现在网页上。

```
  ref = document.createElement('h3') ;
```

此语句将 document.createElement('h3')所返回的新的 h3 元素实例，赋给变量 ref。

```
    ref.innerHTML = 'Great' ;
```

此语句将字符串'Great',赋给变量 ref 当前代表的 h3 元素实例的属性 innerHTML, 使得 h3 元素实例在网页上,呈现出文本'Great'。

```
    ref.id = 'small_block' ;
```

此语句将字符串'small_block',赋给变量 ref 当前代表的 h3 元素实例的属性 id, 使得此 h3 元素实例的属性 id 的数据,成为'small_block'。

```
    with (ref.style)
    {
      color = 'GreenYellow' ;
      background = 'ForestGreen' ;
      width = height = '60px' ;
      textAlign = 'center' ;
      margin = '10px auto' ;
      borderRadius = '10px' ;
      padding = '5px' ;
    }
```

此语法简化了类似【ref.style.color = 'GreenYellow' ;】等语句,成为大括号里【color = 'GreenYellow' ;】等语句,以设置变量 ref 当前代表的 h3 元素实例,在网页上的各个外观。

```
    big_block.appendChild(ref) ;
```

此语句将变量 ref 当前代表的 h3 元素实例,新增至属性 id 的数据为'big_block'的 div 元素实例中,以正式呈现在上述 div 元素实例的范围里。

```
    console.log(small_block.attributes) ;
```

此语句显示出【NamedNodeMap {0: id, 1: style, id: id, style: style, length: 2}】的信息,意味着 small_block.attributes 返回的,是属性 id 的数据为'small_block'的 div 元素实例中的 NamedNodeMap 对象实例。

```
    console.log('') ;
    console.log(small_block.attributes.length) ;
```

small_block.attributes.length 当前返回整数值 2, 意味着在属性 id 的数据为'small_block'的 div 元素实例中,存在两个可对应至 HTML 语法的属性(attribute),也就是属性 id 与 style。

包括 JavaScript 语言的属性 innerHTML 和其他未在本示例中被设置的属性,因为无法对应到特定 HTML 语言的属性,所以并不包含在 small_block.attributes.length 中的属性个数的计算行列中!

```
    console.log('') ;
    console.log(small_block.attributes[0]) ;
```

显示出【id="small_block"】的信息。

```
    console.log(small_block.attributes[0].nodeValue) ;
```

small_block.attributes[0].nodeValue 当前返回字符串'small_block'。其中,nodeValue 的数据,即是代表特定属性的节点(node)实例的值(value)。在此,其 nodeValue 的数据,是属性 id 的数

据字符串'small_block'。

```
console.log('') ;
```

```
console.log(small_block.attributes[1]) ;
```

显示出【style="color: greenyellow; background: forestgreen; height: 60px; width: 60px; text-align: center; margin: 10px auto; border-radius: 10px; padding: 5px;"】的信息。

```
console.log(small_block.attributes[1].nodeValue) ;
```

small_block.attributes[1].nodeValue 当前返回字符串'color: greenyellow; background: forestgreen; height: 60px; width: 60px; text-align: center; margin: 10px auto; border-radius: 10px; padding: 5px;'。

在此，nodeValue 的数据，是属性 style 的数据字符串'color: greenyellow; background: forestgreen; height: 60px; width: 60px; text-align: center; margin: 10px auto; border-radius: 10px; padding: 5px;'。

16.4　使得特定元素实例失去和获取焦点

通过【数据为特定元素实例的变量名称】衔接函数【.focus()】的调用，可使得特定元素实例获取焦点（focus）；通过【数据为特定元素实例的变量名称】衔接函数【.blur()】的调用，则可使得特定元素实例失去焦点。请参看如下示例。

【16-4-^-DOM-blur-and-focus.html】

```html
<!DOCTYPE html>
<html>
  <head>
    <title>DOM blur() and focus()</title>
    <style>
      a, [type=text], [type=button]
      {
        margin: 5px ;
      }
    </style>
  </head>
  <body align="center">
    <a href="#">Link 1</a>
    <a href="#">Link 2</a>
    <a href="#">Link 3</a>
    <a href="#">Link 4</a>
    <a href="#">Link 5</a>

    <p></p>
    <input type="text">
    <input type="text">
    <input type="text">

    <p></p>
    <button type="button">Button 1</button>
    <button type="button">Button 2</button>
```

```
    <button type="button">Button 3</button>
 </body>
 <script>
   let refs = document.querySelectorAll('a, [type=text], button') ;
   let index = 0 ;

   console.log(refs.length) ;

   ///
   let interval_no = setInterval(focus_next, 500) ;

   function focus_next()
   {
    refs[index].focus() ;

    index = index < refs.length - 1 ? index + 1 : 0 ;
   }

   ///
   setTimeout(blur_all, 7000) ;

   function blur_all()
   {
    clearInterval(interval_no) ;

    refs[index - 1].blur() ;
   }
 </script>
</html>
```

【相关说明】

```
 <script>

   let refs = document.querySelectorAll('a, [type=text], button') ;
```

此语句声明了变量 refs，其初始数据为 document.querySelectorAll('a, [type=text], button')所返回的多个元素实例构成的集合。被传入函数 document.querySelectorAll('a, [type=text], button')的参数数据字符串'a, [type=text], button'，是 CSS 选择器名称（selector name）的字符串，在此代表选择了【在当前网页中，元素/标签名称为'a'、属性 type 为'text'，或者元素/标签名称为 button】的所有元素实例。

```
   let index = 0 ;
```

此语句声明了初始值为 0 的变量 index。

```
   console.log(refs.length) ;
```

refs.length 返回整数值 11，意味着在变量 refs 的集合中，内含 5 个作为超链接的元素实例、3 个作为文本字段的 input 元素实例，以及 3 个作为按钮的 button 元素实例，共计 11 个元素实例。

```
   let interval_no = setInterval(focus_next, 500) ;
```

此语句声明了变量 interval_no，其初始数据为 setInterval(focus_next, 500)所返回的整数值（例如 2）。此整数值代表【间接定时调用其名称为 focus_next 的函数】的动作编号。在此语句的等号右侧，调用而执行了内置的全局函数 setInterval(focus_next, 500)，进而每隔 500 毫秒（半秒），就间接调用其名称为 focus_next 的函数一次。

```
function focus_next()
{
    refs[index].focus() ;
```

此语句使得变量 refs 中的第 index + 1 个（索引值为变量 index 的数值）的元素实例，在网页上，获得焦点而呈现出其动态外框。

```
    index = index < refs.length - 1 ? index + 1 : 0 ;
```

此语句使得变量 index 的数值，介于 0 ~ 10 之间。

```
}
```

上述语法定义了函数 focus_next()。

```
setTimeout(blur_all, 7000) ;
```

此语句调用了内置的全局函数 setTimeout()，并在执行此语句的 7000 毫秒（7 秒）之后，再间接调用其名称为 blur_all 的函数。

```
function blur_all()
{
    clearInterval(interval_no) ;
```

此语句调用了内置的全局函数 clearInterval()，进而移除【动作编号为变量 interval_no 的数值】的定时调用动作。

```
    refs[index - 1].blur() ;
```

此语句使得变量 refs 中的第 index - 1 + 1 个（索引值为变量 index 的数值，再减去 1）的元素实例，在网页上【失去】焦点而被去除其动态外框。

```
}
```

上述语法定义了函数 blur_all()。

16.5　访问子节点或子元素的实例

通过【数据为特定元素实例的变量名称】：

- 衔接【.childNodes】，可访问特定元素实例的子节点实例。
- 衔接【.children】，可访问特定元素实例内的子元素实例。
- 衔接【.childElementCount】，可获取特定元素实例内的子元素实例的个数。

关于访问子节点或子元素的实例的综合运用,请参看如下示例。

【16-5-^-child-nodes-and-elements.html】

```html
<!DOCTYPE html>
<html>
  <head>
    <title>child nodes and elements</title>
  </head>
  <body>
    <h1>Hope List</h1>

    <div id="div01">
      <a href="#">Link 1</a>
      <a href="#">Link 2</a>
      <a href="#">Link 3</a>
    </div>

    <div id="div02">
      <ul>
        <li>Item 1</li>
        <li>Item 2</li>
        <li>Item 3</li>
      </ul>

      <ol>
        <li>Step A</li>
        <li>Step B</li>
        <li>Step C</li>
      </ol>
    </div>

  </body>
  <script>
    with (console)
    {
      log(document.body.childElementCount) ;
      log(div01.childElementCount) ;
      log(div02.childElementCount) ;
      log('') ;

      log(document.body.childNodes) ;
      log(div01.childNodes) ;
      log(div02.childNodes) ;
      log('') ;

      log(document.body.children) ;
      log(div01.children) ;
      log(div02.children) ;
      log('') ;

      ///
      // div01.children.item(0).style.textDecoration = 'overline' ;
      div01.children[0].style.textDecoration = 'overline' ;
```

```
      // div01.children.item(1).style.textDecoration = 'line-through' ;
      div01.children[1].style.textDecoration = 'line-through' ;
    }
  </script>
</html>
```

【相关说明】

```
<script>
  with (console)
  {

    log(document.body.childElementCount) ;
```

document.body.childElementCount 会返回 body 元素实例中的子元素（child element）实例的个数 4。

```
    log(div01.childElementCount) ;
```

div01.childElementCount 会返回【在属性 id 的数据为'div01'的 div 元素实例中，其子元素实例】的个数 3。

```
    log(div02.childElementCount) ;
```

div02.childElementCount 会返回【在属性 id 的数据为'div02'的 div 元素实例中，其子元素实例】的个数 2。

```
    log('') ;

    log(document.body.childNodes) ;
```

此语句显示出可展开的【NodeList(8)】的信息，代表 body 元素实例内含 8 个节点实例。document.body.childNodes 会返回 NodeList(节点表)对象实例[text, h1, text, div#div01, text, div#div02, text, script, text]。其中，text 意味着其为文本节点实例；而 script 意味着其为原本被放在 body 元素实例下方的 script 元素实例，并事后被浏览器动态搬移到 body 元素实例内。

在 body 元素实例的起始语句【<body>】到其内部第 1 个子元素实例的起始语句【<h1>】之间，所存在的换行（line feed）和空格（space）字符，构成了 1 个文本节点实例。同样的，第 1 个子元素实例的结束语句【</h1>】到第 2 个子元素实例的开始语句【<div id="div01">】之间，所存在的换行和空格字符，构成了第 2 个文本节点实例，以此类推。

```
    log(div01.childNodes) ;
```

此语句显示出可展开的【NodeList(7)】的信息，意味着在属性 id 的数据为'div01'的 div 元素实例中，存在 7 个节点实例：其中 3 个是元素实例，另外 4 个为【介于元素实例之间】的文本节点实例。

```
    log(div02.childNodes) ;
```

此语句显示出可展开的【NodeList(5)】的信息，意味着在属性 id 的数据为'div02'的 div 元素实例中，存在 5 个节点实例：其中两个是元素实例，另外 3 个是【介于元素实例之间】的文本节点实例。

```
      log('') ;

      log(document.body.children) ;
```

此语句显示出可展开的【HTMLCollection(4)】的信息,意味着在 body 元素实例中,存在 4 个子元素实例。其中,包含 1 个 h1 元素实例、2 个 div 元素实例,以及 1 个 script 元素实例。

```
      log(div01.children) ;
```

此语句显示出可展开的【HTMLCollection(3)】的信息,意味着在属性 id 的数据为'div01'的 div 元素实例中,存在 3 个 a 子元素实例。

```
      log(div02.children) ;
```

此语句显示出可展开的【HTMLCollection(2)】的信息,意味着在属性 id 的数据为'div02'的 div 元素实例中,存在 2 个子元素实例。其中,包含 1 个 ul 元素实例和 1 个 ol 元素实例。

```
      log('') ;

      // div01.children.item(0).style.textDecoration = 'overline' ;
      div01.children[0].style.textDecoration = 'overline' ;
```

div01.children[0]可返回【在属性 id 的数据为'div01'的 div 元素实例中,其第 1 个(索引值为 0)】的 a 元素实例。此语句可设置上述 a 元素实例,具有上画线(overline)的文本装饰(text decoration)。

```
      // div01.children.item(1).style.textDecoration = 'line-through' ;
      div01.children[1].style.textDecoration = 'line-through' ;
```

div01.children[1]可返回【在属性 id 的数据为'div01'的 div 元素实例中,其第 2 个(索引值为 1)】的 a 元素实例。此语句可设置上述 a 元素实例,具有删除线(line-through)的文本装饰(text decoration)。

16.6 访问被应用在特定元素实例的所有 CSS 类名

通过【数据为特定元素实例的变量名称】衔接【.classList】或【.className】,可访问被应用在特定元素实例上的所有 CSS 类名。

通过【数据为特定元素实例的变量名称】衔接函数【.toggle()】的调用,可切换(toggle)被应用在特定元素实例上的 CSS 类名的**存在性**。

关于其综合运用,请参看如下示例。

【16-6-^-DOM-classList-and-className.html】

```html
<!DOCTYPE html>
<html>
  <head>
    <title>DOM classList and className</title>
    <style>
      .color
      {
```

```
      color: GoldenRod ;
      background-color: Gold ;
    }

    .size
    {
      font-size: 0.9em ;
      width: 80px ;
      height: 40px ;
    }

    .alignment
    {
      text-align: center ;
      margin: 10px auto ;
      padding: 3px ;
    }

    .border
    {
      border: 2px Chocolate dashed ;
      border-radius: 5px ;
    }
  </style>
</head>
<body>
  <div id="div01">Good</div>
  <div id="div02">Nice</div>
  <div id="div03">Great</div>
  <div id="div04">Fantastic</div>
  <div id="div05">Tested Block</div>
</body>
<script>
  div01.className = 'color' ;
  div02.className = 'color size' ;
  div03.className = 'color size alignment' ;
  div04.className = 'color size alignment border' ;

  console.log(div01.className) ;
  console.log(div02.className) ;
  console.log(div03.className) ;
  console.log(div04.className) ;
  console.log('') ;

  ///
  div05.className = 'color' ;
  div05.classList.add('size') ;
  div05.classList.add('alignment','border') ;
  div05.classList.remove('color','size') ;

  console.log(div05.classList.contains('size')) ;
  console.log(div05.classList.contains('border')) ;
  console.log('') ;

  ///
```

```
        div05.classList.toggle('size') ;
        div05.classList.toggle('border') ;

        console.log(div05.className) ;
        console.log(div05.classList) ;
        console.log(div05.classList.length) ;
    </script>
</html>
```

【相关说明】

```
<script>

    div01.className = 'color' ;
```

此语句将字符串'color',赋给属性id 的数据为'div01'的 div 元素实例的属性 className。此语句被执行之后，会使得上述 div 元素实例，从【<div id="div01">Good</div>】变成【<div id="div01"**class="color"**>Good</div>】。

请留意，在 JavaScript 源代码中，名称为 className 的属性（property），对应到在 HTML 源代码里，其名称为 class 的属性（attribute）。

其中，【class="color"】的语法，会让上述 div 元素实例，被应用 CSS 类名（class name）为 color 的 CSS 类样式（class style），使得其字体颜色，成为黄花（golden rod）的颜色，而其背景颜色，成为金色（gold）。

```
    div02.className = 'color size' ;
```

此语句将字符串'color size',赋给属性 id 的数据为'div02'的 div 元素实例的属性 className。此语句被执行之后，会使得上述 div 元素实例，从【<div id="div02">Nice</div>】变成【<div id="div02"**class="color size"**>Nice</div>】。其中，【class="color size"】的语法，会让上述 div 元素实例，同时被应用 CSS 类名为 color 和 size 的 CSS 类样式。

```
    div03.className = 'color size alignment' ;
```

此语句将字符串'color size alignment',赋给属性 id 的数据为'div03'的 div 元素实例的属性 className。此语句被执行之后，会使得上述 div 元素实例，从【<div id="div03">Great</div>】变成【<div id="div03"**class="color size alignment"**>Great</div>】。其中，【class="color size alignment"】的语法，会让上述 div 元素实例，同时被应用 CSS 类名为 color、size 与 alignment 的 CSS 类样式。

```
    div04.className = 'color size alignment border' ;
```

此语句将字符串'color size alignment border',赋给属性 id 的数据为'div04'的 div 元素实例的属性 className。此语句被执行之后，会使得上述 div 元素实例，从【<div id="div04">Fantastic</div>】变成【<div id="div04"**class="color size alignment border"**>Fantastic</div>】。其中，【class="color size alignment border"】的语法,会让上述 div 元素实例,同时被应用 CSS 类名为 color、size、alignment 与 border 的 CSS 类样式。

```
    console.log(div01.className) ;
```

div01.className 返回字符串'color'。

```
console.log(div02.className) ;
```

div02.className 返回字符串'color size'。

```
console.log(div03.className) ;
```

div03.className 返回字符串'color size alignment'。

```
console.log(div04.className) ;
```

div04.className 返回字符串'color size alignment border'。

```
console.log('') ;

div05.className = 'color' ;
```

此语句将字符串'color'，赋给属性 id 的数据为'div05'的 div 元素实例的属性 className，使得上述 div 元素实例，变成【<div id="div05"**class="color"**>Tested Block</div>】。

```
div05.classList.add('size') ;
```

此语句将字符串'size'，新增至属性 id 的数据为'div05'的 div 元素实例的 CSS 类列表（class list）里，使得上述 div 元素实例，变成【<div id="div05" class="color **size**">Tested Block</div>】。

```
div05.classList.add('alignment','border') ;
```

此语句将字符串'alignment'与'border'，新增至属性 id 的数据为'div05'的 div 元素实例的 CSS 类列表里，使得上述 div 元素实例，变成【<div id="div05" class="color size **alignment border**">Tested Block</div>】。

```
div05.classList.remove('color','size') ;
```

此语句在属性 id 的数据为'div05'的 div 元素实例的 CSS 类列表中，移除了 CSS 类名 color 与 size，使得上述 div 元素实例，变成【<div id="div05"**class="alignment border"**>Tested Block</div>】。

```
console.log(div05.classList.contains('size')) ;
```

div05.classList.contains('size')返回布尔值 false，意味着在属性 id 的数据为'div05'的 div 元素实例的 CSS 类列表'alignment border'里，当前并不包含（contain） CSS 类名 size。

```
console.log(div05.classList.contains('border')) ;
```

div05.classList.contains('border')返回布尔值 true，意味着在属性 id 的数据为'div05'的 div 元素实例的 CSS 类列表'alignment border'里，当前包含 CSS 类名 border。

```
console.log('') ;

div05.classList.toggle('size') ;
```

此语句在属性 id 的数据为'div05'的 div 元素实例的 CSS 类列表里，进行了【包含／不包含】CSS 类名 size 的切换（toggle）。因为在上述 div 元素实例当前的 CSS 类列表'alignment border'里，并不包含 CSS 类名 size，所以会被切换成为包含 CSS 类名 size，使得上述 div 元素实例，变成【<div id="div05" class="alignment border **size**">Tested Block</div>】。

```
div05.classList.toggle('border') ;
```

此语句在属性 id 的数据为'div05'的 div 元素实例的 CSS 类列表里，进行了【包含／不包含】CSS 类名 border 的切换。因为在上述 div 元素实例当前的 CSS 类列表'alignment border size'里，包含了 CSS 类名 border，所以会被切换成为不包含 CSS 类名 border，使得上述 div 元素实例，变成【<div id="div05" **class**="alignment size">Tested Block</div>】。

```
console.log(div05.className) ;
```

div05.className 在此返回字符串'alignment size'。

```
console.log(div05.classList) ;
```

显示出可展开的【DOMTokenList(2) ["alignment", "size", value: "alignment size"]】的信息。

```
console.log(div05.classList.length) ;
```

div05.classList.length 在此返回整数值 2，意味着在属性 id 的数据为'div05'的 div 元素实例的 CSS 类列表里，当前内含 2 个 CSS 类名，也就是 alignment 与 size。

16.7　模拟鼠标单击特定元素实例的动作

通过【数据为特定元素实例的变量名称】衔接函数【.click()】的调用，可模拟用户借助鼠标，以单击特定元素实例的动作。请参看如下示例。

【16-7-^-DOM-click.html】

```
<!DOCTYPE html>
<html>
  <head>
    <title>DOM click()</title>
    <style>
      input, button, select
      {
        font-size: 1.2em ;
        margin: 5px ;
      }
    </style>
  </head>
  <body>
    <form id="form01" name="form01" style="text-align: center;">
      <h3>个人资料</h3>
      <input type="text" id="username" name="username" placeholder="username" size="16"
        required>
      <input type="password" id="password" name="password" placeholder="password" size="16"
        required>
      <p></p>
      <label>I am not a robot</label>
      <input type="checkbox" id="check01" onmouseover="" onclick="">

      <p></p>
      <button type="submit">Login</button>
```

```
      <button type="reset">Reset</button>
    </form>
  </body>
  <script>
    setTimeout(() => check01.click(), 2000) ;

    check01.onclick = () => console.log('check01 is automatically checked.') ;
  </script>
</html>
```

【相关说明】

```
  <script>

    setTimeout(() => check01.click(), 2000) ;
```

此语句设置了在 2000 毫秒（2 秒）之后，间接调用箭头函数【() => check01.click()】。上述箭头函数被调用时，会间接调用 check01.click()，如同单击（click）了属性 id 的数据为'check01'的 input 元素实例一样，使得上述 input 元素实例所代表的复选框（checkbox）被选中。

```
    check01.onclick = () => console.log('check01 is clicked to be checked.') ;
```

此语句将等号右侧的箭头函数的定义，赋给了属性 id 的数据为'check01'的 input 元素实例的属性 onlick，使得上述 input 元素实例，被单击时，即会调用上述箭头函数，并显示出【check01 is automatically checked.】的信息。

16.8　获取特定元素的尺寸、坐标与可定位的上层元素

通过【数据为特定元素实例的变量名称】：

- 衔接【.clientWidth】、【.clientHeight】、【.offsetWidth】和【.offsetHeight】，可获取特定元素实例的尺寸相关数据。
- 衔接【.clientLeft】、【.clientTop】、【.offsetLeft】和【.offsetTop】，可获取特定元素实例的坐标相关数据。
- 衔接【.offsetParent】，可获取可定位的上层元素实例。

关于其调试，请参看如下示例。

【16-8-^-DOM-coordinates-and-offsetParent.html】

```
<!DOCTYPE html>
<html>
  <head>
    <title>DOM coordinates and offsetParent</title>
    <style>
      #div01
      {
        color: Chocolate ;
        text-align: center ;
```

```
            background-color: Gold ;
            padding: 10px ;
            width: 100px ;
            height: 60px ;
            margin: 50px auto ;
            border: 8px RoyalBlue solid ;
        }
    </style>
</head>
<body>
    <div id="div01">
      Grateful
    </div>
</body>
<script>
    console.log(div01.clientWidth) ;
    console.log(div01.offsetWidth) ;
    console.log('') ;

    console.log(div01.clientHeight) ;
    console.log(div01.offsetHeight) ;
    console.log('') ;

    console.log(div01.clientLeft) ;
    console.log(div01.offsetLeft) ;
    console.log('') ;

    console.log(div01.clientTop) ;
    console.log(div01.offsetTop) ;
    console.log('') ;

    // console.log(div01.parentElement.tagName) ;
    console.log(div01.offsetParent.tagName) ;

    div01.offsetParent.style.backgroundColor = 'YellowGreen' ;
</script>
</html>
```

【相关说明】

```
<script>

    console.log(div01.clientWidth) ;
```

div01.clientWidth 会返回【在属性 id 的数据为'div01'的 div 元素实例中，其客户区域宽度（client width）】的像素个数 120。所谓的客户区域宽度，并不包含滚动条（scroll bar）、边框（border）和外边距（margin）的宽度，但是包含内边距（padding）的宽度。

```
    console.log(div01.offsetWidth) ;
```

div01.offsetWidth 会返回【在属性 id 的数据为'div01'的 div 元素实例中，其偏移宽度（offset width）】的像素个数 136。所谓的偏移宽度，是指【客户区域宽度】加上【边框的宽度】。

```
    console.log('') ;
```

```
console.log(div01.clientHeight) ;
```

div01.clientHeight 会返回【在属性 id 的数据为'div01'的 div 元素实例中,其客户区域高度(client height)】的像素个数 80。所谓的客户区域高度,并不包含滚动条、边框和外边距的高度,但是包含内边距(padding)的高度。

```
console.log(div01.offsetHeight) ;
```

div01.offsetHeight 会返回【在属性 id 的数据为'div01'的 div 元素实例中,其偏移高度(offset height)】的像素个数 96。所谓的偏移高度,是指【客户区域高度】加上【边框的高度】。

```
console.log('') ;
console.log(div01.clientLeft) ;
```

div01.clientLeft 会返回【在属性 id 的数据为'div01'的 div 元素实例中,其客户区域的左上顶点】的坐标 x 值。在此,将上述 div 元素实例的边框外侧的左上顶点,作为坐标原点,以评估出在上述 div 元素实例的客户区域里,其左上顶点的坐标 x 值。

```
console.log(div01.offsetLeft) ;
```

div01.offsetLeft 会返回【在属性 id 的数据为'div01'的 div 元素实例中,其客户区域的左上顶点】的偏移(offset)坐标 x 值。在此,将浏览器窗口的左上顶点,作为坐标原点,以评估出在上述 div 元素实例的边框外侧中,其左上顶点的偏移坐标 x 值。

```
console.log('') ;
console.log(div01.clientTop) ;
```

div01.clientTop 会返回【在属性 id 的数据为'div01'的 div 元素实例中,其客户区域的左上顶点】的坐标 y 值。在此,将上述 div 元素实例的边框外侧的左上顶点,作为坐标原点,以评估出在上述 div 元素实例的客户区域里,其左上顶点的坐标 y 值。

```
console.log(div01.offsetTop) ;
```

div01.offsetTop 会返回【在属性 id 的数据为'div01'的 div 元素实例中,其客户区域的左上顶点】的偏移坐标 y 值。在此,将浏览器窗口的左上顶点,作为坐标原点,以评估出在上述 div 元素实例的边框外侧中,其左上顶点的偏移坐标 y 值。

```
console.log('') ;
// console.log(div01.parentElement.tagName) ;
console.log(div01.offsetParent.tagName) ;
```

div01.offsetParent.tagName 与 div01.parentElement.tagName 有着相同的效果,均会返回【在属性 id 的数据为'div01'的 div 元素实例的上层元素实例中,其代表元素名称 / 标签名称(tag name)】的字符串'BODY'。

对于当前元素实例的父元素实例而言,若其 style.display 的数据为字符串'static',则其属性 parentElement 并不是返回当前元素实例的父元素实例,而是返回【其 style.display 的数据,并非字符串'static'】的更上层的元素实例。

```
div01.offsetParent.style.backgroundColor = 'YellowGreen' ;
```

此语句间接设置了 body 元素实例的背景颜色，成为黄绿色（yellow green）。

16.9　比较两个元素之间的位置关系

通过【其数据为特定元素实例的变量名称】衔接函数【.compareDocumentPosition(数据为另一元素实例的变量名称)】的调用，可在当前网页及其 iframe 元素实例中的子网页里，比较两个特定元素实例之间的**位置关系**。请参看如下带有两个源代码文档的示例。请留意，本示例的源代码文档，必须通过 HTTP 协议，才能调试出其正确的结果。

【16-9-^-DOM-compareDocumentPosition.html】

```html
<!DOCTYPE html>
<html>
  <head>
    <title>DOM compareDocumentPosition()</title>
    <style>
      div
      {
        color: DodgerBlue ;
        background-color: Pink ;
        width: 50px ;
        height: 50px ;
        padding: 5px ;
        text-align: center ;
        margin: 5px auto ;
        display: inline-block ;
      }
    </style>
  </head>
  <body align="center">
    <iframe id="iframe01" src="16-9-^-test-page.html" width="100%" height="70" style="border:
      none"></iframe>

    <div id="div01">1st</div>
    <div id="div02">2nd</div>
    <div id="div03">3rd</div>
    <div id="div04">4th</div>
    <div id="div05">5th</div>
  </body>
  <script>
    let result = false, ref = null ;

    result = div01.compareDocumentPosition(div03) ;

    console.log(result) ;
    console.log('') ;

    result = div03.compareDocumentPosition(div01) ;

    console.log(result) ;
```

```
      console.log('') ;

      ///
      document.onreadystatechange = function (event)
      {
        switch (document.readyState)
        {
          case 'complete':
            ref = iframe01.contentDocument.getElementById('span01') ;

            result = div01.compareDocumentPosition(ref) ;

            console.log(result) ;
            break ;
        }
      } ;
    </script>
</html>
```

【相关说明】

```
    <script>

      let result = false, ref = null ;
```

此语句声明了初始数据为布尔值 false 的变量 result，以及初始数据为空值（null）的变量 ref。

```
      result = div01.compareDocumentPosition(div03) ;
```

div01.compareDocumentPosition(div03)当前返回整数值 4，意味着属性 id 的数据为'div03'的 div 元素实例，存在于属性 id 的数据为'div01'的 div 元素实例的后方。

通过【元素实例 1 的 id】衔接函数【.compareDocumentPosition(元素实例 2 的 id 字符串)】的调用，可返回如下各整数值的组合：

- 1：意味着两个元素实例，不在相同的网页文档里。
- 2：意味着元素实例 2，是在元素实例 1 的前方。
- 4：意味着元素实例 2，是在元素实例 1 的后方。
- 8：意味着元素实例 2，是元素实例 1 的上层（祖先）元素实例。
- 16：意味着元素实例 2，是元素实例 1 的内层（后代）元素实例。
- 32：意味着两个元素实例之间的关系，是被动态实现的。

```
      console.log(result) ;
```

显示出整数值 4 的信息。

```
      console.log('') ;

      result = div03.compareDocumentPosition(div01) ;
```

div03.compareDocumentPosition(div01)当前返回整数值 2，意味着属性 id 的数据为'div01'的 div 元素实例，存在于属性 id 的数据为'div03'的 div 元素实例的前方。

```
      console.log(result) ;
```

显示出整数值 2 的信息。

```
console.log('') ;

document.onreadystatechange = function (event)
{
  switch (document.readyState)
  {
    case 'complete':

      ref = iframe01.contentDocument.getElementById('span01') ;
```

在属性 id 的数据为'iframe01'的 iframe 元素实例的范围里，当前被呈现出外部网页 16-9-^-test-page.html 的内容。在此，iframe01.contentDocument.getElementById('span01')会返回【在外部网页中，其属性 id 的数据为'span01'】的 span 元素实例，并赋给变量 ref。

```
      result = div01.compareDocumentPosition(ref) ;
```

div01.compareDocumentPosition(ref)当前返回整数值 37，也就是 1 + 4 + 32 的总和，因此同时意味着以下状况：

- 属性 id 的数据为'div01'的 div 元素实例，和变量 ref 所代表的属性 id 的数据为'span01'的 span 元素实例，并不存在于相同的网页文档里。
- 变量 ref 所代表的属性 id 的数据为'span01'的 span 元素实例，存在于属性 id 的数据为'div01' 的 div 元素实例的后方。
- 属性 id 的数据为'div01'的 div 元素实例，和变量 ref 所代表的属性 id 的数据为'span01'的 span 元素实例之间的关系，是被动态实现的。

```
      console.log(result) ;
```

显示出整数值 37 的信息。

```
      break ;
  }
} ;
```

上述语法将等号右侧的匿名函数的定义，赋给了代表网页本身的 document 对象实例的属性 onreadystatechange，使得网页当前的加载状态，一旦有变化，则上述匿名函数就会被调用。

【16-9-^-test-page.html】

```
<!DOCTYPE html>
<html>
  <head>
    <title>test page</title>
    <style>
      span
      {
        color: LightGreen ;
        background-color: DodgerBlue ;
        width: 40px ;
        height: 30px ;
```

```
      padding: 5px ;
      border-radius: 5px ;
      text-align: center ;
      margin: 5px auto ;
      display: inline-block ;
    }
  </style>
 </head>
 <body align="center">
  <span id="span01">one</span>
  <span id="span02">two</span>
  <span id="span03">three</span>
  <span id="span04">four</span>
  <span id="span05">five</span>
 </body>
</html>
```

【相关说明】

```
<body align="center">

  <span id="span01">one</span>
  <span id="span02">two</span>
  <span id="span03">three</span>
  <span id="span04">four</span>
  <span id="span05">five</span>
```

属性 id 的数据分别为 'span01' ~ 'span05' 的这些 span 元素实例，是位于网页文档【16-9-^-test-page.html】里。

16.10　判断是否存在特定子元素或可被编辑

通过【document.body】或【数据为特定元素实例的变量名称】衔接函数【.contains()】的调用，可用来判断 body 元素实例或者特定元素实例，是否内含特定子元素实例。而通过【数据为特定元素实例的变量名称】：

- 衔接【.isContentEditable】，可判断特定元素实例，是否可被编辑 (editable)。
- 衔接【.contentEditable】，可动态设置特定元素实例的可被编辑的状态。

对于相关的综合运用，请参看如下示例。

【16-10-^-DOM-contains-and-isContentEditable.html】

```
<!DOCTYPE html>
<html>
  <head>
    <title>DOM contains() and isContentEditable</title>
    <style>
      div
      {
```

```
      text-align: center ;
      color: IndianRed ;
      background-color: Pink ;
      width: 80px ;
      height: 80px ;
      margin: 10px ;
    }

    h3
    {
      line-height: 1em ;
    }

    a
    {
      text-decoration: none ;
      color: Yellow ;
    }
  </style>
</head>
<body>
  <div id="div01">
    <p id="p01">Super</p>
    <a href="#" id="a01">happy time</a>
  </div>
</body>
<script>
  console.log(document.body.contains(div01)) ;
  console.log(document.body.contains(p01)) ;
  console.log(div01.contains(a01)) ;
  console.log('') ;

  console.log(p01.contains(a01)) ;
  console.log('') ;

  ///
  console.log(p01.isContentEditable) ;

  p01.contentEditable = true ;
</script>
</html>
```

【相关说明】

```
<script>

  console.log(document.body.contains(div01)) ;
```

document.body.contains(div01)当前返回布尔值 true，意味着在 body 元素实例中，存在属性 id 的数据为'div01'的 div 元素实例。

```
  console.log(document.body.contains(p01)) ;
```

document.body.contains(p01)当前返回布尔值 true，意味着在 body 元素实例中，存在属性 id 的数据为'p01'的 p 元素实例。

```
console.log(div01.contains(a01)) ;
```

div01.contains(a01)当前返回布尔值 true，意味着在属性 id 的数据为'div01'的 div 元素实例中，存在属性 id 的数据为'a01'的 a 元素实例。

```
console.log('') ;
console.log(p01.contains(a01)) ;
```

p01.contains(a01)当前返回布尔值 false，意味着在属性 id 的数据为'p01'的 p 元素实例中，并不存在属性 id 的数据为'a01'的 a 元素实例。

```
console.log('') ;
console.log(p01.isContentEditable) ;
```

属性 p01 的 p 元素实例，默认为不可直接在网页上，被编辑的元素实例；所以 p01.isContentEditable 的默认值为布尔值 false。

```
p01.contentEditable = true ;
```

此语句将布尔值 true，赋给了属性 id 的数据为'p01'的 p 元素实例的属性 contentEditable，使得上述 p 元素实例被开放了【在网页上的其范围内，用户可直接进行编辑】的功能。换言之，读者可在此示例网页中，单击 p 元素实例内的文本 Super；待键盘光标出现之后，将原始文本，修改成为新的文本。

16.11 访问特定元素实例的文本被书写的方向

通过【数据为特定元素实例的变量名称】衔接【.dir】，可获取或设置特定元素实例上的文本被书写的方向（direction）。请参看如下示例。

【16-11-^-DOM-dir.html】

```html
<!DOCTYPE html>
<html>
  <head>
    <title>DOM dir</title>
    <style>
      #p01
      {
        color: RoyalBlue ;
        background-color: Gold ;
      }
    </style>
  </head>
  <body>
    <p id="p01">
      To live happily ever after...
    </p>
  </body>
```

```
<script>
  // p01.style.direction = 'rtl';
  p01.dir = 'rtl';
</script>
</html>
```

【相关说明】

```
<script>

  // p01.style.direction = 'rtl';
  p01.dir = 'rtl';
```

此语句将字符串'rtl'，赋给了属性id 的数据为'p01'的 p 元素实例的属性 dir，使得上述 p 元素实例上的文本，变更了显示的方向，成为【从右至左 (RTL, right to left)】。实际上，其效果看起来相似于靠右侧对齐而已。

16.12 返回头尾的子节点实例或子元素实例

通过【数据为特定元素实例的变量名称】：
- 衔接【.firstChild】，可返回特定元素实例的第 1 个子**节点**实例。
- 衔接【.firstElementChild】，可返回特定元素实例的第 1 个子**元素**实例。
- 衔接【.lastChild】，可返回特定元素实例的最后 1 个子**节点**实例。
- 衔接【.lastElementChild】，可返回特定元素实例的最后 1 个子**元素**实例。

对于相关的综合调试，请参看如下示例。

【16-12-^-DOM-first-and-last-children.html】

```
<!DOCTYPE html>
<html>
  <head>
    <title>DOM first and last children</title>
  </head>
  <body>

    Inventory
    <div id="div01">
      Fruits
      <p>apple</p>
      <p>banana</p>
      <p>cherry</p>
    </div>

    <div id="div02">
      Vegetables
      <p>burdock</p>
      <p>cabbage</p>
      <p>spinach</p>
```

```
    </div>

    <span>test</span>
  </body>
  <script>
    console.log(document.body.firstChild) ;
    console.log(document.body.firstElementChild) ;
    console.log('') ;

    ///
    console.log(div01.firstChild) ;
    console.log(div01.firstElementChild) ;
    console.log('') ;

    console.log(div01.lastChild) ;
    console.log(div01.lastElementChild) ;
    console.log('') ;

    ///
    console.log(div02.firstChild) ;
    console.log(div02.firstElementChild) ;
    console.log('') ;

    console.log(div02.lastChild) ;
    console.log(div02.lastElementChild) ;
    console.log('') ;
  </script>
</html>
```

【相关说明】

```
<script>

  console.log(document.body.firstChild) ;
```

document.body.firstChild 在此会返回 body 元素实例内的第 1 个子**节点**实例。其中，节点实例可分为**元素**实例、**属性**节点实例与**文本节点**实例。在此，其第 1 个子节点实例，是如下横跨多行的**文本**节点实例：

"

　　Inventory
"

```
  console.log(document.body.firstElementChild) ;
```

document.body.firstElementChild 当前会返回 body 元素实例中的第 1 个子元素实例【<div id="div01">...</div>】。

```
  console.log('') ;

  console.log(div01.firstChild) ;
```

div01.firstChild 会返回属性 id 的数据为'div01'的 div 元素实例中的第 1 个子节点实例，也就是如下横跨多行的文本节点实例：

"
　　Fruits
"

```
    console.log(div01.firstElementChild);
```

div01.firstElementChild 会返回属性 id 的数据为'div01'的 div 元素实例中的第 1 个子元素实例【<p>apple</p>】。

```
    console.log('');
    console.log(div01.lastChild);
```

div01.lastChild 会返回属性 id 的数据为'div01'的 div 元素实例中的最后 1 个子节点实例。此语句会显示出可展开的【#text】的信息。

```
    console.log(div01.lastElementChild);
```

div01.lastElementChild 会返回属性 id 的数据为'div01'的 div 元素实例中的最后 1 个子元素实例【<p>cherry</p>】。

```
    console.log('');
    console.log(div02.firstChild);
```

div02.firstChild 会返回属性 id 的数据为'div02'的 div 元素实例中的第 1 个子节点实例，也就是如下横跨多行的文本节点实例：

"
　　Vegetables
"

```
    console.log(div02.firstElementChild);
```

div02.firstElementChild 会返回属性 id 的数据为'div02'的 div 元素实例中的第 1 个子元素实例【<p>burdock</p>】。

```
    console.log('');
    console.log(div02.lastChild);
```

div02.lastChild 会返回属性 id 的数据为'div02'的 div 元素实例中的最后 1 个子节点实例。此语句会显示出可展开的【#text】的信息。

```
    console.log(div02.lastElementChild);
```

div02.lastElementChild 会返回属性 id 的数据为'div02'的 div 元素实例中的最后 1 个子元素实例【<p>spinach</p>】。

16.13　访问或删除特定元素实例的特定属性

通过【数据为特定元素实例的变量名称】：

- 衔接函数【.getAttribute()】的调用，可获取特定元素实例的特定属性的数据。
- 衔接函数【.getAttributeNode()】的调用，可获取特定元素实例的特定属性对应的**属性**节点实例。
- 衔接函数【.setAttribute()】的调用，可设置特定元素实例的特定属性的数据。
- 衔接函数【.setAttributeNode()】的调用，可设置特定元素实例的特定属性对应的**属性**节点实例。
- 衔接函数【.removeAttribute()】的调用，可删除特定元素实例的特定属性。
- 衔接函数【.removeAttributeNode()】的调用，可删除特定元素实例的特定属性对应的**属性**节点实例。

对于相关的综合调试，请参看如下示例。

【16-13-^-DOM-attribute-access-or-deletion.html】

```
<!DOCTYPE html>
<html>
 <head>
  <title>DOM attributeaccess or deletion</title>
 </head>
 <body>
  <div id="div01" style="color: RoyalBlue ; background-color: Gold">
    Time is money ...
  </div>
 </body>
 <script>
  console.log(div01.getAttribute('id')) ;
  console.log(div01.getAttributeNode('id')) ;
  console.log('') ;

  let a_node = div01.getAttributeNode('id') ;
  console.log(a_node.name) ;
  console.log(a_node.value) ;
  console.log('') ;

  ///
  a_node = document.createAttribute('title') ;

  a_node.value = 'Author: Francesco' ;

  div01.setAttributeNode(a_node) ;

  console.log(div01.title) ;
  console.log('') ;

  ///
```

```
        div01.setAttribute('align','center') ;

        console.log(div01.align) ;
        console.log('') ;

        ///
        div01.removeAttribute('title') ;

        console.log(div01.title) ;
        console.log('') ;

        ///
        a_node = div01.getAttributeNode('align') ;

        div01.removeAttributeNode(a_node) ;

        console.log(div01.align) ;
        console.log('') ;
    </script>
</html>
```

【相关说明】

```
<script>

    console.log(div01.getAttribute('id')) ;
```

div01.getAttribute('id')会返回【在属性 id 的数据为'div01'的 div 元素实例中，其属性 id】的数据字符串'div01'。

```
        console.log(div01.getAttributeNode('id')) ;
```

此语句会显示出【id="div01"】的信息。其中，div01.getAttributeNode('id')会返回【在属性 id 的数据为'div01'的 div 元素实例中，其属性 id】对应的**属性节点**实例。

```
        console.log('') ;

        let a_node = div01.getAttributeNode('id') ;
```

此语句声明了变量 a_node，其初始数据为 div01.getAttributeNode('id')所返回的属性 id 对应的属性节点实例。

```
        console.log(a_node.name) ;
```

a_node.name 当前返回字符串'id'，意味着此属性节点实例的名称为 id。

```
        console.log(a_node.value) ;
```

a_node.value 当前返回字符串'div01'，意味着此属性节点实例的值（value）为 div01。

```
        console.log('') ;

        a_node = document.createAttribute('title') ;
```

此语句将 document.createAttribute('title')所返回的属性 title 对应的属性节点实例，赋给了变量 a_node。

```
            a_node.value = 'Author: Francesco' ;
```

此语句将字符串'Author: Francesco',赋给了变量 a_node.value,使得变量 a_node 所代表的属性节点实例的值,成为了【Author: Francesco】,进而等价于 HTML 语法【title="Author: Francesco"】。

```
            div01.setAttributeNode(a_node) ;
```

此语句将变量 a_node 所代表的属性 title 对应的属性节点实例,设置成为属性 id 的数据是'div01'的 div 元素实例中的属性 title。

```
            console.log(div01.title) ;
```

div01.title 在此返回了字符串'Author: Francesco'。

```
            console.log('') ;
            div01.setAttribute('align','center') ;
```

此语句设置了属性 id 的数据为'div01'的 div 元素实例中的属性 align 及其数据字符串'center',进而等价于 HTML 语法【align="center"】。

```
            console.log(div01.align) ;
```

div01.align 在此返回了字符串'center'。

```
            console.log('') ;
            div01.removeAttribute('title') ;
```

此语句删除了属性 id 的数据为'div01'的 div 元素实例中的属性 title 及其数据。

```
            console.log(div01.title) ;
```

div01.title 在此返回空字符串'',意味着其属性 title 已经被删除了。

```
            console.log('') ;
            a_node = div01.getAttributeNode('align') ;
```

此语句在属性 id 的数据为'div01'的 div 元素实例中,将属性 align 对应的属性节点实例,赋给了变量 a_node。

```
            div01.removeAttributeNode(a_node) ;
```

此语句删除了属性 id 的数据为'div01'的 div 元素实例中的属性 align 及其数据。

```
            console.log(div01.align) ;
```

div01.align 在此返回空字符串'',意味着属性 align 已经被被删除了。

16.14 访问特定元素实例的常见属性的数据

通过【数据为特定元素实例的变量名称】:

- 衔接【.id】，可访问特定元素实例的属性 id 的数据，并代表其身份识别码（identification code）。
- 衔接【.innerHTML】，可访问特定元素实例的属性 innerHTML 的数据，也就是在其内部（inner）带有所有子节点实例的 HTML 源代码片段。
- 衔接【.tagName】，可获取特定元素实例的属性 tagName 的数据，并代表其标签名称（tag name）。
- 衔接【.title】，可访问特定元素实例的属性 title 的数据，并代表【在鼠标指针移入其范围内时，所呈现】的提示文本。
- 衔接【.style】，可访问特定元素实例的各项外观相关的 CSS 属性。

对于相关的综合调试，请参看如下示例。

【16-14-^-DOM-common-attributes.html】

```html
<!DOCTYPE html>
<html>
  <head>
    <title>DOM common attributes</title>
    <style>
      [id^=box]
      {
        width: 80px ;
        height: 80px ;
        color: DodgerBlue ;
        border: 1px RoyalBlue solid ;
        border-radius: 5px ;
        text-align: center ;
        margin: 5px ;
      }
    </style>
  </head>
<body>
  <div id="box01">
    Area one
  </div>

  <div id="box02">
    Area two
  </div>

  <div id="box03">
    Area three
  </div>
</body>
<script>
  with (console)
  {
    log(box01.id) ;
    log(box01.innerHTML) ;
    log(box01.tagName) ;
    log('') ;

    box02.title = 'This area is optional.' ;

    // box03.textContent = '3rd area' ;
```

```
      box03.innerHTML = '3rd area' ;
      box03.style.backgroundColor = 'Gold' ;
      box03.style.textDecoration = 'underline' ;

      let nodes = document.getElementsByTagName('div') ;

      log(nodes.length) ;
    }
  </script>
</html>
```

【相关说明】

```
<script>
  with (console)
  {

    log(box01.id) ;
```

box01.id 会返回【在属性 id 的数据为'box01'的 div 元素实例中,其属性 id】的数据字符串'box01'。

```
    log(box01.innerHTML) ;
```

box01.innerHTML 会返回【在属性 id 的数据为'box01'的 div 元素实例中】的如下内容文本:

Area one

```
    log(box01.tagName) ;
```

box01.tagName 在此会返回【在属性 id 的数据为'box01'的 div 元素实例中',其元素名称 / 标签名称（tag name）】对应的字符串'DIV'。

```
    log('') ;

    box02.title = 'This area is optional.' ;
```

此语句将字符串'This area is optional.'，赋给了属性 id 的数据为'box02'的 div 元素实例中的属性 title，使得鼠标指针移入上述 div 元素实例之后，呈现出提示信息【This area is optional.】。

```
    // box03.textContent = '3rd area' ;
    box03.innerHTML = '3rd area' ;
```

此语句将字符串'3rd area'，赋给了属性 id 的数据为'box03'的 div 元素实例中的属性 innerHTML，使得其内容文本【3rd area】，呈现在上述 div 元素实例的范围内。

```
    box03.style.backgroundColor = 'Gold' ;
```

此语句将字符串'Gold'，赋给了属性 id 的数据为'box03'的 div 元素实例中的子属性 style.backgroundColor，使得上述 div 元素实例，呈现出金色（gold）的背景颜色。

```
    box03.style.textDecoration = 'underline' ;
```

此语句将字符串'underline'，赋给了属性 id 的数据为'box03'的 div 元素实例中的子属性

style.textDecoration，使得上述 div 元素实例，呈现出下画线（underline）的文本装饰 (text decoration)。

```
let nodes = document.getElementsByTagName('div') ;
```

此语句声明了变量 nodes，其初始数据为 document.getElementsByTagName('div')所返回的 3 个 div 元素实例构成的集合。

```
log(nodes.length) ;
```

nodes.length 在此返回整数值 3，意味着在变量 nodes 所代表的集合中，存在 3 个元素实例。

16.15　判断是否存在任何子节点实例

通过【数据为特定元素实例的变量名称】衔接函数【.hasChildNodes()】的调用，可判断在特定元素实例中，是否存在任何子节点（child node）实例。请参看如下示例。

【16-15-^-DOM-hasChildNodes.html】

```
<!DOCTYPE html>
<html>
  <head>
    <title>DOM hasChildNodes()</title>
  </head>
  <body>
    <select id="list01">
      <option>item 1</option>
      <option>item 2</option>
      <option>item 3</option>
    </select>

    <ul id="list02">
      <li>item 1</li>
      <li>item 2</li>
      <li>item 3</li>
    </ul>

    <div id="list03">
      <p id="p01">paragraph 1</p>
      <p id="p02">paragraph 2</p>
      <p id="p03">paragraph 3</p>
      <p id="p04"></p>
      <p id="p05"></p>
    </div>
  </body>
  <script>
    with (console)
    {
      log(list01.hasChildNodes()) ;
      log(list02.hasChildNodes()) ;
      log(list03.hasChildNodes()) ;
      log('') ;
```

```
      log(p01.hasChildNodes()) ;
      log(p04.hasChildNodes()) ;
      log(p05.hasChildNodes()) ;
    }
  </script>
</html>
```

【相关说明】

```
<script>
  with (console)
  {

    log(list01.hasChildNodes()) ;
```

list01.hasChildNodes()返回布尔值 true，意味着在属性 id 的数据为'list01'的 select 元素实例中，存在任意子节点实例。其中，节点实例可分为元素实例、属性节点实例和文本节点实例。

```
    log(list02.hasChildNodes()) ;
```

list02.hasChildNodes()返回布尔值 true，代表在属性 id 的数据为'list02'的 ul 元素实例中，存在任意子节点实例。

```
    log(list03.hasChildNodes()) ;
```

list03.hasChildNodes()返回布尔值 true，代表在属性 id 的数据为'list03'的 div 元素实例中，存在任意子节点实例。

```
    log('') ;
    log(p01.hasChildNodes()) ;
```

p01.hasChildNodes()返回布尔值 true，代表在属性 id 的数据为'p01'的 p 元素实例中，存在任意子节点实例。

```
    log(p04.hasChildNodes()) ;
```

p04.hasChildNodes()返回布尔值 true，代表在属性 id 的数据为'p04'的 p 元素实例中，存在任意子节点实例。请留意，在属性 id 的数据为'p04'的 p 元素实例中，其内含的仅仅 1 个空格字符，仍然被视为 1 个文本节点实例。

```
    log(p05.hasChildNodes()) ;
```

p05.hasChildNodes()返回布尔值 false，代表在属性 id 的数据为'p05'的 p 元素实例中，并不存在任何子节点实例。

16.16 在特定子节点实例之前新增另一子节点实例

通过【数据为特定元素实例的变量名称】衔接函数【.insertBefore()】的调用，可在特定元素实例中的特定子节点实例之前，新增子节点实例。请参看如下示例。

【16-16-^-DOM-insertBefore.html】

```html
<!DOCTYPE html>
<html>
  <head>
    <title>DOM insertBefore()</title>
  </head>
  <body>
    <div id="content">
      <p id="seperator">Default paragraph</p>
    </div>
  </body>
  <script>
    let ref = document.createElement('p') ;

    ref.innerHTML = '1st inserted.' ;

    ref.style.color = 'RoyalBlue' ;

    content.appendChild(ref) ;

    ///
    ref = document.createElement('p') ;

    ref.innerHTML = '2nd inserted..' ;

    ref.style.color = 'Purple' ;

    content.insertBefore(ref, content.children[0]) ;

    ///
    ref = document.createElement('p') ;

    ref.innerHTML = '3rd inserted...' ;

    ref.style.color = 'GreenYellow' ;

    content.insertBefore(ref, seperator) ;
  </script>
</html>
```

【相关说明】

```
<script>

  let ref = document.createElement('p') ;
```

此语句声明了变量ref，其初始数据为document.createElement('p')所返回的新的p元素实例。

```
  ref.innerHTML = '1st inserted.' ;
```

此语句将字符串'1st inserted.'，赋给了变量ref所代表的p元素实例的属性innerHTML，使得p元素实例，在网页上呈现出【1st inserted】的文本。

```
  ref.style.color = 'RoyalBlue' ;
```

此语句将字符串'RoyalBlue'，赋给了变量 ref 所代表的 p 元素实例的属性 style.color，使得 p 元素实例，在网页上呈现出宝蓝色（royal blue）的文本颜色。

```
content.appendChild(ref) ;
```

将变量 ref 所代表的 p 元素实例，新增至属性 id 的数据为'content'的 div 元素实例内的末尾。

```
ref = document.createElement('p') ;
```

此语句将 document.createElement('p')所返回的另一个新的 p 元素实例，赋给了变量 ref。

```
ref.innerHTML = '2nd inserted..' ;
```

此语句将字符串'2nd inserted..'，赋给了变量 ref 所代表的 p 元素实例的属性 innerHTML，使得 p 元素实例，在网页上呈现出【2nd inserted..】的文本。

```
ref.style.color = 'Purple' ;
```

此语句将字符串'Purple'，赋给了变量 ref 所代表的 p 元素实例的属性 style.color，使得 p 元素实例，在网页上呈现出紫色（purple）的文本颜色。

```
content.insertBefore(ref, content.children[0]) ;
```

content.children[0]会返回【在属性 id 的数据为'content'的 div 元素实例中，其第 1 个 (索引值为 0)】子元素实例。所以，此语句在 content.children[0]所代表的子元素实例之前，插入了变量 ref 所代表的 p 元素实例。

```
ref = document.createElement('p') ;
```

将 document.createElement('p')所返回的新的 p 元素实例，赋给变量 ref。

```
ref.innerHTML = '3rd inserted...' ;
```

此语句将字符串'3rd inserted..'，赋给了变量 ref 所代表的 p 元素实例的属性 innerHTML，使得 p 元素实例，在网页上呈现出【3rd inserted..】的文本。

```
ref.style.color = 'GreenYellow' ;
```

此语句将字符串'GreenYellow'，赋给了变量 ref 所代表的 p 元素实例的属性 style.color，使得 p 元素实例，在网页上呈现出绿黄色（green yellow）的文本颜色。

```
content.insertBefore(ref, seperator) ;
```

此语句在属性 id 的数据为'content'的 div 元素实例中，其属性 id 的数据为'seperator'的子元素实例之前，插入了变量 ref 所代表的 p 元素实例。

16.17　判断两个节点实例的内容是否完全相同

通过【数据为特定元素实例的变量名称】：

- 衔接函数【.isEqualNode()】的调用，可判断 2 个特定变量的数据，是否指向同一个内存引址的节点实例。
- 衔接函数【.isSameNode()】的调用，可判断 2 个特定变量的数据所指向的节点实例内容，是否完全相同。

对于相关的综合调试，请参看如下示例。

【16-17-^-DOM-isEqualNode-and-isSameNode.html】

```html
<!DOCTYPE html>
<html>
  <head>
    <title>DOM isEqualNode() and isSameNode()</title>
  </head>
  <body>
    <p>
      Go
      <span>Travelling</span>
    </p>

    <p>
      Go
      <span>Trip</span>
    </p>

    <p>
      Go
      <span>Travelling</span>
    </p>
  </body>
  <script>
    let nodes = document.body.children ;

    console.log(nodes[0].isEqualNode(nodes[1])) ;
    console.log(nodes[0].isEqualNode(nodes[2])) ;
    console.log('') ;

    ///
    console.log(nodes[0].isSameNode(nodes[1])) ;
    console.log(nodes[0].isSameNode(nodes[2])) ;
    console.log('') ;

    ///
    let p_nodes = document.querySelectorAll('p') ;

    console.log(nodes[0].isSameNode(p_nodes[0])) ;
    console.log(nodes[1].isSameNode(p_nodes[1])) ;
    console.log(nodes[2].isSameNode(p_nodes[2])) ;
  </script>
</html>
```

【相关说明】

```
<script>
  let nodes = document.body.children ;
```

此语句声明变量 nodes，其初始数据为 body 元素实例中的所有子元素实例构成的集合。

```
console.log(nodes[0].isEqualNode(nodes[1])) ;
```

nodes[0]会返回【在变量 nodes 的集合中的第 1 个（索引值为 0）】子元素实例，也就是 body 元素实例中的第 1 个子元素实例。nodes[1]会返回【在变量 nodes 的集合中的第 2 个（索引值为 1）】子元素实例，也就是 body 元素实例中的第 2 个子元素实例。

在此，nodes[0].isEqualNode(nodes[1])会返回布尔值 false，意味着【在变量 nodes 的集合中，其第 1 个（索引值为 0）和第 2 个（索引值为 1）】子元素实例的**内容**，存在着**不同**之处。

```
console.log(nodes[0].isEqualNode(nodes[2])) ;
```

nodes[0]会返回【在变量 nodes 的集合中的第 1 个（索引值为 0）】子元素实例，也就是 body 元素实例中的第 1 个子元素实例。nodes[2]会返回【在变量 nodes 的集合中的第 3 个（索引值为 2）】子元素实例，也就是 body 元素实例中的第 3 个子元素实例。

nodes[0].isEqualNode(nodes[2]) 返回布尔值 true，意味着【在变量 nodes 的集合中，第 1 个（索引值为 0）和第 3 个（索引值为 2）】子元素实例的内容，是**完全相同**的。

所谓的完全相同，是指【从两个元素实例的起始语句的位置开始，连细微的 1 个文字，甚至 1 个空格（space）字符的相对位置】，都是**没有差别**的。

```
console.log('') ;

console.log(nodes[0].isSameNode(nodes[1])) ;
```

nodes[0].isSameNode(nodes[1])返回布尔值 false，意味着【在变量 nodes 的集合中，第 1 个（索引值为 0）和第 2 个（索引值为 1）】子元素实例，并不是相同的元素实例，也就是各自占用不同的内存引址。其实，语法【nodes[0].isSameNode(nodes[1])】等价于语法【nodes[0] == nodes[1]】。

```
console.log(nodes[0].isSameNode(nodes[2])) ;
```

nodes[0].isSameNode(nodes[2])也会返回布尔值 false，意味着【在变量 nodes 的集合中的第 1 个（索引值为 0）和第 3 个（索引值为 2）】子元素实例，亦不是相同的元素实例，也就是各自占用不同的内存引址。在此，语法【nodes[0].isSameNode(nodes[2])】等价于语法【nodes[0] == nodes[2]】。

```
console.log('') ;

let p_nodes = document.querySelectorAll('p') ;
```

此语句声明了变量 p_nodes，其初始数据为 document.querySelectorAll('p')返回的多个 p 元素实例构成的集合。

```
console.log(nodes[0].isSameNode(p_nodes[0])) ;
console.log(nodes[1].isSameNode(p_nodes[1])) ;
console.log(nodes[2].isSameNode(p_nodes[2])) ;
```

这 3 个语句均显示出布尔值 true 的信息，意味着如下个别的组合，均被视为同一个元素实例，也就是指向相同的内存引址：

- nodes[0]和 p_nodes[0] 的数据。
- nodes[1]和 p_nodes[1] 的数据。

- nodes[2]和 p_nodes[2] 的数据。

此外在本示例中,语法【nodes[索引值 n].isSameNode(p_nodes[索引值 n])】等价于语法【nodes[索引值 n] == p_nodes[索引值 n]】。

16.18 返回下一个或上一个兄弟节点实例

通过【数据为特定元素实例的变量名称】:
- 衔接【.nextSibling】,可返回特定元素实例的下一个兄弟(next sibling)节点实例。
- 衔接【.previousSibling】,可返回特定元素实例的上一个兄弟(previous sibling)节点实例。

对于相关的调试,请参看如下示例。

【16-18-^-DOM-next-and-previous-siblings.html】

```
<!DOCTYPE html>
<html>
  <head>
    <title>DOM next and previous siblings</title>
  </head>
  <body>
    <p id="p01">Paragraph 1</p>
    <p id="p02">Paragraph 2</p>
    <p id="p03">Paragraph 3</p>
    <p id="p04">Paragraph 4</p>
    <p id="p05">Paragraph 5</p>
  </body>
  <script>
    console.log(p01.nextSibling) ;
    console.log(p01.nextSibling.nextSibling) ;
    console.log('') ;

    console.log(p03.nextElementSibling) ;
    console.log('') ;

    ///
    console.log(p05.previousSibling) ;
    console.log(p05.previousSibling.previousSibling) ;
    console.log('') ;

    console.log(p03.previousElementSibling) ;
  </script>
</html>
```

【相关说明】

```
<script>

  console.log(p01.nextSibling) ;
```

p01.nextSibling 会返回【紧接在属性 id 的数据为'p01'的 p 元素实例】之后的兄弟（sibling）节点实例，也就是指**紧接在后**的**同一层**节点实例。紧接在属性 id 的数据为'p01'的 p 元素实例之后的节点实例，其实是由【换行（line feed）与数个空格（space）字符】所构成的文本节点实例。也因此，此语句会显示出代表文本节点实例的【#text】的信息。

```
console.log(p01.nextSibling.nextSibling) ;
```

因为 p01.nextSibling.nextSibling 重复了两次 nextSibling，所以会返回属性 id 的数据为'p01'的 p 元素实例之后的第 2 个节点实例【<p id="p02">Paragraph 2</p>】。

```
console.log('') ;
console.log(p03.nextElementSibling) ;
```

p03.nextElementSibling 会返回【紧接在属性 id 的数据为'p03'的 p 元素实例】之后的兄弟元素（sibling element）实例，也就是**紧接在后**的**同一层**元素实例【<p id="p04">Paragraph 4</p>】。

```
console.log('') ;
console.log(p05.previousSibling) ;
```

p05.previousSibling 会返回【紧接在属性 id 的数据为'p05'的 p 元素实例】**之前**的兄弟节点（sibling node）实例，也就是**紧接在前**的【同一层】节点实例。紧接在属性 id 的数据为'p05'的 p 元素实例之前的节点实例，其实是由【数个空格（space）和 1 个换行（line feed）字符】所构成的文本节点实例。也因此，此语句会显示出代表文本节点实例的【#text】的信息。

```
console.log(p05.previousSibling.previousSibling) ;
```

因为 p05.previousSibling.previousSibling 重复了 2 次 previousSibling，所以会返回属性 id 的数据为'p05'的 p 元素实例之前的第 2 个节点实例【<p id="p04">Paragraph 4</p>】。

```
console.log('') ;
console.log(p03.previousElementSibling) ;
```

p03.previousElementSibling 会返回【紧接在属性 id 的数据为'p03'的 p 元素实例】之前的兄弟元素实例，也就是**紧接在前**的**同一层**元素实例【<p id="p02">Paragraph 2</p>】。

16.19　返回特定节点实例的相关数据

通过【特定节点实例的变量名称】衔接【.nodeName】、【.nodeType】和【.nodeValue】，可分别返回特定节点实例的节点名称（node name）、节点类型（node type）以及节点值（node value）的数据字符串，也就是特定节点实例的节点数据（node data）。请参看如下示例。

【16-19-^-DOM-node-data.html】

```
<!DOCTYPE html>
<html>
  <head>
```

```
    <title>DOM node data</title>
  </head>
  <body>
    Sample …
    <div id="div01" style="color: RoyalBlue ; background-color: Gold">
      Time is money …
    </div>
  </body>
  <script>
    let node = div01 ;

    console.log(node.nodeName) ;
    console.log(node.nodeType) ;
    console.log(node.nodeValue) ;
    console.log('') ;

    node = div01.getAttributeNode('id') ;

    console.log(node.nodeName) ;
    console.log(node.nodeType) ;
    console.log(node.nodeValue) ;
    console.log('') ;

    node = document.body.firstChild ;

    console.log(node.nodeName) ;
    console.log(node.nodeType) ;
    console.log(node.nodeValue) ;
  </script>
</html>
```

【相关说明】

```
  <script>

    let node = div01 ;
```

此语句声明了变量 node，其初始数据为属性 id 的数据为'div01'的 div 元素实例。

```
    console.log(node.nodeName) ;
```

node.nodeName 返回了字符串'DIV'，代表着变量 node 所代表的节点实例的节点名称（node name）。

```
    console.log(node.nodeType) ;
```

node.nodeType 返回整数值 1，意味着在变量 node 所代表的节点实例中，其节点类型（node type）是元素（element）。

```
    console.log(node.nodeValue) ;
```

node.nodeValue 返回 null，意味着在变量 node 所代表的节点实例中，并不存在对应的节点值（node value）。

```
    console.log('') ;
```

```
            node = div01.getAttributeNode('id') ;
```

此语句将 div01.getAttributeNode('id')所返回的属性 id 的属性节点（attribute node）实例，赋给了变量 node。

```
            console.log(node.nodeName) ;
```

node.nodeName 返回字符串'id'，意味着在变量 node 当前代表的属性节点实例中，其节点名称（node name）为 id。

```
            console.log(node.nodeType) ;
```

node.nodeType 返回整数值 2，意味着在变量 node 代表的节点实例中，其节点类型（node type）为属性（attribute）。

```
            console.log(node.nodeValue) ;
```

node.nodeValue 返回字符串'div01'，意味着在变量 node 代表的节点实例中，其节点值（node value）为 div01。

```
            console.log('') ;
            node = document.body.firstChild ;
```

此语句将 document.body.firstChild 所返回的 body 元素实例中的第 1 个子节点实例，赋给了变量 node。

```
            console.log(node.nodeName) ;
```

node.nodeName 返回字符串'#text'，意味着在变量 node 当前代表的**文本**节点实例中，其节点名称（node name）为#text。

```
            console.log(node.nodeType) ;
```

node.nodeType 返回整数值 3，意味着在变量 node 所代表的节点实例中，其节点类型（node type）为文本（text）。

```
            console.log(node.nodeValue) ;
```

node.nodeValue 在此返回了【在变量 node 所代表的节点实例中，其如下横跨多行的字符串的节点值（node value）:

    Sample。。。

16.20　合并多个相邻的文本子节点实例

通过【数据为特定元素实例的变量名称】衔接函数【.normalize()】的调用，可在特定元素实例中，将多个相邻的文本（text）子节点（child node）实例，合并成为单一文本子节点实例。请参看如下示例。

【16-20-^-DOM-normalizing-text-nodes.html】

```html
<!DOCTYPE html>
<!DOCTYPE html>
<html>
  <head>
    <title>DOM normalizing text nodes</title>
  </head>
  <body>
    <div id="div01"></div>
  </body>
  <script>
    let node = document.createTextNode('Happily') ;

    div01.appendChild(node) ;

    node = document.createTextNode('ever') ;

    div01.appendChild(node) ;

    node = document.createTextNode('after') ;

    div01.appendChild(node) ;

    console.log(div01.childNodes) ;
    console.log(div01.childNodes.length) ;
    console.log(div01.childNodes[0]) ;
    console.log(div01.childNodes[1]) ;
    console.log(div01.childNodes[2]) ;
    console.log('') ;
    /*
    document.body.normalize() ;

    console.log(div01.childNodes) ;
    console.log(div01.childNodes.length) ;
    console.log(div01.childNodes[0]) ;
    */
  </script>
</html>
```

【相关说明】

```
<script>

  let node = document.createTextNode('Happily') ;
```

此语句声明了变量 node，其初始数据为 document.createTextNode('Happily')所返回的内含字符串'Happily'的文本节点实例。

```
    div01.appendChild(node) ;
```

此语句将变量 node 当前所代表的第 1 个文本节点实例，新增至属性 id 的数据为'div01'的 div 元素实例中。

```
            node = document.createTextNode('ever') ;
```

此语句将 document.createTextNode('ever')所返回的内含字符串'ever'的文本节点实例，赋给了变量 node。

```
            div01.appendChild(node) ;
```

此语句将变量 node 当前所代表的第 2 个文本节点实例，新增至属性 id 的数据为'div01'的 div 元素实例中。

```
            node = document.createTextNode('after') ;
```

此语句将 document.createTextNode('after')所返回的内含字符串'after'的文本节点实例，赋给了变量 node。

```
            div01.appendChild(node) ;
```

此语句将变量 node 当前所代表的第 3 个文本节点实例，新增至属性 id 的数据为'div01'的 div 元素实例中。

```
            console.log(div01.childNodes) ;
```

div01.childNodes 返回【在属性 id 的数据为'div01'的 div 元素实例中，其所有子节点实例】构成的集合。此语句显示出【NodeList(3) [text, text, text]】的信息，意味着其所有子节点实例，均为文本（text）节点实例。

```
            console.log(div01.childNodes.length) ;
```

div01.childNodes.length 返回整数值 3，意味着属性 id 的数据为'div01'的 div 元素实例，内含 3 个子节点实例。

```
            console.log(div01.childNodes[0]) ;
```

div01.childNodes[0]会返回【属性 id 的数据为'div01'的 div 元素实例中的第 1 个（索引值为 0）】子节点实例。此语句显示出其第 1 个子节点实例的文本"Happily "。

```
            console.log(div01.childNodes[1]) ;
```

div01.childNodes[1]会返回【属性 id 的数据为'div01'的 div 元素实例中的第 2 个（索引值为 1）】子节点实例。此语句显示出其第 2 个子节点实例的文本"ever "。

```
            console.log(div01.childNodes[2]) ;
```

div01.childNodes[2]会返回【属性 id 的数据为'div01'的 div 元素实例中的第 3 个（索引值为 2）】子节点实例。此语句显示出其第 3 个子节点实例的文本"after"。

```
            console.log('') ;
            /*
```

欲调试时，可先直接调试一次，并观察着显示在浏览器的调试工具【Console】面板里的信息之后，进一步消除这个注释在此处的开始符号【/*】与下方的结束符号【*/】。随后再次调试，并观察着显示的信息，以便深度理解其差异。

```
            document.body.normalize() ;
```

浏览器执行此语句时，会在 body 元素实例中，将所有相邻的数个文本节点实例，合并成为单一文本节点实例。

```
console.log(div01.childNodes) ;
```

此语句显示出【NodeList(1) [text]】的信息，意味着原本 3 个相邻的文本子节点实例，被合并成为单一文本子节点实例了。

```
console.log(div01.childNodes.length) ;
```

div01.childNodes.length 返回整数值 1，意味着其当前只剩下 1 个子节点实例了。

```
console.log(div01.childNodes[0]) ;
```

div01.childNodes[0]会返回【当前在属性 id 的数据为'div01'的 div 元素实例中的第 1 个（索引值为 0）】子节点实例。此语句显示出其**第 1 个子节点实例**当前的文本"Happily ever after"。

16.21 返回父节点实例

通过【数据为特定元素实例的变量名称】：

- 衔接【.parentNode】，可返回特定元素实例的父节点（parent node）实例。
- 衔接【.parentElement】，可返回特定元素实例的父元素（parent element）实例。

对于相关的综合调试，请参看如下示例。

【16-21-^-DOM-parentNode-and-parentElement.html】

```
<!DOCTYPE html>
<html>
  <head>
    <title>DOM parentNode and parentElement</title>
  </head>
  <body>
    <div id="div01">
      <p id="p01">
        <a id="a01" href="#">Link 1</a>
      </p>
    </div>
  </body>
  <script>
    with (console)
    {
      log(a01.parentNode) ;
      log(a01.parentElement) ;
      log('') ;

      log(a01.parentNode.id) ;
      log(a01.parentElement.id) ;
      log('') ;
```

```
        log(div01.parentNode) ;
        log(div01.parentElement) ;
        log('') ;

        log(document.body.parentNode) ;
        log(document.body.parentElement) ;
        log('') ;

        log(document.body.parentNode.parentNode) ;
        log(document.documentElement.parentNode) ;
        log(document.documentElement.parentElement) ;
      }
  </script>
</html>
```

【相关说明】

```
  <script>
    with (console)
    {

      log(a01.parentNode) ;
      log(a01.parentElement) ;
```

在此，a01.parentNode 和 a01.parentElement 均返回元素实例【<p id="p01">...</p>】。其中，a01.parentNode 会返回属性 id 的数据为'a01'的 a 元素实例的父**节点**（parent node）实例，a01.parentElement 则会返回属性 id 的数据为'a01'的 a 元素实例的父**元素**（parent element）实例；只是上述父**节点**实例和上述父**元素**实例，当前刚好对应到同一个元素实例【<p id="p01">...</p>】。

```
      log('') ;

      log(a01.parentNode.id) ;
      log(a01.parentElement.id) ;
```

a01.parentNode.id 与 a01.parentElement.id 当前均返回字符串'p01'。

```
      log('') ;

      log(div01.parentNode) ;
      log(div01.parentElement) ;
```

div01.parentNode 和 div01.parentElement 均返回元素实例【<body>...</body>】。其中，div01.parentNode 会返回属性 id 的数据为'div01'的 div 元素实例的父**节点**实例，div01.parentElement 会返回属性 id 的数据为'div01'的 div 元素实例的父**元素**实例；只是上述父**节点**实例和上述父**元素**实例，当前刚好对应到同一个元素实例【<body>...</body>】。

```
      log('') ;

      log(document.body.parentNode) ;
      log(document.body.parentElement) ;
```

document.body.parentNode 和 document.body.parentElement 均返回元素实例【<html>...</html>】。其中，document.body.parentNode 会返回 body 元素实例的父**节点**实例，document.body.parentElement 会返回 body 元素实例的父**元素**实例；只是上述父**节点**实例和上述父**元素**实例，当前刚好对应到同一

个元素实例【<html>...</html>】。

```
    log('') ;

    log(document.body.parentNode.parentNode) ;
    log(document.documentElement.parentNode) ;
```

document.body.parentNode.parentNode 与 document.documentElement.parentNode 在此均返回 #document。

```
    log(document.documentElement.parentElement) ;
```

document.documentElement.parentElement 返回 null，意味着其并不存在父元素实例，这是因为 document.documentElement 对应的元素实例【<html>...</html>】，已经是当前网页的根元素（root element）实例了。

16.22　删除或替换子节点实例

通过【数据为特定元素实例的变量名称】：

- 衔接函数【.removeChild()】的调用，可在特定元素实例中，删除特定子节点实例。
- 衔接函数【.replacechild()】的调用，可在特定元素实例中，替换特定子节点实例。

对于相关的综合调试，请参看如下示例。

【16-22-^-DOM-removeChild-and-replaceChild.html】

```
<!DOCTYPE html>
<html>
 <head>
    <title>DOM removeChild() and replaceChild()</title>
 </head>
 <body>
    <ul id="ul01">
      <li>Item 1</li>
      <li>Item 2</li>
      <li>Item 3</li>
      <li>Item 4</li>
      <li>Item 5</li>
    </ul>
 </body>
<script>
  let refs = ul01.children ;

  console.log(refs) ;
  console.log('') ;

  ul01.removeChild(refs[2]) ;
  ul01.removeChild(refs[2]) ;
```

```
    console.log(refs) ;
    console.log('') ;

    ///
    let new_item = document.createElement('li') ;

    new_item.innerHTML = 'Special X' ;

    ul01.replaceChild(new_item, refs[1]) ;
  </script>
</html>
```

【相关说明】

```
<script>

    let refs = ul01.children ;
```

此语句声明了变量 refs，其初始数据为【在属性 id 的数据为'ul01'的 ul 元素实例中，其所有子元素实例】构成的集合。

```
    console.log(refs) ;
```

此语句显示出【HTMLCollection(5) [li, li, li, li, li]】的信息，意味着变量 refs 所代表的集合，内含 5 个 li 元素实例。

```
    console.log('') ;
    ul01.removeChild(refs[2]) ;
```

此语句在属性 id 的数据为'ul01'的 ul 元素实例中，移除了其当前第 3 个（索引值为 2）li 元素实例【<li>Item 3</li>】。其中，refs[2]当前代表变量 refs 的集合中的第 3 个元素实例【<li>Item 3</li>】）。

```
    ul01.removeChild(refs[2]) ;
```

此语句在属性 id 的数据为'ul01'的 ul 元素实例中，移除了其当前第 3 个（索引值为 2）li 元素实例【<li>Item 4</li>】。

```
    console.log(refs) ;
```

此语句显示出【HTMLCollection(3) [li, li, li]】的信息，意味着移除了其两个元素实例之后，在变量 refs 所代表的集合里，只剩下 3 个 li 元素实例。

```
    console.log('') ;
    let new_item = document.createElement('li') ;
```

此语句声明了变量 new_item，其初始数据为 document.createElement('li')所返回的新的 li 元素实例。

```
    new_item.innerHTML = 'Special X' ;
```

此语句将字符串'Special X'，赋给了变量 new_item 所代表的 li 元素实例的属性 innerHTML，使得 li 元素实例的内容文本，成为'Special X'。

```
ul01.replaceChild(new_item, refs[1]) ;
```

此语句使得变量 refs 所代表的第 2 个（索引值为 1）li 元素实例，被取代成为变量 new_item 所代表的新的 li 元素实例。也因此，在属性 id 的数据为'ul01'的 ul 元素实例中，其当前第 2 个（索引值为 1）li 元素实例中的文本，从'Item 2'，变成'Special X'。

16.23 获取滚动条的相关数据

通过【数据为特定元素实例的变量名称】：

- 衔接【.scrollLeft】，可返回特定元素实例的水平滚动条的当前滚动刻度。
- 衔接【.scrollTop】，可返回特定元素实例的垂直滚动条的当前滚动刻度。
- 衔接【.scrollWidth】，可返回【在特定元素实例中，其内容可被水平滚动】的宽度。
- 衔接【.scrollHeight】，可返回【在特定元素实例中，其内容可被垂直滚动】的高度。
- 衔接函数【.scrollIntoView()】的调用，可使得特定元素实例，被显示在浏览器的当前窗口（window）或特定窗格（pane）中。

对于相关的综合调试，请参看如下示例。

【16-23-^-DOM-scroll-related-attributes.html】

```
<!DOCTYPE html>
<html>
  <head>
    <title>DOM scroll related attributes</title>
    <style>
      body
      {
        height: 1000px ;
        width: 2000px ;
      }

      #content
      {
        width: 200px ;
        height: 140px ;
        padding: 10px ;
        background-color: GreenYellow ;
        overflow: scroll ;
      }

      #content p
      {
        color: White ;
        padding: 5px ;
        width: 400px ;
        height: 300px ;
        border-radius: 5px ;
      }
```

```
    #p01
    {
      background-color: Orange ;
    }

    #p02
    {
      background-color: Yellow ;
    }

    #p03
    {
      background-color: RoyalBlue ;
    }

    #p04
    {
      background-color: Cyan ;
    }

    #p05
    {
      background-color: Purple ;
    }
  </style>
</head>
<body>
  <p id="message01" style="position: fixed ; left: 20%; top: 50% ; color: RoyalBlue">(0, 0)</p>

  <p id="message02" style="position: fixed ; left: 60%; top: 50% ; color: Teal">(0, 0)</p>

  <button type="button" id="scroll_button">Scroll to Paragraph 3</button>

  <div id="content">
    <p id="p01">Paragraph 1</p>
    <p id="p02">Paragraph 2</p>
    <p id="p03">Paragraph 3</p>
    <p id="p04">Paragraph 4</p>
    <p id="p05">Paragraph 5</p>
  </div>
</body>
<script>
  scroll_button.onclick = function (event)
  {
    p03.scrollIntoView() ;
  } ;

  window.onscroll = function (event)
  {
    with (document.documentElement)
    {
      message01.innerHTML = `(${scrollLeft}, ${scrollTop})` ;
    }
  }
```

```
    content.onscroll = function (event)
    {
     with (event.target)
     {
       message02.innerHTML = `(${scrollLeft}, ${scrollTop})`;
     }
    }

    console.log(document.body.scrollHeight);
    console.log(document.body.scrollWidth);
  </script>
</html>
```

【相关说明】

```
  <script>
    scroll_button.onclick = function (event)
    {

      p03.scrollIntoView();
```

此语句使得属性 id 的数据为'p03'的 p 元素实例，被显示在当前窗口或特定窗格中。

```
    };
```

上述语法将带有参数 event 的匿名函数的定义，赋给了属性 id 的数据为'scroll_button'的 button 元素实例的属性 onclick。所以，在 button 元素实例被单击时，上述匿名函数就会被调用。

```
    window.onscroll = function (event)
    {
     with (document.documentElement)
     {

       message01.innerHTML = `(${scrollLeft}, ${scrollTop})`;
```

此语句使得属性 id 的数据为'message01'的 p 元素实例的内容文本，变成类似【(173, 231)】的文本，代表着浏览器当前窗口的水平滚动条的滚动刻度（例如 173 像素）及其垂直滚动条的滚动刻度（例如 231 像素）。

```
     }
```

在上述关键字 with 开头的整段语法中，子语法 document.documentElement.scrollLeft 及 document.documentElement.scrollTop，在其大括号里，被简化成 scrollLeft 和 scrollTop。

```
    }
```

在上述 window.onscroll 开头的整段语法中，则将带有参数 event 的匿名函数的定义，赋给了代表浏览器当前窗口的 window 对象实例的属性 onscroll。所以，浏览器**窗口**的内容被滚动时，上述匿名函数就会被调用。

```
    content.onscroll = function (event)
    {
     with (event.target)
     {
```

```
            message02.innerHTML = `(${scrollLeft}, ${scrollTop})` ;
```

此语句使得属性 id 的数据为'message02'的 p 元素实例的内容文本，变成类似【(107, 360)】的文本，代表着在属性 id 的数据为'content'的 div 元素实例中，其当前水平滚动条的滚动刻度（例如 107 像素）和垂直滚动条的滚动刻度（例如 360 像素）。

```
    }
  }
```

上述语法将带有参数 event 的匿名函数的定义，赋给属性 id 的数据为'content'的 div 元素实例的属性 onscroll。属性 id 的数据为'content'的 div 元素实例的内容被滚动时，上述匿名函数就会被调用。

```
            console.log(document.body.scrollHeight) ;
```

document.body.scrollHeight 会返回【当前网页可被垂直滚动的内容】的高度，例如 1000 像素。

```
            console.log(document.body.scrollWidth) ;
```

document.body.scrollWidth 会返回【当前网页可被水平滚动的内容】的宽度，例如 2000 像素。

16.24　练　习　题

1. 已知如下 HTML 源代码片段：

```
<div id="div01" class="color02 border01 background03">太阳系</div>
```

试编写 JavaScript 源代码，以动态将上述 div 元素实例的属性 class 的数据，变更为'color03 border02 background01'。

2. 试编写 JavaScript 源代码片段，以获取并显示当前网页中 img 元素实例的个数。

3. 已知在当前网页里，存在如下 JavaScript 源代码片段：

```
function display()
{
  ⋮
}

document.body.onclick = display ;
```

试编写后续的 JavaScript 源代码片段，使得在当前网页中，任意元素实例被单击时，函数 display() 不会再被调用。

4. 试编写 JavaScript 源代码片段，以计算并显示出在当前网页中，a 元素实例对应的超链接的个数。

5. 试编写 JavaScript 源代码片段，以显示出在当前网页的 body 元素实例中，其子元素实例的个数。

第 17 章

BOM 的多个对象实例

在浏览器对象模型（BOM,browser object model）中，特定网页中的 window 对象实例，可被视为根对象（root object）实例。在 window 对象实例中，较常被运用的，有 document、screen、history 与 location 子对象实例。其中，document 子对象实例，可被视为文档对象模型（DOM, document object model）中的根对象（root object）实例。

17.1　window 对象实例

在 JavaScript 语言里，window 对象实例即是对应到容纳当前网页的浏览器**窗口**（window）。在特定网页里，任意 iframe 元素实例，亦可被视为容纳特定子网页的子窗口。

17.1.1　显示多种对话框与搜索特定文本

通过函数 window.alert()或 alert()的调用，可在当前网页上，呈现出警告对话框（alert dialogue box）。通过函数 window.confirm()或 confirm()的调用，可在当前网页上，呈现出显示确认对话框（confirm dialogue box）。通过函数 window.prompt()或 prompt()的调用，则可在当前网页上，呈现出提示对话框（prompt dialogue box）。

通过函数 window.open()的调用，可使得浏览器打开内含另一网页的新窗口（window）。通过函数 window.find()的调用，则可在当前网页中，**突出**被搜寻到的文本。请参看如下示例。

【17-1-1-window-dialogues-and-find.html】

```
<!DOCTYPE html>
<html>
  <head>
    <title>window dialogues and find()</title>
```

```html
</head>
<body>
  <p>
    Lycopene is a symmetrical tetraterpene assembled from eight isoprene units. It is a member
      of the carotenoid family of compounds, and because it consists entirely of carbon and
      hydrogen, is also a carotene.
  </p>
  <button id="b01">display alert dialog</button>
  <button id="b02">display confirm dialog</button>
  <button id="b03">display prompt dialog</button>
  <button id="b04">open a new page</button>
</body>
<script>
  let sentence = 'See ya...' ;
  let isOK = false, isFound = false ;
  let given_data = '' ;
  let win_obj = null ;

  b01.onclick = function (event)
  {
    // window.alert(sentence) ;
    alert(sentence) ;
  } ;

  b02.onclick = function (event)
  {
    // isOK = window.confirm(sentence) ;
    isOK = confirm(sentence) ;

    // window.console.log(`Is OK button pressed? ${isOK}`) ;
    console.log(`Is OK button pressed to close this window? ${isOK}`) ;

    if (isOK)
      // window.close() ;
      close() ;
  } ;

  b03.onclick = function (event)
  {
    // given_data = window.prompt(sentence) ;
    given_data = prompt('Please input your mood','happy') ;

    // window.console.log(`Is OK button pressed? ${isOK}`) ;
    console.log(`Given data: ${given_data}`) ;

    given_data = prompt('Please input the string to be found out.') ;

    console.log(`Given data: ${given_data}`) ;

    isFound = window.find(given_data) ;

    console.log(`Is Given data found on the page? ${isFound}`) ;
  } ;

  b04.onclick = function (event)
```

```
      {
        // win_obj = open(…) ;
        win_obj = window.open('http://www.tup.tsinghua.edu.cn/index.html','publisher', 'menubar,
      location, status, scrollbars, resizable') ;
      } ;
    </script>
  </html>
```

【相关说明】

```
<script>

  let sentence = 'See ya...' ;
```

声明初始数据为字符串'See ya...'的变量 sentence。

```
  let isOK = false, isFound = false ;
```

声明初始数据均为布尔值 false 的变量 isOK 与 isFound。

```
  let given_data = '' ;
```

声明初始数据为空字符串''的变量 given_data。

```
  let win_obj = null ;
```

声明初始数据为空值（null）的变量 win_obj。

```
  b01.onclick = function (event)
  {
    // window.alert(sentence) ;

    alert(sentence) ;
```

执行此语句之后，浏览器会在其窗口中，呈现出【显示变量 sentence 中的数据字符串'See ya...'】的警报（alert）对话框。

```
  } ;
```

上述语法将匿名函数的定义，赋给了属性 id 的数据为'b01'的 button 元素实例的属性 onclick。一旦上述 button 元素实例被单击时，上述匿名函数就会被调用。

```
  b02.onclick = function (event)
  {

    // window.confirm(sentence) ;
    isOK = confirm(sentence) ;
```

执行此语句之后，浏览器会在其窗口中，呈现出【显示变量 sentence 中的数据字符串'See ya...'】的确认（confirm）对话框。当用户单击此对话框内的**确定按钮**时，布尔值 true 就会被赋给变量 isOK。若用户此时按下键盘上的ESC键或者单击此对话框的**取消按钮**，则布尔值 false 会被赋给变量 isOK。

```
      // window.console.log(`Is OK button pressed? ${isOK}`) ;
      console.log(`Is OK button pressed to close this window? ${isOK}`) ;
```

此语句将类似【Is OK button pressed to close this window? false】的信息，显示在浏览器的调试工具【Console】面板里。

```
      if (isOK)
        // window.close() ;
        close() ;
```

若变量 isOK 的数据为布尔值 true，则当前的浏览器窗口会被关闭。

```
    } ;
```

上述语法将匿名函数的定义，赋给了属性 id 的数据为'b02'的 button 元素实例的属性 onclick。一旦上述 button 元素实例被单击时，上述匿名函数就会被调用。

```
    b03.onclick = function (event)
    {
      // window.prompt(sentence) ;
      given_data = prompt('Please input your mood','happy') ;
```

执行此语句之后，浏览器会在其窗口中，呈现出显示信息【Please input your mood】的提示（prompt）对话框，并请求用户在【已存在默认文本'happy'】的输入字段里，填入新的文本。当用户单击此对话框内的确定按钮时，新的文本字符串就会被赋给变量 given_data。

```
      // window.console.log(`Is OK button pressed? ${isOK}`) ;
      console.log(`Given data: ${given_data}`) ;
```

此语句将类似【Given data: delight】的信息，显示在浏览器的调试工具【Console】面板里。

```
      given_data = prompt('Please input the string to be found out.') ;
```

执行此语句之后，浏览器会在其窗口中，呈现出显示信息【Please input the string to be found out.】的提示（prompt）对话框，并请求用户在没有任何默认文本的输入字段里，填入新的文本。当用户单击此对话框内的确定按钮时，新的文本字符串就会被赋给变量 given_data。

```
      console.log(`Given data: ${given_data}`) ;
```

此语句将类似【Given data: carbon】的信息，显示在浏览器的调试工具【Console】面板里。

```
      isFound = window.find(given_data) ;
```

在本示例网页中的文本【Lycopene is a symmetrical tetraterpene assembled from eight isoprene units. It is a member of the carotenoid family of compounds, and because it consists entirely of carbon and hydrogen, is also a carotene.】里，存在子字符串'carbon'。

若变量 given_data 的数据字符串为'carbon'，则 window.find(given_data)会返回布尔值 true，并**突出**在当前网页里的文本'carbon'。此语句最后会将 window.find(given_data)返回的布尔值 false 或者 true，赋给变量 isFound。

```
      console.log(`Is Given data found on the page? ${isFound}`) ;
```

此语句将类似【Is Given data found on the page? true】的信息，显示在浏览器的调试工具【Console】面板里。

```
    } ;
```

上述语法将匿名函数的定义，赋给了属性 id 的数据为'b03'的 button 元素实例的属性 onclick。一旦上述 button 元素实例被单击时，上述匿名函数就会被调用。

```
      b04.onclick = function (event)
      {
         win_obj = window.open('http://www.tup.tsinghua.edu.cn/index.html','publisher', 'menubar,
location, status, scrollbars, resizable') ;
```

在此语句中，函数window.open()被调用时，会将网址【http://www.tup.tsinghua.edu.cn/index.html】对应的网页，加载到具有如下特征的浏览器的新窗口里，并返回名称为 publisher 的新窗口对应的 window 对象实例。上述新窗口会具有菜单栏（menu bar）、网址栏（location bar）、状态栏（status bar）、可被显示的滚动条（scroll bar），以及可被调整（resizable）的窗口尺寸。

在此，函数 window.open()被调用时，会返回上述新窗口对应的 window 对象实例，并赋给变量 win_obj。所以可通过变量 win_obj，再次访问到上述新窗口的 window 对象实例的内置属性。

```
      } ;
```

上述语法将匿名函数的定义，赋给了属性 id 的数据为'b04'的 button 元素实例的属性 onclick。一旦上述 button 元素实例被单击时，上述匿名函数就会被调用。

17.1.2　滚动至坐标或滚动特定距离

通过函数 window.scrollTo()或者 scrollTo()的调用，可在当前网页的内容中，滚动至特定坐标。通过函数 window.scrollBy()或者 scrollBy()的调用，则可在当前网页的内容中，滚动特定距离。请参看如下示例。

【17-1-2-window-scrollTo-and-scrollBy.html】

```html
<!DOCTYPE html>
<html>
  <head>
    <title>window scrollTo() and scrollBy()</title>
    <style>
      #button_block
      {
        position: fixed ;
        left: 30% ;
        top: 30% ;
      }

      #content
      {
        background: linear-gradient(45deg, Gold, YellowGreen, RoyalBlue, Purple) ;
        width: 1500px ;
        height: 1500px ;
      }
    </style>
  </head>
  <body>
    <div id="button_block">
      <button id="scroll01">Scroll To</button>
      <button id="scroll02">Scroll By</button>
    </div>
```

```
    <div id="content"></div>
  </body>
  <script>
    scroll01.onclick = function (event)
    {
      // window.scroll(0, 500) ;
      // scrollTo(300, 500) ;
      window.scrollTo(300, 500) ;
    } ;

    scroll02.onclick = function (event)
    {
      // scrollBy(20, 60) ;
      window.scrollBy(20, 60) ;
    } ;
  </script>
</html>
```

【相关说明】

```
  <script>
    scroll01.onclick = function (event)
    {
      // window.scroll(0, 500) ;
      // scrollTo(300, 500) ;
      window.scrollTo(300, 500) ;
```

window.scrollTo(0, 500)会驱使浏览器, 将当前网页的内容, 滚动到坐标 x 值为 300 像素、y 值为 500 像素的位置。

```
    } ;
```

上述语法将匿名函数的定义, 赋给了属性 id 的数据为'scroll01'的 button 元素实例的属性 onclick。一旦上述 button 元素实例被单击时, 上述匿名函数就会被调用。

```
    scroll02.onclick = function (event)
    {
      // scrollBy(20, 60) ;
      window.scrollBy(20, 60) ;
```

window.scrollBy(20, 60)会驱使浏览器, 将当前网页的内容, 向右滚动 20 像素, 以及向下滚动 60 像素。

```
    } ;
```

上述语法将匿名函数的定义, 赋给了属性 id 的数据为'scroll02'的 button 元素实例的属性 onclick。一旦上述 button 元素实例被单击时, 上述匿名函数就会被调用。

17.2　screen 对象实例

window.screen 对象实例，可被简写为 screen 对象实例，内含当前设备的屏幕（screen）的相关数据。请参看如下示例。

【17-2-^-window-screen.html】

```
<!DOCTYPE html>
<html>
  <head>
    <title>window screen</title>
  </head>
  <body>
  </body>
  <script>
    // console.log(window.screen.width) ;
    console.log(screen.width) ;
    console.log(screen.height) ;
    console.log('') ;

    console.log(screen.availWidth) ;
    console.log(screen.availHeight) ;
    console.log('') ;

    console.log(screen.colorDepth) ;
    console.log(screen.pixelDepth) ;
    console.log('') ;
  </script>
</html>
```

【相关说明】

```
<script>

  // console.log(window.screen.width) ;
  console.log(screen.width) ;
```

screen.width 会返回当前设备的屏幕的宽度，例如 1360 像素。

```
  console.log(screen.height) ;
```

screen.height 会返回当前设备的屏幕的高度，例如 768 像素。

```
  console.log('') ;

  console.log(screen.availWidth) ;
```

screen.availWidth 会返回当前浏览器窗口可被占用的宽度（available width），例如 1360 像素。

```
  console.log(screen.availHeight) ;
```

screen.availHeight 会返回当前浏览器窗口可被占用的高度，例如 728 像素。

```
  console.log('') ;
```

```
console.log(screen.colorDepth) ;
```

screen.colorDepth 会返回当前设备的屏幕的颜色深度（color depth），也就是用来表示特定颜色的比特位（bit）个数，例如 24 比特位。

```
console.log(screen.pixelDepth) ;
```

screen.pixelDepth 会返回当前设备的屏幕的像素深度（pixel depth），也就是 1 个像素占用的比特位（bit）个数，例如 24 比特位。

17.3　history 对象实例

window.history 对象实例，可被简写为 history 对象实例，内含在当前浏览器窗口里，其用户浏览过的各网页网址的历史（history）记录。请参看如下带有 3 个源代码文档的示例。

【17-3-^-window-history.html】

```html
<!DOCTYPE html>
<html>
  <head>
    <title>window history</title>
  </head>
  <body>
    <button id="previous" disabled>Go back</button>
    <button id="next">Go next</button>

    <h1 style="color: Gold">This is main page.</h1>
    <a href="17-3-^-window-history-test-page-a.html">
      Go to page a
    </a>
  </body>
  <script>
    next.onclick = function (event)
    {
      // window.history.forward() ;
      // history.forward() ;
      history.go(1) ;
    } ;
  </script>
</html>
```

【相关说明】

```
<script>
  next.onclick = function (event)
  {
    // window.history.forward() ;
    // history.forward() ;

    history.go(1) ;
```

此语句会驱使浏览器，访问其历史（history）记录中的下一个（1）网页。

```
  } ;
```

上述语法将匿名函数的定义，赋给了属性 id 的数据为'next'的 button 元素实例的属性 onclick。一旦上述 button 元素实例被单击时，上述匿名函数就会被调用。

【17-3-^-window-history-test-page-a.html】

```
<!DOCTYPE html>
<html>
 <head>
   <title>window history test page a</title>
 </head>
 <body>
   <button id="previous">Go back</button>
   <button id="next">Go next</button>

   <h1 style="color: RoyalBlue">This is test page a</h1>
   <a href="17-3-^-window-history-test-page-b.html">
     Go to page b
   </a>
 </body>
 <script>
   previous.onclick = function (event)
   {
     // history.back() ;
     history.go(-1) ;
   } ;

   next.onclick = function (event)
   {
     // history.next() ;
     history.go(1) ;
   } ;
 </script>
</html>
```

【相关说明】

```
<script>
  previous.onclick = function (event)
  {
    // history.back() ;
    history.go(-1) ;
```

此语句会驱使浏览器，访问其历史（history）记录中的上一个（-1）网页。

```
  } ;
```

上述语法将匿名函数的定义，赋给了属性 id 的数据为'previous'的 button 元素实例的属性 onclick。一旦上述 button 元素实例被单击时，上述匿名函数就会被调用。

```
    next.onclick = function (event)
    {
```

```
    // history.next() ;
    history.go(1) ;
```

此语句会驱使浏览器,访问其历史(history)记录中的下一个网页。

```
  } ;
```

上述语法将匿名函数的定义,赋给了属性 id 的数据为'next'的 button 元素实例的属性 onclick。一旦上述 button 元素实例被单击时,上述匿名函数就会被调用。

【17-3-^-window-history-test-page-b.html】

```
<!DOCTYPE html>
<html>
 <head>
  <title>window history test page b</title>
 </head>
 <body>
  <button id="previous">Go back</button>
  <button id="next" disabled>Go next</button>

  <h1 style="color: GreenYellow">This is test page b</h1>
 </body>
 <script>
  previous.onclick = function (event)
  {
    // history.back() ;
    history.go(-1) ;
  } ;
 </script>
</html>
```

【相关说明】

```
<script>
  previous.onclick = function (event)
  {

    // history.back() ;
    history.go(-1) ;
```

此语句会驱使浏览器,访问其历史(history)记录中的上一个(-1)网页。

```
  } ;
```

上述语法将匿名函数的定义,赋给了属性 id 的数据为'previous'的 button 元素实例的属性 onclick。一旦上述 button 元素实例被单击时,上述匿名函数就会被调用。

17.4　location 对象实例

window.location 对象实例,可被简写为 location 对象实例,用来获取或设置浏览器窗口内的当前网页的网址。请参看如下示例。

【17-4-^-window-location.html】

```html
<!DOCTYPE html>
<html>
  <head>
    <title>window location</title>
  </head>
  <body>
    <button id="assign_btn">Assign new URL</button>
  </body>
  <script>
    // console.log(window.location) ;
    console.log(location) ;
    console.log('') ;

    console.log(location.protocol) ;
    console.log(location.hostname) ;
    console.log(location.port) ;
    console.log(location.pathname) ;
    console.log(location.href) ;

    assign_btn.onclick = function (event)
    {
      // location.href = 'http://www.tup.tsinghua.edu.cn/index.html' ;
      location.assign('http://www.tup.tsinghua.edu.cn/index.html') ;
    } ;
  </script>
</html>
```

【相关说明】

```html
<script>

  // console.log(window.location) ;
  console.log(location) ;
```

window.location 或者 location 均会返回【内含当前网页的网址相关数据】的 location 对象实例。此语句当前会显示出类似【Location {replace: *f*, assign: *f*, href: "file:///D:/examples/html/17-4-^-window-location.html", ancestorOrigins: DOMStringList, origin: "file://", …}】的可展开的信息。

```
console.log('') ;

console.log(location.protocol) ;
```

location.protocol 会返回当前网页的网址中的**协议代号**字符串，例如'file:'或'http:'。

```
console.log(location.hostname) ;
```

location.hostname 会返回当前网页的网址中的主机名称（host name），例如'localhost'、'www.local.com'或空字符串""等。若用户并未通过 HTTP 协议，浏览当前网页，则被返回的主机名称会是空字符串"。

```
console.log(location.port) ;
```

location.port 会返回当前网页的网址中的端口号（port number），例如字符串'80'或空字符串""等。

```
console.log(location.pathname) ;
```

location.pathname 会返回当前网页的网址中的路径名称（path name），例如【/D:/examples/html/17-4-^-window-location.html】或【/html/17-4-^- window-location.html】等。

```
console.log(location.href) ;
```

location.href 会返回当前网页的完整网址，例如【file:///D:/examples/html/17-4-^-window-location.html】或者【http://主机名称/html/17-4-^-window-location.html】等。

```
assign_btn.onclick = function (event)
{
    // location.href = 'http://www.tup.tsinghua.edu.cn/index.html' ;
    location.assign('http://www.tup.tsinghua.edu.cn/index.html') ;
```

此语句会驱使浏览器，加载网址【http://www.tup.tsinghua.edu.cn/index.html】对应的网页。

```
} ;
```

上述语法将匿名函数的定义，赋给了属性 id 的数据为'assign-btn'的 button 元素实例的属性 onclick。一旦上述 button 元素实例被单击时，上述匿名函数就会被调用。

17.5　练　习　题

1. 试编写 JavaScript 源代码片段，可在不直接与间接调用函数 assign()的情况下，等价于 JavaScript 语句【location.assign('http://www.tup.tsinghua.edu.cn/index.html') ;】的效果。

2. 试编写 JavaScript 源代码片段，可在不直接与间接调用函数 forward()的情况下，等价于 JavaScript 语句【history.forward() ;】。

第18章

类

在 JavaScript 语言里,类(class)的自定义语法,具有面向对象编程(OOP, object-oriented programming)的理念,使得特定的类,如同具有特定功能的模块(module),可用来创建作为其副本的对象(object)实例。

18.1 类的定义和继承

定义特定的类,必须通过关键字 class 的协助。欲使得特定子类(sub class),继承自特定父类(parent class),则必须通过关键字 extends 的协助。请参看如下示例。

【18-1-^-class-definitions-and-inheritance.js】

```js
class RadiusClass
{
  constructor(r)
  {
    this.r = r ;
  }

  circumference(r = this.r)
  {
    return 2 * Math.PI * r ;
  }

  circle_area(r = this.r)
  {
    return Math.PI * r * r ;
  }
}
```

```
class Cylinder extends RadiusClass
{
  constructor(r, h)
  {
    super(r) ;

    this.r = r ;
    this.h = h ;
  }

  volume(r = this.r, h = this.h)
  {
    return super.circle_area(r) * h ;
  }

  surface_area(r = this.r, h = this.h)
  {
    return 2 * super.circle_area(r) + super.circumference(r) * h ;
  }
}

let rc01 = new RadiusClass(10) ;

console.log(rc01.circumference()) ;
console.log(rc01.circumference(15)) ;
console.log('') ;

let c01 = new Cylinder(20, 30) ;

console.log(c01.volume()) ;
console.log(c01.surface_area()) ;
```

【相关说明】

请参考本节后面的相关说明。

18.1.1 类的定义（ES6）

在类的定义语法中，存在关键字 class、类名（class name），以及一对大括号里的类体（class body）。类体可内含成员函数（member function），例如由关键字 constructor 开头的唯一的构造函数（constructor），以及**不带**关键字 function 开头的多个自定义函数。

欲在类里，定义数据成员（data member）/ 成员变量（member variable），则必须在特定成员函数（例如构造函数）的定义语法中，通过内置的局部变量 this 来实现，例如【this.r = 100 ;】或者【this.r = r ;】等语法。

```
class RadiusClass
```

此语法通过关键字 class，设置了类名 RadiusClass。

```
{
  constructor(r)
  {
```

```
        this.r = r ;
```

此语句将【其构造函数被调用时的参数 r】的数据，赋给了类 RadiusClass 的对象实例的成员变量 this.r。其中，内置的局部变量 this，是用来指向此类的特定对象实例的指针（pointer）变量。

```
    }
```

上述语法在类 RadiusClass 的定义语法里，进一步定义了带有参数 r 的构造函数 constructor()。请留意，在 JavaScript 语言中，调用特定类的构造函数的语法，并不是借助名称 constructor，而是通过其类名，再加上一对带有被传入的参数数据的小括号，例如 RadiusClass(10)。

```
    circumference(r = this.r)
    {
      return 2 * Math.PI * r ;
```

在此语句中，参数 r 的数值，代表特定半径值。所以，此语句返回了特定半径值的圆周长。

```
    }
```

上述语法定义了带有参数 r 的成员函数 circumference()。其中，参数 r 被设置了默认值，所以若成员函数 circumference()被调用时，未被传入参数 r 的数据，则参数 r 的默认值，就会是【在类 RadiusClass 的特定对象实例中，其成员变量 this.r】的数值。

```
    circle_area(r = this.r)
    {
      return Math.PI * r * r ;
```

此语句返回了【以参数 r 的数值为半径值】的圆面积。

```
    }
```

上述语法定义了带有参数 r 的成员函数 circle_area()。其中，参数 r 被设置了默认值。所以若成员函数 circle_area()被调用时，未被传入参数 r 的数据，则参数 r 的默认值，就会是【在类 RadiusClass 的特定对象实例中，其成员变量 this.r】的数值。

```
}
```

上述语法定义了内含 3 个函数的类 RadiusClass。

```
let rc01 = new RadiusClass(10) ;
```

此语句声明了变量 rc01，其初始数据为【类 RadiusClass 的构造函数 constructor(10)被调用时，所返回】的类 RadiusClass 的对象实例。

```
console.log(rc01.circumference()) ;
```

在此，变量 rc01 所代表的类 RadiusClass 的对象实例，内含成员变量 r 的数值 10。所以，rc01.circumference()当前会返回半径值为 10 的圆周长 62.83185307179586。

```
console.log(rc01.circumference(15)) ;
```

rc01.circumference(15)会返回半径值为 15 的圆周长 94.24777960769379。

18.1.2 类的继承（ES6）

在类的继承（inheritance）语法中，主要借助关键字 extends，以及其前方与后方的两个类名。例如在语法【class **Cylinder** extends **RadiusClass** { ...}】中，其前方的类 Cylinder 为子类（sub class），其后方的类 RadiusClass 为父类（parent class）。有了继承关系之后，在子类的函数内，欲访问父类的成员时，可通过代表着父类的对象实例的关键字 super 来达成。

```
class Cylinder extends RadiusClass
```

此语法通过关键字 extends，使得子类 Cylinder，继承自父类 RadiusClass。

```
{
  constructor(r, h)
  {
    super(r) ;
```

super(r)被调用时，等同于调用了父类 RadiusClass 的构造函数 constructor(r)。

```
    this.r = r ;
    this.h = h ;
```

这两个语句将【构造函数 constructor()被调用时，其被传入的参数 r 与 h】的数据，分别赋给了类 Cylinder 的特定对象实例的成员变量 this.r 与 this.h。

```
  }
```

上述语法在类 Cylinder 的定义语法中，进一步定义了带有参数 r 与 h 的构造函数 constructor()。请留意，在 JavaScript 语言中，调用特定类的构造函数的语法，并不是借助名称 constructor，而是通过其类名，再加上一对带有被传入的参数数据的小括号，例如 Cylinder(20, 30)。

```
  volume(r = this.r, h = this.h)
  {
    return super.circle_area(r) * h ;
```

此语句返回了特定圆柱体积。

```
  }
```

上述语法定义了带有参数 r 与 h 的成员函数 volume()。其中，参数 r 与 h 均被设置了默认值。所以若函数 volume()被调用时，未被传入参数 r 与 h 的数据，则参数 r 的默认值，就会是【在类 RadiusClass 的特定对象实例中，其成员变量 this.r】的数值；而参数 h 的默认值，就会是【在类 RadiusClass 的特定对象实例中，其成员变量 this.h】的数值。

```
  surface_area(r = this.r, h = this.h)
  {
    return 2 * super.circle_area(r) + super.circumference(r) * h ;
```

此语句返回了特定圆柱体表面积。

```
  }
```

上述语法定义了带有参数 r 与 h 的成员函数 surface_area()。其中，参数 r 与 h 均被设置了默认值。所以若函数 surface_area() 被调用时，未被传入参数 r 与 h 的数据，则参数 r 的默认值，就会是【在 RadiusClass 类的特定对象实例中，其成员变量 this.r】的数值；而参数 h 的默认值，就会是【在 RadiusClass 类的特定对象实例中，其成员变量 this.h】的数值。

```
}
```

上述语法定义了【内含 3 个函数，并且继承自父类 RadiusClass】的子类 Cylinder。

```
let c01 = new Cylinder(20, 30) ;
```

此语句声明了变量 c01，其初始数据为【类 Cylinder 的构造函数 constructor(20, 30) 被调用时，所返回】的类 Cylinder 的新对象实例。

```
console.log(c01.volume()) ;
```

在变量 c01 所代表的类 Cylinder 的对象实例中，存在成员变量 r 的数值 20 与成员变量 h 的数值 30。所以，c01.volume() 当前会返回半径值为 20、高度值为 30 的圆柱体积 37699.11184307752。

```
console.log(c01.surface_area()) ;
```

c01.surface_area() 会返回半径值为 20、高度值为 30 的圆柱体表面积 6283.185307179587。

18.2 类的静态成员

关键字 static 开头的函数的定义语法，是用来描述类的静态（static）函数（function）/ 方法（method），例如：

```
class Cubic
{
    ┆
    static about()
    {
        ┆
    }
    ┆
}
┆
c01 = new Cubic(•••) ;
```

请留意，调用特定类的静态函数的语法，必须通过其类名，而不是借助【数据为特定类的对象实例】的变量名称。举例来说，已知变量 c01 的数据为类 Cubic 的对象实例，此时调用类 Cubic 的静态函数 about() 的语法为【Cubic.about()】，而不是【c01.about()】。关于类的静态成员的综合运用，请参看如下示例。

【18-2-^-class-static-members.js】

```
class Cubic
{
    constructor(l, w, h)
```

```
  {
    this.length = l ;
    this.width = w ;
    this.height = h ;
  }

  volume(l = this.length, w = this.width, h = this.height)
  {
    return l * w * h ;
  }

  surface_area(l = this.length, w = this.width, h = this.height)
  {
    return 2 * (l * w + w * h + h * l) ;
  }

  static about()
  {
    console.log('This Class is about calculations of cubic volume and surface area.') ;
  }
}

let c01 = new Cubic(15, 23, 37) ;

console.log(c01.volume()) ;
console.log(c01.volume(10, 20, 30)) ;
console.log('') ;

console.log(c01.surface_area()) ;
console.log(c01.surface_area(10, 20, 30)) ;
console.log('') ;

Cubic.about() ;
c01.about() ;
```

【相关说明】

请参看本节后面的相关说明。

18.2.1 静态成员的概念和定义（ES6）

所谓的静态成员（static member）是被共用的成员。在 JavaScript 语言中，对于静态成员，尚未支持静态数据成员（static data member）/ 静态成员变量（static member variable），当前仅限于静态成员函数（static member function）。请参看下面的说明。

```
class Cubic
```

此语法通过关键字 class，设置了类名 Cubic。

```
{
  constructor(l, w, h)
  {
    this.length = l ;
```

```
    this.width = w ;
    this.height = h ;
```

在其构造函数 constructor() 被调用时,这 3 个语句可将参数 l、w 与 h 的数据,分别赋给类 Cubic 的特定对象实例的成员变量 this.length、this.width 与 this.height。其中,内置的局部变量 this,是用来指向类 Cubic 的特定对象实例的指针(pointer)变量。

```
  }
```

上述语法在类 Cubic 的定义语法中,进一步定义了带有参数 r 的构造函数 constructor()。请留意,在 JavaScript 语言中,调用特定类的构造函数的语法,并不是借助名称 constructor,而是通过其类名,再加上一对带有被传入的参数数据的小括号,例如 Cubic(15, 23, 37)。

```
  volume(l = this.length, w = this.width, h = this.height)
  {
    return l * w * h ;
```

此语句返回了特定方体积。

```
  }
```

上述语法定义了带有参数 l、w 与 h 的成员函数 volume()。其中,参数 l、w 与 h 均被设置了默认值。若函数 volume() 被调用时,未被传入参数 l、w 或 h 的数据,则:

- 参数 l 的默认值,就会是【在类 Cubic 特定对象实例中,其成员变量 this.length】的数值。
- 参数 w 的默认值,就会是【在类 Cubic 特定对象实例中,其成员变量 this.width】的数值。
- 参数 h 的默认值,就会是【在类 Cubic 特定对象实例中,其成员变量 this.height】的数值。

```
  surface_area(l = this.length, w = this.width, h = this.height)
  {
    return 2 * (l * w + w * h + h * l) ;
```

上述语法返回了特定方体表面积。

```
  }
```

上述语法定义了带有参数 l、w 与 h 的成员函数 surface_area()。其中,参数 l、w 与 h 均被设置了默认值。若函数 surface_area() 被调用时,未被传入参数 l、w 或 h 的数据,则:

- 参数 l 的默认值,就会是【在类 Cubic 特定对象实例中,其成员变量 this.length】的数值。
- 参数 w 的默认值,就会是【在类 Cubic 特定对象实例中,其成员变量 this.width】的数值。
- 参数 h 的默认值,就会是【在类 Cubic 特定对象实例中,其成员变量 this.height】的数值。

```
  static about()
  {
    console.log('This Class is about calculations of cubic volume and surface area.') ;
  }
```

上述语法定义了静态成员函数 about(),并用来显示出【This Class is about calculations of cubic volume and surface area.】的信息。

```
}
```

上述语法定义了类 Cubic。

18.2.2　静态成员的运用（ES6）

无论特定类的对象实例有多少个，其各个对象实例均是共用所有的静态成员！也就是说，调用其静态成员函数的方式，例如 Cubic.about()，并不是通过【数据为特定类的对象实例】的变量名称（例如 c01），而是借助其类名（例如 Cubic）。请参看如下说明。

```
c01 = new Cubic(15, 23, 37) ;
```

此语句声明了变量 c01，其初始数据为【类 Cubic 的构造函数 constructor(15, 23, 37)被调用时，所返回】的类 Cubic 的对象实例。

```
console.log(c01.volume()) ;
```

变量 c01 所代表的类 Cubic 的对象实例，内含成员变量 l、w 与 h 的个别数值 15、23 与 37。c01.volume()会返回长度为 15、宽度为 23、高度为 37 的方体积 12765。

```
console.log(c01.volume(10, 20, 30)) ;
```

c01.volume(10, 20, 30)则会返回长度为 10、宽度为 20、高度为 30 的方体积 6000。

```
console.log(c01.surface_area()) ;
```

c01.surface_area()会返回长度为 15、宽度为 23、高度为 37 的方体表面积 3502。

```
console.log(c01.surface_area(10, 20, 30)) ;
```

c01.surface_area(10, 20, 30)则会返回长度为 10、宽度为 20、高度为 30 的方体表面积 2200。

```
Cubic.about() ;
```

Cubic.about()会显示出【This Class is about calculations of cubic volume and surface area.】的信息。因为函数 about()被定义成为类 Cubic 的静态成员函数，所以调用函数 about()的语法，并不是通过【数据为类 Cubic 的特定对象实例】的变量名称 c01，例如【c01.about()】，而是借助类名 Cubic，例如【Cubic.about()】。

```
c01.about() ;
```

若通过变量名称 c01，来调用类 Cubic 的静态成员函数 about()，则会产生【Uncaught TypeError: c01.about is not a function】的错误信息。

18.3　类的设置器和取得器

类的设置器（setter）和取得器（getter），可分别用来设置或取得【经过动态评估之后】的数据。其设置器和取得器的定义语法，除了开头的关键字 set 和 get 之外，其余部分均相同于不带关键字 function 的函数的定义语法。请参看如下示例。

【18-3-^-class-setter-and-getter.js】

```javascript
class Person
{
  constructor(name = '?', age = '?', gender = '?', department = '?')
  {
    this._name = name ;
    this._age = age ;
    this._gender = gender ;
    this._department = department ;
  }

  set name (name)
  {
    this._name = name ;
    console.log('Name is edited!') ;
  }

  get name ()
  {
    console.log('Name is got!') ;
    return this._name ;
  }
}

let p01 = new Person() ;

console.log(p01.name) ;
console.log('') ;

p01.name = 'Jasper' ;

console.log(p01.name) ;
console.log(p01._name) ;
```

【相关说明】
请参看本节后面的相关说明。

18.3.1 设置器和取得器的概念和定义（ES6）

设置器如同是【会动态设置特定成员变量的数据】的函数，取得器则如同是【会动态返回特定成员变量的数据】的函数。请参看如下说明。

```javascript
class Person
{
  constructor(name = '?', age = '?', gender = '?', department = '?')
  {
    this._name = name ;
    this._age = age ;
    this._gender = gender ;
    this._department = department ;
```

在其构造函数 constructor()被调用时，这 4 个语句可分别将其参数 name、age、gender 与 department 的数据，赋给类 Person 的特定对象实例的个别成员变量 this._name、this._age、this._gender 与 this._department。其中，内置的局部变量 this，是用来指向类 Person 的特定对象实例的指针 (pointer) 变量。

```
    }
```

上述语法在类 Person 的定义语法中，进一步定义了带有参数 name、age、gender 与 department 的构造函数 constructor()。请留意，在 JavaScript 语言中，调用特定类的构造函数的语法，并不是借助名称 constructor，而是通过其类名，再加上一对带有被传入的参数数据的小括号，例如 Person() 或者 Person('Jasper', 28, 'male', 'IT')等。其中，参数 name、age、gender 与 department 均被设置了默认数据为字符'?'。若其构造函数 constructor()被调用时，未被传入上述各参数的数据，则上述各参数的默认数据，就会是字符'?'。

```
    set name (new_name)
    {
      this._name = new_name ;
      console.log('Name is edited!') ;
    }
```

此语法通过关键字 set，成为了类 Person 的 name 设置器。其中，name 设置器被访问（调用）时，必须省略其小括号。例如语句【p01.name = 'Jasper' ;】，可使得字符串'Jasper'，间接被赋给变量 p01 的对象实例的成员变量 this._name。

```
    get name ()
    {
      console.log('Name is got!') ;
      return this._name ;
    }
```

此语法通过关键字 get，成为了类 Person 的 name 取得器。其中，name 取得器被访问（调用）时，也必须省略其小括号。例如语法【p01.name】，可返回【在变量 p01 的对象实例中，其成员变量 this._name】的数据字符串'Jasper'。

```
}
```

上述语法定义了类 Person。

18.3.2　设置器和取得器的运用（ES6）

在特定类的特定对象实例中，特定设置器和取得器的访问语法，均如同于特定成员变量的访问语法。例如：

- 已知语句【let p01 = new Person() ;】是用来声明【数据为类 Person 的对象实例】的变量 p01。
- 若其 name 设置器的定义语法为【set name (new_name) { ⋯ }】，则 name 设置器的访问语法，就会是【p01.name】。
- 若其 name 取得器的定义语法为【get name () { ⋯ }】，则 name 取得器的访问语法，就会是【p01 – 'Jasper' ;】。

```
let p01 = new Person();
```

此语句声明了变量 p01，其初始数据为【类 Person 的构造函数 constructor() 被调用时，所返回】的类 Person 的对象实例。

```
console.log(p01.name);
```

p01.name 会访问到类 Person 的 name 取得器，间接返回【在变量 p01 的对象实例中，其成员变量 this._name】的默认数据字符'?'。

```
p01.name = 'Jasper';
```

此语句会访问到类 Person 的 name 设置器，间接设置【在变量 p01 的对象实例中，其成员变量 this._name】的数据，成为字符串'Jasper'。

```
console.log(p01.name);
```

p01.name 会再次访问到类 Person 的 name 取得器，间接返回【在变量 p01 的对象实例中，其成员变量 this._name】的当前数据字符串'Jasper'。

```
console.log('');
console.log(p01._name);
```

p01._name 则会直接返回【在变量 p01 的对象实例中，其属性_name】的数据字符串'Jasper'。

18.4 练习题

1. 编写 JavaScript 源代码，以定义 1 个内含如下函数的类（class）：

- 用来计算并返回 $\sum_{x=1}^{n} x$, where $\{n \in \mathbb{N} \mid n \geq 2\}$ 函数 sum(n)。
- 用来计算并返回 $\sum_{x=1}^{n} x^2$, where $\{n \in \mathbb{N} \mid n \geq 2\}$ 的函数 square_sum(n)。
- 用来计算并返回 $\sum_{x=1}^{n} x^3$, where $\{n \in \mathbb{N} \mid n \geq 2\}$ 的函数 cube_sum(n)。

2. 编写 JavaScript 源代码，以定义 1 个内含如下函数的类（class）：

- 用来计算并返回 $\sum_{x=1}^{n} (-1)^x \cdot x$, where $\{n \in \mathbb{N} \mid n \geq 2\}$ 的函数 alternate_sum(n)。
- 用来计算并返回 $\sum_{x=1}^{n} (-1)^x \cdot x^2$, where $\{n \in \mathbb{N} \mid n \geq 2\}$ 的函数 alternate_square_sum(n)。
- 用来计算并返回 $\sum_{x=1}^{n} (-1)^x \cdot x^3$, where $\{n \in \mathbb{N} \mid n \geq 2\}$ 的函数 alternate_cube_sum(n)。

3. 试改写第 2 题的类，以继承自第 1 题的类。
4. 请在 JavaScript 语言的各个内置的对象里，找出至少 7 个静态函数。

5. 已知如下 JavaScript 源代码片段：

```
class Shape
{
  display()
  {
    console.log('display Shape message...') ;
  }
}

class Triangle extends Shape
{
  display()
  {
    console.log('display Triangle message...') ;
  }
}

let t01 = new Triangle() ;
```

试编写 JavaScript 源代码的语法：

（1）以通过变量 t01，调用到类 Triangle 的函数 display()。

（2）以通过变量 t01，调用到类 Shape 的函数 display()。

（3）进而改写类 Triangle 的定义语法，使得调用类 Triangle 的函数 display()时，可连带调用到类 Shape 的函数 display()。

第 19 章

错 误 处 理

在支持特定编程语言的集成开发环境中，执行开发中的应用程序时，若发生任何异常错误（exception error），则汇编器（assembler）、编译器（compiler）或解释器（interpreter），应该显示特定的错误信息，以供编程人员进行调试。

JavaScript 引擎属于解释器的一种，一边解释（interpret）特定源代码片段，一边立即执行对应的任务。在特定应用程序里，若潜藏着未被发觉的特定异常错误，则可能会导致用户在操作时，看见无法理解的相关错误信息。

欲使得本来无法被理解的相关错误信息，成为用户可以理解或者不会看见的错误信息，则编程人员必须加以编写，用来处理特定异常错误的相关源代码片段才行。

19.1 异常错误的种类

在 JavaScript 语言中，常见的异常错误种类，大致如下：

- 语法（syntax）错误
- 数据类型（data type）错误
- 范围（range）错误
- 引用（reference）错误
- 网址（URI / URL, uniform resource identifier / uniform resource locator）的编码（encoding）或者解码（decoding）错误
- 逻辑（logical）错误

其中，逻辑错误的主要原因，是编程人员思考得不够周全而导致的，所以汇编器、编译器和解释器均难以察觉。

19.1.1 语法错误

JavaScript 引擎无法顺利解释特定语法（syntax）时，便会产生语法的错误（syntax error）。请参看如下示例。

【19-1-1-SyntaxError.js】

```
let v01 = 100, v02 = 230 ;
let result = 0 ;

try
{
  eval('result = v01 -- v02 ;') ;
}
catch (err)
{
  with (console)
  {
    log(err) ;
    log('') ;

    log(typeof err) ;
    log(err instanceof SyntaxError) ;
    log('') ;

    log(err.name) ;
    log(err.message) ;
    log(err.stack) ;
  }
}
```

【相关说明】

```
let v01 = 100, v02 = 230 ;
```

声明初始值为不同整数值的变量 v01 与 v02。

```
let result = 0 ;
```

声明初始值为 0 的变量 result。

```
try
{

  eval('result = v01 -- v02 ;') ;
```

在此语句中，调用内置函数 eval() 时，传入了代表一个表达式的数据字符串'result =v01 -- v02;'。在上述表达式里，被视为操作数的变量名称 v01 和 v02 之间，放入递减运算符【--】，是不被支持的。所以，此语句被执行时，会发生 SyntaxError 异常错误，而跳转至其下方【catch(err)】语句的大括号里。

```
}
```

通过 try 语句，可使得其大括号里的各个子语句，例如【eval('result = v01 -- v02 ;') ;】，一旦被监听到任何异常错误时，其下方 catch(err)语句之大括号里的源代码，就会被执行。

```
catch (err)
{
  with (console)
```

此语法简化了【conole.log(err)】，成为下方大括号里的【log(err)】。

```
  {
    log(err) ;
```

通过内含错误信息（例如【SyntaxError: Unexpected identifier】）相关数据的参数 err，可得知究竟发生了什么异常错误。请留意，调用【console.log(err)】，虽然可返回参数 err 的数据信息；但是，参数 err 的数据类型，并非 string（字符串）。

```
    log('') ;
    log(typeof err) ;
    log(err instanceof SyntaxError) ;
```

因为【typeof err】返回了字符串'object'，再加上【err instanceof SyntaxError】返回了布尔值 true 的缘故，可得知参数 err 的数据类型，是 SyntaxError 对象（object）实例。

```
    log('') ;
    log(err.name) ;
```

err.name 当前返回了字符串'SyntaxError'。

```
    log(err.message) ;
```

err.message 当前返回了字符串'Unexpected identifier'。

```
    log(err.stack) ;
```

err.stack 当前返回了'SyntaxError: Unexpected identifier'。

```
  }
}
```

在上述 try 语句的大括号里，一旦被监听到发生任何异常错误时，其【catch(err)】语句的大括号里的源代码，就会被执行。在其小括号里的 err，可被视为内含特定异常错误相关数据的参数。

19.1.2 数据类型错误

JavaScript 引擎无法顺利转换特定操作数（operand）的数据类型（data type）时，便会产生数据类型的错误（data type error）。请参看如下示例。

【19-1-2-TypeError.js】

```
let v01 = 100, v02 = 230 ;
let result = 0 ;
```

```
try
{
  result = v01.replace('0', '1') ;

  console.log(result) ;
}
catch (err)
{
  with (console)
  {
    log(err) ;
    log('') ;

    log(typeof err) ;
    log(err instanceof TypeError) ;
    log('') ;

    log(err.name) ;
    log(err.message) ;
    log(err.stack) ;
  }
}
```

【相关说明】

```
let v01 = 100, v02 = 230 ;
```

声明初始值为不同整数值的变量 v01 与 v02。

```
let result = 0 ;
```

声明初始值为 0 的变量 result。

```
try
{

  result = v01.replace('0', '1') ;
```

因为变量 v01 当前的数据是数值，并不支持函数 replace()，所以上述语句被执行时，会发生 TypeError 异常错误，进而使得其下方【catch(err)】语句的大括号里的源代码，被执行。

```
  console.log(result) ;
```

一旦上一个语句【result = v01.replace('0', '1') ;】发生了异常错误，则这个语句就不会被执行了。

```
}
```

通过上述 try 语句，可使得其大括号里的各个子语句，例如【result = v01.replace('0', '1') ;】，一旦被监听到发生任何异常错误时，其下方【catch(err)】语句的大括号里的源代码，就会被执行。

```
catch (err)
{

  with (console)
```

此语法简化了【conole.log(err)】，成为下方大括号里的【log(err)】。

```
  {
```

```
        log(err) ;
```

通过内含错误信息（例如【TypeError: v01.replace is not a function】）相关数据的参数 err，可得知究竟发生了什么异常错误。请留意，调用【console.log(err)】，虽然可返回参数 err 的数据信息；但是，参数 err 的数据类型，并非 string（字符串）。

```
        log('') ;
        log(typeof err) ;
        log(err instanceof TypeError) ;
```

【typeof err】返回了字符串'object'，再加上【err instanceof TypeError】返回了布尔值 true，可得知参数 err 的数据类型是 TypeError 对象（object）实例。

```
        log('') ;
        log(err.name) ;
```

err.name 当前返回了字符串'TypeError'。

```
        log(err.message) ;
```

err.message 当前返回了代表错误信息的字符串'v01.replace is not a function'。

```
        log(err.stack) ;
```

err.stack 当前返回了代表更完整的错误信息的字符串'TypeError: v01.replace is not a function'。

```
    }
}
```

在上述 try 语句的大括号里，一旦被监听到发生任何异常错误时，【catch(err)】语句的大括号里的源代码，就会被执行。在其小括号里的 err，可被视为内含特定异常错误相关数据的参数。

19.1.3 评估错误

JavaScript 引擎现今只能通过内置函数【eval(代表特定语句的字符串)】或者语句【throw new EvalError(代表特定错误信息的字符串);】，才会产生评估错误（evaluation error）。请参看如下示例。

【19-1-3-EvalError.js】

```
try
{
  throw new EvalError('Equation evaluation error occurs!') ;
}
catch (err)
{
 with (console)
  {
    log(err) ;
    log('') ;
```

```
    log(typeof err) ;
    log(err instanceof EvalError) ;
    log('') ;

    log(err.name) ;
    log(err.message) ;
    log(err.stack) ;
   }
 }
```

【相关说明】

```
try
{

  throw new EvalError('Equation evaluation error occurs!') ;
```

此语句会直接产生带有自定义错误信息的字符串'Equation evaluation error occurs!'的 EvalError 对象实例，进而使得其下方【catch(err)】语句的大括号里的源代码，间接被执行。

```
}
```

通过上述 try 语句，可使得其大括号里的各个子语句，例如【throw new EvalError('Equation evaluation error occurs!') ;】，一旦被监听到发生任何异常错误时，其下方【catch(err)】语句的大括号里的源代码，就会被执行。

```
catch (err)
{
  with (console)
  {

    log(err) ;
```

通过内含错误信息（例如【EvalError: Equation evaluation error occurs!】）相关数据的参数 err，可得知究竟发生了什么异常错误。请留意，调用【console.log(err)】，虽然可返回参数 err 的数据信息，但是参数 err 的数据类型，并非 string（字符串）。

```
    log('') ;

    log(typeof err) ;
    log(err instanceof EvalError) ;
```

【typeof err】返回了字符串'object'，再加上【err instanceof EvalError】返回了布尔值 true，可得知参数 err 的数据类型是 EvalError 对象（object）实例。

```
    log('') ;

    log(err.name) ;
```

err.name 当前返回了字符串'EvalError'。

```
    log(err.message) ;
```

err.message 当前返回了代表错误信息的字符串'Equation evaluation error occurs!'。

```
    log(err.stack) ;
```

err.stack 当前返回了代表更完整的错误信息的字符串'EvalError: Equation evaluation error occurs!'。

```
    }
}
```

在上述 try 语句的大括号里,一旦被监听到发生任何异常错误时,【catch(err)】语句的大括号里的源代码,就会被执行。在其小括号里的 err,可被视为内含特定异常错误相关数据的参数。

19.1.4 范围错误

JavaScript 引擎评估到特定数据范围(data range)以外的数据时,便会产生数据范围错误(data range error)。请参看如下示例。

【19-1-4-RangeError.js】

```
let value = 200 ;
let list = null ;
let result = 0 ;

try
{
  /*
  // result = value.toFixed(101) ;
  result = value.toPrecision(101) ;
  */
  list = new Array(-3) ;

  console.log(result) ;
}
catch (err)
{
  with (console)
  {
    log(err) ;
    log('') ;

    log(typeof err) ;
    log(err instanceof RangeError) ;

    log('') ;
    log(err.name) ;
    log(err.message) ;
    log(err.stack) ;
  }
}
```

【相关说明】

```
let value = 200 ;
let list = null ;
let result = 0 ;
```

这 3 个语句分别声明了变量 value、list 与 result。

```
try
{
 /*
 // result = value.toFixed(101) ;
 result = value.toPrecision(101) ;
 */
```

内置函数 toFixed()或 toPrecision()并不支持被传入代表高达 101 个小数位的整数值 101。所以此语句被执行时，会发生 RangeError 异常错误，进而使得其下方【catch(err)】语句的大括号里的源代码，间接被执行。

```
 list = new Array(-3) ;
```

Array 对象的构造函数 Array()，并不支持被传入负整数-3。所以此语句被执行时，会发生 RangeError 异常错误，进而使得其下方【catch(err)】语句的大括号里的源代码，间接被执行。

```
 console.log(result) ;
```

因为前面的语句已产生了异常错误，所以此语句就不会被执行了。

```
}
```

通过上述 try 语句，可使得大括号里的各个子语句，例如【list = new Array(-3) ;】，一旦被监听到发生任何异常错误时，其下方【catch(err)】语句的大括号里的源代码，就会间接被执行。

```
catch (err)
{
 with (console)
 {
  log(err) ;
```

通过内含错误信息（例如【RangeError: Invalid array length】）相关数据的参数 err，可得知究竟发生了什么**异常错误**。请留意，通过调用【console.log(err)】，虽然可返回参数 err 的数据信息，但是参数 err 的数据类型，并非 string（字符串）。

```
  log('') ;
  log(typeof err) ;
  log(err instanceof RangeError) ;
```

【typeof err】返回了字符串'object'，再加上【err instanceof RangeError】返回了布尔值 true，可得知参数 err 的数据类型是 RangeError 对象（object）实例。

```
  log('') ;
  log(err.name) ;
```

err.name 当前返回了字符串'RangeError'。

```
  log(err.message) ;
```

err.message 当前返回了代表错误信息的字符串'Invalid array length'。

```
   log(err.stack) ;
```

err.stack 当前返回了代表更完整的错误信息的字符串'RangeError: Invalid array length'。

```
  }
}
```

在上述 try 语句的大括号里，一旦被监听到发生任何异常错误时，其下方【catch(err)】语句的大括号里的源代码，就会间接被执行。在其小括号里的 err，可被视为内含特定异常错误相关数据的参数。

19.1.5　引用错误

尚未被声明的变量名称，被引用时，会产生引用错误（reference error）。请参看如下示例。

【19-1-5-ReferenceError.js】

```
let v01 = 100, v02 = 150 ;
let result = 0 ;

try
{
  result = v01 + v03 ;

  console.log(result) ;
}
catch (err)
{
  with (console)
  {
    log(err) ;
    log('') ;

    log(typeof err) ;
    log(err instanceof ReferenceError) ;
    log('') ;

    log(err.name) ;
    log(err.message) ;
    log(err.stack) ;
  }
}
```

【相关说明】

```
let v01 = 100, v02 = 150 ;
let result = 0 ;
```

这两个语句声明了变量 v01、v02 与 result。

```
try
{
```

```
    result = v01 + v03 ;
```

v03 当前**尚未**被声明成为变量,所以此语句被执行时,会发生 ReferenceError 异常错误,使得其下方【catch(err)】语句的大括号里的源代码,间接被执行。

```
    console.log(result) ;
```

因为前面的语句已产生了异常错误,所以此语句就不会被执行了。

```
  }
```

通过上述 try 语句,可使得其大括号里的各个子语句,例如【result = v01 + v03 ;】,一旦被监听到发生任何异常错误时,其下方【catch(err)】语句的大括号里的源代码,就会被执行。

```
  catch (err)
  {
    with (console)
    {
      log(err) ;
```

通过内含错误信息(例如【ReferenceError: v03 is not defined】)相关数据的参数 err,可得知究竟发生了什么异常错误。请留意,通过调用【console.log(err)】,虽然可返回参数 err 的数据信息,但是参数 err 的数据类型,并非 string (字符串)。

```
      log('') ;
      log(typeof err) ;
      log(err instanceof ReferenceError) ;
```

【typeof err】返回了字符串'object',再加上【err instanceof ReferenceError】返回了布尔值 true,可得知参数 err 的数据类型是 ReferenceError 对象(object)实例。

```
      log('') ;
      log(err.name) ;
```

err.name 当前返回了字符串'ReferenceError'。

```
      log(err.message) ;
```

err.message 当前返回了代表错误信息的字符串'v03 is not defined'。

```
      log(err.stack) ;
```

err.stack 当前返回了代表更完整的错误信息的字符串'ReferenceError: v03 is not defined'。

```
    }
  }
```

在上述 try 语句的大括号里,一旦被监听到任何异常错误时,其下方【catch(err)】语句的大括号里的源代码,就会间接被执行。在其小括号里的 err,可被视为内含特定异常错误相关数据的参数。

19.1.6　网址在编码或解码上的错误

JavaScript 引擎无法顺利对特定网址（URI / URL, uniform resource identifier / uniform resource locator）进行编码（encode）或解码（decode）时，便会产生 URI 错误（URI Error），引申为网址在编码或解码上的错误。请参看如下示例。

【19-1-6-URIError.js】

```
try
{
  decodeURIComponent('https://tw.dictionary.yahoo.com/dictionary?
    p=%E8%81%96%E8%AA%95%E7%AF%FF') ;
}
catch (err)
{
  with (console)
  {
    log(err) ;
    log('') ;

    log(typeof err) ;
    log(err instanceof URIError) ;
    log('') ;

    log(err.name) ;
    log(err.message) ;
    log(err.stack) ;
  }
}
```

【相关说明】

```
try
{
  decodeURIComponent('https://tw.dictionary.yahoo.com/dictionary?
    p=%E8%81%96%E8%AA%95%E7%AF%FF') ;
```

在被传入函数 decodeURIComponent() 的参数数据字符串 'https://tw.dictionary.yahoo.com/dictionary?p=%E8%81%96%E8%AA%95%E7%AF%FF' 里，其中【%E8%81%96%E8%AA%95%E7%AF%FF】的部分，潜藏着 URI 的编码错误。所以，此语句被执行时，会发生 URIError 异常错误，进而使得其下方【catch(err)】语句的大括号里的源代码，间接被执行。

```
}
```

通过上述 try 语句，可使得大括号里的各个子语句，例如【decodeURIComponent('https://tw.dictionary.yahoo.com/dictionary?p=%E8%81%96%E8%AA%95%E7%AF%FF') ;】，一旦被监听到发生任何异常错误时，其下方【catch (err)】语句的大括号里的源代码，就会间接被执行。

```
catch (err)
{
  with (console)
  {
```

```
        log(err) ;
```

通过内含错误信息（例如【URIError: URI malformed at decodeURIComponent (<anonymous>)】）相关数据的参数 err，可得知究竟发生了什么异常错误。请留意，调用 console.log(err)，虽然可返回参数 err 的数据信息，但是参数 err 的数据类型并非 string（字符串）。

```
        log('') ;
        log(typeof err) ;
        log(err instanceof URIError) ;
```

【typeof err】返回了字符串'object'，再加上【err instanceof URIError】返回了布尔值 true，可得知参数 err 的数据类型是 URIError 对象（object）实例。

```
        log('') ;
        log(err.name) ;
```

err.name 当前返回了字符串'URIError'。

```
        log(err.message) ;
```

err.message 当前返回了代表错误信息的字符串'URI malformed'。

```
        log(err.stack) ;
```

err.stack 当前返回了代表更完整的错误信息的字符串'URIError: URI malformed at decodeURIComponent (<anonymous>)'。

```
    }
}
```

在上述 try 语句的大括号里，一旦被监听到任何异常错误时，其下方【catch(err)】语句的大括号里的源代码，就会间接被执行。在其小括号里的 err，可被视为内含特定异常错误相关数据的参数。

19.1.7　逻辑错误

JavaScript 引擎无法察觉，编程人员思考得不够周全而导致的逻辑错误（logical error）。请参看如下示例。

【19-1-7-logical-errors.js】

```
let num_list = Array.from(new Array(11), (value, index) => parseInt(100 * Math.random())) ;

let result = null ;

result = Math.min(... num_list) ;

console.log(num_list) ;
console.log(result) ;
```

```
    // message with logical error
    console.log(`The maximum value is ${result}.`) ;
    console.log('') ;

    ///
    let radius = 55 ;

    // result with logical error
    result = (3 / 4 * Math.PI * radius ** 3).toFixed(3) ;

    console.log(`The sphere volume of radius ${radius} is ${result}`) ;
```

【相关说明】

```
    let num_list = Array.from(new Array(11), (value, index) => parseInt(100 * Math.random())) ;
```

此语句声明了变量 num_list，其初始数据为内含 11 个介于 0～100 之间的随机整数值的数组实例，例如[28, 84, 45, 83, 24, 45, 84, 45, 41, 55, 66]。

```
    let result = null ;
```

声明初始数据为空值（null）的变量 result。

```
    result = Math.min(... num_list) ;
```

Math.min(... num_list)会返回【在变量 num_list 的数组实例中，其最小】的整数值，例如 24。

```
    console.log(num_list) ;
```

显示出变量 num_list 的数组实例，例如[28, 84, 45, 83, 24, 45, 84, 45, 41, 55, 66]。

```
    console.log(result) ;
```

显示出变量 num_list 数组实例中的最小整数值，例如 24。

```
    // message with logical error
    console.log(`The maximum value is ${result}.`) ;
```

此语句显示出类似【The maximum value is 24.】的信息，透露出【在变量 num_list 的数组实例中，其最大】的整数值为 24。

然而，在变量 num_list 的数组实例（例如[28, 84, 45, 83, 24, 45, 84, 45, 41, 55, 66]）中，其最大的整数值并不是 24，而是 84。这种显示出不正确信息的错误，并非语法上的错误，而是编程人员因为思考**不周全**所造成的逻辑错误。

```
    console.log('') ;

    let radius = 55 ;
```

声明初始值为 55 的变量 radius。

```
    // result with logical error
    result = (3 / 4 * Math.PI * radius ** 3).toFixed(3) ;
```

在球体积公式【3 / 4 * Math.PI * radius ** 3】中，其子表达式【3 / 4】存在谬误，其正确的应该是【4 / 3】。这种不正确结果值的错误，也是编程人员思考不周全所造成的逻辑错误。

```
console.log(`The sphere volume of radius ${radius} is ${result}`) ;
```

此语句显示出类似【The sphere volume of radius 55 is 392011.858】的信息。

19.2　处置特定异常错误

在特定应用程序中，难免会潜藏着未被发觉的错误，进而导致用户在操作时，看见无法理解的相关错误信息。

欲转换原本无法被理解的相关错误信息，成为用户可以理解或不会看见的错误信息，编程人员必须在其源代码中，撰写错误处理的相关部分。请参看如下示例。

【19-2-^-try-catch-finally.js】

```
try
{
 //throw new URIError("Defined error message here...") ;
 throw "Logical error occurs." ;
}
catch (err)
{
 console.log(err.name) ;
 console.log(err.message) ;
 console.log(err.stack) ;
 console.log('') ;
 
 console.log(`Special: ${err}`) ;
 console.log('') ;
}
finally
{
 console.log('Message after exception handling.') ;
}
```

【相关说明】

请参看本节下面的相关说明。

19.2.1　试验与捕获特定异常错误

在 JavaScript 语言中，可通过【try {...} catch {...} finally {...}】的语法达成试验（try）与捕获（catch）特定异常错误，最后（finally）综合处置（handling）的机制。

```
try
{
 |
}
```

通过 try 语句，可使得其大括号里的各个子语句，一旦被监听到发生任何异常错误时，其下方

【catch(err)】语句的人括号里的源代码，就会间接被执行。

```
catch (err)
{
  console.log(err.name) ;
```

此语句会显示出内置的异常错误的对象名称，例如 SyntaxError（语法错误）、TypeError（类型错误）、EvalError（评估错误）、RangeError（范围错误）、ReferenceError（参考错误）、URIError（URI 错误）等。

```
  console.log(err.message) ;
  console.log(err.stack) ;
```

err.name、err.message、err.stack 在此均返回原始常量 undefined（未被定义）。

```
  console.log('') ;
  console.log(`Special: ${err}`) ;
```

显示出【Special: Logical error occurs.】的信息。

```
  console.log('') ;
}
```

在前面的 try 语句的大括号里，一旦被监听到任何异常错误时，上述【catch(err)】语句的大括号里的源代码，就会间接被执行。在其小括号里的 err，可被视为内含特定异常错误相关数据的参数。

```
finally
{
  console.log('Message after exception handling.') ;
```

此语句显示出【Message after exception handling.】的信息。

```
}
```

无论在前面的 try 语句的大括号里，其源代码有无异常错误，后续在上述 finally 语句的大括号里的源代码，都会间接被执行。

19.2.2 抛出自定义的异常错误

通过关键字 throw 开头的语句，可使得 JavaScript 引擎抛出（throw）特定自定义的异常错误（exception error）的对象实例（object instance）。请参看示例的如下片段。

```
try
{
  //throw new URIError("Defined error message here...") ;
  throw "Logical error occurs." ;
```

此语句会直接产生带有自定义错误信息'Logical error occurs.'的异常错误的对象实例,进而使得其下方【catch(err)】语句的大括号里的源代码,间接被执行。

19.3 调 试 机 制

JavaScript 引擎支持如下两种调试机制:

- 通过带有字符串字面量的语句【"use strict";】,启动 JavaScript 引擎的严格模式(strict mode)。
- 通过语句【debugger;】,在特定源代码的位置上,设置调试用途的断点(breakpoint)。

19.3.1 严格模式

JavaScript 引擎处于严格模式的情况下,对于源代码的语法的要求,是相当严格(strict)的!请参看如下示例。

【19-3-1-strict-mode.js】

```
"use strict" ;
try
{
  let v01 = 100 ;
  var v02 = 250 ;
  v03 = 600 ;
}
catch (err)
{
  console.log(err) ;
}

///
function func01()
{
  radius = 10 ;

  return 4 * Math.PI * radius ** 2 ;
}
try
{
  func01() ;
}
catch (err)
{
  console.log(err) ;
}

///
```

```
try
{
  eval('\
    with (Math)\
    {\
      console.log(PI) ;\
    }') ;
}
catch (err)
{
  console.log(err) ;
}

///
try
{
  let num01 = 0 ;

  eval('num01 = 0377 ;') ;
}
catch (err)
{
  console.log(err) ;
}

///
try
{
  let person = {name: 'Jasper', gender: 'male', age: '33'} ;

  eval('delete person ;') ;
}
catch (err)
{
  console.log(err) ;
}
```

【相关说明】

```
"use strict" ;
```

此语句启动了严格模式，使得 JavaScript 引擎以特别高的标准，严格检视源代码的语法。

```
try
{
  let v01 = 100 ;
  var v02 = 250 ;

  v03 = 600 ;
```

在严格模式中，v03 被访问之前，尚未被定义成为变量，所以语句【v03 = 600 ;】会产生 ReferenceError 异常错误。

```
}
```

通过上述 try 语句，可使得其大括号里的各个子语句，例如【v03 = 600 ;】，一旦被监听到发生任何异常错误时，其下方的【catch(err)】语句的大括号里的源代码，就会间接被执行。

```
catch (err)
{
  console.log(err) ;
```

此语句显示出【ReferenceError: v03 is not defined】的信息。

```
}
```

在上述 try 语句的大括号里，一旦被监听到任何异常错误时，【catch(err)】语句的大括号里的源代码，就会被执行。在其小括号里的 err，可被视为内含特定异常错误相关数据的参数。

```
function func01()
{
  radius = 10 ;

  return 4 * Math.PI * radius ** 2 ;
}
```

上述语法定义了函数 func01()。

```
try
{
  func01() ;
```

于严格模式中，在函数 func01() 被调用之后，及其内部的语句【radius = 10 ;】被执行之前，radius 尚未被定义成为变量，所以【radius = 10 ;】被执行时，就会产生 ReferenceError 异常错误。

```
}
catch (err)
{
  console.log(err) ;
```

此语句显示出【ReferenceError: radius is not defined】的信息。

```
}
try
{
  eval('\
    with (Math)\
    {\
      console.log(PI) ;\
    }') ;
```

在严格模式中，并不能通过 with 语句，简化特定对象实例的成员的访问语法，否则会产生 SyntaxError 异常错误。

有些 SyntaxError 异常错误的发生，无法被捕获，导致其下方【catch (err)】语句的大括号里的源代码，并不会被执行。所以在此，必须额外将其原始语法，以字符串的形式，传入内置函数 eval()，才能顺利间接进行调试。

```
      }
      catch (err)
      {
         console.log(err) ;
```

此语句显示出【SyntaxError: Strict mode code may not include a with statement】的信息。

```
      }
      try
      {
         let num01 = 0 ;
```

声明初始值为 0 的变量 num01。

```
         eval('num01 = 0377 ;') ;
```

在严格模式下，不能通过【0】加上【八进制】数码的旧语法，来表示八进制数值，否则会产生 SyntaxError 异常错误。

有些 SyntaxError 异常错误的发生，无法被捕获，导致其下方【cath(err)】语句的大括号里的源代码。不会被执行。所以在此，必须额外将其原始语法，以字符串的形式，传入内置函数 eval()，才能顺利间接进行调试。

```
      }
      catch (err)
      {
         console.log(err) ;
```

此语句显示出【SyntaxError: Octal literals are not allowed in strict mode.】的信息。

```
      }
      try
      {
         let person = {name: 'Jasper', gender: 'male', age: '33'} ;
```

声明初始数据为对象实例的变量 person。

```
         eval('delete person ;') ;
```

在严格模式下，不能通过关键字 delete 的语句，来删除特定对象实例，例如上述变量 person 的对象实例。

有些 SyntaxeError 异常错误的发生，无法被捕获，使得其下方【catch (err)】语句的大括号里的源代码，不会被执行。所以在此，必须额外将其原始语法，以字符串的形式，传入内置函数 eval()，才能顺利间接进行调试。

```
      }
      catch (err)
      {
         console.log(err) ;
```

此语句显示出【SyntaxError: Delete of an unqualified identifier in strict mode.】的信息。

19.3.2　源代码的断点设置和逐句执行

在 JavaScript 源代码里，可安排多个调试用途的断点 (breakpoint) 的语句【debugger ;】，以便 JavaScript 引擎执行到特定断点时，可暂停并等待编程人员进一步的调试动作，比如逐句执行、跳至下一个断点等。请参看如下示例。

【19-3-2-debugging-mechanism.js】

```
let radius = 10, height = 15 ;
let result = 0 ;

result = Math.PI * (radius ** 2) * height ;

console.log(result) ;
console.log('') ;

debugger ;

result = 2 * Math.PI * radius ** 2 + 2 * Math.PI * radius * height ;

console.log(result) ;
console.log('') ;

debugger ;

result = 4 / 3 * Math.PI * radius ** 3 ;

console.log(result) ;
console.log('') ;

debugger ;

result =  4 * Math.PI * radius ** 2 ;

console.log(result) ;
```

【相关说明】

```
let radius = 10, height = 15 ;
let result = 0 ;
```

这两个语句分别声明了初始值为不同整数值的变量 radius、height 与 result。

```
result = Math.PI * (radius ** 2) * height ;
```

此语句将半径为 10、高度为 15 的圆柱体积 4712.38898038469，赋给了变量 result。

```
console.log(result) ;
```

此语句显示出圆柱体积 4712.38898038469 的信息。

```
console.log('') ;
```

```
debugger ;
```

此语句除了会间接启动浏览器的调试工具【Sources】面板之外，也会使得 JavaScript 引擎，执行源代码的动作，暂停在此语句上，并等待调试人员后续的手动操作，例如：

- 恢复脚本执行 (resume script execution)
- 跳转至下一个函数的调用完成（step over next function call）
- 进入下一个函数的调用开始（step into next function call）
- 跳出至当前函数的调用完成（step out of current function）

```
result = 2 * Math.PI * radius ** 2 + 2 * Math.PI * radius * height ;
```

此语句将半径为 10、高度为 15 的圆柱体表面积 1570.7963267948967，赋给了变量 result。

```
console.log(result) ;
```

此语句显示出圆柱体表面积 1570.7963267948967 的信息。

```
console.log('') ;
```

```
debugger ;
```

若调试人员手动进行【恢复脚本执行】，则 JavaScript 引擎会从上一个【debugger ;】的位置，往下执行，并暂停于当前语句上。

```
result = 4 / 3 * Math.PI * radius ** 3 ;
```

此语句将半径为 10、高度为 15 的球体积 4188.790204786391，赋给了变量 result。

```
console.log(result) ;
```

此语句显示出球体积 4188.790204786391 的信息。

```
console.log('') ;
```

```
debugger ;
```

若调试人员手动进行【恢复脚本执行】，则 JavaScript 引擎会从上一个【debugger ;】的位置，往下执行，并暂停于当前语句上。

```
result = 4 * Math.PI * radius ** 2 ;
```

此语句将半径为 10、高度为 15 的球体表面积 1256.6370614359173，赋给变量 result。

```
console.log(result) ;
```

此语句显示出球体积 1256.6370614359173 的信息。

19.4 练 习 题

1. 已知如下 JavaScript 源代码片段：

```
let num01 = 155, num02 = 250 ;
```

```
let result ;

result = v01 -- v02 ;
```

试描述如上源代码片段被执行时，会发生什么种类的异常错误。

2. 已知如下 JavaScript 源代码片段：

```
let num03 = 10, num04 = 30 ;
let result ;

result = num03++ + ++num04 ;
```

试描述如上源代码片段被执行时，会发生什么种类的异常错误。

3. 已知如下 JavaScript 源代码语句：

```
eval('let num05 = 123.456. ;') ;
```

试描述如上源代码片段被执行时，会发生什么种类的异常错误。

4. 已知如下 JavaScript 源代码语句：

```
eval('let num06 = 789 ;') ;
eval('let num07 = num06 * num08 ;') ;
```

试描述如上源代码片段被执行时，会发生什么种类的异常错误。

5. 编写 JavaScript 源代码，以完成如下任务：

（1）声明数值可为整数 1、2 与 3 的变量 error_code。

（2）根据变量 error_code 的如下数值，使得 JavaScript 引擎主动抛出对应的异常错误的对象实例：

- 其数值为 1 时，抛出自定义异常错误的类 InputError 的对象实例。
- 其数值为 2 时，抛出自定义异常错误的类 DivisionError 的对象实例。
- 其数值为 3 时，抛出自定义异常错误的类 TimeoutError 的对象实例。

（3）捕捉以上 3 种自定义异常错误的类的对象实例，并显示对应的特定信息。

第 20 章

数据的验证与传输

在网页上的数据验证（data validation）机制，可分为客户端（client side）和服务器端（server side）的版本，以确保用户输入的数据是可用的。

待客户端的数据，以客户端的验证机制，确认为可用的之后，才可通过特定形式的数据传输机制，传递至服务器端，并以服务器端的验证机制，进行最后的核实。

20.1 HTML 表单的内置验证

于 HTML 语言中，可通过如下各个属性，来达成内置在表单中的各数据的验证机制：

- required（必填的）
- pattern（模式）
- min（最小值）
- max（最大值）
- maxLength（字符个数的上限）
- type（类型）

20.1.1 必填验证

若在特定网页的 HTML 源代码中，已知存在如下代表文本字段（text field）的【属性 id 的数据为'username'】的 input 元素实例：

```
<input type="text" id="username" name="username" placeholder="username" size="16">
```

欲让上述 input 元素实例的数据文本，具有必备（required）的特征，其方式大致如下之一：

- 在对应的 HTML 源代码语法里，加上 1 个空格字符，以及无须带有任何数据的属性 required：

```
<input type="text" id="username" name="username" placeholder="username" size="16" required>
```

- 在后续的 script 元素实例中，加上【username.required = true ;】的 JavaScript 语句：

```
<script>
  ┆
  username.required = true ;
  ┆
</script>
```

对于相关的综合运用，请参看如下示例。

【20-1-1-required-validations.html】

```
<!DOCTYPE html>
<html>
  <head>
    <title>required validations</title>
    <style>
      input, button, select
      {
        font-size: 1.2em ;
        margin: 5px ;
      }
    </style>
  </head>
  <body>
    <form id="form01" name="form01" style="text-align: center;">
      <h3>个人资料</h3>
      <input type="text" id="username" name="username" placeholder="username" size="16" required>
      <input type="password" id="password" name="password" placeholder="password" size="16"
        required>
      <p></p>
      <select id="select01" name="select01">
        <option value="">choice of day-off</option>
        <option value="0">Sunday</option>
        <option value="1">Monday</option>
        <option value="2">Tuesday</option>
        <option value="3">Wednesday</option>
        <option value="4">Thursday</option>
        <option value="5">Friday</option>
        <option value="6">Saturday</option>
      </select>
      <input type="search" id="search" name="search" placeholder="job category.." size="9">

      <p></p>
      <button type="submit">Login</button>
      <button type="reset">Reset</button>
    </form>
  </body>
  <script>
    select01.required = true ;
```

```
    search.required = true ;
  </script>
</html>
```

【相关说明】

```
    <input type="text" id="username" name="username" placeholder="username" size="16" required>
    <input type="password" id="password" name="password" placeholder="password" size="16"
        required>
```

这两个 HTML 语法定义了属性 id 的数据分别为'username'和'password'的 input 元素实例，而且均带有属性 required，意味着其数据文本是必备（required）的。换句话说，在用户按下可启动提交（submit）机制的 Login 按钮之前，这两个 input 元素实例的数据是是必须被填写的。

```
    <button type="submit">Login</button>
```

这个 button 元素实例在网页上，呈现为带有文本【Login】的按钮。此外，因为其属性 type 是 submit，所以同时具备可启动提交（submit）机制的特征。

```
    <button type="reset">Reset</button>
  </form>
</body>
<script>

  select01.required = true ;
```

此语句将布尔值 true，赋给了属性 id 的数据为'select01'的 select 元素实例的属性 required。在用户按下可启动提交（submit）机制的 Login 按钮之前，属性 id 的数据为'select01'的 select 元素实例的数据是必备的，引申为必须被选定的。

```
    search.required = true ;
```

此语句将布尔值 true 赋给，属性 id 的数据为'search'的 input 元素实例的属性 required。在用户按下可启动提交机制的 Login 按钮之前，属性 id 的数据为'search'的 input 元素实例的数据是必备的，引申为必须被填写的。

可启动提交机制的 Login 按钮被按下时，若缺少任何必备的元素实例的数据，则浏览器会在网页上，其第 1 个缺少数据的元素实例的附近，呈现出对应的提示信息，并暂停提交的动作。

20.1.2　字符个数和数值范围的验证

在 HTML 语言中，验证特定文本字段（text field）里的字符个数（character amount）和数值范围（number-value range），主要可通过如下属性来达成：

- type（类型）
- maxLength（字符个数的上限）
- min（最小值）
- max（最大值）

对于相关的综合运用，请参看如下示例。

【20-1-2-character-amount-and-number-range-validations.html】

```html
<!DOCTYPE html>
<html>
  <head>
    <title>character amount and number range validations</title>
    <style>
      input, button, select
      {
        font-size: 1.2em ;
        margin: 5px ;
      }
    </style>
  </head>
  <body>
    <form id="form01" name="form01" style="text-align: center;">
      <h3>个人资料</h3>
      <!-- <input type="text" id="username" name="username" placeholder="username" size="16"
        maxlength="12" required> -->
      <input type="text" id="username" name="username" placeholder="username" size="16" required>

      <input type="password" id="password" name="password" placeholder="password" size="16"
        required>
      <p></p>

      <label>Age: </label>
      <!-- <input type="number" id="age" name="age" min="5" max="150"> -->
      <input type="number" id="age" name="age" placeholder="5 ~ 150.." style="width: 100px">

      <p></p>
      <label>Phone No.: </label>
      <input type="text" id="phone_no" name="phone_no" size="10" placeholder="09XX-XXX-XXX"
        pattern="09\d{2}(-\d{3}){2}">

      <p></p>
      <label>Email address: </label>
      <input type="text" id="email_addr" name="email_addr" size="24"
        pattern="^\w+(\.\w+)?@\w+(\.\w+){1,3}$" required>

      <p></p>
      <input type="search" id="search" name="search" placeholder="job category.." size="9"
        novalidate>

      <p></p>
      <button type="submit">Login</button>
      <button type="reset">Reset</button>
    </form>
  </body>
  <script>
    username.maxLength = 12 ;

    age.min = 5 ;
    age.max = 150 ;

    search.required = true ;
```

```
    </script>
</html>
```

【相关说明】

```
<script>
  username.maxLength = 12 ;
```

此语句使得属性 id 的数据为'username'的 input 元素实例，最多只可被填入 12 个字符的文本。

```
  age.min = 5 ;
```

此语句使得属性 id 的数据为'age'的 input 元素实例，必须被填入至少为 5 的整数值。

```
  age.max = 150 ;
```

此语句使得属性 id 的数据为'age'的 input 元素实例，必须被填入最多为 150 的整数值。

```
  search.required = true ;
```

此语句将布尔值 true，赋给了属性 id 的数据为'search'的 input 元素实例的属性 required。在用户按下可启动提交机制的 Login 按钮之前，属性 id 的数据为'search'的 input 元素实例的数据是必备的，引申为必须被填写的。

可启动提交机制的 Login 按钮被按下时，若缺少任何必备或者格式正确的元素实例的数据，则浏览器会在网页上，其第 1 个缺少【必备或者格式正确】的数据的元素实例附近，呈现出对应的提示信息，并暂停提交的动作。

20.2　自定义的验证

在 HTML 语言中，特定文本字段中的模式验证（pattern validation），主要可通过属性 type（数据类型）和 pattern（模式）来达成。请参看如下示例。

【20-2-^-validation-implementations-by-JavaScript.html】

```
<!DOCTYPE html>
<html>
  <head>
    <title>validation implementations by JavaScript</title>
    <style>
      input, button, select
      {
        /*font-size: 1.2em ;*/
        margin: 5px ;
      }

      h4, p
      {
        margin: 0 ;
      }
    </style>
```

```html
</head>
<body>
  <form id="form01" name="form01" style="text-align: center;">
    <h4>个人资料</h4>
    <!-- <input type="text" id="username" name="username" placeholder="username" size="16"
      maxlength="12" required> -->
    <input type="text" id="username" name="username" placeholder="username" size="16">

    <input type="password" id="password" name="password" placeholder="password" size="16">
    <p></p>

    <!-- <input type="number" id="age" name="age" min="5" max="150"> -->
    <input type="number" id="age" name="age" placeholder="(age) 5 ~ 150.." style="width: 100px">

    <p></p>
    <input type="text" id="phone_no" name="phone_no" size="28" placeholder="(phone number)
      12X-XXXX-XXXX">

    <p></p>
    <input type="text" id="email_addr" name="email_addr" size="24" placeholder="email address..">

    <p></p>
    <input type="search" id="search" name="search" placeholder="job category.." size="9">

    <p></p>
    <button type="button" id="check">Check</button>
    <p></p>
    <button type="submit">Login</button>
    <button type="reset">Reset</button>
  </form>
</body>
<script>
  username.required = true ;
  username.maxLength = 12 ;

  password.required = true ;

  age.min = 5 ;
  age.max = 150 ;

  search.required = true ;

  phone_no.pattern = "12\\d-\\d{4}-\\d{4}" ;

  email_addr.required = true ;
  email_addr.pattern = "^\\w+(\\.\\w+)?@\\w+(\\.\\w+){1,3}$" ;

  ///
  let refs, ref ;

  check.onclick = function (event)
  {
    refs = form01.elements ;

    console.log(refs) ;
```

```
        console.log('') ;

    for (ref of refs)
    {
      if (ref.tagName == 'BUTTON') continue ;
      else if (! ref.checkValidity())
      {
        ref.style.border = '2px IndianRed solid' ;

        for (let prop in ref.validity)
        {
          if (ref.validity[prop])
          {
            console.log(`Error message: (${prop}) ${ref.validationMessage}`) ;
            // ref.setCustomValidity('Please fill the text field or correct the format. Orz') ;
          }
        }
      }
      else ref.style.border = '2px YellowGreen solid' ;
    }
  } ;
</script>
</html>
```

【相关说明】

请参看本节下面的相关说明。

20.2.1　文本字段的模式验证

在 HTML 语言中，文本字段的模式验证，必须借助属性 pattern 和其代表【正则表达式字面量（regular-expression literal）】的数据，才能达成。例如，加入属性 pattern 的语法【pattern="12\\d-\\d{4}-\\d{4}"】，至如下 input 元素实例里：

<input type="text" id="phone_no" name="phone_no" size="28" placeholder="(phone number) 12X-XXXX-XXXX">

进而成为如下语句：

<input type="text" id="phone_no" name="phone_no" size="28" placeholder="(phone number) 12X-XXXX-XXXX" **pattern="12\\d-\\d{4}-\\d{4}"**>

对于文本字段的模式验证的综合运用，请继续参看下面的相关说明。

【相关说明】

```
<input type="text" id="phone_no" name="phone_no" size="28" placeholder="(phone number)
 12X-XXXX-XXXX" pattern="12\\d-\\d{4}-\\d{4}">
```

在这个 input 元素实例中，借助属性 pattern 和其代表【正则表达式字面量"12\\d-\\d{4}-\\d{4}"】的数据，来验证其文本字段中的电话号码的格式。

```
<p></p>
<input type="text" id="email_addr" name="email_addr" size="24" placeholder="email address.."
  pattern="^\\w+(\\.\\w+)?@\\w+(\\.\\w+){1,3}$">
```

在这个 input 元素实例中，借助属性 pattern 和其代表【正则表达式字面量"^\\w+(\\.\\w+)?@\\w+(\\.\\w+){1,3}$"】的数据，来验证其文本字段中的电子邮件地址的格式。

20.2.2　JavaScript 源代码实现的验证

原本通过 HTML 语法而实现的验证机制，均可被改写成为 JavaScript 语法的版本。举例来说，已知如下 HTML 的语法：

<input type="text" id="phone_no" name="phone_no" size="28" placeholder="(phone number) 12X-XXXX-XXXX" pattern="12\\d-\\d{4}-\\d{4}">

如上 HTML 语法，可被改写成为如下 HTML 和 JavaScript 语法互相搭配的版本：

  <input type="text" id="phone_no" name="phone_no" size="28" placeholder="(phone number) 12X-XXXX-XXXX">
  ︙
<script>
  ︙
  phone_no.pattern = "12\\d-\\d{4}-\\d{4}" ;
  ︙
</script>

原本在 HTML 语法中的属性 pattern 和其代表【正则表达式字面量"12\\d-\\d{4}-\\d{4}"】的数据，被改写成为【在 script 元素实例中，具有相同效果】的 JavaScript 语法。对于 JavaScript 源代码实现的验证，请继续参看下面的相关说明。

【相关说明】

```
<input type="text" id="phone_no" name="phone_no" size="28" placeholder="(phone number)
  12X-XXXX-XXXX">
```

在这个 HTML 语法中，原本存在属性 pattern 和其代表【正则表达式字面量"12\\d-\\d{4}-\\d{4}"】的数据，当前则被改写成为【在 script 元素实例中，具有相同效果】的 JavaScript 语法。

```
<p></p>
<input type="text" id="email_addr" name="email_addr" size="24" placeholder="email address.."
  pattern="^\\w+(\\.\\w+)?@\\w+(\\.\\w+){1,3}$">
```

在这个 HTML 语法中，原本存在属性 pattern 和其代表【正则表达式字面量"^\\w+(\\.\\w+)?@\\w+(\\.\\w+){1,3}$"】的数据，当前则被改写成为【在 script 元素实例中，具有相同效果】的 JavaScript 语法。

```
<script>
    username.required = true ;
```

此语句将布尔值 true，赋给了属性 id 的数据为'username'的 input 元素实例的属性 required。在用户按下可启动提交机制的 Login 按钮之前，属性 id 的数据为'username'的 input 元素实例的数据是必备的，引申为必须被填写的。

```
    username.maxLength = 12 ;
```

此语句使得属性 id 的数据为'username'的 input 元素实例，最多只可被填入 12 个字符。

```
    password.required = true ;
```

此语句将布尔值 true，赋给了属性 id 的数据为'password'的 input 元素实例的属性 required。在用户按下可启动提交机制的 Login 按钮之前，属性 id 的数据为'password'的 input 元素实例的数据是必备的，引申为必须被填写的。

```
    age.min = 5 ;
```

此语句使得属性 id 的数据为'age'的 input 元素实例，被填入的整数值至少为 5。

```
    age.max = 150 ;
```

此语句使得属性 id 的数据为'age'的 input 元素实例，被填入的整数值最多为 150。

```
    search.required = true ;
```

此语句将布尔值 true，赋给了属性 id 的数据为'search'的 input 元素实例的属性 required。在用户按下可启动提交机制的 Login 按钮之前，属性 id 的数据为'search'的 input 元素实例的数据是必备的，引申为必须被填写的。

```
    phone_no.pattern = "12\\d-\\d{4}-\\d{4}" ;
```

此语句使得属性 id 的数据为'phone_no'的 input 元素实例，被填入的手机号码必须类似于【122-3456-7890】，才是正则表达式字面量【12\d-\d{4}-\d{4}】对应的正确的手机号码格式。此外，将正则表达式字面量【12\d-\d{4}-\d{4}】赋给属性 id 的数据为'phone_no'的属性 pattern 时，必须重复其反斜杠符号【\】，成为【12\\d-\\d{4}-\\d{4}】，才不至于产生 JavaScript 语法上的问题。

```
    email_addr.required = true ;
```

此语句将布尔值 true，赋给了属性 id 的数据为'email_addr'的 input 元素实例的属性 required。在用户按下可启动提交机制的 Login 按钮之前，属性 id 的数据为'email_addr'的 input 元素实例的数据是必备的，引申为必须被填写的。

```
    email_addr.pattern = "^\\w+(\\.\\w+)?@\\w+(\\.\\w+){1,3}$" ;
```

在此语句中，将匹配电子邮件地址的正则表达式字面量【^\w+(\.\w+)?@\w+(\.\w+){1,3}$】，赋给属性 id 的数据为'email_addr'的 input 元素实例的属性 pattern 时，必须重复其反斜杠符号【\】，成为【^\\w+(\\.\\w+)?@\\w+(\\.\\w+){1,3}$】，才不至于产生 JavaScript 语法上的问题。

```
    let refs, ref ;
```

此语句分别声明了变量 refs 与 ref。

```
    check.onclick = function (event)
    {

        refs = form01.elements ;
```

form01.elements 会返回【在属性 id 的数据为'form01'的 form 元素实例中，其 9 个子元素实例】构成的集合，并赋给变量 refs。

```
        console.log(refs) ;
```

此语句显示出可被展开的【HTMLFormControlsCollection(9) [input#username, input#password, input#age, input#phone_no, input#email_addr, input#search, button#check, button, button, username: input#username, password: input#password, age: input#age, phone_no: input#phone_no, email_addr: input#email_addr, …]】的信息。

```
        console.log('') ;
        for (ref of refs)
```

在变量 refs 的集合中，内含 9 个**子元素**实例，所以这个【for...of { ...}】循环语句会迭代 9 次。

```
        {
            if (ref.tagName == 'BUTTON') continue ;
```

此语法用来在变量 ref 当前对应的元素实例中，判断其元素名称 / 标签名称（tag name）是否为'BUTTON'。若其为'BUTTON'，则【ref.tagName == 'BUTTON'】会返回布尔值 true，使得子语句【continue ;】被执行，从而直接跳转至上述循环语句的下一次迭代。

```
            else if (! ref.checkValidity())
```

此语法用来检验（check）变量 ref 的数据的有效性（validity）。若变量 ref 的数据，存在以下任何一个问题，则 ref.checkValidity() 会返回布尔值 false：

- 其数据是必备（required）的，却未被填入任何数据。
- 其数据不符合特定格式或者数值范围。

```
            {
                ref.style.border = '2px IndianRed solid' ;
```

此语句使得变量 ref 当前对应的【其数据有问题】的元素实例，在网页上呈现出 2 像素（2px）宽的印度红色（Indian red）的实线（solid）边框（border）。

```
                for (let prop in ref.validity)
```

在此语法中，ref.validity 会返回 ValidityState 对象实例，例如【ValidityState {valueMissing: false, typeMismatch: false, patternMismatch: false, tooLong: false, tooShort: false, …}】。也因此，变量 prop 在 for 循环语句的每次迭代中，均会被赋给代表特定属性名称的字符串'valueMissing'、'typeMismatch'、'patternMismatch'、'tooLong'和'tooShort'等。

```
                {
```

```
            if (ref.validity[prop])
```

此语法用来判断 ref.validity[prop] 是否返回布尔值 true。若变量 prop 的数据字符串为 'valueMissing'，则语法【ref.validity[prop]】等价于语法【ref.validity['valueMissing']】和语法【ref.validity.valueMissing】。

若【ref.validity[prop]】返回布尔值 true，则代表变量 ref 对应的元素实例的数据，出现了 valueMissing（值的缺失）的问题，也就是缺少（missing）必备的（required）数据值（data value）。

```
            {
                console.log(`Error message: (${prop}) ${ref.validationMessage}`);
```

此语句会显示出类似【Error message: (valueMissing) 请填写此字段。】或【Error message: (patternMismatch) 请与所请求的格式保持一致。】的信息。

```
                // ref.setCustomValidity('Please fill the text field or correct the format. Orz');
            }
        }
    }

            else ref.style.border = '2px YellowGreen solid';
```

此语法使得变量 ref 当前对应的【其数据无问题】的元素实例，在网页上呈现出 2 像素宽的黄绿色（yellow green）的实线边框。

```
        }
    };
```

上述语法将匿名函数的定义，赋给了属性 id 的数据为'check'的 button 元素实例的属性 onclick。一旦上述 button 元素实例被单击时，上述匿名函数就会被调用。

20.3 异步数据传输

在 JavaScript 语言中，异步数据传输的过程，是倚赖 JavaScript 引擎支持的【异步 JavaScript 和 XML (AJAX，asynchronous JavaScript and XML)】的技术，才得以实现的。然而，AJAX 技术现今不再只限于传递 XML 文档的数据，也可用来传递纯文本（plain text）或 JSON 文本（JSON text, JavaScript object notation text）等。

在网页上的异步数据传输的过程中，客户端和服务器端不用等待对方的回应（response），即可继续当前的操作，并互相传递特定数据，进而完成以下任务：

- 传递数据至特定服务器。
- 接收来自特定服务器的数据。
- 在无须重新加载网页文档的情况下，更新其页面的部分内容。

20.3.1 AJAX 的工作原理

AJAX 技术的工作原理，大致由如下 7 个步骤来实现：

（1）在客户端的当前网页中，JavaScript 引擎处理已经发生的特定事件。
（2）在客户端的当前网页中，JavaScript 引擎创建 XMLHttpRequest 对象实例。
（3）在客户端的当前网页中，JavaScript 引擎将【内含特定数据的 XMLHttpRequest 对象实例】的请求（request），传递至特定网络服务器。
（4）网络服务器处理来自客户端的请求。
（5）网络服务器将处理了请求之后的响应（response）数据，传递至客户端。
（6）在客户端的当前网页中，JavaScript 引擎读取来自网络服务器的响应数据。
（7）在客户端的当前网页中，JavaScript 引擎根据来自网络服务器的响应数据，加以更新其页面的部分内容。

对于 AJAX 的工作原理的综合理解，请参看如下带有 2 个源代码文档的示例。

【20-3-1-AJAX-working-principles.html】

```html
<!DOCTYPE html>
<html>
  <head>
    <title>AJAX working principles</title>
    <style>
      select, button
      {
        font-size: 1.2em ;
      }
    </style>
  </head>
  <body align="center">
    <form id="form01" method="get" action="20-3-1-server-responses.php">
      <select id="fruit" name="fruit">
        <option value="">Choose a fruit</option>
        <option value="apple">Apple</option>
        <option value="durian">Durian</option>
        <option value="grape">Grape</option>
        <option value="kiwifruit">Kiwifruit</option>
        <option value="mango">Mango</option>
      </select>

      <p></p>
      <button type="submit">Confirm</button>
    </form>
  </body>
  <script>
  </script>
</html>
```

【相关说明】

首先，用户需要在正式或者调试用途的网络服务器的公共（public）目录里，放置本示例的网页文档【20-3-1-AJAX-working-principles.html】和另一动态网页文档【20-3-1-server-responses.php】。

若在客户端，通过 file 协议，浏览本示例网页，则其异步数据传输的机制，会无法正常运作；必须通过 HTTP 协议，才能正常进行异步数据的传输。

在本示例网页中，form 元素实例的起始语法【<form id="form01" method="get" action="20-3-1-server-responses.php">】里，设置了：

- 属性 method 的数据为'get'或'post'。
- 属性 action 的数据为另一动态网页文档的【名称】或者【所在路径 + 名称】的字符串，例如'20-3-1-server-responses.php'。

因为本示例的网页文档【20-3-1-AJAX-working-principles.html】和动态网页文档【20-3-1-server-responses.php】，被放置于特定网络服务器的相同目录里，所以字符串'20-3-1-server-responses.php'并未带有其所在路径。

当用户在属性 name 的数据为'fruit'的 select 元素实例中，选择了特定项目（例如数据为'apple'的项目 Apple），并按下可启动提交机制的 Confirm 按钮之后，客户端的浏览器便会将用户选择的项目的相关数据（例如冗长的字符串'The apple tree (Malus pumila, commonly and erroneously called Malus domestica) ..., Greek and European Christian traditions.'），传递至网络服务器，并按照动态网页文档【20-3-1-server-responses.php】中的 PHP 源代码的指示，进一步将响应数据，回复给客户端的浏览器。最后，客户端的浏览器，便可进一步将网络服务器回复的响应数据，显示在浏览器窗口的网页上。

【20-3-1-server-responses.php】

```php
<?php
    $fruits['apple'] = 'The apple tree (Malus pumila, commonly and erroneously called Malus
        domestica) is a deciduous tree in the rose family best known for its sweet, pomaceous fruit,
        the apple. It is cultivated worldwide as a fruit tree, and is the most widely grown species
        in the genus Malus. The tree originated in Central Asia, where its wild ancestor, Malus
        sieversii, is still found today. Apples have been grown for thousands of years in Asia and
        Europe, and were brought to North America by European colonists. Apples have religious and
        mythological significance in many cultures, including Norse, Greek and European Christian
        traditions.' ;

    $fruits['durian'] = 'The durian (/ djʊəriən/) or / dʊriən/ is the fruit of several tree species
        belonging to the genus Durio. There are 30 recognised Durio species, at least nine of which
        produce edible fruit, and over 300 named varieties in Thailand, 102 in Indonesia, and 100
        in Malaysia. Durio zibethinus is the only species available in the international market:
        other species are sold in their local regions.' ;

    $fruits['grape'] = 'A grape is a fruit, botanically a berry, of the deciduous woody vines of
        the flowering plant genus Vitis. Grapes can be eaten fresh as table grapes or they can be
        used for making wine, jam, juice, jelly, grape seed extract, raisins, vinegar, and grape
        seed oil. Grapes are a non-climacteric type of fruit, generally occurring in clusters.' ;

    $fruits['kiwifruit'] = 'Kiwifruit (often abbreviated as kiwi) or Chinese gooseberry is the
        edible berries of several species of woody vines in the genus Actinidia.[1][2] The most common
        cultivar group of kiwifruit ("Hayward") is oval, about the size of a large hen\'s egg (5
        - 8 cm (2.0 - 3.1 in) in length and 4.5 - 5.5 cm (1.8 - 2.2 in) in diameter). It has a fibrous,
        dull greenish-brown skin and bright green or golden flesh with rows of tiny, black, edible
```

```
    seeds. The fruit has a soft texture and a sweet but unique flavor. It is a commercial crop
    in several countries, such as China, Italy, New Zealand, Chile, Greece, and France.' ;

  $fruits['mango'] = 'Mangoes are juicy stone fruit (drupe) from numerous species of tropical
    trees belonging to the flowering plant genus Mangifera, cultivated mostly for their edible
    fruit.The majority of these species are found in nature as wild mangoes. The genus belongs
    to the cashew family Anacardiaceae. Mangoes are native to South Asia,from where the "common
    mango" or "Indian mango", Mangifera indica, has been distributed worldwide to become one of
    the most widely cultivated fruits in the tropics. Other Mangifera species (e.g. horse mango,
    Mangifera foetida) are grown on a more localized basis. It is the national fruit of India,
    Pakistan, and the Philippines, and the national tree of Bangladesh.' ;

  if ($_REQUEST['fruit'] == '' || $_REQUEST['fruit'] == 'undefined' ||
    ! isset($_REQUEST['fruit']))
  {
    exit("No fruit is chosen...") ;
  }
  else
  {
    echo $fruits[$_REQUEST['fruit']] . "\n" ;
  }
?>
```

【相关说明】

```
<?php
  $fruits['apple'] = 'The apple tree (Malus pumila, commonly and erroneously called Malus
    domestica) is a deciduous tree in the rose family best known for its sweet, pomaceous fruit,
    the apple. It is cultivated worldwide as a fruit tree, and is the most widely grown species
    in the genus Malus. The tree originated in Central Asia, where its wild ancestor, Malus
    sieversii, is still found today. Apples have been grown for thousands of years in Asia and
    Europe, and were brought to North America by European colonists. Apples have religious and
    mythological significance in many cultures, including Norse, Greek and European Christian
    traditions.' ;

  $fruits['mango'] = 'Mangoes are juicy stone fruit (drupe) from numerous species of tropical
    trees belonging to the flowering plant genus Mangifera, cultivated mostly for their edible
    fruit.The majority of these species are found in nature as wild mangoes. The genus belongs
    to the cashew family Anacardiaceae. Mangoes are native to South Asia,from where the "common
    mango" or "Indian mango", Mangifera indica, has been distributed worldwide to become one of
    the most widely cultivated fruits in the tropics. Other Mangifera species (e.g. horse mango,
    Mangifera foetida) are grown on a more localized basis. It is the national fruit of India,
    Pakistan, and the Philippines, and the national tree of Bangladesh.' ;
```

在如上数个源代码片段里，设置了变量$fruits 的初始数据，成为【内含冗长字符串的属性 apple、durian、grape 和 kiwifruit】构成的对象实例。

```
  if ($_REQUEST['fruit'] == '' || $_REQUEST['fruit'] == 'undefined' ||
    ! isset($_REQUEST['fruit']))
```

在网页文档【20-3-1-AJAX-working-principles.html】中，存在属性 name 的数据为'fruit'的 select 元素实例。也因此，内置变量$_REQUEST['fruit']的数据，会对应到【属性 name 的数据为'fruit'的 select 元素实例】的数据，例如数据字符串'apple'。

在上述语法里，若内置变量$_REQUEST['fruit']的数据，为如下之一，则意味着在【属性 name

的数据为'fruit'的 select 元素实例中，没有选择任何项目】的情况下，用户就按下了可启动提交机制的 Confirm 按钮：

- 空字符串"。
- 原始常量 undefined（未被定义）。
- 完全不存在（也就是表达式【! isset($_REQUEST['fruit'])】所代表的含义）。

```
{
    exit("No fruit is chosen...") ;
```

此语句可立即终止后续源代码的执行，并将响应数据'No fruit is chosen ...' 回复给客户端的浏览器。

```
}
else
{
    echo $fruits[$_REQUEST['fruit']] . "\n" ;
```

若内置变量$_REQUEST['fruit']当前内含字符串'apple'，则此语句会将冗长的响应'The apple tree (Malus pumila, commonly and erroneously called Malus domestica) ..., including Norse, Greek and European Christian traditions.
\n'，回复给客户端的浏览器，以供显示在浏览器窗口的网页上。

20.3.2 AJAX 的编程方式

现今较为流行的 AJAX 的编程方式，大致可分为：

- **纯粹**通过 JavaScript 语法。
- 主要通过 jQuery 框架。
- 主要通过 AngularJS、React、Vue.js 等框架。

关于**纯粹**通过 JavaScript 语法的编程方式，请参看如下示例。

【20-3-2-e1-AJAX-by-pure-JavaScript.html】

```
<!DOCTYPE html>
<html>
  <head>
    <title>AJAX by pure JavaScript</title>
    <style>
      select, button
      {
        font-size: 1.2em ;
      }

      #message_box
      {
        color: RoyalBlue ;
      }
    </style>
```

```
      </head>
      <body align="center">
        <!-- <form id="form01" method="get" action="20-3-1-server-responses.php"> -->
        <form id="form01">
          <select id="fruit" name="fruit">
            <option value="">Choose a fruit</option>
            <option value="apple">Apple</option>
            <option value="durian">Durian</option>
            <option value="grape">Grape</option>
            <option value="kiwifruit">Kiwifruit</option>
            <option value="mango">Mango</option>
          </select>

          <p></p>
          <!-- <button type="submit" id="confirm_button">Confirm</button> -->
          <button type="button" id="confirm_button">Confirm</button>

          <p></p>
          <div id="message_box"></div>
        </form>
      </body>
      <script>
        confirm_button.onclick = function (event)
        {
          load_data() ;
        } ;

        function load_data()
        {
          let req = new XMLHttpRequest() ;
          let url = '20-3-1-server-responses.php' ;
          let args = '?fruit=' + fruit.value ;

          req.onreadystatechange = function(event)
          {
            if (this.readyState == 4 && this.status == 200)
            {
              message_box.innerHTML = this.responseText ;
            }
          } ;

          req.open("GET", url + args, true) ;
          req.send() ;
        }
      </script>
    </html>
```

【相关说明】

```
    <!-- <form id="form01" method="get" action="20-3-1-server-responses.php"> -->
    <form id="form01">
```

欲通过 AJAX 技术，进行数据传输，则可在 JavaScript 源代码里，编写达成如上原本的 HTML 语法片段【method="get" action="20-3-1-server-responses.php"】的等价语法。

```html
        <select id="fruit" name="fruit">
         <option value="">Choose a fruit</option>
         <option value="apple">Apple</option>
         <option value="durian">Durian</option>
         <option value="grape">Grape</option>
         <option value="kiwifruit">Kiwifruit</option>
         <option value="mango">Mango</option>
        </select>

        <p></p>

        <!-- <button type="submit" id="confirm_button">Confirm</button> -->
        <button type="button" id="confirm_button">Confirm</button>
```

欲通过 AJAX 技术，进行数据传输，则也可进一步在 JavaScript 源代码里，改写达成 HTML 如上原本的 HTML 语法片段【type="submit"】所实现的提交机制。在此，将其属性 type 的数据，设置成为 "button"，使得上述 button 元素实例，变成一般按钮即可。

```html
        <p></p>
        <div id="message_box"></div>
      </form>
    </body>
    <script>
      confirm_button.onclick = function (event)
      {

        load_data() ;
```

此语句调用了函数 load_data()，从而启动异步数据传输的机制。

```
      } ;
```

上述语法将匿名函数的定义，赋给了属性 id 的数据为'confirm_button'的 button 元素实例的属性 onclick。一旦上述 button 元素实例被单击时，上述匿名函数就会被调用。

```javascript
      function load_data()
      {

        let req = new XMLHttpRequest() ;
```

此语句声明了变量 req，其初始数据为 XMLHttpRequest()所返回的 XMLHttpRequest 对象实例。

```javascript
        let url = '20-3-1-server-responses.php' ;
```

此语句声明了变量 url，其初始数据为代表【文档路径 + 文档名称】的字符串 '20-3-1-server-responses.php'。因为上述 php 文档和本示例的 html 文档，位于相同的公共目录里，所以无须设置其文档路径的部分。

```javascript
        let args = '?fruit=' + fruit.value ;
```

此语句声明了变量 args，其初始数据为类似'?fruit=apple'的字符串。

```javascript
        req.onreadystatechange = function(evnt)
        {

          if (this.readyState == 4 && this.status == 200)
```

在这个语法中，内置的局部变量 this，是指向变量 req。若表达式【this.readyState == 4】返回布尔值 true，则意味着网络服务器已经处理了此次的请求（request），并将特定响应数据，回复给客户端的浏览器了。若表达式【this.status == 200】返回布尔值 true，则意味着网络服务器已经接受此次的请求了。

```
        {
            message_box.innerHTML = this.responseText ;
```

网络服务器回复的冗长的响应数据，会被存储于 this.responseText 里，并赋给属性 id 的数据为 'message_box' 的 div 元素实例的属性 innerHTML。换言之，也就是将网络服务器回复的冗长的响应数据，例如 'The apple tree (Malus pumila, commonly and erroneously called Malus domestica) ..., including Norse, Greek and European Christian traditions.'，呈现在属性 id 的数据为 'message_box' 的元素实例上。

```
        }
    } ;
```

上述语法将等号右侧的匿名函数的定义，赋给了变量 req 所代表的 XMLHttpRequest 对象实例的属性 onreadystatechange，使得网页当前的加载状态，一旦有了改变时，上述匿名函数就会被调用。

```
    req.open("GET", url + args, true) ;
```

在此，语句【req.open("GET", url + args, true) ;】当前等价于更为明确的语句【req.open("GET", "20-3-1-server-responses.php?fruit=apple", true) ;】，并启动了变量 req 的 XMLHttpRequest 对象实例所代表的请求程序。

函数 open() 中第 1 个参数数据为 "GET"，意味着在请求程序中，欲传递至网络服务器的数据字符串，会被附加于动态网页文档对应的网址里的末尾之处，例如网页文档对应的网址【20-3-1-server-responses.php?fruit=apple】里的【?fruit=apple】。

函数 open() 第 3 个参数数据为布尔值 true，意味着其数据传输是采用异步（asynchronous）方式，使得在数据传输的同时，用户还可以继续操作当前网页上的其他功能。若其第 3 个参数数据，被设置为布尔值 false，则意味着其数据传输是采用同步（synchronous）方式，进而导致在数据传输的期间内，无法操作当前网页上的其他功能。

```
    req.send() ;
```

此语句在【变量 req 的 XMLHttpRequest 对象实例所代表的请求程序】中，使得其相关数据字符串，正式被传递到网络服务器。

```
}
```

上述语法定义了函数 load_data()。

接下来，关于主要通过 **jQuery** 框架的编程方式，请参看如下示例。

【20-3-2-e2-AJAX-by-jQuery.html】

```
<!DOCTYPE html>
```

```html
<html>
  <head>
    <title>AJAX by jQuery</title>
    <style>
      select, button
      {
        font-size: 1.2em ;
      }

      #message_box
      {
        color: RoyalBlue ;
      }
    </style>
  </head>
  <body align="center">
    <form id="form01">
      <select id="fruit" name="fruit">
        <option value="">Choose a fruit</option>
        <option value="apple">Apple</option>
        <option value="durian">Durian</option>
        <option value="grape">Grape</option>
        <option value="kiwifruit">Kiwifruit</option>
        <option value="mango">Mango</option>
      </select>

      <p></p>
      <button type="button" id="confirm_button">Confirm</button>

      <p></p>
      <div id="message_box"></div>
    </form>

  </body>
  <script src="https://ajax.googleapis.com/ajax/libs/jquery/3.2.1/jquery.min.js">
  </script>
  <script>
    $("#confirm_button").click(load_data) ;

    function load_data()
    {
      let url = '20-3-1-server-reponses.php' ;

      $.ajax(
      {
        url: url,
        data: {fruit: fruit.value},
        success: function(responseText)
        {
          $("#message_box").html(responseText) ;
        }
      } ) ;
    }
  </script>
</html>
```

【相关说明】

```
<script src="https://ajax.googleapis.com/ajax/libs/jquery/3.2.1/jquery.min.js"></script>
```

运用 jQuery 框架的语法和函数库之前，必须通过 script 元素实例和其属性 src 的协助，使得其属性 src 的数据，成为在特定网址中，其特定版本的 jQuery 框架对应的文档名称，例如 "https://ajax.googleapis.com/ajax/libs/jquery/3.2.1/jquery.min.js"。

```
</body>
<script>

    $("#confirm_button").click(load_data) ;
```

此语句融入了 jQuery 框架的语法，使得属性 id 的数据为'confirm_button'的 button 元素实例，一旦被单击时，则名称为 load_data 的函数，就会被调用。

```
    function load_data()
    {

      let url = '20-3-1-server-responses.php' ;
```

此语句声明了变量 url，其初始数据为代表【文档路径 + 文档名称】的字符串 '20-3-1-server-responses.php'。因为上述 php 文档和本示例的 html 文档，位于相同的公共目录里，所以无须设置其文档路径的部分。

```
      $.ajax(
      {

        url: url,
```

在这个语法中，其冒号左侧的 url，是指特定对象实例的属性（property）名称，其冒号右侧的 url，则是指【数据为字符串'20-3-1-server-responses.php'】的变量 url。也因此，此语法等价于【url: '20-3-1-server-responses.php',】。

```
        data: {fruit: fruit.value},
```

在这个语法中，其冒号左侧的 data，是指特定对象实例的属性名称，其冒号右侧的 {fruit: fruit.value}，则是一个子对象实例。在这个子对象实例的语法中，其冒号左侧的 fruit，是这个子对象实例的一个属性名称，其右侧的 fruit.value，则会返回【属性 id 的数据为'fruit'的 select 元素实例】的数据字符串，例如'apple'。若 fruit.value 返回字符串'apple'，则上述语法等价于【data: {fruit: 'apple'},】。

```
        success: function(responseText)
        {

          $("#message_box").html(responseText) ;
```

这个语句和 JavaScript 语句【message_box.innerHTML = responseText ;】是等价的。在这个语句中，网络服务器回复的冗长的响应数据，会被存储于参数 responseText 里，并赋给属性 id 的数据为'message_box'的 div 元素实例的属性 innerHTML。换言之，也就是将网络服务器回复的冗长的响应数据，例如'The apple tree (Malus pumila, commonly and erroneously called Malus domestica) ..., including Norse, Greek and European Christian traditions.', 呈现在属性 id 的数据为'message_box'的元素实例上。

```
        }
```

在上述语法里, 其冒号左侧的 success, 可被视为对象实例中的函数名称, 这是因为其冒号右侧的语法, 是 1 个带有参数 resposeText 的匿名函数的定义。

```
    } ) ;
```

上述语法为融入了 jQuery 框架的语法, 一旦 jQuery 框架的函数【$.ajax(...)】被调用时, 则【内含属性 url、data 与函数 success(...)的定义】的对象 (object) 实例, 就会成为了被传入上述函数【$.ajax(...)】的参数数据。

```
}
```

上述语法完成了函数 load_data()的定义。

关于主要通过 AngularJS 框架的编程方式, 请参看如下示例。

【20-3-2-e3-AJAX-by-AngularJS.html】

```html
<!DOCTYPE html>
<html>
  <head>
    <title>AJAX by AngularJS</title>
    <style>
      select, button
      {
        font-size: 1.2em ;
      }

      #message_box
      {
        color: RoyalBlue ;
      }
    </style>
  </head>
  <body align="center">
    <div ng-app="my_app" ng-controller="my_ctrl">
      <form id="form01">

        <select ng-model="fruit">
          <option value="">Choose a fruit</option>
          <option value="apple">Apple</option>
          <option value="durian">Durian</option>
          <option value="grape">Grape</option>
          <option value="kiwifruit">Kiwifruit</option>
          <option value="mango">Mango</option>
        </select>

        <p></p>
        <button type="button" id="confirm_button" ng-click="load_data()">Confirm</button>

        <p></p>
```

```
      <div id="message_box">
        <div>{{message}}</div>
      </div>

    </form>
  </div>
</body>
<script src="https://ajax.googleapis.com/ajax/libs/angularjs/1.6.4/angular.min.js"></script>
<script>
  let spa = angular.module('my_app', []) ;

  spa.controller('my_ctrl', my_controller) ;

  function my_controller($scope, $http)
  {
    $scope.url = '20-3-1-server-responses.php' ;

    $scope.load_data = function (event)
    {
      $scope.args = '?fruit=' + $scope.fruit ;

      console.log($scope.args) ;

      let response_object = $http.get($scope.url + $scope.args) ;

      response_object.then((response) => $scope.message = response.data, (response) =>
        $scope.message = 'Error occurred...') ;
    } ;
  }
</script>
</html>
```

【相关说明】

```
<body align="center">

  <div ng-app="my_app"ng-controller="my_ctrl">
```

在这个 HTML 语法中，通过属性 ng-app 和其数据，使得 div 元素实例，成为了代表名称为 my_app 的 AngularJS 应用程序（application program）。通过属性 ng-controller 和其数据，使得 AngularJS 应用程序，内含了名称为 my_ctrl 的 AngularJS 控制器（controller）。

```
    <form id="form01">

      <select ng-model="fruit">
```

在这个 HTML 语法中，通过属性 ng-model 和其数据，使得这个 select 元素实例的数据，对应到【被附加至 AngularJS 对象实例$scope 中，其名称为 fruit】的 AngularJS 变量【$scope.fruit】的数据。

```
        <option value="">Choose a fruit</option>
        <option value="apple">Apple</option>
        <option value="durian">Durian</option>
        <option value="grape">Grape</option>
        <option value="kiwifruit">Kiwifruit</option>
```

```
        <option value="mango">Mango</option>
    </select>

    <p></p>

    <button type="button" id="confirm_button" ng-click="load_data()">Confirm</button>
```

在这个 HTML 语法中，属性 ng-click 和其数据，可在这个 button 元素实例被按下时，调用【被附加至 AngularJS 对象实例$scope 中，其名称为 load_data】的 AngularJS 函数【$scope.load_data()】。

```
    <p></p>
    <div id="message_box">

        <div>{{message}}</div>
```

AngularJS 支持的语法【{{message}}】，可使得这个 div 元素实例的数据，对应到【被附加至 AngularJS 对象实例$scope 中，其名称为 message】的 AngularJS 变量【$scope.message】。

```
        </div>

    </form>
  </div>
</body>

<script src="https://ajax.googleapis.com/ajax/libs/angularjs/1.6.4/angular.min.js"></script>
```

运用 AngularJS 框架的语法和函数库之前，必须通过 script 元素实例和其属性 src 的协助，设置其属性 src 的数据，成为在特定网址中，其特定版本的 AngularJS 框架对应的文档名称，例如 "https://ajax.googleapis.com/ajax/libs/angularjs/1.6.4/angular.min.js"。

```
    <script>
    let spa = angular.module('my_app', []) ;
```

此语句声明了变量 spa，其初始数据为 AngularJS 函数【angular.module('my_app', [])】所返回的 AngularJS 模块（module）实例。其中，【属性 ng-app 的数据为'my_app'】的 div 元素实例，会被视为 AngularJS 应用程序（application program）。

```
    spa.controller('my_ctrl', my_controller) ;
```

此语句在变量 spa 所代表的 AngularJS 模块实例中，使得名称为 my_controller 的函数，成为了其控制器（controller）函数。

```
    function my_controller($scope, $http)
    {

        $scope.url = '20-3-1-server-responses.php' ;
```

此语句将字符串'20-3-1-server-responses.php'，赋给了新的 AngularJS 变量【$scope.url】。

```
        $scope.load_data = function (event)
        {

            $scope.args = '?fruit=' + $scope.fruit ;
```

在 HTML 语法中，属性 ng-model 的数据为'fruit'的 select 元素实例，会对应到 AngularJS 变量 $scope.fruit。所以，上述 select 元素实例被选定的数据字符串（例如'apple'），会成为 AngularJS 变量$scope.fruit 当前的数据字符串。一旦上述 select 元素实例被重新选定新的项目时，则新的项目所代表的数据字符串，就会成为 AngularJS 变量$scope.fruit 的当前数据字符串。也因此，这个语句将类似'?fruit=apple'的字符串，赋给了 AngularJS 变量$scope.args。

```
console.log($scope.args) ;
```

此语句在浏览器的调试工具【Console】面板里，显示出【?fruit=apple】的信息。

```
let response_object = $http.get($scope.url + $scope.args) ;
```

在上述 select 元素实例中，若数据字符串为'apple'的 Apple 项目被选中时，则语法【$http.get($scope.url + $scope.args)】等价于【$http.get('20-3-1-server-responses.php?fruit=apple')】，并且使得客户端的浏览器，正式向网络服务器，传递其网址为【20-3-1-server-responses.php?fruit=apple】的请求。一旦网络服务器接受并完成了客户端的请求程序之后，就会将响应数据，回复给客户端的浏览器。

也因此，这个语句声明了变量 response_object，其初始数据为函数【$http.get($scope.url + $scope.args)】所返回的响应对象（response object）实例，并且内含网络服务器回复的响应数据。

```
response_object.then((response) => $scope.message = response.data, (response) =>
    $scope.message = 'Error occurred...') ;
```

在这个语句中，通过变量 response_object 内含的响应对象实例，驱使了 JavaScript 引擎，加以判断网络服务器的请求程序，是否顺利完成了。

若其请求程序已经顺利完成，则上述**第 1 个箭头函数**【(response) => $scope.message = response.data】会被调用，并且在当前网页上，显示出类似【The apple tree (Malus pumila, commonly and erroneously called Malus domestica) ..., Greek and European Christian traditions.】的信息。

若其**请求**程序以失败收场，则**第 2 个箭头函数**【(response) => $scope.message = 'Error occurred..'】会被调用，并且在当前网页上，显示出【Error occurred...】的信息。

```
} ;
```

在上述语法的等号右侧中，将带有参数 event 的匿名函数的定义，赋给了$scope.load_data，进而使得 load_data 被视为 AngularJS 函数。

20.4 练 习 题

1. 已知在特定网页中，存在如下 HTML 源代码片段所代表的 input 元素实例：

```
<input type="text" id="username" name="username" value="用户名称...">
```

试编写 JavaScript 源代码片段，使得上述 input 元素实例中的数据，在被传递至网络服务器之前，进行如下各项验证的动作：

- 其数据为必备的。
- 其数据的字符个数，最少为 6 个、最多为 16 个。
- 其数据中的各字符，必须是下画线、字母或者数字的组合。
- 其数据不可以是数字开头的，必须是以下画线或者字母开头的。

2. 已知在特定网页中，存在如下 HTML 源代码片段所代表的 input 元素实例：

```
<input id="salary" name="salary" value="工资数值...">
```

试编写 JavaScript 源代码片段，使得上述 input 元素实例中的数据，在被传递至网络服务器之前，进行如下各项验证的动作：

- 其数据为必备的。
- 其数据必须是整数。
- 其数据所代表的整数，必须至少为 2300。

3. 已知在特定网页中，存在如下 HTML 源代码片段所代表的 input 元素实例：

```
<input type="text" id="employee_no" name="employee_no" value="员工编号...">
```

试编写 JavaScript 源代码片段，使得上述 input 元素实例中的数据，在被传递至网络服务器之前，进行如下各项验证的动作：

- 其数据为必备的。
- 其数据必须是以 2 个字母衔接 1 个减号【-】开头的。
- 其数据在减号【-】之后，再衔接 6 个数字。

4. 试编写 JavaScript 源代码，以通过 AJAX 技术，实现以下功能。

- 在特定网页被加载完成之后，便传递用户端的操作系统和浏览器的版本相关的数据，至网址为 https://www.qzmp.wxno.com 的网络服务器，进而请求通过 opt/client_info.php，以处理来自用户端的数据。
- 将网络服务器回复的响应数据，赋给变量 message。

5. 试编写 JavaScript 源代码，以通过 AJAX 技术，实现以下功能。

- 在当前网页中，每隔 10 秒钟，便传递代表文本字段且属性 id 的数据为 'question' 的 input 元素实例的数据，至网址为 https://www.qzmp.wxno.com 的网络服务器，进而请求通过 opt/answers.php，以处理来自用户端的数据。
- 将网络服务器回复的响应数据，赋给变量 message。

第 21 章

响应式机制

响应式机制（responsive mechanism）即是指响应式网页设计（RWD, responsive web design）的技术。现今因为 JavaScript 语言的完善，RWD 的技术也可单独通过 JavaScript 语法，来完整实现！

21.1 通过 CSS 语法的版本

通过 CSS 语法来实现响应式机制的方式，即是标准、但并不是唯一的途径。浏览器先行解释当前网页的 HTML 语法的元信息（meta data），再解释其 CSS 语法的媒体查询（media query），即可达成 RWD 的技术，让特定网页内容，可以顺应不同移动设备（mobile device）和台式电脑（desktop computer）的屏幕（screen），进而被呈现于屏幕上不同尺寸的窗口（window）里。对于通过 CSS 语法，达成 RWD 的技术的综合理解，请参看如下示例。

【21-1-^-responsive-by-CSS-syntax.html】

```
<!DOCTYPE html>
<html>
  <head>
    <meta charset="utf-8">
    <meta name="viewport" content="width=device-width, initial-scale=1.0">
    <title>responsive by CSS syntax</title>
    <style>
      header, aside, main, footer
      {
        padding: 5px ;
        box-sizing: border-box ;
      }

      header
      {
```

```css
      color: RoyalBlue ;
      background-color: PowderBlue ;
      height: 80px ;
    }

    aside
    {
      color: SeaGreen ;
      background-color: YellowGreen ;
      width: 20% ;
      height: 500px ;
      display: inline-block ;
    }

    main
    {
      color: Chocolate ;
      background-color: SkyBlue ;
      width: 80% ;
      height: 500px ;
      display: inline-block ;
      /*position: absolute ;*/
      right: 0 ;
    }

    @media (max-width: 768px)
    {
      aside, main
      {
        width: 100% ;
      }

      aside
      {
        background-color: LightGreen ;
      }

      main
      {
        background-color: DeepSkyBlue ;
      }
    }

    footer
    {
      color: LightSkyBlue ;
      background-color: Lavender ;
      height: 50px ;
      text-align: center ;
    }
  </style>
</head>
<body>
  <header>Header</header>
  <aside>Menu</aside><main>Content</main>
```

```
    <footer>Footer</footer>
  </body>
  <script>
  </script>
</html>
```

【相关说明】

请参考本节下面的相关说明。

21.1.1　页面的元信息

在网页文档中的 HTML 源代码里，和响应式机制有关的元信息（meta data），主要为视口（viewport）相关特征的设置语法，例如：

```
<meta name="viewport" content="width=device-width, initial-scale=1.0">
```

其中：

- 语法【width=device-width】使得浏览器根据当前装置（device）的屏幕宽度（screen width），来呈现当前网页的内容。
- 语法【initial-scale=1.0】使得浏览器根据 100%（1.0）的初始缩放比例（initial scale），设置在视口（viewport）中，呈现特定网页内容的显示比例。

21.1.2　媒体查询

所谓的媒体查询（media query），即是在特定网页中，依据不同媒体类型（media type）的设备（device），对特定元素（element）实例，应用不同的 CSS 规则，使得这个元素实例，在不同设备的屏幕窗口里，呈现出不同的外观。

```
aside
{
  color: SeaGreen ;

  background-color: YellowGreen ;
  width: 20% ;
```

这两个语句在 aside 元素实例上，分别设置了其一开始的背景颜色，成为黄绿色（yellow green），以及其宽度，成为父元素实例的总宽度的 20%。

```
  height: 500px ;
  display: inline-block ;
}

main
{
  color: Chocolate ;
  background-color: SkyBlue ;

  width: 80% ;
```

```
            height: 500px ;
```

这两个语句用来在 main 元素实例中，设置其一开始的背景颜色，成为巧克力色（chocolate），以及其宽度，成为父元素实例的总宽度的 80%。

```
        display: inline-block ;
        /*position: absolute ;*/
        right: 0 ;
    }

    @media (max-width: 768px)
```

若浏览器窗口的内容区域的宽度，未超过 768 像素，则其下方大括号里的 CSS 源代码，就会被应用。

```
    {
        aside, main
        {
            width: 100% ;
        }
```

此语法用来设置 aside 与 main 元素实例的宽度，成为其父元素实例（body 元素实例）的总宽度的 100%。

```
        aside
        {
            background-color: LightGreen ;
        }

        main
        {
            background-color: DeepSkyBlue ;
        }
```

这两段语法，用来改变 aside 和 main 元素实例的背景颜色。

```
    }
```

- 若浏览器窗口的内容区域的宽度，未超过 768 像素，则上述语法将使得 header、aside、main 与 footer 元素实例在网页上，除了占用其总宽度之外，均为由上至下地纵向排列。
- 若浏览器窗口的内容区域的宽度，超过 768 像素，虽然纵向排列的 header 与 footer 元素实例，依然占用其总宽度；但是，横向排列的 aside 元素实例，占用其总宽度的 20%，而 main 元素实例，则占用其总宽度的 80%。

通过上述 CSS 源代码，使得本示例网页，具有响应式（responsive）机制，进而在浏览器窗口的内容区域的宽度，超过与**未**超过 768 像素时，分别被呈现出不同的外观。

21.2　通过 JavaScript 语法的版本

原本由 CSS 语法所实现的响应式机制，可以全部或者部分被改写，成为等价的 JavaScript 语法。对于通过 JavaScript 语法，来取代原本由 CSS 语法所实现的响应式机制，请参看如下示例。

【21-2-^-responsive-by-JavaScript.html】

```html
<!DOCTYPE html>
<html>
  <head>
    <meta charset="utf-8">
    <meta name="viewport" content="width=device-width, initial-scale=1.0">
    <title>responsive by JavaScript</title>
    <style>
      header, aside, main, footer
      {
        padding: 5px ;
        box-sizing: border-box ;
      }

      header
      {
        color: RoyalBlue ;
        background-color: PowderBlue ;
        height: 80px ;
      }

      aside
      {
        color: SeaGreen ;
        background-color: YellowGreen ;
        /*width: 20% ;*/
        height: 500px ;
        display: inline-block ;
      }

      main
      {
        color: Chocolate ;
        background-color: SkyBlue ;
        /*width: 80% ;*/
        height: 500px ;
        display: inline-block ;
      }
      /*
      @media (max-width: 768px)
      {
        aside, main
        {
          width: 100% ;
        }
```

```
      aside
      {
        background-color: LightGreen ;
      }

      main
      {
        background-color: DeepSkyBlue ;
      }
    }
    */
    footer
    {
      color: LightSkyBlue ;
      background-color: Lavender ;
      height: 50px ;
      text-align: center ;
    }
  </style>
</head>
<body>
  <header>Header</header>
  <aside>Menu</aside><main>Content</main>
  <footer>Footer</footer>
</body>
<script>
  window.onresize = update_size ;

  update_size() ;

  function update_size()
  {
    if (window.matchMedia('(max-width: 768px)').matches)
    {
      with (document.querySelector('aside').style)
      {
        width = '100%' ;

        backgroundColor = 'LightGreen' ;
      }

      with (document.querySelector('main').style)
      {
        width = '100%' ;

        backgroundColor = 'DeepSkyBlue' ;
      }
    }
    else
    {
      with (document.querySelector('aside').style)
      {
        width = '20%' ;

        backgroundColor = 'YellowGreen' ;
```

```
        }
      with (document.querySelector('main').style)
      {
        width = '80%' ;
        backgroundColor = 'SkyBlue' ;
      }
    }
  </script>
</html>
```

【相关说明】

```
    aside
    {
      color: SeaGreen ;
      background-color: YellowGreen ;

      /*width: 20% ;*/
```

在此，将**被改写**的部分，变成注释，以供读者观察和比较。

```
      height: 500px ;
      display: inline-block ;
    }

    main
    {
      color: Chocolate ;
      background-color: SkyBlue ;

      /*width: 80% ;*/
```

在此，也将**被改写**的部分，变成注释。

```
      height: 500px ;
      display: inline-block ;
    }

    /*
    @media (max-width: 768px)
    {
      aside, main
      {
        width: 100% ;
      }

      aside
      {
        background-color: LightGreen ;
      }

      main
      {
```

```
        background-color: DeepSkyBlue ;
      }
    }
    */
```

在此，亦将**冗长**而**被改写**的部分，变成注释。

```
<script>

  window.onresize = update_size ;
```

此语句将名称为 update_size 的函数的定义，赋给了代表浏览器窗口的 window 对象实例的属性 onresize。一旦其窗口当前的宽度或者高度被变更时，则名称为 update_size 的函数，就会被调用。

```
  update_size() ;
```

在浏览器开启本示例网页时，此语句会使得名称为 update_size 的函数被调用，以修正当前网页的外观。

```
  function update_size()
  {
    if (window.matchMedia('(max-width: 768px)').matches)
```

window.matchMedia('(max-width: 768px)')会返回在 CSS 媒体查询语法中的条件【(max-width: 768px)】里，其代表相关数据的 MediaQueryList 对象实例，例如【MediaQueryList {media: "(max-width: 768px)", matches: true, onchange: null}】。

在上述语法里，window.matchMedia('(max-width: 768px)').**matches** 则会返回布尔值 false 或者 true。若浏览器窗口的内容区域的总宽度，未超过 768 像素，则 window.matchMedia('(max-width: 768px)').matches 会返回布尔值 true，并执行其下方大括号里的源代码。

```
    {
      with (document.querySelector('aside').style)
```

document.querySelector('aside').style 是指 aside 元素实例的属性 style。通过此语法，可简化【document.querySelector('aside').style.width】和【document.querySelector('aside').style.backgroundColor】，成为其下方大括号里的【width】和【backgroundColor】。

```
      {
        width = '100%' ;
```

此语句设置了 aside 元素实例的宽度，成为父元素实例的总宽度（100%）。

```
        backgroundColor = 'LightGreen' ;
```

此语句设置了 aside 元素实例的背景颜色，成为浅绿色（light green）。

```
      }
      with (document.querySelector('main').style)
```

document.querySelector('main').style 是指 main 元素实例的属性 style。通过此语法，可简化

【document.querySelector('main').style.width】和【document.querySelector('main').style.backgroundColor】，成为其下方大括号里的【width】和【backgroundColor】。

```
        {
            width = '100%' ;
```

此语句设置了 main 元素实例的宽度，成为浏览器窗口的内容区域的总宽度（100%）。

```
            backgroundColor = 'DeepSkyBlue' ;
```

此语句设置了 main 元素实例的背景颜色，成为深天蓝色（deep sky blue）。

```
        }
    }
    else
```

若浏览器窗口的内容区域的总宽度，已超过 768 像素，则 window.matchMedia('(max-width: 768px)').matches 会返回布尔值 false，并执行其下方大括号里的源代码。

```
    {
        with (document.querySelector('aside').style)
        {
            width = '20%' ;
```

此语句设置了 aside 元素实例的宽度，成为上述内容区域的总宽度的 20%。

```
            backgroundColor = 'YellowGreen' ;
```

此语句设置了 aside 元素实例的背景颜色，成为黄绿色（yellow green）。

```
        }
        with (document.querySelector('main').style)
        {
            width = '80%' ;
```

此语句设置了 main 元素实例的宽度，成为上述内容区域的总宽度的 80%。

```
            backgroundColor = 'SkyBlue' ;
```

此语句设置了 main 元素实例的背景颜色，成为天蓝色（sky blue）。

```
        }
    }
}
```

上述语法完成了函数 update_size() 的定义。

上述示例的源代码，可进一步被改写，成为在下一小节中，更加简洁的版本。

21.2.1 简易判断窗口尺寸的版本

欲改写 CSS 媒体查询语法所对应的 JavaScript 语法，成为更加简洁的 JavaScript 语法，可将语法【if (window.matchMedia('(max-width: 768px)').matches)】，改写成为【if (window.innerWidth <= 768)】即可。请参看如下相关【函数名称】有所变动的示例。

【21-2-1-version-of-simply-checking-window-size.html】

```
<!DOCTYPE html>
<html>
  <head>
    <meta charset="utf-8">
    <meta name="viewport" content="width=device-width, initial-scale=1.0">
    <title>version of simply checking window size</title>
    <style>
      header, aside, main, footer
      {
        padding: 5px ;
        box-sizing: border-box ;
      }

      header
      {
        color: RoyalBlue ;
        background-color: PowderBlue ;
        height: 80px ;
      }

      aside
      {
        color: SeaGreen ;
        background-color: YellowGreen ;
        /*width: 20% ;*/
        height: 500px ;
        display: inline-block ;
      }

      main
      {
        color: Chocolate ;
        background-color: SkyBlue ;
        /*width: 80% ;*/
        height: 500px ;
        display: inline-block ;
      }

      footer
      {
        color: LightSkyBlue ;
        background-color: Lavender ;
        height: 50px ;
        text-align: center ;
      }
    </style>
```

```html
  </head>
  <body>
    <header>Header</header>
    <aside>Menu</aside><main>Content</main>
    <footer>Footer</footer>
  </body>
  <script>
    window.onresize = change_size ;

    change_size() ;

    function change_size()
    {
      // if (window.matchMedia('(max-width: 768px)').matches)
      if (window.innerWidth <= 768)
      {
        with (document.querySelector('aside').style)
        {
          width = '100%' ;

          backgroundColor = 'LightGreen' ;
        }

        with (document.querySelector('main').style)
        {
          width = '100%' ;

          backgroundColor = 'DeepSkyBlue' ;
        }
      }
      else
      {
        with (document.querySelector('aside').style)
        {
          width = '20%' ;

          backgroundColor = 'YellowGreen' ;
        }

        with (document.querySelector('main').style)
        {
          width = '80%' ;

          backgroundColor = 'SkyBlue' ;
        }
      }
    }
  </script>
</html>
```

【相关说明】

```html
  <script>

    window.onresize = change_size ;
```

此语句将名称为 change_size 的函数的定义，赋给了代表浏览器窗口的 window 对象实例的属性 onresize。一旦窗口当前的宽度或高度被变更时，名称为 change_size 的函数就会被调用。

```
change_size() ;
```

在浏览器开启本示例网页时，此语句会使得名称为 change_size 的函数被调用，以修正当前网页的外观。

```
function change_size()
{
    // if (window.matchMedia('(max-width: 768px)').matches)
    if (window.innerWidth <= 768)
```

此语法可用来判断【浏览器窗口的宽度，扣除垂直滚动条的宽度之后，其剩余的宽度，是否小于或等于 768 像素】。若是，则执行其下方大括号里的源代码。

```
    {
        with (document.querySelector('aside').style)
```

document.querySelector('aside').style 是指 aside 元素实例的属性 style。通过此语法，可简化【document.querySelector('aside').style.width】和【document.querySelector('aside').style.backgroundColor】，成为其下方大括号里的【width】和【backgroundColor】。

```
        {
            width = '100%' ;
```

此语句设置了 aside 元素实例的宽度，成为浏览器窗口的内容区域的总宽度（100%）。

```
            backgroundColor = 'LightGreen' ;
```

此语句设置了 aside 元素实例的背景颜色，成为浅绿色（light green）。

```
        }
        with (document.querySelector('main').style)
```

document.querySelector('main').style 是指 main 元素实例的属性 style。通过此语法，可简化【document.querySelector('main').style.width】和【document.querySelector('main').style.backgroundColor】，成为其下方大括号里的【width】和【backgroundColor】。

```
        {
            width = '100%' ;
```

此语句设置了 main 元素实例的宽度，成为浏览器窗口的内容区域的总宽度（100%）。

```
            backgroundColor = 'DeepSkyBlue' ;
```

此语句设置了 main 元素实例的背景颜色，成为深天蓝色（deep sky blue）。

```
        }
    }
```

```
            else
```

若浏览器窗口的内容区域的宽度，大于 768 像素，则执行其下方大括号里的源代码。

```
            {
              with (document.querySelector('aside').style)
              {
                width = '20%' ;
```

此语句设置了 aside 元素实例的宽度，成为上述内容区域的总宽度的 20%。

```
                backgroundColor = 'YellowGreen' ;
```

此语句设置了 aside 元素实例的背景颜色，成为黄绿色（yellow green）。

```
              }
              with (document.querySelector('main').style)
              {
                width = '80%' ;
```

此语句设置了 main 元素实例的宽度，成为上述内容区域的总宽度的 80%。

```
                backgroundColor = 'SkyBlue' ;
```

此语句设置了 main 元素实例的背景颜色，成为天蓝色（sky blue）。

```
              }
            }
          }
```

上述语法完成了函数 change_size() 的定义。

上述示例的源代码，可进一步被改写，成为下一小节中的示例源代码。

21.2.2 直接变更 CSS 规则的版本

欲改写操控 CSS 内联样式（inline style）的 JavaScript 语法，成为【操控 CSS 内部样式（inner style）】的 JavaScript 语法，需要通过 JavaScript 语法【document.styleSheets[0].cssRules】来达成。请参看如下示例。

【21-2-2-version-of-directly-modifying-CSS-rules.html】

```
<!DOCTYPE html>
<html>
  <head>
    <meta charset="utf-8">
    <meta name="viewport" content="width=device-width, initial-scale=1.0">
    <title>version of directly modifying CSS rules</title>
    <style>
      header, aside, main, footer
      {
```

```
      padding: 5px ;
      box-sizing: border-box ;
    }

    header
    {
      color: RoyalBlue ;
      background-color: PowderBlue ;
      height: 80px ;
    }

    aside
    {
      color: SeaGreen ;
      background-color: YellowGreen ;
      width: 20% ;
      height: 500px ;
      display: inline-block ;
    }

    main
    {
      color: Chocolate ;
      background-color: SkyBlue ;
      width: 80% ;
      height: 500px ;
      display: inline-block ;
    }

    footer
    {
      color: LightSkyBlue ;
      background-color: Lavender ;
      height: 50px ;
      text-align: center ;
    }
  </style>
</head>
<body>
  <header>Header</header>
  <aside>Menu</aside><main>Content</main>
  <footer>Footer</footer>
</body>
<script>
  window.onresize = modify_rules ;

  let rules = document.styleSheets[0].cssRules ;

  console.log(rules) ;
  console.log('') ;

  console.log(rules[2].style.width) ;
  console.log(rules[3].style.width) ;
  console.log('') ;
```

```
      modify_rules() ;

    function modify_rules()
    {
      if (window.innerWidth <= 768)
      {
       with (rules[2].style)
       {
         width = '100%' ;

         backgroundColor = 'LightGreen' ;
       }

       with (rules[3].style)
       {
         width = '100%' ;

         backgroundColor = 'DeepSkyBlue' ;
       }
      }
      else
      {
       with (rules[2].style)
       {
         width = '20%' ;

         backgroundColor = 'YellowGreen' ;
       }

       with (rules[3].style)
       {
         width = '80%' ;

         backgroundColor = 'SkyBlue' ;
       }
      }
    }
  </script>
</html>
```

【相关说明】

```
  <script>

    window.onresize = modify_rules ;
```

此语句将名称为 modify_rules 函数的定义，赋给了代表浏览器窗口的 window 对象实例的属性 onresize。一旦浏览器窗口当前的宽度或高度被变更时，名称为 modify_rules 的函数就会被调用。

```
    let rules = document.styleSheets[0].cssRules ;
```

document.styleSheets[0] 会返回【在当前网页里，代表第一个 CSS 样式表（style sheet）】的 CSSStyleSheet 对象实例，其实就是对应到当前网页里的第一个 style 元素实例【<style> ... </style>】。
document.styleSheets[0].cssRules 则会返回上述 CSSStyleSheet 对象实例中的 CSSRuleList 子对

象实例,其实就是对应到【在当前网页里的第 1 个(索引值为 0)style 元素实例中,其各个 CSS 规则】的列表(rule list)实例,例如由【header, aside, main, footer {...}】、【header {...}】、【aside {...}】、【main {...}】与【footer {...}】5 个 CSS 规则所组成的列表实例。

也因此,上述语句声明了变量 rules,其初始数据为 document.styleSheets[0].cssRules 所返回的 CSSRuleList 对象实例。

```
    console.log(rules) ;
    console.log('') ;

    console.log(rules[2].style.width) ;
```

rules[2].style.width 当前会返回字符串'20%'。

```
    console.log(rules[3].style.width) ;
```

rules[3].style.width 当前会返回字符串'80%'。

```
    console.log('') ;

    modify_rules() ;
```

此语句使得名称为 modify_rules 的函数,会被调用,以修正当前网页的外观。

```
function modify_rules()
{
    if (window.innerWidth <= 768)
```

此语法用来判断【浏览器窗口的宽度,扣除垂直滚动条的宽度之后,其剩余的宽度,是否小于或等于 768 像素】。若是,则执行其下方大括号里的源代码。

```
    {
        with (rules[2].style)
```

rules[2].style 是对应到当前网页里的第 1 个(索引值为 0)style 元素实例中的第 3 个(索引值为 2)CSS 规则实例【aside {...}】。通过此语法,可简化【rules[2].style.width】,成为其下方大括号里的【width】。

```
        {
            width = '100%' ;
```

此语句设置了 aside 元素实例的宽度,成为浏览器窗口的内容区域的总宽度(100%)。

```
            backgroundColor = 'LightGreen' ;
```

此语句设置了 aside 元素实例的背景颜色,成为浅绿色(light green)。

```
        }

        with (rules[3].style)
```

rules[3].style 是对应到当页网页里的第 1 个 style 元素实例中的第 4 个(索引值为 3)CSS 规则实例【main {...}】。通过此语句,可简化【rules[3].style.width】,成为其下方大括号里的【width】。

```
            {
                width = '100%' ;
```

此语句设置了 main 元素实例的宽度，成为浏览器窗口的内容区域的总宽度（100%）。

```
                backgroundColor = 'DeepSkyBlue' ;
```

此语句设置了 main 元素实例的背景颜色，成为深天蓝色（deep sky blue）。

```
            }
        }
        else
```

若浏览器窗口的内容区域的宽度，大于 768 像素，则执行其下方大括号里的源代码。

```
        {
            with (rules[2].style)
            {
                width = '20%' ;
```

此语句设置了 aside 元素实例的宽度，成为上述内容区域的总宽度的 20%。

```
                backgroundColor = 'YellowGreen' ;
```

此语句设置了 aside 元素实例的背景颜色，成为黄绿色（yellow green）。

```
            }
            with (rules[3].style)
            {
                width = '80%' ;
```

此语句设置了 main 元素实例的宽度，成为上述内容区域的总宽度的 80%。

```
                backgroundColor = 'SkyBlue' ;
```

此语句设置了 main 元素实例的背景颜色，成为天蓝色（sky blue）。

```
            }
        }
    }
```

上述语法完成了函数 modify_rules() 的定义。

21.3　源代码的加密

在网页里的 HTML、CSS 和 JavaScript 源代码，其实是无法被彻底加密的。但是，编程人员可以先将 HTML 和 CSS 的源代码，改写成为等价的 JavaScript 源代码；然后，再将其所有 JavaScript

的源代码，加密成为难以被理解的 JavaScript 源代码。JavaScript 源代码的加密操作，大致存在以下各项步骤：

（1）替换各个名称，成为失去原本含义的新名称。
（2）替换各个常量，成为等价且带有 2^n 进制字面量的表达式。
（3）替换各个字符串，成为等价的十六进制数码所表示的连续字符。
（4）删除原本所有的注释文本，以及多余的空白字符，包括空格、制表符和换行符等。
（5）进行仅有制造业者本身才知道的更高级的编码处理。

21.3.1　改写 HTML 与 CSS 成为 JavaScript 源代码

欲达成加密任务，首先必须将特定网页中的所有 HTML 和 CSS 的源代码，尽可能地改写成为等价的 JavaScript 源代码。请参看如下示例。

【21-3-1-recodings-from-HTML-and-CSS-to-JavaScript.html】

```
<!DOCTYPE html>
<html><body></body></html>
<script>
  // Produce CSS style sheet.
  let ele = document.createElement('style') ;

  document.head.appendChild(ele) ;

  ele.sheet.insertRule('header, aside, main, footer {padding: 5px ; box-sizing: border-box ;}', 0) ;

  ele.sheet.insertRule('header {color: RoyalBlue ; background-color: PowderBlue ; height: 80px ;}', 1) ;

  ele.sheet.insertRule('aside {color: SeaGreen ; background-color: YellowGreen ; height: 500px ; display: inline-block ;}', 2) ;

  ele.sheet.insertRule('main {color: Chocolate ; background-color: SkyBlue ; height: 500px ; display: inline-block ;}', 3) ;

  ele.sheet.insertRule('footer {color: LightSkyBlue ; background-color: Lavender ; height: 50px ; text-align: center ;}', 4) ;
  ///
  // Produce document-head elements with Javascript.
  let ref = document.createElement('meta') ;

  ref.setAttribute('charset', 'utf-8') ;

  document.head.appendChild(ref) ;

  ///
  ref = document.createElement('meta') ;

  ref.setAttribute('name', 'viewport') ;

  ref.setAttribute('content', 'width=device-width, initial-scale=1.0') ;
```

```
  document.head.appendChild(ref) ;

  // Produce <title>...</title>
  document.title = 'Rewrite HTML to Javascript source code.' ;

  ///
  ref = document.createElement('header') ;

  ref.innerHTML = 'Header' ;

  document.body.appendChild(ref) ;

  ref = document.createElement('aside') ;

  ref.innerHTML = 'Menu' ;

  document.body.appendChild(ref) ;

  ref = document.createElement('main') ;

  ref.innerHTML = 'Content' ;

  document.body.appendChild(ref) ;

  ref = document.createElement('footer') ;

  ref.innerHTML = 'Footer' ;

  document.body.appendChild(ref) ;
  ///
  console.log(document.head) ;

  window.onresize = change_size ;

  change_size() ;

  function change_size()
  {
    // if (window.matchMedia('(max-width: 768px)').matches)
    if (window.innerWidth <= 768)
    {
      with (document.querySelector('aside').style)
      {
        width = '100%' ;
        backgroundColor = 'LightGreen' ;
      }

      with (document.querySelector('main').style)
      {
        width = '100%' ;
        backgroundColor = 'DeepSkyBlue' ;
      }
    }
```

```
    else
    {
      with (document.querySelector('aside').style)
      {
        width = '20%' ;
        backgroundColor = 'YellowGreen' ;
      }
      with (document.querySelector('main').style)
      {
        width = '80%' ;
        backgroundColor = 'SkyBlue' ;
      }
    }
  }
</script>
```

【相关说明】

```
<!DOCTYPE html>

<html><body></body></html>
```

原本在 body 元素实例中的各子元素实例对应的 HTML 源代码，均会被改写成为等价的 JavaScript 源代码。所以在上述 HTML 语法里的 body 元素实例中，可看出并未内含任何子元素实例。

```
<script>
  // Produce CSS style sheet.
  let ele = document.createElement('style') ;
```

此语句声明了变量 ele，其初始数据为 document.createElement('style')所返回的新的 style 元素实例。其中，style 元素实例即是对应到各个 CSS 规则所构成的样式表（style sheet）。

```
  document.head.appendChild(ele) ;
```

此语句将变量 ele 所代表的 style 元素实例，新增至当前网页里的 head 元素实例中，以调整当前网页的外观。

```
  ele.sheet.insertRule('header, aside, main, footer {padding: 5px ; box-sizing: border-box ;}', 0) ;
```

此语句将第 1 个（索引值为 0）CSS 规则实例【header, aside, main, footer {padding: 5px ; box-sizing: border-box ;}】，新增至变量 ele 所代表的 style 元素实例中的样式表（style sheet）里。

```
  ele.sheet.insertRule('header {color: RoyalBlue ; background-color: PowderBlue ; height: 80px ;}', 1) ;
```

此语句将第 2 个（索引值为 1）CSS 规则实例【header {color: RoyalBlue ; background-color: PowderBlue ; height: 80px ;}】，新增至变量 ele 所代表的 style 元素实例中的样式表里。

```
  ele.sheet.insertRule('aside {color: SeaGreen ; background-color: YellowGreen ; height: 500px ; display: inline-block ;}', 2) ;
```

此语句将第 3 个（索引值为 2）CSS 规则实例【aside {color: SeaGreen ; background-color:

YellowGreen ; height: 500px ; display: inline-block ;}】，新增至变量 ele 所代表的 style 元素实例中的样式表里。

```
ele.sheet.insertRule('main {color: Chocolate ; background-color: SkyBlue ; height: 500px ;
   display: inline-block ;}', 3) ;
```

此语句将第 4 个（索引值为 3）CSS 规则实例【main {color: Chocolate ; background-color: SkyBlue ; height: 500px ; display: inline-block ;}】，新增至变量 ele 所代表的 style 元素实例中的样式表里。

```
ele.sheet.insertRule('footer {color: LightSkyBlue ; background-color: Lavender ; height:
   50px ; text-align: center ;}', 4) ;
```

此语句将第 5 个（索引值为 4）CSS 规则实例【footer {color: LightSkyBlue ; background-color: Lavender ; height: 50px ; text-align: center ;}】，新增至变量 ele 所代表的 style 元素实例中的样式表里。

```
///
// Produce document-head elements with Javascript.
let ref = document.createElement('meta') ;
```

此语句声明了变量 ref，其初始数据为 document.createElement('meta')所返回的新的 meta 元素实例。

```
ref.setAttribute('charset', 'utf-8') ;
```

此语句使得变量 ref 所代表的 meta 元素实例，被设置了属性 charset 和其数据字符串'utf-8'，进而如同建立了 meta 元素实例【<meta charset="utf-8">】。

```
document.head.appendChild(ref) ;
```

此语句将变量 ref 所代表的 meta 元素实例，新增至当前网页里的 head 元素实例中，进而设置了当前网页的编码字符集（character set）为 utf-8，以正常显示各国文本。

```
///
ref = document.createElement('meta') ;
```

此语句将 document.createElement('meta')所返回的新的 meta 元素实例，赋给了变量 ref。

```
ref.setAttribute('name', 'viewport') ;
```

此语句使得变量 ref 所代表的 meta 元素实例，被设置了属性 name 和其数据字符串'viewport'，进而如同建立了 meta 元素实例【<meta name="viewport">】。

```
ref.setAttribute('content', 'width=device-width, initial-scale=1.0') ;
```

此语句使得变量 ref 所代表的 meta 元素实例，被设置了另一属性 content 和其数据字符串'width=device-width, initial-scale=1.0'，进而如同建立了 meta 元素实例【<meta name="viewport" content="width=device-width, initial-scale=1.0">】，以驱使浏览器加以调整【当前网页呈现在其窗口可视范围／视口（viewport）】时的如下特征：

- 依据**设备**（device）的屏幕宽度（width），设置【当前网页在上述视口（viewport）里】的宽度（width）。
- 依据 100%（1.0）的初始比例（initial scale），设置浏览器加载当前网页时的显示比例。

```
document.head.appendChild(ref) ;
```

此语句将变量 ref 所代表的 meta 元素实例,新增至当前网页的 head 元素实例中,进而设置了浏览器加载当前网页的显示比例。

```
// Produce <title>...</title>
document.title = 'Rewrite HTML to Javascript source code.' ;
```

此语句将字符串'Rewrite HTML to Javascript source code.',赋给了代表当前网页的 document 对象实例的属性 title,使得当前网页的标题,成为了【Rewrite HTML to Javascript source code.】。

```
///
ref = document.createElement('header') ;
```

此语句将 document.createElement('header')所返回的新的 header 元素实例,赋给了变量 ref。

```
ref.innerHTML = 'Header' ;
```

此语句将字符串'Header',赋给了变量 ref 当前所代表的 header 元素实例的属性 innerHTML,使得其 header 元素实例带有【Header】的文本,例如【<header>Header</header>】。

```
document.body.appendChild(ref) ;
```

此语句将变量 ref 当前所代表的 header 元素实例,新增至 body 元素实例内。

```
ref = document.createElement('aside') ;
```

此语句将 document.createElement('aside')所返回的新的 aside 元素实例,赋给了变量 ref。

```
ref.innerHTML = 'Menu' ;
```

此语句将字符串'Menu',赋给了变量 ref 当前所代表的 aside 元素实例的属性 innerHTML,使得 aside 元素实例带有【Menu】的文本,例如【<aside>Menu</aside>】。

```
document.body.appendChild(ref) ;
```

此语句将变量 ref 当前所代表的 aside 元素实例,新增至 body 元素实例内。

```
ref = document.createElement('main') ;
```

此语句将 document.createElement('main')所返回的新的 main 元素实例,赋给了变量 ref。

```
ref.innerHTML = 'Content' ;
```

此语句将字符串'Content',赋给了变量 ref 当前所代表的 main 元素实例的属性 innerHTML,使得其 main 元素实例带有【Content】的文本,例如【<main>Content</main>】。

```
document.body.appendChild(ref) ;
```

此语句将变量 ref 当前所代表的 main 元素实例,新增至 body 元素实例内。

```
ref = document.createElement('footer') ;
```

此语句将 document.createElement('footer')所返回的新的 footer 元素实例,赋给了变量 ref。

```
ref.innerHTML = 'Footer' ;
```

此语句将字符串'Footer',赋给了变量 ref 当前所代表的 footer 元素实例的属性 innerHTML,使得 footer 元素实例带有【Footer】的文本，例如【<footer>Footer</footer>】。

```
document.body.appendChild(ref) ;
```

此语句将变量 ref 当前所代表的 footer 元素实例，新增至 body 元素实例内。

```
///
console.log(document.head) ;
```

此语句显示出可展开的【<head>...</head>】的信息。

21.3.2　JavaScript 源代码的全数加密

在互联网上，不少网站提供了 JavaScript 源代码的加密应用。读者可以操作如下步骤，以获得加密之后的 JavaScript 源代码：

- 将前一示例的网页文档，另存成为不同档名的新网页文档。
- 将【在新网页文档里，其特定 script 元素实例中】的 JavaScript 源代码，全数复制并粘贴至【特定网站页面（例如 https://javascriptobfuscator.com/Javascript-Obfuscator.aspx）里，其加密应用】的左侧文本区域中，再单击上述网站页面里的加密按钮（例如 Obfuscate 按钮）。
- 单击加密按钮之后，若无任何错误信息，则将【在上述网站页面里的右侧文本区域中，已经加密完成】之后的 JavaScript 源代码，粘贴并覆盖至原本的 JavaScript 源代码。
- 存储并调试新的网页文档。

【21-3-2-encrypting-JavaScript-source-codes.html】及其【相关说明】

```
<!DOCTYPE html>
<html><body></body></html>
<script>
  var _0xc6f8=["\x73\x74\x79\x6C\x65","\x63\x72\x65\x61\x74\x65\x45\x6C\x65\x6D\x65\x6E\x74","\x61\x70\x70\x65\x6E\x64\x43\x68\x69\x6C\x64","\x68\x65\x61\x64","\x68\x65\x61\x64\x65\x72\x2C\x20\x61\x73\x69\x64\x65\x2C\x20\x6D\x61\x69\x6E\x2C\x20\x66\x6F\x6F\x74\x65\x72\x20\x7B\x70\x61\x64\x64\x69\x6E\x67\x3A\x20\x35\x70\x78\x20\x3B\x20\x62\x6F\x78\x2D\x73\x69\x7A\x69\x6E\x67\x3A\x20\x62\x6F\x72\x64\x65\x72\x2D\x62\x6F\x78\x20\x3B\x7D","\x69\x6E\x73\x65\x72\x74\x52\x75\x6C\x65","\x73\x68\x65\x65\x74","\x68\x65\x61\x64\x65\x72\x20\x7B\x63\x6F\x6C\x6F\x72\x3A\x20\x52\x6F\x79\x61\x6C\x42\x6C\x75\x65\x20\x3B\x20\x61\x63\x6B\x67\x72\x6F\x75\x6E\x64\x2D\x63\x6F\x6C\x6F\x72\x3A\x20\x50\x6F\x77\x64\x65\x72\x42\x6C\x75\x65\x20\x3B\x20\x68\x65\x69\x67\x68\x74\x3A\x20\x38\x30\x70\x78\x20\x3B\x7D","\x61\x73\x69\x64\x65\x20\x7B\x63\x6F\x6C\x6F\x72\x3A\x20\x53\x65\x61\x47\x72\x65\x65\x6E\x20\x3B\x20\x62\x61\x63\x6B\x67\x72\x6F\x75\x6E\x64\x2D\x63\x6F\x6C\x6F\x72\x3A\x20\x59\x65\x6C\x6C\x6F\x77\x47\x72\x65\x65\x6E\x20\x3B\x20\x68\x65\x69\x67\x68\x74\x3A\x20\x35\x30\x30\x70\x78\x20\x3B\x20\x64\x69\x73\x70\x6C\x61\x79\x3A\x20\x69\x6E\x6C\x69\x6E\x65\x2D\x62\x6C\x6F\x63\x6B\x20\x3B\x7D","\x6D\x61\x69\x6E\x20\x7B\x63\x6F\x6C\x6F\x72\x3A\x20\x43\x68\x6F\x63\x6F\x6C\x61\x74\x65\x20\x3B\x20\x62\x61\x63\x6B\x67\x72\x6F\x75\x6E\x64\x2D\x63\x6F\x6C\x6F\x72\x3A\x20\x53\x6B\x79\x42\x6C\x75\x65\x20\x3B\x20\x68\x65\x69\x67\x68\x74\x3A\x20\x35\x30\x30\x70\x78\x20\x3B\x20\x64\x69\x73\x70\x6C\x61\x79\x3A\x20\x69\x6E\x6C\x69\x6E\x65\x2D\x62\x6C\x6F\x63\x6B\x20\x3B\x7D","\x66\x6F\x6F\x74\x65\x72\x20\x7B\x63\x6F\x6C\x6F\x72\x3A\x20\x4C\x69\x67\x68\x74\x53\x6B\x79\x42\x6C\x75\x65\x20\x3B\x20\x62\x61\x63\x6B\x67\x72\x6F\x75\x6E\x64\x2D\x63\x6F\x6C\x6F\x72\x3A\x20\x4C\x61\x76\x65\x6E\x64\x65\x72\x20\x3B\x20\x68\x65\x69\x67\x68\x74\x3A\x20\x35\x30\x70\x78\x20\x3B\x20\x74\x65\x78\x74\x2D\x61\x6C\x69\x67\x6E\x3A\x20\x63\x65\x6E\x74\x65\x72\x20\x3B\x7D","\x6D\x65
```

```
x74\x61","\x63\x68\x61\x72\x73\x65\x74","\x75\x74\x66\x2D\x38","\x73\x65\x74\x41\x74
\x74\x72\x69\x62\x75\x74\x65","\x6E\x61\x6D\x65","\x76\x69\x65\x77\x70\x6F\x72\x74","\
x63\x6F\x6E\x74\x65\x6E\x74","\x77\x69\x64\x74\x68\x3D\x64\x65\x76\x69\x63\x65\x2D\x7
7\x69\x64\x74\x68\x2C\x20\x69\x6E\x69\x74\x69\x61\x6C\x2D\x73\x63\x61\x6C\x65\x3D\x31
\x2E\x30","\x74\x69\x74\x6C\x65","\x52\x65\x77\x72\x69\x74\x65\x20\x48\x54\x4D\x4C\x2
0\x74\x6F\x20\x4A\x61\x76\x61\x73\x63\x72\x69\x70\x74\x20\x73\x6F\x75\x72\x63\x65\x20
\x63\x6F\x64\x65\x2E","\x68\x65\x61\x64\x65\x72","\x69\x6E\x6E\x65\x72\x48\x54\x4D\x4
C","\x48\x65\x61\x64\x65\x72","\x62\x6F\x64\x79","\x61\x73\x69\x64\x65","\x4D\x65\x6E
\x75","\x6D\x61\x69\x6E","\x43\x6F\x6E\x74\x65\x6E\x74","\x66\x6F\x6F\x74\x65\x72","\
x46\x6F\x6F\x74\x65\x72","\x6F\x6E\x72\x65\x73\x69\x7A\x65","\x69\x6E\x6E\x65\x72\x57
\x69\x64\x74\x68","\x31\x30\x30\x25","\x4C\x69\x67\x68\x74\x47\x72\x65\x65\x6E","\x71
\x75\x65\x72\x79\x53\x65\x6C\x65\x63\x74\x6F\x72","\x44\x65\x65\x70\x53\x6B\x79\x42\x
6C\x75\x65","\x32\x30\x25","\x59\x65\x6C\x6C\x6F\x77\x47\x72\x65\x65\x6E","\x38\x30\x
25","\x53\x6B\x79\x42\x6C\x75\x65"];let ele=document[_0xc6f8[1]]
(_0xc6f8[0]);document[_0xc6f8[3]][_0xc6f8[2]](ele);ele[_0xc6f8[6]][_0xc6f8[5]](_0xc6f
8[4],0);ele[_0xc6f8[6]][_0xc6f8[5]](_0xc6f8[7],0);ele[_0xc6f8[6]][_0xc6f8[5]](_0xc6f8
[8],0);ele[_0xc6f8[6]][_0xc6f8[5]](_0xc6f8[9],0);ele[_0xc6f8[6]][_0xc6f8[5]](_0xc6f8[
10],0);let
ref=document[_0xc6f8[1]](_0xc6f8[11]);ref[_0xc6f8[14]](_0xc6f8[12],_0xc6f8[13]);
document[_0xc6f8[3]][_0xc6f8[2]](ref);ref= document[_0xc6f8[1]](_0xc6f8[11]);
ref[_0xc6f8[14]](_0xc6f8[15],_0xc6f8[16]);ref[_0xc6f8[14]](_0xc6f8[17],_0xc6f8[18]);d
ocument[_0xc6f8[3]][_0xc6f8[2]](ref);document[_0xc6f8[19]]= _0xc6f8[20];ref=
document[_0xc6f8[1]](_0xc6f8[21]);ref[_0xc6f8[22]]= _0xc6f8[23];document[_0xc6f8[24]]
[_0xc6f8[2]](ref);ref= document[_0xc6f8[1]](_0xc6f8[25]);ref[_0xc6f8[22]]=_0xc6f8[26];
document[_0xc6f8[24]][_0xc6f8[2]](ref);ref= document[_0xc6f8[1]](_0xc6f8[27]);
ref[_0xc6f8[22]]= _0xc6f8[28];document[_0xc6f8[24]][_0xc6f8[2]](ref);ref= document
[_0xc6f8[1]](_0xc6f8[29]);ref[_0xc6f8[22]]= _0xc6f8[30];document[_0xc6f8[24]]
[_0xc6f8[2]](ref);window[_0xc6f8[31]]= change_size;change_size();function change_size()
{if(window[_0xc6f8[32]]<= 768){with(document[_0xc6f8[35]](_0xc6f8[25])[_0xc6f8[0]])
{width= _0xc6f8[33];backgroundColor= _0xc6f8[34]};with(document[_0xc6f8[35]]
(_0xc6f8[27])[_0xc6f8[0]]){width= _0xc6f8[33];backgroundColor= _0xc6f8[36]}}else
{with(document[_0xc6f8[35]](_0xc6f8[25])[_0xc6f8[0]]){width= _0xc6f8[37];
backgroundColor= _0xc6f8[38]};with(document[_0xc6f8[35]](_0xc6f8[27])[_0xc6f8[0]])
{width= _0xc6f8[39];backgroundColor= _0xc6f8[40]}}} ;
```

以上冗长的 JavaScript 源代码，是经过加密之后的 JavaScript 源代码，并且可以实现【与前一个示例的网页文档】相同的功能和画面！

`</script>`

21.4 练 习 题

1. 编写必要的 HTML 和 CSS 源代码，以实现如下布局的网页：

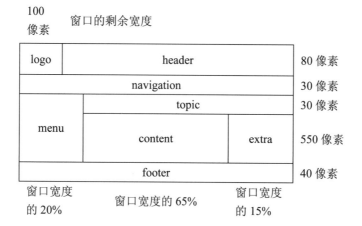

其中，在上述布局中，其各个元素实例的属性 id 的数据（例如 footer），即是如同以上各块内的文本（例如【<div id="footer">...</div>】）。

2. 进一步通过 CSS 媒体查询的语法，改写第 1 题的源代码，使得浏览器窗口的宽度，不足 768 像素时，则在当前网页上，除了属性 id 的数据为'logo'的 div 元素实例之外，请**垂直排列**均占用浏览器窗口的总宽度的**其余**各块元素实例。

3. 试改写第 2 题中的 CSS 媒体查询（media query）相关的源代码，成为等价的 JavaScript 源代码。

4. 试改写第 3 题中的剩余的 CSS 源代码，成为具有同样效果的 JavaScript 源代码。

附录

练习题答案

第 1 章

1.
9 个版本。

2.
HTML 和 CSS。

3.
browser 或 web browser。

4.
ActionScript 和 JScript。

5.
机器代码、源代码、编程语言、语法、子程序、变量、常量和调用。

6.
汇编器、编译器和解释器。

7.
```
color = 'Gold' ;
backgroundColor = 'DodgerBlue' ;
fontSize = '1.5em' ;
```

8.
```
let i = 0 ;
i < 10 ;
i++
count++ ;
sum += count * i ** 2 ;
```

9.
message、subject、greeting 和 object。

10.

常量、变量、函数返回值和子表达式。

11.

!、()、>、||、>= 和 &&。

12.

15、30、75、90、180、2、5 和 770。

13.

'z1 = x1 ^ 2 + y1 * 3 + 6' 和 /[a-zA-Z]\d/g。

14.

```
result = sphere(50) ;
```

15.

仅有 catchup。

16.

按下快捷键 Ctrl + Shift + I 或 Ctrl + Shift + J。

第 2 章

1.

770、Math.PI、'Nice Day'、"good"、/\w\s\d/g 和 /.\w.\d/ 均是常量的一种。请特别留意其大小写字母！剩余的项目均不是内置的常量。

2.

```
const love_your_forever = 201314 ;
```

3.

```
const love_me_longer = 2591.8 ;
```

4.

全局变量：value01、value02、str01、str02。

局部变量：data、identity、result、amount、price、output。

5.

3 个主要表达式：【sphere_volume = r => 4 / 3 * Math.PI * Math.pow(r, 3)】、【Math.PI】和【Math.pow(r, 3)】。

2 个不同的左侧表达式：【sphere_volume =】和【Math.】。

1 个箭头函数表达式：【r => 4 / 3 * Math.PI * Math.pow(r, 3)】。

4 个算术表达式：【4 / 3 * Math.PI * Math.pow(r, 3)】、【4 / 3 * Math.PI】、【Math.PI * Math.pow(r, 3)】、【4 / 3】。

6.
```
function trapezoid_area(upper_base, lower_base, height)
{
  let result ;

  result = (upper_base + lower_base) * height / 2 ;

  return result ;
}
```

或

```
function trapezoid_area(upper_base, lower_base, height)
{
  return (upper_base + lower_base) * height / 2 ;
}
```

或

```
trapezoid_area = (upper_base, lower_base, height) => (upper_base + lower_base) * height / 2 ;
```

7.
```
function ellipse_area(major_axe, minor_axe)
{
  let result ;

  result = Math.PI * major_axe * minor_axe / 4 ;

  return result ;
}
```

或

```
function ellipse_area(major_axe, minor_axe)
{
  return Math.PI * major_axe * minor_axe / 4 ;
}
```

或

```
ellipse_area = (major_axe, minor_axe) => Math.PI * major_axe * minor_axe / 4 ;
```

8.

整数 13。

9.

整数 27。

10.

整数 1036。

11.

整数 51。

12.
```
rating = score < 60 ? 'failed' : score < 80 ? 'passed' : score <= 100 ? 'nice' : 'error' ;
```

或

```
rating =
  score < 60 ? 'failed' :
  score < 80 ? 'passed' :
  score <= 100 ? 'nice' : 'error' ;
```

13.

此处列举出 6 种：number、string、object、boolean、undefined 和 function。

第 3 章

1.

二进制的语法：(25.75).toString(2)。

八进制的语法：(25.75).toString(8)。

十六进制的语法：(25.75).toString(16)。

或

先编写【let num01 = 25.75 ;】的语句，再编写如下相关语法：

- 二进制的语法：num01.toString(2)。
- 八进制的语法：num01.toString(8)。
- 十六进制的语法：num01.toString(16)。

2.

```
0b110111011.toString(16)
```

或

```
parseInt('110111011', 2).toString(16)
```

或

```
Number('0b110111011').toString(16)
```

3.

```
(0b101100011100 + 0o1275 + 0x51cf).toString(2)
```

4.
$3 * x ** 2 + 2 * ((x - 1) ** 2) * y + 2 * x * y ** 2 + 5 * y ** 3$

5.

```
var result = product.item01[1] + product.item02[1] + product.item03[1] + product.item04[1] +
  product.item05[1] ;
```

或

```
var result = 0 ;

for (item in product)
{
  result += product[item][1] ;
}
```

6.

NaN：代表【并非数值 (not a number)】的含义。

Infinity：代表【无穷值】的含义。

undefined：代表特定名称【未 (被) 定义】的含义。

7.

303。

8.

```
var price = 72583000 ;
var cnf_cn = Intl.NumberFormat('cn', {style: 'currency', currency: 'cny'}) ;

console.log(cnf_cn.format(price)) ;
```

9.

先编写【let x = 768 ;】，再编写下面其中一个语句：

```
console.log(x ** (1 / 3)) ;
console.log(Math.pow(x, 1 / 3)) ;
```

10.

```
! price < 300 && ! amount >= 10
! (price < 300 || amount >= 10)
```

11.

访问 25：arr[1][0][0]。

访问 40：arr[1][1][1]。

访问 60：arr[2][1][1]。

12.

访问'browser'：【obj.product.en】或【obj['product']['en']】。

访问'谷歌'：【obj.developer.cn】或【obj['developer']['cn']】。

访问'free'：【obj.price.en】或【obj['price']['en']】。

13.

```
let message = `${profile.name} now lives on ${profile.planet}.` ;
```

14.

```
console.log('Apple: 11\nBanana: 15\nGuava: 23') ;
```

15.

```
console.log(Date()) ;
```

16.
```
let components = new Set() ;
components.add('window').add('pane').add('dialogue').add('button').add('scrollbar') ;
```

17.
```
let devices = new Map() ;
devices.set('mobile phone', 10).set('tablet PC', 7).set('notebook PC', 3).set('desktop PC', 20) ;
```

18.
```
let num = 201314 ;
let message = `变量num 的：\n  十进制数值 = ${num}\n  二进制数码 = ${num.toString(2)}\n  八进制数码 = ${num.toString(8)}\n  十六进制数码 = ${num.toString(16)}` ;

console.log(message) ;
```

第 4 章

1.

if 语句的版本：

```
let dtype = ['integer', 'float', 'alphabet'][parseInt(3 * Math.random())] ;

if (dtype == 'integer')
  value = parseInt(100 * Math.random()) ;
else if (dtype == 'float')
  value = (100 * Math.random()).toFixed(3) ;
else if (dtype == 'alphabet')
  value = String.fromCharCode(65 + parseInt(26 * Math.random())) ;
else
  value = null ;

console.log(value) ;
```

switch 语句的版本：

```
let dtype = ['integer', 'float', 'alphabet'][parseInt(3 * Math.random())] ;

switch (dtype)
{
 case 'integer':
   value = parseInt(100 * Math.random()) ;
   break ;
 case 'float':
   value = (100 * Math.random()).toFixed(3) ;
   break ;
 case 'alphabet':
   value = String.fromCharCode(65 + parseInt(26 * Math.random())) ;
   break ;
 default:
```

```
            value = null ;
    }

    console.log(value) ;
```

2.

for 语句的版本：

```
let num_list = [10, 33, 21, 56, 77, 64, 82, 98, 6] ;
let result = 0 ;

for (let i = 0; i < num_list.length; i++)
{
    result += num_list[i] ;
}

console.log(result) ;
```

for…of 语句的版本：

```
let num_list = [10, 33, 21, 56, 77, 64, 82, 98, 6] ;
let result = 0 ;

for (let value of num_list)
{
    result += value ;
}

console.log(result) ;
```

for…in 语句的版本：

```
let num_list = [10, 33, 21, 56, 77, 64, 82, 98, 6] ;
let result = 0 ;

for (let key in num_list)
{
    result += num_list[key] ;
}

console.log(result) ;
```

3.

```
let chairs = {wood: 15, metal: 23, plastic: 37, others: 60} ;

for (let key in chairs)
{
    console.log(`${key}: ${chairs[key]}`) ;
}
```

4.

```
let str = 'Happily ever after.' ;
let output = '' ;
```

```
for (let i = str.length - 1; i > -1; i--)
{
  output += str[i] ;
}

console.log(output) ;
```

或

```
let str = 'Happily ever after.' ;
console.log(str.split('').reverse().join('')) ;
```

5.

（1）

```
function * weekday(language = 'cn')
{
  let weekday_obj = {cn: ['日', '一', '二', '三', '四', '五', '六'], en: ['Sun', 'Mon', 'Tues',
    'Wednes', 'Thurs', 'Fri', 'Sat']} ;

  for (let day of weekday_obj[language])
  {
    yield language == 'cn' ? '星期' + day : day + 'day' ;
  }
}
```

（2）

```
for (let day of weekday())
{
  console.log(day) ;
}

for (let day of weekday('en'))
{
  console.log(day) ;
}
```

6.

for 语句的版本：

```
let base_list = [7, 12, 21, 30, 40, 55] ;
let result = 0 ;

for (let i = 0; i < base_list.length; i++)
{
  result += base_list[i] ** 2 ;
}

console.log(result) ;
```

while 语句的版本：

```
let base_list = [7, 12, 21, 30, 40, 55] ;
let result = 0 ;
```

```
let i = 0 ;

while (i < base_list.length)
{
  result += base_list[i] ** 2 ;

  i++ ;
}

console.log(result) ;
```

do…while 语句的版本：

```
let base_list = [7, 12, 21, 30, 40, 55] ;
let result = 0 ;
let i = 0 ;

do
{
  result += base_list[i] ** 2 ;

  i++ ;
} while (i < base_list.length) ;

console.log(result) ;
```

第 5 章

1.
```
let timing = function ()
{
  console.log(new Date().toTimeString()) ;
} ;
```

或

```
let timing = () => console.log(new Date().toTimeString()) ;
```

2.
```
let date_obj =
{
  today()
  {
    console.log(new Date().toDateString()) ;
  }
} ;
```

或

```
let date_obj =
{
  today: function ()
  {
    console.log(new Date().toDateString()) ;
```

```
  }
} ;
```

3.
```
let monster_base = (name = 'new', HP = 100, MP = 50) =>
{
  console.log('About this monster:') ;
  console.log(`  name: ${name}`) ;
  console.log(`  HP: ${HP}`) ;
  console.log(`  MP: ${MP}`) ;
} ;
```

4.
```
function f02()
{
  console.log('Hello, Solar System!') ;

  f01() ;

  function f01()
  {
    console.log('Hello, Earth!') ;
  }
}

function f03()
{
  console.log('Hello, Galaxy!') ;
}
```

5.
```
(function ()
{
  alert('G\'day, and welcome to our official website.') ;
}) () ;
```

或

```
void function ()
{
  alert('G\'day, and welcome to our official website.') ;
} () ;
```

或

```
(() => alert('G\'day, and welcome to our official website.')) () ;
```

6.
```
function sum_of_square(n)
{
  if (n == 1) return 1 ;
  else return n ** 2 + sum_of_square(n - 1) ;
}
```

```
// sum_of_square(10) ;
```

或

```
function sum_of_square(n)
{
  if (n > 1) return n ** 2 + sum_of_square(n - 1) ;
  else return 1 ;
}

// sum_of_square(10) ;
```

或

```
function sum_of_square(current, n)
{
  if (current == n) return n ** 2 ;
  else return current ** 2 + sum_of_square(current + 1, n) ;
}

// sum_of_square(1, 10) ;
```

或

```
function sum_of_square(current, n)
{
  if (current < n) return current ** 2 + sum_of_square(current + 1, n) ;
  else return n ** 2 ;
}

// sum_of_square(1, 10) ;
```

7.
生成器的版本：

```
let zodiac01 =
{
  zodiac_list: ['Rat', 'Ox', 'Tiger', 'Hare', 'Dragon', 'Snake', 'Horse', 'Sheep', 'Monkey',
    'Rooster', 'Dog', 'Pig'],

  *[Symbol.iterator]()
  {
    for(let i = 0; i < 10; i++)
    {
      yield this.zodiac_list[i] ;
    }
  }
}

for (let one of zodiac01)
{
  console.log(one) ;
}
```

生成器函数的版本：

```
function * zodiac02()
```

```
{
  let zodiac_list = ['Rat', 'Ox', 'Tiger', 'Hare', 'Dragon', 'Snake', 'Horse', 'Sheep',
    'Monkey', 'Rooster', 'Dog', 'Pig'] ;

  for(let i = 0; i < 10; i++)
  {
    yield zodiac_list[i] ;
  }
}

for (let one of zodiac02())
{
  console.log(one) ;
}
```

8.

```
let textfield = document.createElement('input') ;

textfield.id = 't01' ;

textfield.style.fontSize = '2em' ;
textfield.style.paddingLeft = '5px' ;
textfield.style.color = 'RoyalBlue' ;

document.body.innerHTML += '<p></p>' ;

document.body.appendChild(textfield) ;

let message_box = document.createElement('span') ;

message_box.id = 'mbox' ;
message_box.style.fontSize = '1.5em' ;
message_box.style.marginLeft = '15px' ;

document.body.appendChild(message_box) ;

textfield.focus() ;

///
let message_promise ;

function promise_executor(resolve, reject)
{
  let current_value = Number(t01.value) ;

  console.log(current_value) ;

  if (t01.value == '')
  {
    reject('超过 6 秒，未被输入资料，真是遗憾...') ;
  }

  if (current_value < 1 || current_value > 100)
```

```
      {
        reject('输入值超出范围 Orz') ;
      }
      else
      {
        resolve('输入正确，谢谢您！') ;
      }
    }

    function positive(message)
    {
      mbox.style.color = 'ForestGreen' ;
      mbox.innerHTML = message ;
    }

    function negative(message)
    {
      mbox.style.color = 'Pink' ;
      mbox.innerHTML = message ;
    }

    function check_error()
    {
      message_promise = new Promise(promise_executor) ;

      message_promise.then(positive, negative) ;
    }

    setTimeout(check_error, 6000) ;
```

第 6 章

1.
其最大正整数：$1.7976931348623157 \times 10^{308}$。
其最大安全整数：$2^{53} - 1$，也就是 9007199254740991。

2.
浮点数。

3.
$2.220446049250313 \times 10^{-16}$。

4.
```
    let num01 = 2013.5625 ;

    console.log(num01.toString(2)) ;
    console.log(num01.toString(8)) ;
    console.log(num01.toString(16)) ;
```

5.
```
    let num02 = 0xff285c ;
```

```
console.log(num02.toString(2)) ;
console.log(num02.toString(8)) ;
```

6.
```
let num03 = Math.PI ;

// 方式 1
console.log(num03.toFixed(3)) ;
// 方式 2
console.log(num03.toPrecision(4)) ;
```

7.
```
let str01 = '0x57fc', str02 = '0b111011' ;
let result = new Array(3) ;

// 方式 1
result[0] = Number(str01) + Number(str02) ;
// 方式 2
result[1] = parseInt(str01) + Number(str02) ;
// 方式 3
result[2] = Math.round(str01) + Math.round(str02) ;

console.log(result) ;
```

8.
```
let value01 = 1 + parseInt(99 * Math.random()) ;
let value02 = 1 + parseInt(9 * Math.random()) ;

console.log(value01, value02) ;
console.log(Number.isInteger(value01 / value02)) ;
```

9.
```
let x = parseInt(10 + 60 * Math.random()) ;

// 方式 1
console.log(Math.sqrt(x)) ;
// 方式 2
console.log(Math.pow(x, 0.5)) ;
// 方式 3
console.log(x ** 0.5) ;
```

10.
```
let hypotenuse = 150 ;
let angle = 30 ;
let radian = angle * (Math.PI / 180) ;

let side01 = Math.abs(hypotenuse * Math.cos(radian)) ;
let side02 = Math.abs(hypotenuse * Math.sin(radian)) ;

console.log(side01, side02) ;
// check
console.log(Math.sqrt(side01 ** 2 + side02 ** 2)) ;
```

11.

```
let num = -100 + parseInt(200 * Math.random()) ;

console.log(num) ;
```

第 7 章

1.

分别返回 7、23 和 19。可参考如下源代码：

```
let str01 = '今晨明曦暖春风，午后炙意似晴空，夕刻彩晖遍地羞，夜来寂阑愁思浓。' ;

let index01 = str01.indexOf('，') ;
let index02 = str01.lastIndexOf('，') ;
let index03 = str01.indexOf('晖') ;

console.log(index01) ;
console.log(index02) ;
console.log(index03) ;
```

2.

```
let str02 = '建筑千层塔，通达万里跋，绽露盈满华，顶天立宏霸。' ;
let count5 = str02.match(/[\u4E00-\u9FA5]{5}(，|。)/g) ;
let count7 = str02.match(/[\u4E00-\u9FA5]{7}(，|。)/g) ;

if (count5 == null) count5 = [] ;

if (count7 == null) count7 = [] ;

let answer =
  count5.length == 4 ? '五言绝句' :
  count5.length == 8 ? '五言律诗' :
  count7.length == 4 ? '七言绝句' :
  count7.length == 8 ? '七语律诗' :
  '无法辨识' ;

console.log(answer) ;
```

3.

```
let str03 = '上底：15，下底：25，高度：35' ;
let parts = str03.split('，') ;
let trapezoid_obj = {} ;
let temp ;

for (let str of parts)
{
  temp = str.split('：') ;

  switch(temp[0])
  {
    case '上底':
```

```
      trapezoid_obj.upper_base = Number(temp[1]) ;
      break ;
    case '下底':
      trapezoid_obj.lower_base = Number(temp[1]) ;
      break ;
    case '高度':
      trapezoid_obj.height = Number(temp[1]) ;
      break ;
  }
}

with (trapezoid_obj)
{
  temp = (upper_base + lower_base) * height / 2 ;
}

console.log(temp) ;
```

4.
```
let str04 = 'zelda, peter, paul, jasper, alex, daisy, eric, john, stella, tommy, adam, betty,
  sean, frank, kevin, sammy, julia, yolanda, william' ;

str04 = str04.replace(/ /g, '') ;

let parts = str04.split(',') ;

parts.sort() ;

let output = '' ;

for (let name of parts)
{
  name = name.charAt(0).toUpperCase() + name.slice(1) ;

  output += name + ', ' ;
}

output = output.slice(0, -2) ;

console.log(output) ;
```

5.
```
let count ;

for (count = 1; count < 1000; count++)
{
  console.log(`no. ${count.toString().padStart(4, '0')}`) ;
}
```

第 8 章

1.

```
let digit_list = '20131452053077520099';
let digit_array;

// 方式 1
digit_array = digit_list.split('');

console.log(digit_array);

// 方式 2
digit_array = [... digit_list];

console.log(digit_array);
```

2.

```
let juices = ['orange', 'apple', 'guava', 'pineapple', 'passion fruit', 'lemon', 'kiwi fruit',
 'watermelon', 'grapefruit', 'grape'];

let random_index, output = [];

while (juices.length > 0)
{
  random_index = parseInt(juices.length * Math.random());

  output.push(... juices.splice(random_index, 1));
}

console.log(output);
```

第 9 章

1.

```
let profile01 = {name: 'Alex Lee', age: 27, gender: 'male', married: false};
let profile02 = {name: 'Alex Lee', profession: 'software engineer', annual_salary: 15000,
 company: 'Weird Alien'};
let profile03 = {name: 'Alex Lee', country: 'China', place: 'Beijing', phone_number: 'secret'};

let profiles = {};

Object.assign(profiles, profile01, profile02, profile03);

console.log(profiles);
```

2.

```
let product = {name: 'computer', price: 5000};

product = Object.create(product, {manufacturer: {value: '~O_O~', writable: false}});
```

```
product = Object.create(product, {model: {value: '+^v^+', writable: false}});
```

或

```
let product = {name: 'computer', price: 5000};

Object.defineProperty(product, 'manufacturer', {value: '~O_O~', writable: false});

Object.defineProperty(product, 'model', {value: '+^v^+', writable: false});
```

或

```
let product = {name: 'computer', price: 5000};

Reflect.defineProperty(product, 'manufacturer', {value: '~O_O~', writable: false});

Reflect.defineProperty(product, 'model', {value: '+^v^+', writable: false});
```

第 10 章

1.

```
// Assume someone's birthday was 1980/06/27 14:18
// Be ware! 5 means June, not May.
let birth_time = new Date(1980, 5, 27, 14, 18) ;

let current_time = new Date() ;

let second_count = (Date.parse(current_time) - Date.parse(birth_time)) / 1000 ;

let day_count = second_count / (60 * 60 * 24) ;
let day_passed = parseInt(day_count) ;

console.log(day_passed) ;
```

2.

```
let current = new Date();

current.setDate(current.getDate() + 100);

console.log(current.toLocaleString());
```

或

```
let current = new Date();

current.setTime(current.getTime() + 100 * 24 * 60 * 60 * 1000);

console.log(current.toLocaleString());
```

第 11 章

1.
方式 1：

```
<button id="login_btn">Login</button>

<script>
  function display()
  {
    
  }

  login_btn.onclick = display ;
</script>
```

方式 2：

```
<button id="login_btn">Login</button>

<script>
  function display()
  {
    
  }

  login_btn.addEventListener('click', display) ;
</script>
```

2.

```
window.location.href = 'file:///○○○/html/6-3-6-onscroll.html#list01' ;

function display_message()
{
  if (location.hash == '#list02')
    console.log('特定信息写在此处...') ;
}

document.body.onhashchange = display_message ;
```

3.
方式 1：

```
function input_focus()
{
  username.focus() ;
}

window.onload = input_focus ;
```

方式 2：

```
window.onload = function ()
{
```

```
  username.focus() ;
} ;
```

方式 3：

```
window.onload = () => username.focus() ;
```

方式 4：

```
function input_focus()
{
  username.focus() ;
}

window.addEventListener('load', input_focus) ;
```

方式 5：

```
window.addEventListener('load', () => username.focus()) ;
```

第 12 章

1.

```
let point01 = {x: 0, y: 0}, point02 = {x: 0, y: 0} ;
let distance ;

let ref = document.createElement('div') ;

ref.id = 'div01' ;
ref.draggable = true ;

with (ref.style)
{
  width = height = '100px' ;
  backgroundColor = 'RoyalBlue' ;
  borderRadius = '10px' ;
  maring = '0px' ;
  position = 'relative' ;
}

document.body.appendChild(ref);
document.body.ondragover = (event) => event.preventDefault() ;

div01.ondragstart = function (event)
{
  point01.x = event.clientX ;
  point01.y = event.clientY ;
} ;

div01.ondragend = function (event)
{
  point02.x = event.clientX ;
  point02.y = event.clientY ;
```

```
    distance = parseInt(Math.sqrt((point02.x - point01.x) ** 2 + (point02.y - point01.y) ** 2)) ;

    console.log(distance) ;
} ;

document.body.ondrop = function (event)
{
    div01.style.left = event.clientX - 8 + 'px' ;
    div01.style.top = event.clientY - 8 + 'px' ;
} ;
```

2.

```
let ref01 = document.createElement('div') ;

ref01.id = 'message_box' ;
ref01.innerHTML = '特定信息 ...' ;

with (ref01.style)
{
    width = '200px' ;
    height = '50px' ;
    textAlign = 'center' ;
    padding = '10px' ;
    borderRadius = '10px' ;

    color = 'RoyalBlue' ;
    backgroundColor = 'gold' ;

    position = 'absolute' ;
    left = Math.round(window.innerWidth / 2 - 100) + 'px' ;
    top = Math.round(window.innerHeight / 2 - 30) + 'px' ;
}

document.body.appendChild(ref01) ;

let ref02 = document.createElement('video') ;

ref02.id = 'video' ;
ref02.src = '...' ; // please specify your video source.
ref02.type = 'video/mp4' ;
ref02.controls = true ;
ref02.style.display = 'block' ;
ref02.style.width = '400px' ;
ref02.style.margin = 'auto' ;

document.body.appendChild(ref02) ;

video.onpause = function (event)
{
    message_box.style.visibility = 'visible' ;
} ;

video.onplay = function (event)
{
```

```
    message_box.style.visibility = 'hidden' ;
  } ;
```

3.

```
let result_obj = {click_count: 0, character_count: 0, scale_count: 0} ;

window.onclick = function (event)
{
  result_obj.click_count++ ;
} ;

window.onkeyup = function (event)
{
  result_obj.character_count++ ;
} ;

window.onwheel = function (event)
{
  result_obj.scale_count++ ;
} ;

function display_result()
{
  with (result_obj)
    console.log(`点击的次数：${click_count}\n 输入字符的个数：${character_count}\n 滚动滚轮的总刻度：
    ${scale_count}`) ;
}

setTimeout(display_result, 5000) ;
```

第 13 章

1.
```
Reflect.defineProperty(profile, 'department', {value: 'IT', writable: true, enumerable: true,
  configurable: false}) ;
```

2.
```
let result = 'department' in profile ;
```

3.
```
let result = Object.keys(profile) ;
```

第 14 章

1.
```
function get_weekday_string(obj, attr)
{
  let result = obj[attr] ;
```

```javascript
    if (attr == 'weekday')
    {
      let num = parseInt(obj[attr]) ;
      let weekday_string = '一二三四五六日' ;

      result = (num > 0 && num < 8) ? '星期' + weekday_string[num - 1]:'只能是整数值1~7。';
    }

    return result ;
}
function set_weekday(obj, attr, value)
{
  let final_value = value ;

  if (attr == 'weekday')
  {
    final_value = (value > 0 && value < 8) ? value : 0;
  }

  obj[attr] = final_value ;
}

var one_day = {type: 'chinese', weekday: 0} ;

var proxy_handler = {get: get_weekday_string, set: set_weekday } ;

var new_day = new Proxy(one_day, proxy_handler) ;

new_day.weekday = 100 ;

console.log(new_day.weekday) ;

new_day.weekday = 5 ;

console.log(new_day.weekday) ;
```

2.

```javascript
function get_weekday_string(obj, attr)
{
  let result = obj[attr] ;

  if (attr == 'weekday')
  {
    let num = parseInt(obj[attr]) ;
    let weekday_string = '一二三四五六日' ;
    let weekday_array = ['Mon', 'Tues', 'Wednes', 'Thurs', 'Fri', 'Satur', 'Sun'];

    if (obj['type'] == 'chinese')
    {
      result = (num > 0 && num < 8) ? '星期'+weekday_string[num - 1]:'只能是整数值1~7。';
    }
    else if (obj['type'] == 'english')
    {
```

```
      result=(num>0 && num<8)?weekday_array[num-1]+'day':'Canonly betheinteger 1to 7.';
    }
  }

  return result ;
}

function set_weekday(obj, attr, value)
{
  let final_value = value ;

  if (attr == 'weekday')
  {
    final_value = (value > 0 && value < 8) ? value : 0 ;
  }

  obj[attr] = final_value ;
}

var one_day = {type: 'chinese', weekday: 0} ;

var proxy_handler = {get: get_weekday_string, set: set_weekday } ;

var new_day = new Proxy(one_day, proxy_handler) ;

new_day.type = 'english' ;

new_day.weekday = 100 ;

console.log(new_day.weekday) ;

new_day.weekday = 6 ;

console.log(new_day.weekday) ;
```

第 15 章

1.
```
function display(event)
{
    |
}

form01.onclick = display ;
```

2.
```
current_ref = div01 ;
```

3.
```
with (username.style)
{
  color = 'RoyalBlue' ;
```

```
backgroundColor = 'GoldenRod' ;
padding = '5px' ;
margin = '10px' ;
fontSize = '30px' ;
}
```

第 16 章

1.
```
div01.className = 'color03 border02 background01' ;
```
或
```
div01.classList = 'color03 border02 background01' ;
```
或
```
div01.setAttribute('class', 'color03 border02 background01') ;
```

2.
```
console.log(document.images.length) ;
```
或
```
console.log(document.querySelectorAll('img').length) ;
```

3.
```
document.body.onclick = null ;
```
其中，上述的原始常量 null，可被置换成原始常量 undefined 或者空字符串""。

4.
```
console.log(document.querySelectorAll('a[href]').length) ;
```
请留意，并不是所有 a 元素实例，均为超链接！举例来说，HTML 语法【<a name="…"></a>】是锚点，而 HTML 语法【<a href="…">…</a>】才是超链接。

5.
```
console.log(document.body.children.length) ;
```

第 17 章

1.
```
location.href = 'http://www.tup.tsinghua.edu.cn/index.html' ;
```
或
```
location = 'http://www.tup.tsinghua.edu.cn/index.html' ;
```

2.
```
history.go(1) ;
```

第 18 章

1.
```
class Summation
{
  sum(n)
  {
    return (1 + n) * n / 2 ;
  }

  square_sum(n)
  {
    let output ;

    for (output = 1; n > 1; n--)
      output += n ** 2 ;

    return output ;
  }

  cube_sum(n)
  {
    let output ;

    for (output = 1; n > 1; n--)
    {
      output += n ** 3 ;
    }

    return output ;
  }
}

let s01 = new Summation() ;

// for testing
console.log(s01.sum(10)) ;
console.log(s01.square_sum(10)) ;
console.log(s01.cube_sum(10)) ;
```

2.
```
class Alternation
{
  alternate_sum(n)
  {
    let output ;

    for (output = -1; n > 1; n--)
      output += ((-1) ** n) * n ;

    return output ;
  }
```

```
  alternate_square_sum(n)
  {
    let output ;

    for (output = -1; n > 1; n--)
      output += ((-1) ** n) * n ** 2 ;

    return output ;
  }

  alternate_cube_sum(n)
  {
    let output ;

    for (output = -1; n > 1; n--)
    {
      output += ((-1) ** n) * n ** 3 ;
    }

    return output ;
  }
}

let a01 = new Alternation() ;

// for testing
console.log(a01.alternate_sum(7)) ;
console.log(a01.alternate_square_sum(7)) ;
console.log(a01.alternate_cube_sum(7)) ;
```

3.
```
class Summation
{
  sum(n)
  {
    return (1 + n) * n / 2 ;
  }

  square_sum(n)
  {
    let output ;

    for (output = 1; n > 1; n--)
      output += n ** 2 ;

    return output ;
  }

  cube_sum(n)
  {
    let output ;

    for (output = 1; n > 1; n--)
    {
```

```
      output += n ** 3 ;
    }

    return output ;
  }
}

let s01 = new Summation() ;

class Alternation extends Summation
{
  alternate_sum(n)
  {
    let output ;

    for (output = -1; n > 1; n--)
      output += ((-1) ** n) * n ;

    return output ;
  }

  alternate_square_sum(n)
  {
    let output ;

    for (output = -1; n > 1; n--)
      output += ((-1) ** n) * n ** 2 ;

    return output ;
  }

  alternate_cube_sum(n)
  {
    let output ;

    for (output = -1; n > 1; n--)
    {
      output += ((-1) ** n) * n ** 3 ;
    }

    return output ;
  }
}

let a02 = new Alternation() ;

// for testing
console.log(a02.sum(9)) ;
console.log(a02.square_sum(9)) ;
console.log(a02.cube_sum(9)) ;
console.log(a02.alternate_sum(9)) ;
console.log(a02.alternate_square_sum(9)) ;
console.log(a02.alternate_cube_sum(9)) ;
```

4.

Object.keys()、Object.values()、Object.create()、Array.from()、Array.of()、Array.isArray()、Number.isInteger()、Number.isNaN()、Number.isSafeInteger()、String.fromCharCode()、String.fromCodePoint() 等。

5.

（1）

语句【t01.__proto__.__proto__.display();】可调用到类 Shape 的函数 display()。

（2）

语句【t01.display();】可调用到类 Triangle 的函数 display()。

（3）

改写成为如下类 Triangle 的定义语法：

```
class Triangle extends Shape
{
  display()
  {
    super.display() ;

    console.log('display Triangle message...') ;
  }
}
```

第 19 章

1.

在代表运算符【--】之处，会导致语法错误（syntax error）。

2.

不会发生任何异常错误！

3.

在第 2 个符号【.】之处，会导致语法错误（syntax error）。

4.

在名称【num08】之处，会导致参考错误（reference error），因为 num08 尚未被定义。

5.

```
class InputError
{
  constructor(message)
  {
    this.message = message ;
  }
}

class DivisionError
{
  constructor(message)
  {
```

```
      this.message = message ;
   }
}

class TimeoutError
{
   constructor(message)
   {
      this.message = message ;
   }
}

let error_code = 1 ; // or 2 or 3

try
{
   switch (error_code)
   {
      case 1:
         throw new InputError('Input error...') ;
         break ;
      case 2:
         throw new DivisionError('Division error...') ;
         break ;
      case 3:
         throw new TimeoutError('Timeout error...') ;
   }
}
catch (err)
{
   if (err instanceof InputError)
   {
      console.log(err.message) ;
   }
   else if (err instanceof DivisionError)
   {
      console.log(err.message) ;
   }
   else if (err instanceof TimeoutError)
   {
      console.log(err.message) ;
   }
}
```

第 20 章

1.
```
username.required = true ;
username.pattern = '[_a-zA-Z]\\w{5,15}' ;
```

2.
```
salary.required = true ;
```

```
salary.type = 'number' ;
salary.min = 2300 ;
```

3.

```
salary.required = true ;
salary.pattern = '[a-zA-Z]{2}-\d{6}' ;
```

4.

```
let message ;

window.onload = function (event)
{
  send_client_data() ;
} ;

function send_client_data()
{
  let req = new XMLHttpRequest() ;
  let url = 'https://www.qzmp.wxno.com/opt/client_info.php' ;

                      // or navigator.userAgent
  let args = '?client_data=' + navigator.appVersion ;

  req.onreadystatechange = function(event)
  {
    if (this.readyState == 4 && this.status == 200)
    {
      message = this.responseText ;
    }
  } ;

  req.open("GET", url + args, true) ;
  req.send() ;
}
```

5.

```
let message ;

setInterval(send_question_data, 10000) ;

function send_question_data()
{
  let req = new XMLHttpRequest() ;
  let url = 'https://www.qzmp.wxno.com/opt/answers.php' ;
  let args = '?question_data=' + question.value ;

  req.onreadystatechange = function(event)
  {
    if (this.readyState == 4 && this.status == 200)
    {
      message = this.responseText ;
    }
  } ;
```

```
    req.open("GET", url + args, true) ;
    req.send() ;
}
```

第 21 章

1.
```
<!DOCTYPE html>
<html>
  <head>
    <meta charset="utf-8">
    <meta name="viewport" content="width=device-width, initial-scale=1.0">
    <title>exercise 1 in chapter 21</title>
    <style>
      div
      {
        color: White ;
        float: left ;
        text-align: center ;
      }

      #logo
      {
        width: 100px ;
        height: 80px ;
        background-color: Orange ;
      }

      #header
      {
        width: 100% ;
        height: 80px ;
        background-color: Gold ;
      }

      #navigation
      {
        background-color: RoyalBlue ;
        width: 100% ;
        height: 30px ;
      }

      #menu
      {
        width: 20% ;
        height: 580px ;
        background-color: Cyan ;
      }

      #topic
      {
        width: 80% ;
```

```css
      height: 30px ;
      background-color: Violet ;
    }

    #content
    {
      width: 65% ;
      height: 550px ;
      background-color: YellowGreen ;
    }

    #extra
    {
      width: 15% ;
      height: 550px ;
      background-color: GoldenRod ;
    }

    #footer
    {
      width: 100% ;
      height: 40px ;
      background-color: Lime ;
    }
  </style>
</head>
<body>
  <div id="header">
    header
  <div id="logo">logo</div>
  </div>
  <div id="navigation">navigation</div>
  <div id="menu">menu</div>
  <div id="topic">topic</div>
  <div id="content">content</div>
  <div id="extra">extra</div>
  <div id="footer">footer</div>
</body>
</html>
```

2.

```html
<!DOCTYPE html>
<html>
  <head>
    <meta charset="utf-8">
    <meta name="viewport" content="width=device-width, initial-scale=1.0">
    <title>exercise 2 in chapter 21</title>
    <style>
      div
      {
        color: White ;
        float: left ;
        text-align: center ;
      }
```

```css
#logo
{
  width: 100px ;
  height: 80px ;
  background-color: Orange ;
}


#header
{
  width: 100% ;
  height: 80px ;
  background-color: Gold ;
}

#navigation
{
  background-color: RoyalBlue ;
  width: 100% ;
  height: 30px ;
}

#menu
{
  width: 20% ;
  height: 580px ;
  background-color: Cyan ;
}

#topic
{
  width: 80% ;
  height: 30px ;
  background-color: Violet ;
}

#content
{
  width: 65% ;
  height: 550px ;
  background-color: YellowGreen ;
}

#extra
{
  width: 15% ;
  height: 550px ;
  background-color: GoldenRod ;
}

#footer
{
  width: 100% ;
  height: 40px ;
  background-color: Lime ;
```

```
        }

      @media (max-width: 767px)
      {
        div:not(#logo)
        {
          width: 100% ;
        }
      }
    </style>
  </head>
  <body>
    <div id="header">
      header
      <div id="logo">logo</div>
    </div>
    <div id="navigation">navigation</div>
    <div id="menu">menu</div>
    <div id="topic">topic</div>
    <div id="content">content</div>
    <div id="extra">extra</div>
    <div id="footer">footer</div>
  </body>
</html>
```

3.

```
<!DOCTYPE html>
<html>
  <head>
    <meta charset="utf-8">
    <meta name="viewport" content="width=device-width, initial-scale=1.0">
    <title>exercise 3 in chapter 21</title>
    <style>
      div
      {
        color: White ;
        float: left ;
        text-align: center ;
      }

      #logo
      {
        width: 100px ;
        height: 80px ;
        background-color: Orange ;
      }

      #header
      {
        /*width: 100% ;*/
        height: 80px ;
        background-color: Gold ;
      }
```

```css
    #navigation
    {
      background-color: RoyalBlue ;
      /*width: 100% ;*/
      height: 30px ;
    }

    #menu
    {
      /*width: 20% ;*/
      height: 580px ;
      background-color: Cyan ;
    }

    #topic
    {
      /*width: 80% ;*/
      height: 30px ;
      background-color: Violet ;
    }

    #content
    {
      /*width: 65% ;*/
      height: 550px ;
      background-color: YellowGreen ;
    }

    #extra
    {
      /*width: 15% ; */
      height: 550px ;
      background-color: GoldenRod ;
    }

    #footer
    {
      /*width: 100% ;*/
      height: 40px ;
      background-color: Lime ;
    }
    /*
    @media (max-width: 767px)
    {
      div:not(#logo)
      {
        width: 100% ;
      }
    }
    */
  </style>
</head>
<body>
  <div id="header">
    header
```

```
    <div id="logo">logo</div>
  </div>
  <div id="navigation">navigation</div>
  <div id="menu">menu</div>
  <div id="topic">topic</div>
  <div id="content">content</div>
  <div id="extra">extra</div>
  <div id="footer">footer</div>
</body>
<script>
  let refs = document.querySelectorAll('div:not(#logo)') ;

  window.onresize = update_size ;

  update_size() ;

  function update_size()
  {
    if (window.matchMedia('(max-width: 767px)').matches)
    {
      for (let i = 0; i < refs.length; i++)
      {
        refs[i].style.width = '100%' ;
      }
    }
    else
    {
      header.style.width = '100%' ;
      navigation.style.width = '100%' ;
      menu.style.width = '20%' ;
      topic.style.width = '80%' ;
      content.style.width = '65%' ;
      extra.style.width = '15%' ;
      footer.style.width = '100%' ;
    }
  }
</script>
</html>
```

4.
```
<!DOCTYPE html>
<html>
  <head>
    <meta charset="utf-8">
    <meta name="viewport" content="width=device-width, initial-scale=1.0">
    <title>exercise 4 in chapter 21</title>
    <!-- <style>
      div
      {
        color: White ;
        float: left ;
        text-align: center ;
      }
```

```css
#logo
{
  width: 100px ;
  height: 80px ;
  background-color: Orange ;
}

#header
{
  /*width: 100% ;*/
  height: 80px ;
  background-color: Gold ;
}

#navigation
{
  background-color: RoyalBlue ;
  /*width: 100% ;*/
  height: 30px ;
}

#menu
{
  /*width: 20% ;*/
  height: 580px ;
  background-color: Cyan ;
}

#topic
{
  /*width: 80% ;*/
  height: 30px ;
  background-color: Violet ;
}

#content
{
  /*width: 65% ;*/
  height: 550px ;
  background-color: YellowGreen ;
}

#extra
{
  /*width: 15% ; */
  height: 550px ;
  background-color: GoldenRod ;
}

#footer
{
  /*width: 100% ;*/
  height: 40px ;
  background-color: Lime ;
}
```

```
    /*
    @media (max-width: 767px)
    {
      div:not(#logo)
      {
        width: 100% ;
      }
    }
    */
  </style> -->
</head>
<body>
  <div id="header">
    header
    <div id="logo">logo</div>
  </div>
  <div id="navigation">navigation</div>
  <div id="menu">menu</div>
  <div id="topic">topic</div>
  <div id="content">content</div>
  <div id="extra">extra</div>
  <div id="footer">footer</div>
</body>
<script>
  let ref = document.createElement('style') ;

  document.head.appendChild(ref) ;

  ref.sheet.insertRule('div { color: White ; float: left ; text-align: center ; }', 0) ;

  ref.sheet.insertRule('#logo {width: 100px; height: 80px; background-color:Orange;}', 1);

  ref.sheet.insertRule('#header { height: 80px ; background-color: Gold ; }', 2) ;

  ref.sheet.insertRule('#navigation { background-color: RoyalBlue ; height: 30px ; }', 3) ;

  ref.sheet.insertRule('#menu { height: 580px ; background-color: Cyan ; }', 4) ;

  ref.sheet.insertRule('#topic { height: 30px ; background-color: Violet ; }', 5) ;

  ref.sheet.insertRule('#content { height: 550px ; background-color: YellowGreen ; }', 6) ;

  ref.sheet.insertRule('#extra { height: 550px ; background-color: GoldenRod ; }', 7) ;

  ref.sheet.insertRule('#footer { height: 40px ; background-color: Lime ; }', 8) ;

  ///
  let refs = document.querySelectorAll('div:not(#logo)') ;

  window.onresize = update_size ;

  update_size() ;

  function update_size()
  {
```

```
      if (window.matchMedia('(max-width: 767px)').matches)
      {
        for (let i = 0; i < refs.length; i++)
        {
          refs[i].style.width = '100%';
        }
      }
      else
      {
        header.style.width = '100%';
        navigation.style.width = '100%';
        menu.style.width = '20%';
        topic.style.width = '80%';
        content.style.width = '65%';
        extra.style.width = '15%';
        footer.style.width = '100%';
      }
    }
  </script>
</html>
```